INTRODUCTORY OCEANOGRAPHY

INTRODUCTORY OCEANOGRAPHY

Fifth Edition

Harold V. Thurman

Mount San Antonio College

Merrill Publishing Company
A Bell & Howell Information Company
Columbus Toronto London Melbourne

Published by Merrill Publishing Company
A Bell & Howell Information Company
Columbus, Ohio 43216

This book was set in Usherwood

Administrative Editor: David Gordon
Developmental Editor: Wendy Jones
Production Coordinator: Carol Sykes
Art Coordinator: James Hubbard
Cover Designer: Cathy Watterson
Text Designer: Cynthia Brunk

Library of Congress Catalog Card Number: 87-62654
International Standard Book Number: 0-675-20855-6
Printed in the United States of America
1 2 3 4 5 6 7 8 9—92 91 90 89 88

Cover illustration: Robert W. Tope/Merrill Publishing

Color Plate Credits: All photos and illustrations copyrighted by individuals or companies listed. Plates 1, 7, 8, 23, 24, 25, 26, 27A, Christopher Newbert; plate 2A, Scripps Institute; plate 2B, Dr. Fred N. Spiess/Scripps Institute; plates 3, 9, 10, 11, 15, 16, 18B, 19B, 20B, 28, Robert W. Tope/Merrill Publishing; plate 4, from C. L. Parkinson, J. C. Comiso, H. J. Zwally, D. J. Cavalieri, P. Gloersen, and W. J. Campbell, 1987, *Arctic Sea Ice, 1973–1976: Satellite Passive-Microwave Observations,* NASA SP-489, National Aeronautics and Space Administration, Washington, D.C., 296 pp.; plate 5, Dr. Charles McLain/Rosensteil School of Marine and Atmospheric Science, University of Miami; plates 6, 21, from H. R. Gordon, "Phytoplankton Pigments from *Nimbus 7* Coastal Zone Scanner: Comparisons with Surface Measurements," *Science* 210: 63–66, 1980; plate 12, Marie Tharp; plate 13, UNESCO; plate 14, Jet Propulsion Laboratory; plate 17, NASA, Goddard Space Flight Center; plates 18A, 19A, 20A, NOAA/NESDIS; plate 22, Paul C. Fiedler/National Marine Fisheries Service; plate 27B, Gregory Silber/EarthViews; plate 29A, Craig Matkin/EarthViews; plate 29B, Bernie Tershy and Craig Strong/EarthViews.

Figure Credits: Figures 3–10 and 3–11 from "Remote Acoustic Detection of Turbidity Current Surge," Hay, A. E., et al., *Science* Vol. 217, pp. 833–835, 27 August 1982. Copyright 1982 by the AAAS. Figures 5–10 and 5–11B from "Hardened Subtidal Stromatolites, Bahamas," Dravis, J. J., *Science* Vol. 219, pp. 385–386, 28 January 1983. Copyright 1983 by the AAAS. Table 11–1 from Table 3, "Average Beach Face Slopes Compared to Sediment Diameters" (p. 127) from SUBMARINE GEOLOGY, third edition by Francis P. Shepard. Copyright © 1948, 1963, 1973 by Francis P. Shepard. Reprinted by permission of Harper & Row, Publishers, Inc. Figure 14–10 from "Nitrogen Fixation by Floating Diatom Mats: A Source of New Nitrogen to Oligotrophic Ocean Waters," Martinez, L., et al., *Science* Vol. 221, pp. 152–154, 8 July 1983. Copyright 1983 by the AAAS. Figures 15–4B, 15–4C, 15–4D, and 16–26 courtesy of James M. King, Graphic Impressions, P.O. Box 21626, Santa Barbara, CA 93121. Figure 5–5C courtesy of Dr. Howard Spiro. Appendix V from *Marine Biology,* Thurman, H., and Webber, Merrill Publishing Co., 1984.

Dedicated to the memory of Kip Sears, who presented my first effort as a writer to the Merrill Publishing Company in 1972. From that time on he was a most positive influence in my life. He had a pleasant manner that radiated self-assurance, and he could make difficult decisions and never look back. Kip was an excellent example of how one can make the most of each day in this life. He will always be missed.

PREFACE

In the first edition of this text, my stated objective was to "help the student develop an appreciation for the oceans through an understanding of their physical and biological processes." I had not, however, presented this very worthwhile goal with any great sense of urgency.

But by the second edition, the 1973 oil embargo conducted by the OPEC nations created strong concern about the development of potential oil reserves lying beneath the continental margin. In a more pressing tone, I discussed the investigation of non-polluting, renewable energy sources such as ocean winds, currents, waves, tides, and temperature differences. Political activities were aimed at clearing the way for the mining of manganese nodules from the deep-ocean floor and the establishment of a 200-mile Exclusive Economic Zone by more and more nations. I pointed out that the Third United Nations Conference on Law of the Sea had been meeting at least once a year since 1973 in an effort to provide a solution to all of these problems. I increased coverage of renewable energy resources of the marine environment in the chapters dealing with physical motions of the ocean and added a chapter to cover in greater depth the problems of the Law of the Sea, fisheries, mariculture, minerals, marine pollution, and ocean research. My objective had intensified from "appreciation" to "understanding" so that we could make sound decisions to proceed with the development of the oceans.

In the third and fourth editions, I continued a comprehensive and intelligible discussion of the concepts necessary for a broad understanding of the physical and biological phenomena of the oceans. I wanted to provide an up-to-date picture of the problems arising from the development of marine resources as well as the degree of progress being achieved toward their solutions.

At present, most of the problem areas mentioned are still with us. A good many of the problems continue to elude solution because of their entanglement in the political confrontation existing between the developing and developed nations of the world. Thus, if we are to take a positive position relative to the possible solution of the problems discussed, we can do so only in the knowledge that there exists a relatively high level of public awareness and understanding of them.

My primary goal with this fifth edition—as it has been with all previous revisions—is to impart the new knowledge about the oceans and our means of studying them. Of course, we cannot cover this new information in depth. But the text does attempt to show clearly the essential concepts involved and explain their significance.

Most of the new information has been incorporated into the general text, but there are eight new features to complement the eighteen of the previous edition. These features explore recent discoveries to promote students' appreciation of the oceans and how we learn about them. The features range from interviews with experts in salmon ranching and deep-ocean benthos to overviews of complex phenomena such as the Southern Oscillation–El Niño system, ophiolites, and remote satellite and underwater observation.

I have made a special effort to reorganize the presentation based on suggestions from readers. The discussion of basic chemistry has been moved from chapter 6, "Properties of Water," to chapter 2, "The Origin of the Earth, Its Oceans, and Life in the Oceans." In its new location, this discussion can help give students some basic understanding of chemistry before they take on the chemical considerations discussed in regard to the origin of life in the oceans. In addition, the discussion of isostasy has been moved from chapter 4, "The Origin of Ocean Basins: Theory of Global Plate Tectonics," to chapter 3, "Marine Provinces." This was suggested because the subject seems to be more related to the fundamental difference between continents and ocean basins than to the plate tectonic process. To assist students in developing their knowledge of word roots, I have added an appendix on roots, prefixes, and suffixes. Improvements in the readability of the text have been made throughout the book.

A major focus of this revision is the improvement of illustrations. Many of the line illustrations have been redrawn, and captions have been improved to make them more useful to the students. The color insert sections have been expanded, and the plates have been arranged to follow four central themes: an introductory overview, marine geology, physical oceanography, and marine biology. Most of the images are new, and they have been carefully integrated into a narrative that explains the significance of each as it relates to the central theme. Both this narrative approach and the placement of the inserts at appropriate locations in the text provide the student with a context for understanding the relevance of each photograph to the specific subject matter.

We anticipate that these changes will be helpful and believe the book will meet the needs of more teachers and students of introductory oceanography than ever before.

I would like to extend a special thank-you for the valuable suggestions made by those professors who reviewed the manuscript for the fifth edition of *Introductory Oceanography:* Ernest Angino, University of Kansas; James Conkin, University of Louisville; William T. Fox, Williams College; David M. Karl, University of Hawaii, C. Ernest Knowles, North Carolina State University; Ted Loder, University of New Hampshire; James M. McWhorter, Miami Dade Community College; Larry L. Malinconico, Southern Illinois University; Edward D. Stroup, University of Hawaii; and Mel Zucker, Skyline College.

CONTENTS

INTRODUCTORY
OCEANOGRAPHY

PLATE 1

We humans are drawn to the ocean by our attraction to our close relatives, whales and other marine mammals, living at its surface. The humpback whale has aroused an especially large amount of interest with its "songs" and spectacular breaching behavior.

Some of the whales, all of which are marvelously streamlined, make remarkably deep dives and flaw-lessly track prey in water that is totally dark.

Sperm whales are known to dive to depths in excess of 2 km (1.2 mi). They are aided in making these long, deep, dives by a much greater volume of blood per unit of body mass than other mammals.

Many whales, including the sperm whale, locate and track their prey in the deep, dark waters by using echolocation—bouncing sound waves off the prey.

PLATE 2

A

B

But we also are driven to learn more of our distant relatives, invertebrates, that inhabit the ocean to its greatest depths. Of particular interest are relatively large bottom-dwelling animals living in association with hot-water springs found along the rift valleys near the axes of submarine mountain ranges. These animals are independent of the microscopic plant life that lives in the sunlit waters and supports the vast majority of marine animals.

These vent communities are supported by bacteria that produce food in total darkness by extracting chemical energy from the hydrogen sulfide gas that is dissolved in the vent water.

The first of these vent communities was discovered in 1977. Off the coast of Ecuador and near the equator, they were recorded by a photographic sled towed through the region of a temperature anomaly at the ocean floor. This initial find was in a rift valley at the axis of a submarine mountain range—the Galapagos Rift.

The more prominent members of this deep assemblage are large tube worms living in tubes over 1 m (3.3 ft) long, as well as clams and mussels up to 25 cm (9 in) long. None of these larger animals possess guts, and they may receive most of their nutrition from a symbiotic relationship with the chemosynthetic bacteria living within their tissue.

Other members of the hydrothermal vent communities may filter the bacteria from the water or graze on bacterial mats growing on the rocky ocean floor.

Subsequent investigations have located vents called black smokers that spew out dark clouds of metallic compounds at temperatures of more than 300°C

(572°F) and biological communities associated with numerous hydrothermal vents in the Pacific and Atlantic oceans.

Cold-water seeps have also been found to support biological communities based on a basic bacterial food supply. The cold-water communities have all been located near the margins of continents. They are known to exist at the base of the Florida Escarpment in the Gulf of Mexico, off the coast of Oregon, and near Japan.

Seeping hydrocarbons support a biological community on the continental slope off the coast of Louisiana.

In addition to the unusual biological phenomena associated with the hydrothermal vents, these emissions precipitate metallic deposits that may be the ultimate source of metal-rich ores mined on the continents.

After the metal deposits form at the axes of the submarine mountain ranges, they are transported away by a slowly moving ocean floor. Eventually, the ocean floor may encounter a continent and descend beneath it. During this process, the metallic compounds are distilled out of the oceanic rocks and emplaced in the overlying rocks of the continent.

PLATE 3

PLATE 4

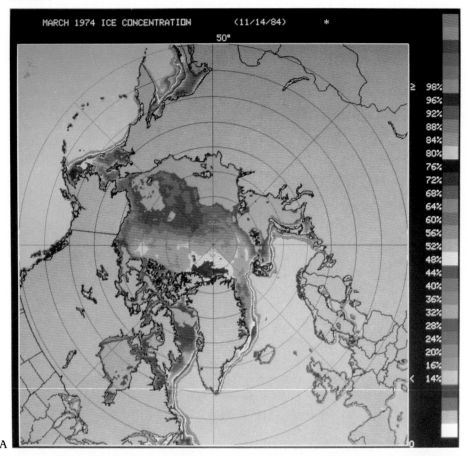

A

MARCH 1974 ICE CONCENTRATION   (11/14/84)   *
50°

≥ 98%
96%
92%
88%
84%
80%
76%
72%
68%
64%
60%
56%
52%
48%
44%
40%
36%
32%
28%
24%
20%
16%
< 14%

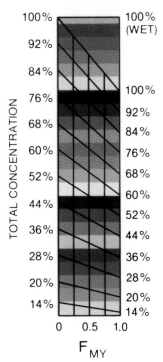

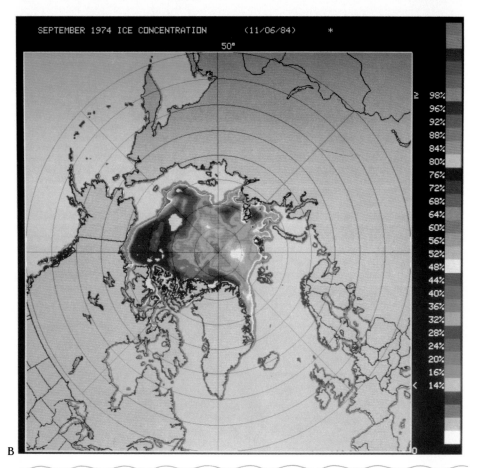

B

SEPTEMBER 1974 ICE CONCENTRATION   (11/06/84)   *
50°

≥ 98%
96%
92%
88%
84%
80%
76%
72%
68%
64%
60%
56%
52%
48%
44%
40%
36%
32%
28%
24%
20%
16%
< 14%

D*uring recent years, much attention has been directed at the polar oceans, where sea ice covers great expanses of ocean surface each winter and melts away in summer months. The seaward edge of this ice is a zone of intense biological activity.*

A permanent polar ice accumulation exists in the Arctic Ocean. During the winter, sea ice in the form of pack ice forms on the open sea to extend the polar ice to the south. Fast ice also forms along the shores to the south. Each summer the pack ice and fast ice melt away.

These 1974 images showing percent of sea-ice concentration are accurate to within 15% for first-year sea ice (pack ice and fast ice) and 25% for multiyear ice (polar ice cap). The scales of ice concentration along the right margins of each image are for first-year ice. The scale above can be used for first-year ice (left side) or multiyear ice (right side). The images were developed from the ESMR (Electrically Scanning Microwave Radiometer) aboard the Nimbus 5 *satellite.*

Sea-surface temperature data gathered by a NOAA satellite was processed to produce this false-color image of the northwest Atlantic Ocean. The warm Gulf Stream waters are shown as orange and red. The colder nearshore waters are blue and purple. Warm water from south of the Gulf Stream is transferred to the north as warm core rings (yellow) surrounded by cooler (blue and green) water. Cold nearshore water spins off to the south of the Gulf Stream as cold core rings (green) surrounded by warmer (yellow and red) water.

The rings form when meanders close to trap warm or cold water within them. The warm rings contain shallow, bowl-shaped masses of warm water about 1 km (0.62 mi) deep with diameters of about 100 km (62 mi). The cold rings have cones of cold water that extend to the ocean floor. They may be more than 500 km (310 mi) across at the surface. The diameter of the cone increases with depth.

PLATE 5

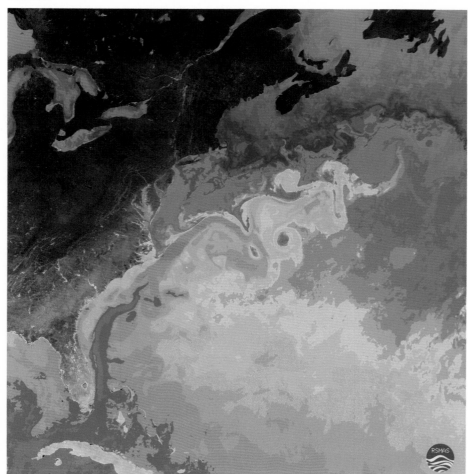

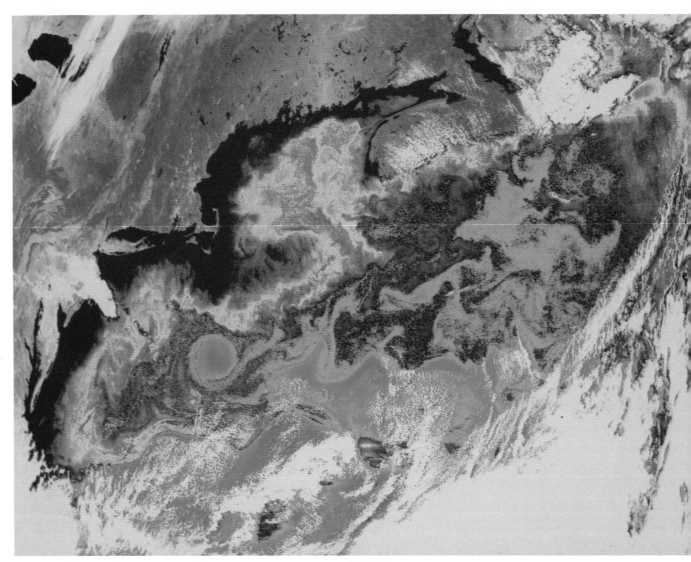

PLATE 6

I n shallow coastal waters, where biological activity reaches its greatest development, there is a correlation between decreased temperature of surface water and high levels of microscopic plant production. Remote sensing by satellites may eventually make it possible to develop a global view of this relationship.

This false-color image of chlorophyll pigment concentration in the northwest Atlantic Ocean shows the low level of biological productivity of the Gulf Stream in blue. Increasing levels of productivity are shown as green, yellow, red, and dark red.

A striking feature of this image is the low level of productivity indicated by the blue color of the warm core ring south of Cape Cod. Note that it is surrounded by more productive green nearshore water. The white is cloud cover.

PLATE 7

oral reefs serve as biological oases in the relatively unproductive
tropical ocean. The coral reef environment includes a great variety of marine life with the most colorful and varied marine communities found in the oceans. Here we see the tentacles of coral polyps
extended to remove microscopic food particles from ocean water.

PLATE 8

oral reefs are found only in warm water, and reef-building corals have microscopic algae living in their tissues to aid them in secreting the calcium carbonate structure that supports them.

However, more than half of the limestone of coral reefs is produced by calcium carbonate-secreting algae that thrive in the nutrient-rich waters of the reef. Actually, the plant mass of a coral reef is three times as great as the animal mass.

The clown fish, which lives unharmed among the stinging tentacles of the sea anemone, may pay for this protection by bringing food to the anemone. This symbiotic relationship that benefits both participants is called mutualism. Two other types of symbiotic relationships that may be found in coral reefs and the ocean as a whole are commensalism and parasitism. In commensalism, one organism benefits while the other is unaffected. Parasitism involves the benefit to one at the expense of the other.

One of the most striking features of the coral reef is the great variety of brilliantly colored fish. The reason for such an audacious display of color has not been discovered. But for whatever reason it exists, no visual experience on this planet will ever produce the thrill and wonderment of one's first view of this display.

CHAPTER ONE
HISTORY OF OCEANOGRAPHY

Our present knowledge of the oceans has many gaps that remain to be filled in the future, but to better understand what is now known about the oceans, we need to take a glimpse into the past. What we know today is an accumulation of advances made in various scientific disciplines, and we will profit from surveying the early history of these investigations.

THE CONCEPT OF GEOGRAPHY DEVELOPS

If human beings evolved in Africa some four million years ago, as present evidence indicates, it is obvious that the vast extent of the ocean was no barrier to their movement from this early home to all parts of the habitable world. When the European explorers set out at the end of the fifteenth century to see what lay beyond the horizon, they discovered that cultures unknown to them had preceded them to many of these far-off places, from Tierra del Fuego in the south to the island of Greenland in the north. Explorers found that not only were the newly discovered continents of the world inhabited, but also populated were the small islands in the Pacific Ocean separated from the mainlands by perhaps thousands of kilometers of ocean expanse. We know little of how the inhabitants reached these islands or by what routes their migrations were made.

Our first concern with the ocean probably was as a source of food. Later in the development of civilization, people built vessels to travel by sea, making it a wide avenue over which distant societies could interact.

Early Ideas about the Land and Sea

Five thousand years ago, the Babylonians thought the earth a mountain surrounded by ocean. The Hebrews of Palestine modified this concept, viewing the earth as a plane with an underground water source to feed the streams and seas, which were thought to cover only one-seventh of the earth's surface.

During the development of ancient Greek culture, the concept of the earth and its oceans evolved significantly. In the eighth century B.C., it is clear from

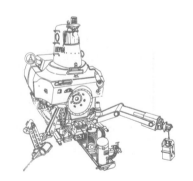

From the *Challenger* report, Great Britain, 1895.

Homer's *Odyssey* that the Greeks saw the earth as a flat or slightly concave disc surrounded by the river Oceanus. During the sixth century B.C., Anaximander hypothesized that the earth stood alone in space at the center of the universe. Anaximander, who made the first known map, thought the earth was a cylinder with a height equal to one-third of its diameter. Land was a large island in the center of the cylinder; this island was divided by the Mediterranean Sea into Europe and Asia. Parmenides was the first to mention a spherical earth, at the beginning of the fifth century B.C.

The world as it was viewed by the Greeks is presented for us on a map constructed in 450 B.C. by the geographer Herodotus. In it he depicts the Greek civilization centered around the Mediterranean Sea (fig. 1–1). To the north, east, and south lie the three continents—Europe, Asia, and Libya—bordered by three major seas: to the west, Mare Atlanticum; to the south, Mare Australis; and to the southeast, Mare Erythraeum. The north and northeastern margins of the continents of Europe and Asia are indicated as unknown, but the Greeks believed that the oceans surrounded all three continents.

The Greek astronomer-geographer Pytheas sailed northward to Iceland in 325 B.C. and worked out a method for determining latitude (fig. 1–2). Using astronomical measurements, he also proposed that the tides were a product of lunar influence.

Eratosthenes (276–192 B.C.), librarian at Alexandria, was the first person known to have determined the world's circumference. He discovered that during the summer solstice (the time when the sun is directly overhead at the Tropic of Cancer) the sun was at zenith over Syene (now called Aswan). Eratosthenes observed that a vertical well in Syene was totally illuminated at noon on the day of the solstice and made the following assumptions (fig. 1–3):

1. Because of the great distance between the earth and the sun, the sun's rays falling on both Syene and Alexandria to the north could be considered as parallel.
2. The distance between Alexandria and Syene was 5000 stadia, 1 stadium equaling approximately 0.16 km (0.1 mi).
3. Alexandria and Syene lay on a north-south line, a *meridian*, perpendicular to the equator.

With these three assumptions and the knowledge that a straight line intersecting two parallel lines creates corresponding angles that are equal, Eratosthenes was prepared to measure the circumference of the planet on which he lived.

By observing the shadow of a vertical wall in Alexandria at noon on the day of the summer solstice, Eratosthenes could measure the angle defined by a line connecting the top of the wall and the top of the wall's shadow on the earth's surface. This angle was measured as 7.2°. The meridian on which he had assumed Syene and Alexandria lay could be continued in a full circle around the surface of the earth. This circle contains a total of 360°. The angle corresponding to the one that Eratosthenes measured is the an-

FIGURE 1–1 The World of Herodotus.

The world according to Herodotus, 450 B.C. From the *Challenger* report, Great Britain, 1895.

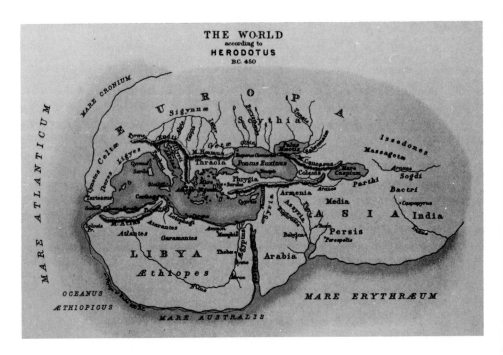

FIGURE 1–2 Determining Latitude.

The method of determining latitude used by Pytheas was to measure the angle between the horizon and the North Star, a star that is directly above the North Pole. Latitude north of the equator is the angle between the two sightings. The present North Star is Polaris.

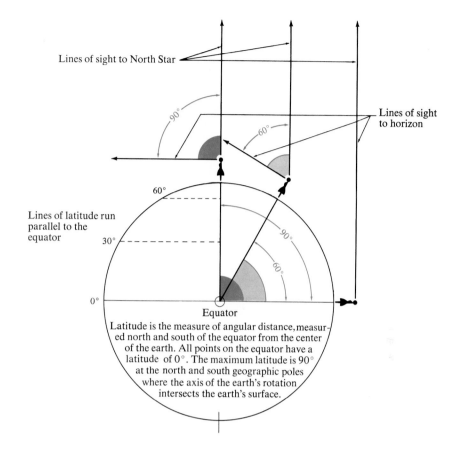

Lines of sight to North Star

Lines of sight to horizon

Lines of latitude run parallel to the equator

Equator

Latitude is the measure of angular distance, measured north and south of the equator from the center of the earth. All points on the equator have a latitude of 0°. The maximum latitude is 90° at the north and south geographic poles where the axis of the earth's rotation intersects the earth's surface.

FIGURE 1–3 Eratosthenes' Determination of the Circumference of the Earth (circa 200 B.C.).

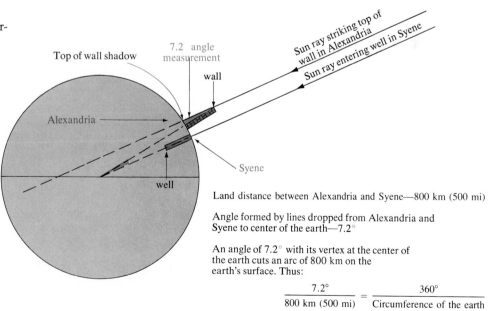

Top of wall shadow

7.2 angle measurement

wall

Sun ray striking top of wall in Alexandria

Sun ray entering well in Syene

Alexandria

Syene

well

Land distance between Alexandria and Syene—800 km (500 mi)

Angle formed by lines dropped from Alexandria and Syene to center of the earth—7.2°

An angle of 7.2° with its vertex at the center of the earth cuts an arc of 800 km on the earth's surface. Thus:

$$\frac{7.2°}{800 \text{ km (500 mi)}} = \frac{360°}{\text{Circumference of the earth}}$$

Circumference of the earth = 40,000 km (24,840 mi)

gle defined at the center of the earth by extending a line along the wall's surface to that point where it intersected the sun's ray that passed through Syene. Since 7.2° represents one-fiftieth of the number of degrees that are in a full circle, the distance defined along the circumference of the circle by an angle of 7.2° must be one-fiftieth of the total circumference of the earth. Therefore, if 5000 stadia represent 800 km (500 mi), the circumference of the earth is 40,000 km (24,850 mi). This figure compares very well with the 40,032 km (24,875 mi) meridional circumference determined by the more precise methods in use today.

Contributions of the Romans

Two Roman contemporaries of Jesus Christ, one a geographer named Strabo (63 B.C.–A.D. 24) and the other a rhetorician named Seneca (54 B.C.–A.D. 39), added greatly to our understanding of the world. Strabo had observed the volcanic activity that is characteristic of the Mediterranean area and concluded that, as a result of this activity, the land periodically sank and rose, causing the sea to invade and then recede from the continents. He further observed that the rivers flowing across the continent erode material from the surface and transport it to the oceans.

Seneca introduced the concept of the *hydrologic cycle* (fig. 1–4), part of which is the flow of the rivers into the oceans. He explained that the level of the oceans did not rise as a result of the continual inflow provided by the rivers because of the process of water evaporating as water vapor into the atmosphere. Some of this vapor is transported over the continents,

where it condenses and is released as precipitation onto the continents. It then returns as runoff to the oceans.

In approximately A.D. 150, Ptolemy produced a world map that indicates Roman knowledge at that time. He introduced lines of longitude and latitude on his map. Before this time mariners could determine the latitude of any point on the surface of the earth using the method introduced by Pytheas, but it was impossible for them to accurately determine longitude. Ptolemy's map indicated, as did the earlier Greek maps, the continents Europe, Asia, and Africa and showed the Indian Ocean to be surrounded by a partly unknown landmass. All oceanic bodies were generally considered to be seas similar to the Mediterranean, having boundaries defined by unknown landmasses.

The Middle Ages

After the fall of the Roman Empire the Mediterranean area was dominated by Arab influence, and the writing of the Greeks and Romans passed into the hands of the Arabs to be forgotten by the Christians in Europe. Subsequently, the Western concept of geography degenerated considerably; one notion envisioned the world as a black disc with Jerusalem at the center.

The post–Roman Empire concept of geography, as held by much of the Western world, is represented by the map of the sixth-century navigator Cosmas (fig. 1–5). His map depicts the earth as a rectangle measuring 20,000 km by 10,000 km (12,420 mi by 6210 mi).

FIGURE 1–4 The Hydrologic Cycle.

Evaporation and plant transpiration send water vapor into the atmosphere. The vapor condenses and returns to the earth's surface as precipitation. That which falls onto the land is evaporated, runs to the oceans through river systems, or soaks into the ground to become groundwater.

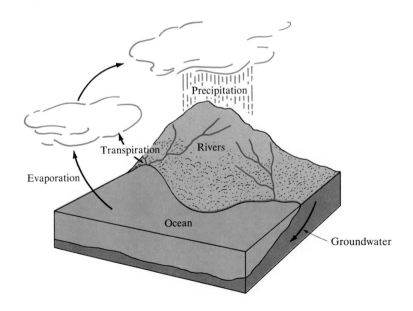

FIGURE 1–5 The World of Cosmas.

The map of Cosmas Indicopleustes shows how the southern European concept of the world had deteriorated by the sixth century. From the *Challenger* report, Great Britain, 1895.

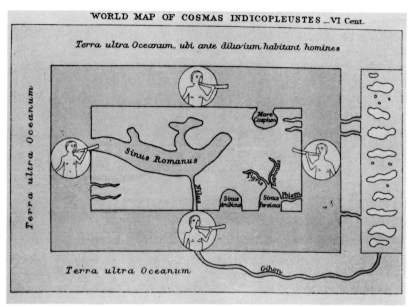

The Arabs, meanwhile, were trading extensively with east Africa, southeast Asia, and India and had learned the secret of the monsoons. They took advantage of these winds, making their trade voyages easier. During the summer, when the monsoon winds blew from the southwest, ships laden with goods for trade would leave the Arabian ports and sail eastward across the Indian Ocean. The return voyage would be timed to take advantage of the northeasterly trade winds that occurred during the winter, making the transit relatively simple.

Major Arabian contributions to the advancement of knowledge during this period include laying the foundation of modern trigonometry; introduction of Arabic numerals, which facilitate calculations; invention of the magnetic compass box with a windrose; and invention of the astrolabe. Celestial navigation is also believed to have been highly developed by the Arabs.

In Europe, the nautical inactivity of the southern Europeans was offset by the vigorous exploration of the Vikings of Scandinavia. Late in the ninth century, not shackled by considerations of Christian theology and aided by a period of climatic warming, the Vikings conquered Iceland. In 982 Eric the Red sailed westward from Greenland and discovered Baffin Island. In 1000 Eric's son, Leif Ericson, discovered what was then called Vinland, and he spent the winter in that portion of North America we call Newfoundland.

By the beginning of the thirteenth century, the climate began to cool and the northern Atlantic became clogged with ice throughout most of the year, isolating the Viking colonies in Greenland. Owing to this deterioration in climate the explorations of the Vikings were curtailed, and they were unable to exploit their newly discovered territories.

Age of Discovery

During a thirty-year period from 1492 to 1522, known as the Age of Discovery, the Western world came to fully realize the vastness of the earth's water-covered surface. The continents of North and South America were "discovered." The globe was circumnavigated, and it was learned that human populations existed elsewhere in the world. Human cultures were found throughout the newly discovered continents and islands, and they were vastly different from those with which the voyagers were familiar.

Precipitating these voyages was the capture of Constantinople, the capital of eastern Christendom, in 1453 by Sultan Mohammed II. This event isolated the Mediterranean ports from the riches of the East Indies and caused the Western world to search for a new trading route to this area. As a result of the Arab occupation of Constantinople, the ancient knowledge of the Greeks and Romans was carried out of that city and became available to the powers of southern Europe, who were interested in reestablishing trade connections with the East Indies.

Prince Henry "the Navigator" of Portugal had established a marine observatory to improve Portuguese sailing skills, but for years attempts to reestablish trade by ocean routes met with failure.

One of the greatest obstacles to an alternate trade route was the need to travel around the tip of Africa. The Cape of Good Hope was finally rounded by Bartholomeu Díaz in 1486. He was followed in 1498 by Vasco da Gama, who continued his trip around the tip of Africa to India.

The idea for a voyage such as the one Columbus undertook was initiated by the Florentine astronomer

Toscanelli. He wrote to the king of Portugal, suggesting that a course be charted to the west in an attempt to reach the East Indies by crossing the Atlantic Ocean. Columbus later contacted Toscanelli and was given a copy of the information indicating that India could be reached at a longitude just west of the continent of North America.

After his well-known difficulties in originating the voyage, Columbus set sail with 88 men and three ships August 3, 1492, from the Canary Islands. During the morning of October 12, 1492, the first American land was sighted, and it was generally believed to have been Watling Island in the Bahamas. Recent research by the National Geographic Society has convinced them Columbus landed at Samana Cay, 120 km (75 mi) to the southeast. Because Columbus had greatly underestimated the distance to the East Indies via the Atlantic Ocean, he was convinced that he had arrived at these Islands. Upon his return to Spain and the announcement of his discovery, additional voyages were planned, and the Spanish and Portuguese continued to explore the coasts of North and South America. In 1505, after his fourth voyage to the "New World," Columbus still had an erroneous view of the earth. He said, "The world is small. Of its parts, six are occupied by land and only the seventh is covered with water."

The Atlantic Ocean became familiar to the European explorers, but the Pacific Ocean was not seen by them until 1513 when Vasco Nuñez de Balboa attempted a land crossing of the Isthmus of Panama and sighted the Pacific Ocean from atop a mountain.

The culmination of this period of discovery is marked by the circumnavigation of the globe by Ferdinand Magellan (fig. 1–6). In September of 1519, Magellan left Sanlucar de Barrameda, Spain, and traveled through a passage to the Pacific at 52° S latitude, now named the Straits of Magellan. After discovering the Philippines on March 15, 1521, Magellan was killed in a fight with the inhabitants of these islands. Juan Sebastian del Caño completed the circumnavigation by taking one of the ships, *Victoria*, across the Indian Ocean and back to Spain in 1522. On this trip Magellan had attempted to measure the depth of the Pacific Ocean by a weighted line, but he was unable to reach bottom.

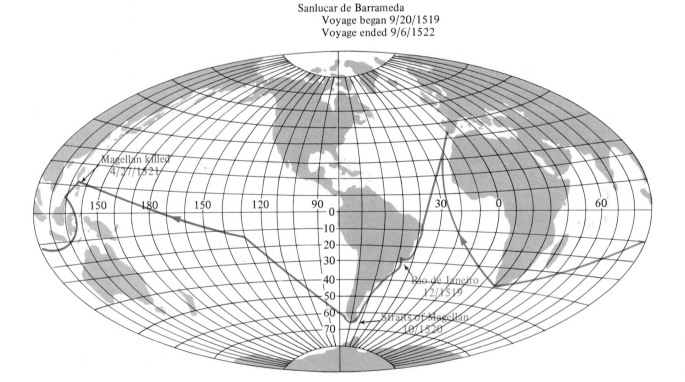

FIGURE 1–6 Voyage of Magellan.
Searching for a western route to the East Indies, Magellan culminated the Age of Discovery by beginning a voyage that ended for him with death in the Philippines. Juan Sebastian del Caño returned to Spain in the *Victoria*, one of five ships that began the voyage, to complete the first circumnavigation of the earth. Base map courtesy of National Ocean Survey.

After these voyages, the Spanish initiated many others for the purpose of obtaining the gold of the Aztec and Inca cultures in Mexico and South America. While the Spanish concentrated on plundering the Aztecs and Incas, the English and Dutch plundered the Spanish. The political dominance of Spain came to an end with the defeat of the Spanish Armada by the British in 1588. With the squelching of this attempted invasion of the British Isles, the English became the dominant maritime power, and they remained so until early in the twentieth century.

THE SEARCH TO INCREASE SCIENTIFIC KNOWLEDGE OF THE OCEANS

Captain James Cook

The next major interest in the oceans was more scientific, as the English determined that increasing their knowledge of the oceans would help them maintain their maritime superiority. The more successful of the early voyages that were initiated to learn more about the physical nature of the oceans were conducted by the English navigator James Cook (1728–79), son of a farm laborer (fig. 1–7).

In 1768, Cook, then a lieutenant, had command of H.M.S. *Endeavour*, which was assigned to conduct observations of the transit of Venus. After completing his tasks in the South Pacific, the ship sailed south to 40° latitude in search of a continent, Terra Australis, thought to exist in the high southern latitudes. Cook did not observe any land, so he sailed to what is now New Zealand and charted those islands. He also charted the eastern coast of Australia, nearly losing his ship in crossing the Great Barrier Reef, before returning home from his first major voyage into the Pacific.

Cook's second voyage, commissioned to continue the search for a continent in the southern ocean, left England on July 13, 1772. The expedition included two ships, the *Resolution* and the *Adventure*. With this voyage, Cook established that if any continent existed in the southern oceans it must be beyond the ice fields and possibly covered by them. Before completing the voyage, Captain Cook discovered and charted South Georgia and the South Sandwich Islands in the South Atlantic. He returned home July 30, 1775.

Following this, his greatest voyage, Cook was honored for preserving the health of his seamen by requiring them to eat sauerkraut and other foods regularly, thus preventing the dread disease scurvy that

FIGURE 1–7 Captain James Cook, RN (1728–79). Courtesy U.S. Navy.

plagued sailors who spent long periods of time at sea. This diet prevented the vitamin C deficiency that was the cause of this disease.

Cook volunteered to take command of a third voyage to examine the northern Pacific coast of America in search of a northwest passage. He visited New Zealand, Tonga, and the Society Islands before sailing north on December 8, 1777, and discovered the Hawaiian group on January 18, 1778. After reaching the coast of North America, he sailed along it and through the Bering Strait as far as 70° 44′ N latitude before being stopped by pack ice. Returning to Hawaii for the winter, Cook anchored in Kealakekua Bay from January 17 to February 4, 1779. Shortly after leaving port, the ships had to put back in to repair a top mast, and on February 14, Cook was killed in a skirmish with the natives ashore.

Cook's expeditions added greatly to the scientific knowledge of the oceans. He determined the outline of the world's largest ocean and was the first person known to cross the Antarctic Circle. Cook also led the way in sampling subsurface temperatures, measuring winds and currents, sounding, and collecting important data on coral reefs. By proving the value of John Harrison's chronometer as a means of determining longitude, Cook made possible the first accurate maps of the earth's surface (*see* fig. 1–8).

An American contemporary of Captain Cook was an early contributor to greater understanding of the ocean's surface currents. Benjamin Franklin, deputy postmaster general for the Colonies, determined that

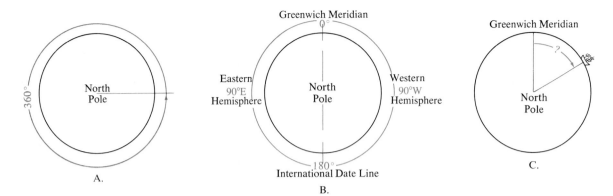

FIGURE 1-8 Determination of Longitude.

View of earth from outer space, looking down on North Pole. *A*, as the earth turns on its rotational axis, it moves through 360° of angle per 24 h. *B*, a prime meridian that runs through Greenwich, England, was selected as the reference meridian, and the earth was divided into a Western and Eastern hemisphere. After John Harrison's chronometer was developed, many ships carried it showing the time on the Greenwich Meridian—Greenwich time. *C*, since the earth rotates through 15° of angle, longitude, per h (360° ÷ 24 h = 15°/h), a ship's captain could easily determine his longitude each day at noon. For example, a ship sets sail west across the Atlantic Ocean, checking its longitude each day at noon (when the sun crosses the meridian running directly overhead).
One day when the sun is at the noon position, the captain checks the chronometer. It reads 16:18 h (0:00 is midnight and 12:00 is noon). What is the ship's longitude? Longitude solution:

16:18 h = 4:18 P.M.

The earth rotates through $\frac{1}{4}°$ (15′) of angle per minute of time.

(One degree of arc is divided into 60 min)

4 h × 15°/h = 60° of longitude

18 minutes of time × 15 minutes of angle/minute of time = 270 minutes of arc

270 min ÷ 60 min/degree (°) = 4.5° of longitude

60° + 4.5° = 64.5° W longitude

it took mail ships coming from Europe by a northerly route two weeks longer to reach the colonies than it did ships that came by a more southerly route. To find out why this occurred, he asked for information concerning the movement of surface waters from captains who came into port and concluded that there was a significant current moving in a northerly path along the eastern coast of the United States. The current moved out in a more easterly path across the North Atlantic. Franklin concluded that this east-flowing current, which the ships traveling a northerly route from Europe had to combat, was responsible for increasing the time of their voyages. He subsequently published in 1777 a map of the Gulf Stream based on these observations (*see* page 190).

Matthew Fontaine Maury

An even greater contribution was made by Matthew Fontaine Maury (1806–73). This career officer in the U.S. Navy was placed in charge of the Depot of Naval Charts and Instruments after suffering an injury early in his career. In the depot were log books containing a large amount of information about currents and

weather conditions in various parts of the oceans. Maury's systematic analysis of these logs produced a compilation of wind and current patterns that proved very useful to the navigators of the nineteenth century. Maury helped organize the first International Meteorological Conference in Brussels in 1853 for the purpose of establishing uniform methods of making nautical and meteorological observations at sea. This standardization greatly increased the dependability of such data, which Maury summarized in *The Physical Geography of the Sea* (1855). Maury is often referred to as the father of oceanography.

Charles Darwin

In the early nineteenth century, an English naturalist named Charles Darwin (1809–82) entered the scientific scene. Darwin's interest, investigating the whole of nature, led him to make one of the most outstanding contributions to the field of biology. Much of the background on which he based his conclusions about evolution by natural selection was gained from observations made during his voyage aboard H.M.S. *Beagle*.

The *Beagle* sailed from Devensport on December 27, 1831, under the command of Caption Robert Fitzroy, with the major objective of completing the survey of the coast of South America.

Darwin spent the next five years aboard the *Beagle*, which continued to sail around the world while making chronometrical measurements (fig. 1–9). The voyage allowed the young naturalist the opportunity to study the plants and animals throughout the world. He studied closely the differences in the plant and animal populations living in different environments and concluded that populations change slowly over the immense period of time represented by the geologic past.

Darwin's observations led him to conclude that birds and mammals must have evolved from the reptiles and that the similar skeletal framework of the human, the bat, the horse, the giraffe, the elephant, the porpoise, and other vertebrates requires that they be grouped together. Darwin felt that the superficial differences that could be observed between populations were the result of adaptation to different environments and modes of existence.

In 1859, Darwin published his controversial book, *The Origin of Species*, which presented his theory of the evolution of living things into their many forms. Although Darwin's ideas were highly controversial, many can be supported by an impressive array of scientific observations.

The Rosses—Sounders of the Deep

Two of the earliest successful investigators of deep oceans were Englishmen, Sir John Ross and his nephew Sir James Clark Ross. Sir John Ross explored Baffin Bay in Canada in 1817 and 1818 and took depth measurements. Ross also collected samples of bottom-dwelling organisms and sediments with a device of his own design called a *deep-sea clamm*. He recovered sea stars and worms in mud samples from a depth of 1.8 km (1.1 mi).

Sir James Clark Ross extended the soundings to greater depths on voyages to the Antarctic from 1839 to 1843. A 7-km (4⅓-mi) sounding line was used on

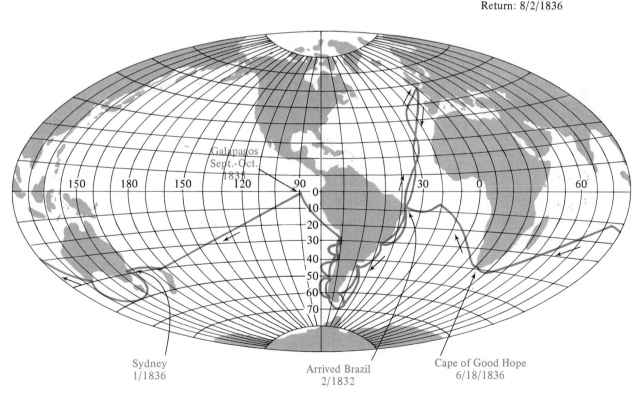

FIGURE 1–9 Voyage of the *Beagle*.
Sailing as naturalist aboard H.M.S. *Beagle*, Charles Darwin gathered the evidence that enabled him to develop his theory of biological evolution through natural selection. Base map courtesy of National Ocean Survey.

these voyages, and even this great length was at times insufficient to reach the bottom of the ocean. Sir James observed that the animals he recovered from the cold waters of the Antarctic were the same species his uncle had recovered from the Arctic area and that they were very sensitive to temperature increase. This discovery led him to the conclusion that the waters making up the deep ocean must be of uniformly low temperature.

Edward Forbes

Also making an important contribution to the biology of the oceans was a British naturalist, Edward Forbes (1815–54). Forbes was interested in determining the vertical distribution of life in the ocean, and after repeated observations, he divided the sea into specific life-depth zones. He came to the conclusion that plant life is limited to the zone near the surface. Forbes also found that the animal concentration was greater near the surface and decreased with increasing depth until only a small trace of life, if any, would remain in the deepest waters.

Forbes's followers either misinterpreted his statement of the condition of life's distribution in the ocean or decided to apply their own logic to it, and they concluded that no life existed in the deep ocean, as life would be impossible under the conditions of high pressure and absence of light and oxygen in this environment. Ignoring the findings of the Rosses in the high southern and northern latitudes, they persisted in their belief that no life existed in the abyssal ocean.

The *Challenger* Expedition

The controversy over the distribution of life in the ocean was one of the factors that helped stimulate interest in the first large-scale voyage with the express purpose of studying the subject.

In 1871, the Royal Society recommended that funds be raised from the British government for an expedition to investigate the following subjects.

1. physical conditions of the deep sea in the great ocean basins
2. the chemical composition of seawater at all depths in the ocean
3. the physical and chemical characteristics of the deposits of the sea floor and the nature of their origin
4. the distribution of organic life at all depths in the sea, as well as on the sea floor

A staff of six scientists under the direction of C. Wyville Thompson left England in December 1872 aboard H.M.S. *Challenger* (fig. 1–10). A 2306-ton corvette that had been refitted to conduct scientific investigation, the *Challenger* returned in May 1876, after having traversed large portions of the Atlantic and Pacific oceans (fig. 1–11). The achievements of the *Challenger* expedition, which covered 127,500 km (79,223 mi), included 492 deep-sea soundings, 133 bottom dredges, 151 open-water trawls, and 263 serial water temperature observations. The more outstanding aspects of the voyage were the netting and classification of 4717 new species of marine life and the measurement of a water depth of 8185 m (26,850 ft) in the Marianas Trench. Seventy-seven samples of ocean water collected during the *Challenger* expedition were analyzed in 1884 by C. R. Dittmar. This first refined analysis contributed greatly to our understanding of ocean salinity, which will be discussed in chapter 6.

Fridtjof Nansen

One of the most unusual of the voyages was initiated by a Norwegian, Fridtjof Nansen (1861–1930). Dr. Nansen developed a great interest in exploring the North Atlantic and the Arctic area. To aid him in his planning, he compiled the results of earlier attempts to explore the Arctic area during the latter part of the nineteenth century. Of particular interest to Nansen was the ill-fated American voyage of George Washington DeLong and his crew on the *Jeanette* in 1879. This expedition had attempted to sail through the Bering Strait to Wrangell Island, from which it planned to go overland to the pole. Scientists believed that Wrangell Island was possibly the southern tip of a peninsula extending southward from the great, but yet undiscovered, Arctic continent.

The *Jeanette* became stuck in the floating pack ice on September 6, 1879, and drifted north of Wrangell Island, proving that the island was not a peninsula of the Arctic continent. After two years of drifting, the *Jeanette* sank off the New Siberian Islands. DeLong and many of his crew perished in the Lena Delta region of eastern Siberia.

In 1884, a number of articles that could be traced to the *Jeanette* were found frozen in the pack ice off the eastern coast of Greenland. On the basis of this evidence, Nansen considered that the articles must have drifted from the wreck of the *Jeanette* to the observed location by following a path that would have taken them over or near the North Pole. He believed

FIGURE 1–10 H.M.S. *Challenger.*

From the *Challenger* report, Great Britain, 1895.

the ice flow that transported these articles could be used to transport an expedition across the North Pole as well.

Eventually Nansen was able to raise sufficient funds for building the *Fram*, a ship that would be so designed that the expanding ice would not crush it, but rather would force it up to the surface, free of the grip of the growing ice sheets. It was assumed that the journey would take a period of approximately three years, as this was the time that had been re-

quired for the *Jeanette* articles to follow the same course.

Provisions for 13 men for five years were stored away on the small ship, and the crew set sail from Oslo on June 24, 1893. The objective was to sail as far as the New Siberian Islands, then attempt to plow in among the ice sheets as far as they could to assure getting into the proper current. On September 21, after fighting their way along the northern cost of Siberia, the men had not yet sighted the New Siberian

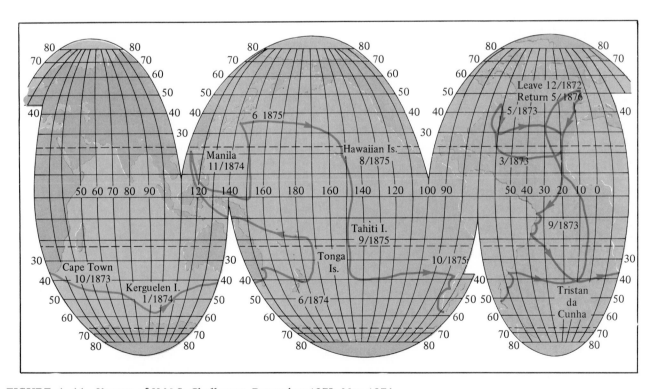

FIGURE 1–11 Voyage of H.M.S. *Challenger*, December 1872–May 1876.

The route traveled by H.M.S. *Challenger* during the first major oceanographic voyage. Base map courtesy of National Ocean Survey.

Islands. The ice captured them at a latitude of 78° 30′ N, over 1100 km (683 mi) from the North Pole. The 13-man crew had now begun what would be a long and lonely endeavor (fig. 1–12).

The *Fram*, on November 15, 1895, reached its most northerly point, 394 km (244 mi) from the pole. On August 13, 1896, the little *Fram* broke ice and was again unbound in the open sea. She had drifted a total of 1658 km (1028 mi) during her three-year entrapment. The early part of the drifting journey was accomplished at a slow rate, with an increasing rate of speed being attained during the last two years of the voyage.

The drift of the *Fram* proved that no continent existed in the Arctic Sea and that the ice that covered the polar area throughout the year was not of glacial origin, but a freely moving pack ice accumulation that had been formed directly upon the ocean surface. During the voyage, the crew had found that the depth of this ocean was in excess of 3000 m (9840 ft) and that there was a rather surprising body of relatively warm water with temperatures as high as 1.5°C (35°F) between the depths of 150 and 900 m (about 500 and 3000 ft). Nansen correctly described this water as being a mass of Atlantic water that had sunk below the less saline Arctic water.

Nansen's realization of the need for more accurate measurements of salinity and temperature led him to develop the once widely used bottle for sampling the ocean depths, the Nansen bottle. His observations of the direction of ice drift relative to the wind direction helped V. Walfrid Ekman, a Scandinavian physicist, to develop the mathematical explanation of this phenomenon. Nansen, who was awarded the Nobel Peace Prize in 1922, Ekman, and other Scandinavian scientists led the way in the development of the field of physical oceanography in the early twentieth century.

TWENTIETH-CENTURY OCEANOGRAPHY

Voyage of the *Meteor*

Oceanography entered a new era with the voyage of the *Meteor*, 1925–27. This German expedition was the first to use an echo sounder, making it possible to obtain a continuous depth recording as a vessel proceeded along its course. This method represented a great advance over reliance on scattered soundings and interpolation of water depth between the points of known depths. The *Meteor* gathered data for 25 months and concentrated on increasing the knowledge of the South Atlantic Ocean. It was this expedition that first revealed the true ruggedness of the ocean floor.

FIGURE 1–12 Voyage of the *Fram*.

The course of the *Fram* after becoming frozen in ice near the New Siberian Islands, September 21, 1893. On August 13, 1896, the little ship was released by the ice off Spitsbergen.

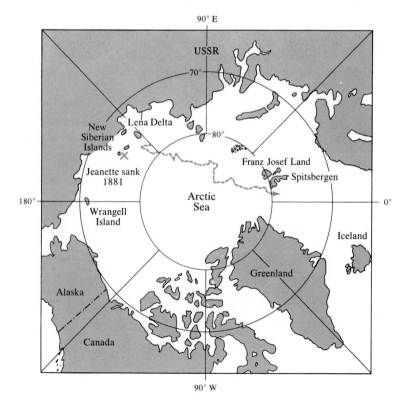

Many contributions from a number of nations have been made to the understanding of the oceans after the voyage of the *Meteor*. We will not discuss the details of these voyages, but we must emphasize that our knowledge of the world's oceans is still far from complete. Efforts to bring it to a more complete state will, of necessity, continue for years.

Oceanography in the United States

Although the first large-scale contribution to oceanography by a citizen of the United States resulted from the efforts of Lieutenant Matthew Maury of the U.S. Navy and culminated in his book *The Physical Geography of the Sea*, it was not until later in the nineteenth century that the United States became active in organizing voyages for the purpose of increasing our knowledge of the ocean. This phase of involvement actually began in 1877 with the voyages of the *Blake* under the direction of Alexander Agassiz. This young scientist, the son of the Swiss naturalist Louis Agassiz, contributed greatly to the development of oceanographic research in the United States. Since then, many cruises have been initiated with the prime objectives of increasing the scientific knowledge of the world's ocean. In the following paragraphs, we will list and discuss briefly the major institutions in the United States that are concerned with studying the world's ocean and some of their more significant investigations.

The U. S. Navy Oceanographic Office has the primary objective of improving the combat readiness of the fleet, and basic research performed toward this objective is a part of its assignment. The results of their research, as well as that of the Office of Naval Research and the U. S. Coast Guard, are of value and basic importance to all oceanographers.

On February 10, 1807, the U.S. Coast and Geodetic Survey was created by an act of Congress after President Jefferson requested a survey of the coast of the United States. This organization, now the National Ocean Survey, has primary responsibility for bathymetric (ocean depth) surveys in the waters adjacent to the United States and its territories. Although the major work done by this agency is underwater mapping, the vessels that are involved are outfitted to conduct a wider scope of oceanographic research, gathering a wide range of oceanographic data.

The Commerce Department's National Oceanic and Atmospheric Administration (NOAA) works to ensure wise use of ocean resources through its National Ocean Survey, National Marine Fisheries Service, and Office of Sea Grant.

FIGURE 1–13 Dr. Maurice Ewing.
Dr. Maurice Ewing (1906–74) pioneered seismic techniques for studying sediments on the ocean floor. He was director of Lamont-Doherty Geological Observatory at Columbia University from 1949 to 1972. Photo courtesy of Columbia University.

Three major marine laboratories have pioneered much of the marine research in the United States. The oldest of these institutions is Scripps Institution of Oceanography of the University of California at San Diego. Originated at La Jolla, California, in 1903, Scripps investigates a wide spectrum of marine problems.

Woods Hole Oceanographic Institution originated at Woods Hole, Massachusetts, in 1930 as a private nonprofit organization devoted to the scientific study of the world ocean. The ship *Atlantis II* operated by this institution was one of the first modern ships designed to conduct oceanographic research.

The Lamont-Doherty Geological Observatory of Columbia University was founded at Torrey Cliffs Palisades, New York, in 1949 under the direction of Dr. Maurice Ewing (fig. 1–13). He was one of the great pioneers of geophysical methods for investigation of the geology of the ocean bottom.

These institutions have led the way in twentieth-century oceanographic exploration. Their development of new devices for the study of the ocean depths has been sparked by the desire to unravel more of the mysteries hidden there. One of these imaginative devices is the Floating Instrument Platform (FLIP) shown in figure 1–14. It is designed for making acoustical and other open-ocean measurements that require a stable platform. Camera systems described in chapter 17 and magnetometer instru-

FIGURE 1–14 Floating Instrument Platform (FLIP).

FLIP, a 108-m (454-ft) Floating Instrument Platform, designed and developed by Marine Physical Laboratory at Scripps Institution of Oceanography. In the vertical position (inset) it gives scientists an extremely stable platform from which to carry out underwater acoustic and other types of oceanographic research. FLIP has no motive power of its own and must be towed to a research site in the horizontal position. Once on station, the ballast tanks are flooded and the vessel flips vertically, leaving 17 m (56 ft) of the platform above water. The work completed, water is forced from the tanks and FLIP resumes the horizontal position. Courtesy of Scripps Institution of Oceanography, University of California, San Diego.

ment capsules (fig. 1–15) have been developed for free-fall emplacement on the ocean floor. They can be left to operate for days or months. Automatic timers or acoustical signals release these devices to float to the surface for recovery.

The need for cooperative efforts was recognized in the Deep Sea Drilling Project, which was carried out under the leadership of the Scripps Institution of Oceanography, with the cooperation of a number of educational institutions that have developed oceanographic programs. In November 1963, the director of the National Science Foundation announced the establishment of a national program for sediment coring. Scripps Institution of Oceanography, the Rosenstiel School of Atmospheric and Oceanic Studies at the University of Miami, Florida, Lamont-Doherty Geological Observatory of Columbia University, and Woods Hole Oceanographic Institution in Massachusetts united to form the Joint Oceanographic Institutions for Deep Earth Sampling (JOIDES). They were later joined by the departments of oceanography of the University of Washington, Texas A&M, University of Hawaii, Oregon State University, and the University of Rhode Island.

A ship capable of drilling the ocean bottom at 6000 m (3.7 mi) while floating at the ocean surface had to be designed specifically to carry on this program. The *Glomar Challenger* was constructed and launched in 1968 (fig. 1–16). The first leg of the Deep Sea Drilling Project commenced that year. The *Glomar Challenger* is 122 m (400 ft) long and 20 m (66 ft) in beam, and its displacement is 10,500 tn loaded. The vessel is a

FIGURE 1–15 Magnetometer Capsule.

The free-falling magnetometer capsule developed at Scripps Institution of Oceanography is dropped to the floor of the ocean from a research vessel and can remain there for several months to record variations in the earth's magnetic field. It is recalled to the surface by acoustical commands transmitted from the ship when it returns to the site. Courtesy of Scripps Institution of Oceanography, University of California, San Diego.

FIGURE 1–16 Old and New *Challengers.*

Models of H.M.S. *Challenger* and D/V *Glomar Challenger* are shown with the tracks of Phase 1 of the Deep Sea Drilling Project in the background. For propulsion, H.M.S. *Challenger* relied on her sails and auxiliary steam power of 1234 hp. D/V *Glomar Challenger* produced 8800 continuous or 10,000 intermittent hp for propulsion and to operate the drilling equipment. Where H.M.S. *Challenger* shortened sails and used steam power to remain over a site, D/V *Glomar Challenger* used dynamic positioning. The first *Challenger* was 61 m (200 ft) long and displaced 2306 tn. *Glomar Challenger* is 122 m (400 ft) long and displaces 10,500 tn. The scientific impact of both vessels has been revolutionary. While H.M.S. *Challenger* is generally thought of as beginning the science of deep-ocean marine geology, the results of D/V *Glomar Challenger's* drilling and coring program have truly revolutionized the science. Courtesy of the Deep Sea Drilling Project, Scripps Institution of Oceanography, University of California, San Diego.

self-sustained unit capable of remaining at sea for 90 days.

Initially the oceanographic research program was financed by the U.S. government, but in 1975 it became international. Financial and scientific support is now received from West Germany, France, Japan, the United Kingdom, and the USSR. In 1983 the Deep Sea Drilling Project became the Ocean Drilling Program (ODP), with the broader objective of also drilling the continental margins. This new phase of operations is supervised by Texas A&M University. Accompanying this change in emphasis was the decommissioning of *Glomar Challenger* after a highly successful, significant 15-year role in which it made possible a major advancement in the field of earth science—confirmation of sea-floor spreading. In this process, new ocean

floor is being created along submarine mountain ranges. Following in the wake of the *Glomar Challenger* as the ODP drill ship is the much larger *JOIDES Resolution*, which conducted its first scientific cruise in January of 1985.

The Geochemical Ocean Section (GEOSECS) program has been operating since 1972. This project involves sampling and analyzing water from all oceans to study ocean circulation patterns, mixing processes, and biogeochemical cycling.

For examples of the use of submersibles in research, see the special feature *Alvin Explores Spreading-Center Median Valleys* (pp. 81–87).

Many examples of ingenious uses of today's technology in the study of the oceans are discussed in later chapters. They include the use of sound waves

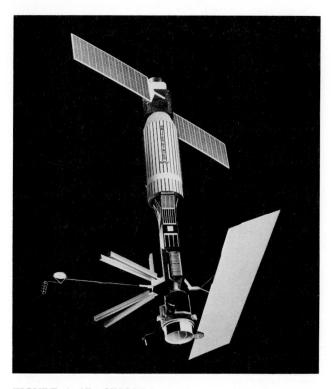

FIGURE 1–17 *SEASAT A.*

SEASAT A, the first oceanographic satellite, was launched July 7, 1978. It shorted out October 10, 1978, but provided sufficient data to prove the value of satellite remote sensing to the study of the oceans. Courtesy Jet Propulsion Laboratory, NASA.

to image horizontal slices of ocean temperature by use of ocean acoustical tomography. This technique is similar in design to the use of X rays to image slices through the human body with computerized axial tomography or CAT scans (see p. 196).

A rapidly developing area of ocean research involves the use of remote sensing from aircraft and satellites. Such techniques are especially suitable for worldwide measurement of surface phenomena; they are also useful in obtaining data about the ocean floor (fig. 1– 17) (see *SEASAT Maps the Ocean Floor,* pp. 52–54, *SEASAT Measures the Ocean Surface,* pp.180–182, and *Satellites and Plankton,* pp. 378–380).

A newly developed Remote Underwater Manipulator (RUM III), about the size of a compact car, will provide deep-ocean research capability previously available only through the costly use of manned submersibles (fig. 1–18). Weighing 363 kg (800 lb) in water, it is light enough to maneuver using its canvas tracks across the softest sediments. Operated from a computerized console aboard ship, it will be towed by steel-armored coaxial cable. RUM III has a manipulator arm that can lift samples and experimental equipment weighing up to 100 kg (220 lb), thrusters for positioning above and on the ocean floor, and television cameras. Two side-looking sonar instruments will be able to map the terrain of the ocean floor. RUM III began sea trials early in 1986 and is designed to work at depths of 6000 m (20,000 ft).

A.

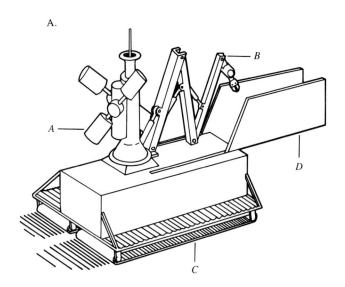

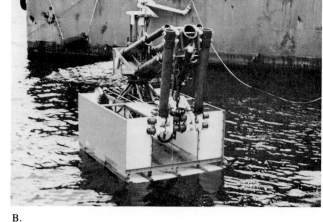

B.

FIGURE 1–18 RUM III: Remote Underwater Manipulator.

A, diagram of the RUM III showing *(A)* thrusters, *(B)* manipulator arm, *(C)* canvas tracks, and *(D)* counterweights to help stabilize RUM III when heavy objects are lifted. *B,* RUM III during sea trials in January 1986. Visuals supplied by The Marine Physical Laboratory of the Scripps Institution of Oceanography.

THE *SEABEAM/ARGO-JASON* REMOTE SYSTEM

A new remote underwater observation system has been designed to replace manned submersibles: a dual vehicle system, *Argo and Jason* is used in conjuction with the *Seabeam* acoustical mapping system.

Currently, detailed deep-sea observation by the manned submersible *Alvin* can be carried out to depths of 3500 m (11,480 ft) using a slow and expensive procedure that has three phases.

1. Using sonar to make accurate maps of the ocean floor.
2. Towing an unmanned camera sled to obtain more detailed reconnaissance profiles of the ocean floor.
3. After identifying prime targets for detailed observation and sampling, sending down the manned submersible.

Following the lead of the offshore oil industry, which has found unmanned remote machines much more cost-effective than manned submersibles, scientists believe such systems will become increasingly more important in deep-ocean research.

While the above three-step procedure may enable the exploration of tens of square kilometers per month, the use of the *Seabeam/Argo-Jason* system will allow the exploration of hundreds of square kilometers per month (fig. 1A). The system will work in the following fashion: *Seabeam*, a complex sonar system connected to a computer and operated on a tow ship, can make instantaneous bathymetric maps of the ocean floor (fig. 1B) over which the *Argo-Jason* system will pass. The data from adjacent paths of the *Seabeam* data can be converted by the computer into a three-dimensional model of the ocean floor. This image will aid in determining at what depth *Argo* should be towed.

Argo will be towed by a coaxial cable and will contain an array of sonar and television camera systems that will give observing scientists on the ship (and even on land) an instant wide view of the strip of ocean floor it passes over. Each television image will cover 4 acres; acoustical images, created by bouncing sound waves off the ocean floor, will include 150 acres of sea floor.

Jason (a tethered, self-propelled vehicle) will be "garaged" under *Argo*. When a particularly interesting area is identified on the *Argo* images, *Jason* can be sent to investigate further. *Jason* will have two mechanical arms and stereo color television cameras.

Seabeam is already perfected and in use. *Argo* provided the first images of the *Titanic* since her sinking in 1912—video images during her first sea trials on September 1, 1985. *Jason*—in the prototype form of *Jason Junior*—provided additional images of the *Titanic* in July 1986. Highly successful tests of this version of *Jason* provided pictures from inside the sunken ship (fig. 1C).

Dr. Robert Ballard (fig. 1D) of Woods Hole Oceanographic Institution, a driving force behind the use of manned submersibles and the more advanced remote systems such as *Argo-Jason*, says, "The ultimate challenge will be to integrate the entire *Seabeam/Argo-Jason* system into a single operational reality."

The development of a more efficient means of deep-ocean exploration is a welcome event. However, even with the use of such remote systems that allow more rapid investigation of the ocean floor, it will be many years before we know as much about this surface as we do the far side of the moon.

FIGURE 1A *Seabeam/Argo-Jason.*
Courtesy of Woods Hole Oceanographic Institution.

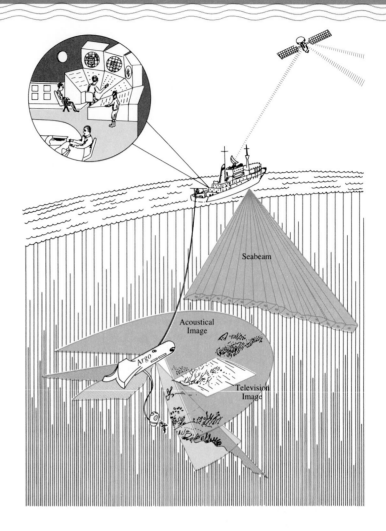

FIGURE 1B The *Seabeam/Argo-Jason* Remote System.

A computer-produced map of the ocean floor produced by the sophisticated *Seabeam* mapping system shows portions of two small volcanic peaks up to 207 m (679 ft) high. The ocean floor mapped is about 3300 m (10,824 ft) deep at a location approximately 2900 km (1798 mi) east of Tahiti. Courtesy of Scripps Institution of Oceanography.

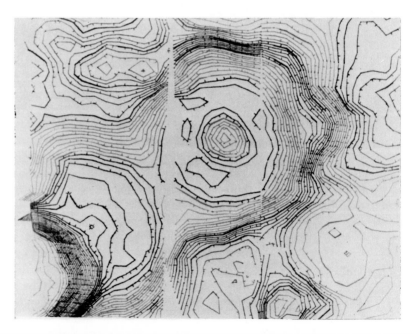

FIGURE 1C R.M.S. *Titanic* Is Photographed.

R.M.S. *Titanic,* unseen by human eyes since it sank 73 years ago 600 km (373 mi) south of Newfoundland, is at last visible in television and photographic images. The ship was discovered during sea trials of the new deep-sea investigative instrument, *Argo,* on September 1, 1985. The two photos were taken by a towed photographic unit, ANGUS (Acoustically Navigated Underwater Survey System), that takes high-resolution still photographs. *A,* anchor chains, capstans, and windlasses. *B,* loading cranes on the bow of the *Titanic. C,* electric winch on boat deck photographed by *Jason Jr. D, Jason Jr.* photographed from *Alvin. Jason Jr's.* "garage" on *Alvin* is in foreground. Courtesy of Woods Hole Oceanographic Institution.

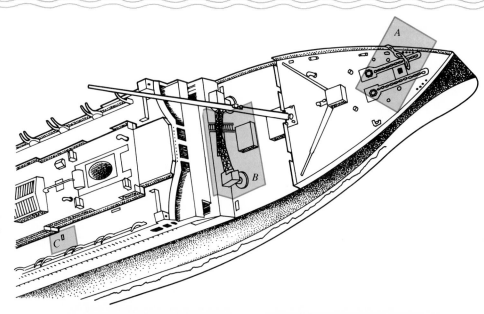

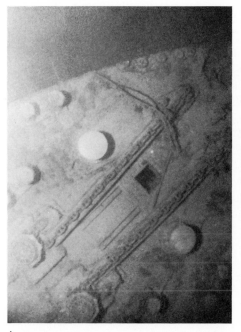

A.

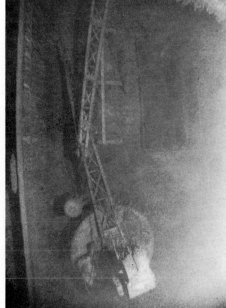

B.

C.

D.

FIGURE 1D Dr. Robert Ballard.

Dr. Robert D. Ballard in the personnel sphere of the research submersible *Alvin.* Courtesy of National Geographic Society.

Another remote system, which made the spectacular discovery of the *Titanic* and surveyed the wreckage, is the *Seabeam/Argo-Jason* Remote System described in the special feature on pages 17–19.

SUMMARY

After human beings evolved, they migrated from a probable early home in Africa to all parts of the earth before the time of written history. Many of these migrations seem to have required long ocean voyages about which we know nothing.

We do know that in the Western world, the **Phoenicians** were the first great navigators. Later the **Greeks, Romans,** and **Arabs** made significant contributions that lay hidden behind the curtain of the Middle Ages for over one thousand years. The **Age of Discovery** renewed the Western world's interest in exploring the unknown. It began with the voyage of **Columbus** in 1492 and ended in 1522 with the circumnavigation of the earth by a voyage initiated by **Ferdinand Magellan.**

Captain James Cook, who sailed in the late eighteenth century, was one of the first navigators to sail the oceans with the primary purpose of learning their natural history. He also made the first accurate maps, using the **chronometer** developed by **John Harrison.** **Matthew Fontaine Maury** was the first American to make a large-scale contribution to the study of the oceans with publication of **The Physical Geography of the Sea** in 1855. By the mid-nineteenth century, the English biologists **Charles Darwin, Sir John Ross, Sir James Ross,** and **Edward Forbes** had made significant contributions to increasing our knowledge of life in the oceans.

The voyage of the H.M.S. *Challenger* from December 1872 through May 1876 represented the first major full-scale oceanographic voyage. A feeling for the significance of this exploration can be gained by contemplating the fact that 4717 new species of marine life were identified during the voyage. An unusual voyage aboard the *Fram,* which spent three years entrapped in Arctic ice, was initiated by the Norwegian explorer **Fridtjof Nansen** in 1893. It was with this voyage that the first knowledge of the oceanography of the Arctic Ocean was gained. From 1925 through 1927, during a period of 25 months, the German ship *Meteor* criss-crossed the South Atlantic Ocean, recording the first large-scale continuous depth record with an echo sounder. This voyage first revealed the true ruggedness of the ocean floor.

Recent oceanographic investigations have been numerous. In the United States, they were conducted early by government agencies and three pioneering oceanographic institutions—**Scripps Institution of Oceanography, Woods Hole Oceanographic Institution,** and **Lamont-Doherty Geological Observatory.** They have since been joined by many educational institutions. A major focus of this widespread involvement in marine research has been the **Deep Sea Drilling Project** and the subsequent **Ocean Drilling Program.** The **Geochemical Ocean Sections (GEOSECS)** program has provided a worldwide chemical analysis of ocean water.

The use of submersibles such as *Alvin* made it possible for oceanographers to see the phenomena of the deep ocean first hand. In the future, much of the deep-ocean exploration will likely be done by use of less expensive and safer unmanned remote systems such as Remote Underwater Manipulator III and the *Seabeam/Argo-Jason.*

The use of remote sensors mounted on airplanes and satellites is beginning to show promise of helping us rapidly expand our knowledge of the oceans.

QUESTIONS AND EXERCISES

1. Construct a time line that includes the major events of human history that have resulted in a greater understanding of our planet in general and the oceans in particular.

2. List the Mediterranean cultures that appear on your time line in the probable order of their dominance in marine navigation.

3. Describe the method used by Pytheas to determine latitude in the Northern Hemisphere.

4. State the plane geometry relationship used by Eratosthenes to calculate the circumference of the earth. Determine the circumference of the earth from the proportion stated in figure 1–3.

5. How did the Greek concept of geography provided by Herodotus differ from that of the Roman geographer Ptolemy?

6. While the Arabs dominated the Mediterranean region during the Middle Ages, what were the most significant events taking place in northern Europe?

7. List the important events from the fall of Constantinople through the Age of Discovery.

8. List some of the major achievements of Captain James Cook.

9. If a ship's captain observes a Greenwich Meridian time of 1530 hours on a chronometer at local noon, what is the ship's longitude? What is the longitude if the chronometer reads 900 hours at local noon?

10. What was Matthew Fontaine Maury's major contribution to an increased knowledge of the oceans?

11. Define in your own words the process of evolution by natural selection.

12. What was the controversy that developed as a result of the observations of the Rosses and Edward Forbes?

13. List some major achievements of the voyage of H.M.S. *Challenger.*

14. What new knowledge of the Arctic polar region did the voyage of the *Fram* provide?

15. The voyage of which ship first revealed the true relief of the ocean floor?

16. List the names and locations of three pioneering oceanographic research institutions.

17. Describe the functions of the Floating Instrument Platform (FLIP) and the Remote Underwater Manipulator III (RUM III).

18. Name the major ocean research programs designed to sample the ocean floor and study ocean circulation.

REFERENCES

Bailey, H. S., Jr. 1953. The voyage of the *Challenger. Scientific American* 188:5, 88–94.

Buchanan, J. Y. 1919. *Accounts rendered of work done and things seen.* New York: Cambridge University Press.

Cromie, W. J. 1962. *Exploring the secrets of the sea.* Englewood Cliffs, N.J.: Prentice-Hall.

Duxbury, A. C. 1971. *The earth and its oceans.* Reading, Mass.: Addison-Wesley.

Engel, L. 1961. *The sea.* New York: Time-Life Books.

Oceanography from space. 1981. *Oceanus* 24:3, 1–75.

The *Titanic:* Lost and found. 1985. *Oceanus* 28:4, 1–112.

Weyl, P. K. 1970. *Oceanography: An introduction to the marine environment.* New York: John Wiley & Sons.

SUGGESTED READING

Sea Frontiers

Baker, S. 1981. The continent that wasn't there. 27:2, 108–14.
A history of the search for Terra Australis Incognita, the unfound southern land.

Browning, M. A. 1973. Stick charting. 19:1, 34–44.
A description of the methods of navigation used by Marshallese and the stick charts that served as maps.

Charlier, R. H., and Charlier, P. A. 1970. Matthew Fontaine Maury, Cyrus Field, and *The physical geography of the sea.* 16:5, 272–81.
A biography of Maury with emphasis on his accomplishments as superintendent of the Department of Charts and Instruments. His role in laying the first transatlantic cable and a discussion of his most famous work, *The physical geography of the sea*, are included.

Denzel, J. F. 1976. Edward Forbes and the birth of marine ecology. 22:1, 16–32.
    The studies conducted by Edward Forbes in the *Mediterranean Sea* and seas
    around the British Isles and his division of living communities within the seas into
    eight zones are discussed.

Gaunt, A. 1975. Marking time for three hundred years. 21:6, 322–30.
    An information-rich history of Greenwich Observatory near London, England.

Heidorn, K. C. 1974. The weather luck of Christopher Columbus. 20:5, 302–11.
    A discussion of Columbus's luck in arriving in the Caribbean during hurricane
    season in 1492 without encountering a storm. Also described are his later encoun-
    ters with storms.

Hennesey, H. 1982. Full circle for *Vema*. 28:2, 92–95
    The history of the famous research vessel *Vema*.

Rice, A. L. 1972. H.M.S. *Challenger*—midwife to oceanography. 18:5, 291–305.
    A summary of the voyage of H.M.S. *Challenger* that was undertaken in December
    1872 and laid the foundation for most branches of marine science.

Schuessler, R. 1984. Ferdinand Magellan: The greatest voyage of them all. 30:5, 299–
    307.
    A brief history of the voyage initiated by Magellan to circumnavigate the globe.

Scientific American

Bailey, H. S. 1953. The voyage of the *Challenger*. 188:5, 88–94.
    A summary of the accomplishments of the English oceanographic expedition of
    1872.

Herbert, S. 1986. Darwin as a geologist. 254:5, 116–123.
    Darwin's early interest in the developing science of geology aided him in the de-
    velopment of his theory of biological evolution.

Revelle, R. 1969. The ocean. 221:3, 54–65.
    A discussion of how humans use the ocean and the geological creation of its pres-
    ent basins.

Before considering the development of the earth's oceans and the origin of life in them, we should try to understand what is known of the relationship of the earth to the cosmos. Much speculation is involved in making these considerations, but science has not made advances by restricting its concerns to what is already known. We should, however, begin our discussion with the aspects of the universe that we can readily observe and describe. Then we will discuss some theories about the origin of the universe and our planet.

CHAPTER TWO
THE ORIGIN OF THE EARTH, ITS OCEANS, AND LIFE IN THE OCEANS

THE UNIVERSE WE SEE

As astronomers looked into the sky surrounding earth, they gradually developed an understanding of earth's position in what is now called the solar system (fig. 2–1). The solar system is composed of the sun and its nine planets, which occupy a disc-shaped portion of the universe about 13 billion km (42.6 billion mi) in diameter. This structure seems enormous, but it represents a very minute portion of the total universe. Our sun is one of perhaps 100 billion billion, or 10^{20}, stars that constitute the known universe. In terms of distances with which we are familiar, the solar system is incomprehensibly isolated. The distance from the sun to the nearest star is 44 million billion km (144.3 million billion mi).

While the planets revolve around the sun, the entire system is moving at about 280 km/s (174 mi/s) around the center of the Milky Way, the galaxy or the group of stars to which the solar system belongs. Estimates of the dimensions of the Milky Way indicate that it is a disc-shaped accumulation of stars about 100,000 light-years in diameter and from 10,000 to 15,000 light-years thick near the center, thinning gradually toward the edges. A *light-year* is a unit of measure used by astronomers to describe distances within the universe and is equal to the distance traveled by light at a speed of 300,000 km/s (186,000 mi/s) during a period of one year—almost 10 trillion (9.8×10^{12}) km. Within the Milky Way are some 100 billion (10^{11}) stars, tens of millions of which very probably possess families of planets, millions of which may be inhabited by intelligent creatures, according to the estimates of some astronomers.

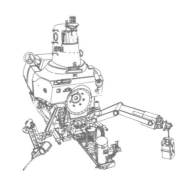

FIGURE 2–1 The Solar System.

A, the orbits of the planets of the solar system are drawn to scale. *B,* relative sizes of the sun and the planets are shown. The distance scale is not maintained in *B.*

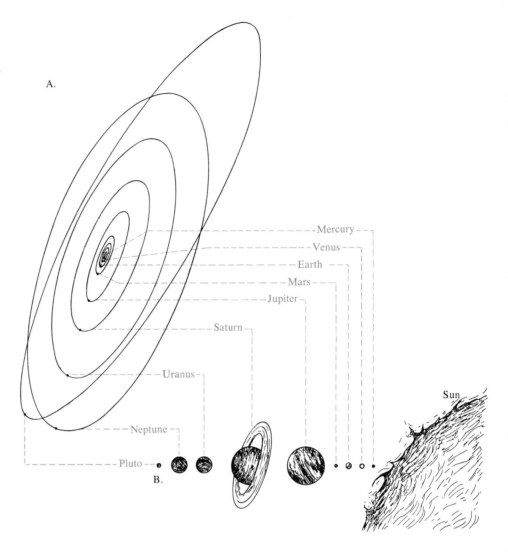

All the stars you see with your naked eye belong to the Milky Way (fig. 2–2). If you have exceptional vision and live in the Northern Hemisphere, you may see a hazy patch of light in the constellation Andromeda that represents another galaxy within the universe. This galaxy is some 1.5 million light-years away. It appears that galaxies are exceedingly numerous, and within a distance of a billion light-years from our galaxy there are at least 100 million galaxies large enough to be observed by our telescopes.

By observing light energy that radiates from these distant galaxies, astronomers have been able to determine that most are moving away from us. This conclusion is based on a spectral shift toward the red end of the visible light spectrum of absorption lines in the light they emit (fig. 2–3). The red shift indicates an *apparent* lower frequency for the light energy that the galaxies radiate. This apparent decrease in the frequency is a result of their moving at high velocities away from the observer. There is proportionality ob-

served between the amount of red shift and the distance between the earth and the galaxies. The largest red shift that has been observed is from a body that is 12.4 billion light-years away, which gives us a minimum size for the visible universe.

The velocity with which the galaxies recede from our system is also proportional to their distance from it. Velocities of over 250,000 km/s (155,250 mi/s) have been calculated for the most distant galaxies as they recede from the Milky Way. This is in excess of four-fifths of the speed of light!

One may be left with the impression that all of the galaxies are being repelled by our solar system, but a more reasonable consideration would be that galaxies are moving away from one another as would fragments that result from an explosion. The universe appears to be rapidly expanding, its component galaxies moving ever farther apart. If all of these galaxies are moving away from one another in a manner similar to that of fragments created by explosion, we

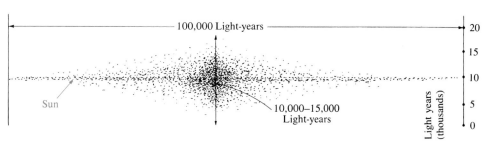

FIGURE 2—2 The Milky Way. The sun and its planets are located about two-thirds the distance from the center of the Milky Way to the edge. The Milky Way completes a rotation every 200 million years. To maintain its position in the Milky Way, the solar system travels around the center of the galaxy at a velocity of about 280 km/s (174 mi/s).

might ask if they all belonged to one large mass. We, of course, do not know but evidence in support of this condition continues to increase. If that is the case, the time required for the galaxies to have reached their present distribution would have been about 14.5 billion years.

ORIGIN OF THE EARTH

As insignificant as our planet is in the overall composition of the cosmos, it is nonetheless of interest to us to learn what we can of its origin. Although no complete understanding of the origin of the earth has yet been developed, we can outline some of the probable sequences that led to its present form. As galactic matter revolved around its center of rotation, the stars began to form as individual glowing masses due to the concentration of space particles under the force of gravity. These concentrations may have occurred in local eddies and over a vast period of time taken their present form. Many of the resulting stars may have developed assemblages of planets that revolve around them.

Such a general sequence may have occurred with the formation of our sun, and in its early stages it may have had a size at least equal to the diameter of

the solar system that we observe today. As this large accumulation of gas and space dust began to contract under the force of gravity, a small percentage of it was left behind.

This remaining material flattened itself into a disc that became increasingly more dense. Because of the increase in density of this disc, it became gravitationally unstable and broke into separate small clouds. These were protoplanets, which later consolidated into the present planets.

Protoearth was a huge mass, perhaps 1000 times greater in diameter than the earth today and 500 times more massive. The heavier constituents of the planet earth, as with all the other protoplanets, were moving toward the center to form the heavy core surrounded by lighter materials. Similarly, the satellites we observe around the planets, such as our moon, began to form.

During this early formation of the protoplanets and their satellites, the sun eventually began to "shine" as it condensed into such a large mass with such high temperature that forces were set up within its interior, creating energy through a process known as the *carbon cycle*, or fusion reaction. When temperatures reach tens of millions of degrees, hydrogen atoms, in the presence of carbon, are converted to helium at-

Constellation in which galaxy may be seen	Velocity km/s (mi/s)	Distance from sun (light-years)	Shift of absorption lines	
			Violet	Red
Virgo	1,200 (745)	43,000,000		
Corona Borealis	21,500 (13,351)	728,000,000		
Hydra	61,000 (37,881)	1,960,000,000		

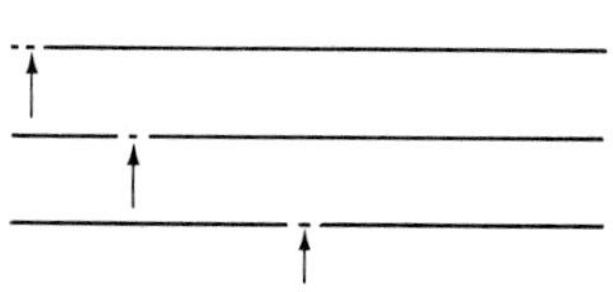

FIGURE 2—3 Galaxies—Velocity, Distance, and Spectral Shift.
Most galaxies are moving away from the Milky Way. This figure gives the characteristics of three such galaxies. A relationship can be seen among the velocity of the galaxy, its distance, and the observed amount of shift toward the red end of the light spectrum. Radiation from distant galaxies is absorbed so that dark lines (gaps) appear in the violet part of the spectrum. The greater the distance to the galaxies, the greater the shift of the dark lines toward the red end of the spectrum.

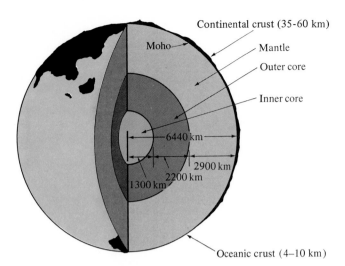

FIGURE 2–4 The Earth's Structure.
The three major subdivisions of the earth—core, mantle, and crust—are shown in the cross section.

oms, and huge amounts of energy are released. In addition to the light that is so obviously emitted by the sun, there is an intense emission of ionized, electrically charged, particles. In the early stages of the development of our solar system, this emission of ionized particles served to blow away the nebular gas that remained from the formation of the planets and their satellites.

As the protoplanets were being warmed by the solar radiation, their "atmospheres" of hydrogen and helium began to boil away. The combination of emission of ionized solar particles and the internal warming of the planets resulted in a drastic decrease in the size of the planets closer to the sun. Since the gas envelope that surrounded most of these planets was possibly composed mostly of hydrogen and helium (we may consider this to be the composition of the earth's first atmosphere), the energy that was available easily allowed atoms of small size to escape from the gravitational fields of the planets. As the protoplanets continued to contract, the heat produced within the planet as a product of spontaneous disintegration of atoms (radioactivity) became more intense.

The evidence indicates that the earth ultimately became molten, allowing the heavier components to concentrate in the core. The lighter of the materials segregated according to their densities in concentric spheres around this core. Surrounding the core, containing abundant iron and nickel, is a zone of material composed of heavy iron and magnesium silicates—the mantle. A thin crust formed at the surface.

On the basis of data gained from the observation of seismic waves passed through the earth's interior, scientists have constructed a model of the earth that shows a solid inner core with a radius of approximately 1300 km (807 mi) composed of iron and nickel and other heavy elements. The solid core gradually changes to a liquid outer core with a thickness of approximately 2200 km (1366 mi). Surrounding the core is a 2900-km (1800-mi) thick mantle. The mantle is composed of silicates of iron and magnesium and is overlain by a very thin crust made up of silicates of iron, magnesium, and aluminum. The crust ranges in thickness from 4 to 60 km (2.5 to 37 mi) and is separated from the mantle by a distinct density change called the Mohorovicic Discontinuity (Moho), after the Yugoslavian seismologist who discovered its existence (fig. 2–4).

ORIGIN OF THE ATMOSPHERE AND OCEANS

The solidification of the earth's crust marks the beginning of geologic history, and evidence indicates that this event occurred about 4.6 billion years ago (fig. 2–5). At that time the earth had lost all but a small fraction of its original gas envelope because of the heat radiating from it as it cooled.

The lighter a molecule and the higher the velocity of that molecule, the greater is the probability that a gas composed of such molecules will escape from

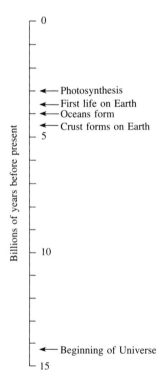

FIGURE 2–5 Major Events in the Evolution of the Earth.

the earth's surface. The radiant energy enabled light gas molecules to escape the earth's gravitational field. It has been calculated that when the average molecular velocity in an atmosphere exceeds one-fourth of the planet's escape velocity, which for the earth is 11.4 km/s (7.1 mi/s), the atmosphere will be lost into space within a relatively short period of time. At present hydrogen and helium are rapidly lost from the earth's atmosphere, while gases such as oxygen and nitrogen, composed of heavier molecules, are retained.

If the earth has in the past been much hotter than it is today, there should be evidence of this fact. We can assume that if the earth did have a considerably higher temperature, then it would have lost the volatile (gaseous) material that once surrounded it. We can compare the relative abundance of the noble (inert) gases that are found on earth to that of the stars to test this theory. The noble gases rarely combine chemically and always remain in the gaseous state, so if we compare the abundance of these volatiles with that of an element that becomes chemically bound to the earth, we can infer temperatures of the past.

A very abundant element that fits this latter condition is silicon. As our theory indicates, the earth formed from the same material as did the sun, so there was probably a ratio of the noble gases to silicon in the material that converged under the force of gravity to produce protoearth similar to the ratio that exists in the sun. If the earth's surface was heated during its early development, we will see that there has been a loss of volatile components, represented by the noble gases, in comparison to silicon, which is held chemically to earth. Studies related to the depletion of noble gases show that neon, one of the lighter noble gases, was depleted by a very large factor and that the depletion factor decreases with the increasing mass of the atoms of the various noble gases. This evidence suggests that the earth's surface may have been much hotter than it is today.

If the earth lost much of its argon, with an atomic weight of 40, we can very well assume that it also lost whatever water vapor, nitrogen, and oxygen it may have contained since they have molecular weights of 18, 28, and 32, respectively. On the basis of this evidence, we can assume that the present atmosphere could not have formed at that time.

The Atmosphere Forms

Where did the material to make up the present atmosphere and ocean come from? Although the earth's surface had become relatively cool by this stage in its development, there was still much volcanic activity according to the geologic record. As volcanic gases bubbled up from the earth's upper mantle, they formed a new atmosphere as soon as the earth had cooled enough to retain one. This atmosphere was far different from the one that the earth had lost and from the one it presently has. It contained little free oxygen and was composed of large percentages of carbon dioxide and water vapor. As the earth cooled to the point at which it could retain these gases, a vast envelope of clouds surrounded the planet, causing the earth to reflect about 60% of all the sunlight that was directed toward it. Much of the energy it absorbed from the sun remained in the thick envelope of gases.

Water vapor released from the surface of the earth was captured in the thick envelope and was forced to remain there as a vapor until sufficient cooling of the earth's crust would allow it to condense and begin to accumulate as a liquid. Eventually, cooling of the crust allowed the water vapor to condense and fall under the force of gravity to the earth's surface. Finally, with further decrease in the surface temperature, permanent accumulation of water began in depressions on the earth's irregular surface (fig. 2–6).

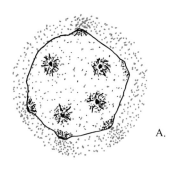

A.

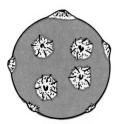

B.

FIGURE 2–6 How Water Accumulated on Earth.

A, widespread volcanic activity releases H_2O and smaller quantities of CO_2, Cl_2, N_2, and H_2, which produced a water vapor atmosphere that also contained carbon dioxide (CO_2), methane (CH_4), and ammonia (NH_3). *B,* as the earth cooled, the water vapor condensed and fell to the earth's surface. There it accumulated to form the oceans. Size of volcanoes is exaggerated.

The Oceans Evolve

These accumulations were the initial oceans on planet earth, and it is assumed that this water has been permanently in existence since its formation. The oceans were probably somewhat less saline than the present oceans, but the primary crystalline rocks that made the earth's crust were weathered and eroded. These processes were aided by the high concentration of carbonic acid (H_2CO_3) resulting from the combination of carbon dioxide and water molecules in the atmosphere. The inorganic compounds of which the rocks were composed were dissolved in the waters that fell relentlessly upon their surfaces and were carried into the newly forming oceans.

The process of cooling the earth's surface, which allowed the water vapor to condense and accumulate to form the earth's first oceans, was accelerated by water's property of high heat capacity and latent heat of vaporization. The heat energy that prevented the water from accumulating on the earth's surface was being absorbed by that massive cloud of water vapor and carbon dioxide that surrounded the earth's solid surface. As the molecules of water absorbed heat at the earth's surface and moved through the cloud to its outer edges, where they were frozen into ice crystals, large quantities of heat were released into outer space.

THE SOURCE

We have obviously been proceeding under the assumption that the materials that make up the atmosphere and the ocean came from within the earth's interior and were brought to the surface by the process of volcanism. Is this a valid assumption? Let us consider the water of the oceans. The material that is brought to the earth's surface as lava must have its source in the lower crust or the upper mantle of the earth. The mantle of the earth has been calculated to have a volume of 10^{27} cm^3 and a density that averages 4.5 g/cm^3. This amounts to a total mass for the mantle of 4.5×10^{27} g. The water of the oceans has a mass of 1.4×10^{24} g. Therefore, the mantle must have lost 0.031% of its mass as water to have produced the earth's oceans. (*See* calculations in appendix VI.)

For meteorites that contain silicates and are believed to resemble the composition of the earth's mantle, we can obtain a water content from which we can estimate a water content for the mantle. The average water content of these stony meteorites is about 0.5%, which is about 16 times the loss required from the mantle to account for the present oceans. If

our assumption is correct, the mantle would contain more than enough water to create the oceans, and therefore the mantle could very adequately serve as a source for the waters of the earth's oceans.

THE CHEMICAL BALANCE SHEET

Because chemical weathering of the primary rocks that composed the original earth's crust was made possible by the appearance of water, the process of chemical weathering needs to be considered in some detail. As a result of chemical weathering, the elements that were contained in the primary crystalline rock were freed, dissolved in the water, and carried off to accumulate in the ocean. As the water carried these dissolved elements and compounds into the ocean, it also carried a great load of particulate material that had been weathered and eroded from the crystalline rocks and deposited this material as sediment. The components of the primary crystalline rock that were freed by chemical weathering must be found dissolved in that water in the ocean or as components of the earth's atmosphere. Some of the components will remain chemically bound with the sediments that have been carried into the sea by the rivers (*see* fig. 2–7).

Because mass must be conserved, we must find in the products of chemical weathering a mass equal to the total mass of weathered crystalline rock. Disre-

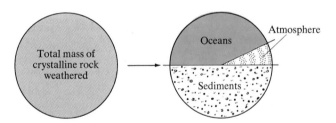

The total mass of crystalline rock components freed by chemical weathering must be found as components of the atmosphere, oceans, or sediments.

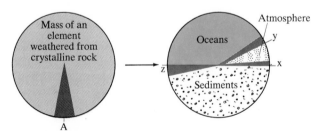

When considering one element represented by the dark wedge in the "pie" to the left, (A), the sum of the masses of that element found in the atmosphere, (x), oceans, (y), and sediments, (z), should equal the mass weathered from the crystalline rock, A = x + y + z

FIGURE 2–7   Geochemical Balances.

AN ALTERNATE THEORY—OCEAN WATER FROM SPACE

Although some oceanographers can't believe he is serious, Louis Frank, one of the world's top space scientists, believes the ocean's water may have come to earth as cometlike balls of ice about 9 m (30 ft) in diameter. Frank believes they enter the earth's atmosphere at a rate of 20 per minute.

The theory evolved from the need to explain some features on images of the earth's atmosphere produced from data obtained from an ultraviolet photometer aboard the *Dynamic Explorer I,* a polar-orbiting satellite that reaches an altitude of 23,350 km (14,500 mi). The image is of dayglow created by the absorption and reradiation of solar energy by atomic oxygen 290 km (180 mi) above sea level. The dayglow, which makes the earth look like a yellow ball, had a number of short-lived dark spots (fig. 2A). These dark "holes" in the atmosphere were 48 km (30 mi) across and existed for up to 3 minutes. Because of the absorption spectra of these dark spots, they can be explained only by the cometlike balls of ice breaking up and vaporizing 1600–3200 km (1000–2000 mi) above the earth's surface.

At the observed rate of occurrence, the earth would receive 0.0025 mm (0.0001 in) of water per year. Four billion years of such bombardment would give us enough water to fill the oceans to their present volume.

FIGURE 2A Image of Earth's Dayglow.

Recorded by ultraviolet photometer aboard the satellite *Dynamic Explorer I.* The dark spot or "hole," thought to be caused by the vaporization of cometlike balls of ice, is shown in the inset. The circular emission at the top is the northern auroral oval. Courtesy L. A. Frank, University of Iowa.

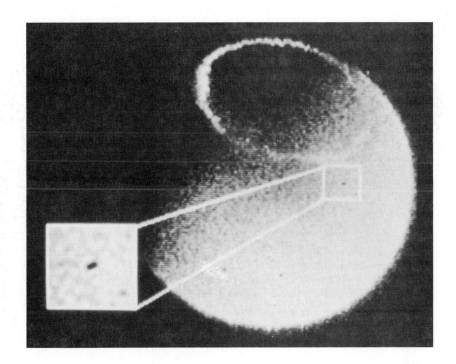

garding those heavy elements that are radioactive, we must be able to find for the total mass of each chemical element that is taken from primary crystalline rock an equivalent mass of that element in the products of chemical weathering. The mass of an element in crystalline rock is equal to the total mass of the rock times the average concentration of that element in it.

In looking for the equivalent mass of an element in the products of chemical weathering, we must estimate the total mass of sediments on the earth, the total mass of the earth's ocean, and the total mass of the atmosphere and multiply these masses by the average concentration of that element in each. The sums of these computations should equal the mass of that element that was computed to have been present in the volume of crystalline rock that has been chemically weathered.

The estimates that have been made for these various volumes will not be exact, but the estimate for the ocean is probably one of the most accurate, since the ocean volume is relatively well known and its composition is comparatively uniform because it is well mixed. The estimates are less accurate for crystalline rock, which is quite heterogeneous, and for the sediments, which are of various types. Nevertheless, geochemists have drawn up balance sheets for the various elements and those calculated by different individuals working on the problem agree fairly well. For those elements which are most common in the crystalline rocks, the geochemical equations balance well within the limits set by the accuracy of the estimates (fig. 2–7).

Excess Volatiles

There are some components that do not balance in any of the attempts that have been made to do so.

These elements are much more abundant in the atmosphere, the ocean, and the sediments than can be accounted for by the volume of crystalline rocks which have been estimated to have undergone chemical weathering. All of these substances that are found in excessive amounts compared to the products of chemical weathering are volatile at or slightly above the average temperatures found at the earth's surface and are therefore called *excess volatiles*. Most abundant of the gaseous substances that make up the excess volatiles are water vapor and carbon dioxide; also important are quantities of chlorine, nitrogen, sulfur, hydrogen, and fluorine. (*See* table 2–1.)

Volcanic activity is presently producing these gases and venting them into the earth's atmosphere. A great percentage of the gases that surface as a result of volcanic activity may simply be recycled volatiles that have been dissolved in groundwater and carried down from the earth's surface, but it has been determined that a small percentage of the total volatiles is always composed of new gases that have for the first time been released from the crystalline rock of the earth's crust. When molten magma cools, solid minerals are formed, and some of the gaseous constituents of the magma are freed. These gases, freed during the crystallization process, represent the new material coming to the earth's surface from the mantle.

We have geological evidence in the form of fossilized algae, assumed to have grown in a marine environment, that the oceans have existed on the earth's surface for well over 3 billion years. If we consider the rate of discharge of excess volatiles by hot springs in the United States to be equal to the average rate of production of excess volatiles throughout the last 3 billion years, we find that enough water vapor would have been produced to fill the oceans to more than 100 times their present volume. We must remember that much of this water that comes to the surface is

TABLE 2–1 Excess Volatiles. The volatile elements and compounds that are found in the atmosphere, ocean, and sedimentary rocks in far greater amounts than have been made available by chemical weathering of primary crystalline rocks are called *excess volatiles*. They are believed to have been released from the interior of the earth by volcanic activity. Note the absence of free oxygen.

	Mass of Elements or Compounds in 10^{20} g					
	H_2O	C as CO_2	Cl	N	S	H, B, Br, Ar, F, etc.
Found in:						
Atmosphere and ocean	14,600	1.5	276	39	13	1.7
Buried in sedimentary rocks	2,100	920	30	4	15	15
Total	16,700	921.5	306	43	28	16.7
Supplied by:						
Weathering of crystalline rocks	130	11	5	0.6	6	3.5
Excess volatiles (unaccounted for by rock weathering)	16,570	910.5	301	42.4	22	13.2

Courtesy of the Geological Society of America. After Rubey, 1951

water that is being recycled; it does not represent new water released from the crystallization of magma. Only 1% of this water would require such an origin to account for the present volume of the earth's oceans from this source.

DEVELOPMENT OF THE OCEANS AND THEIR BASINS

When the earth began to form its solid crust about 4.6 billion years ago, the mass of the oceans was zero, while at present it is 1.4×10^{24} g. We do not know what the rate was at which volatiles escaped the earth's mantle via volcanic activity and hydrothermal springs throughout this period of 4.6 billion years. If we assume it to be at least equal to the present rate, we can visualize the formation of the earth's oceans as a gradual process.

Does continual outgassing mean that the oceans are gradually covering more and more of the earth's surface? Not necessarily, since the area that the oceans cover is directly determined by the volume that the basin in which they form is able to accept. The crust, which we now describe as continental, may not have been present during the initial solidification of the earth's crust and may well have formed gradually as the oceans themselves were formed. Since the volume of the ocean basins is directly related to the difference in thickness between the oceanic and the continental crusts, and if we assume that the continental material has gradually accumulated, we can see that the capacity of the oceans' basins could have gradually increased along with the volume of water that was being produced over this period of time. Thus, it is quite possible that the oceans have had a distribution very similar in total surface area to that of the present oceans for a long period of geologic time and that the major change in the oceans' character has been an increase in depth. Throughout the geologic time during which the oceans and continents were increasing their volumes, the ocean basins were getting deeper to accommodate the increasing volume of water that was being transferred from the underlying mantle to the earth's surface. In fact, the area covered by the oceans may be decreasing as a result of continued growth of the continents. This process will be discussed further in chapter 4.

Considering the history of ocean salinity, we might certainly ask if the oceans have possessed a relatively uniform salinity throughout their history or if they are getting more or less saline. By far the most important component of salinity is the chlorine ion, Cl^-, which is produced by volcanic action, the same process that produces the water vapor forming the oceans. So our considerations boil down to one question: Has the relative amount of water vapor to chlorine ion production remained constant throughout geologic time, or has it varied? We have no indication that there has been any fluctuation in this ratio throughout geologic time. We must then consider, on the basis of the present evidence, that the oceans' salinity has been relatively constant, while their volume and the volume of the crystalline rock that makes up the continental crust has increased throughout the 4.6 billion years since the earth's crust first began to form.

A CHEMICAL BACKGROUND

Before going further in our discussion, we need to define the following terms. An *element* is one of a number of substances each of which is comprised entirely of like particles, atoms, that cannot be broken into smaller particles by chemical means. The *atom* is the smallest particle of an element that can combine with similar particles of other elements to produce compounds. The periodic table of elements, appendix III, lists the elements and describes their atoms. A *compound* is a substance containing two or more elements combined in fixed proportions. A *molecule* is the smallest particle of an element or compound that, in the free state, retains the characteristics of the substance.

As an illustration of these terms, consider Sir Humphrey Davey's use of electrolysis to break the compound water into its component elements, hydrogen and oxygen. Atoms of the elements hydrogen (H) and oxygen (O) combine in the proportion 2 to 1, respectively, to produce molecules of the compound water (H_2O). As an electric current is passed through the water, the molecules dissociate into hydrogen atoms that collect near the cathode (negatively charged electrode) and oxygen atoms that collect near the anode (positively charged electrode). Here they combine to form the diatomic gaseous molecules of the elements hydrogen (H_2) and oxygen (O_2). Because there are twice as many hydrogen atoms as oxygen atoms in a given volume of water, twice as many molecules of hydrogen gas (H_2) as oxygen gas (O_2) are formed. Further, because under identical conditions of temperature and pressure the volume of gas is proportional to the number of gas particles (molecules) present, two volumes of hydrogen gas are produced for each volume of oxygen.

A Look at the Atom

Building upon earlier discoveries, the Danish physicist Niels Bohr (1884–1962) developed his theory of the atom as a small "solar system" in which a positively charged nucleus takes the place of the sun and the "planets" that orbit around it are represented by negatively charged electrons. Although this theory has since been diagrammatically altered, it is still commonly used to demonstrate the arrangement of electrons and nuclear particles in the atom.

Bohr's earliest concern was with the atom of hydrogen, which he considered to consist of a single positively charged *proton* in its nucleus orbited by a single negatively charged *electron*. Since the mass of the electron is only about 1/1840 the mass of the proton, it will be considered negligible in our coming discussion of atomic weights.

According to Bohr, the number of protons, or units of positive charge in the nucleus, coincides with the *atomic number* of the element. Thus hydrogen, with an atomic number of 1, has a nucleus containing a single proton. Helium, the next heavier element, having an atomic number of 2, contains two nuclear protons, and so forth. The atomic number will also indicate the number of electrons in a normal atom of any element, because this number is equal to the number of protons in the nucleus (fig. 2–8).

An *isotope* is an atom of an element that has a different atomic weight than other atoms of the same element. As we will later see in some detail, chemical properties of atoms are determined by the electron arrangement that surrounds the nucleus. This arrangement, in turn, is determined by the number of protons in the nucleus. Because there are isotopes that have different atomic weights but identical chemical characteristics, we suspect that the proton

is not the only nuclear particle that can influence the atomic weight of an atom. There must be a nuclear particle that has no effect on the atom's chemical properties, in that it does not affect the electron structure surrounding the nucleus (fig. 2–9).

Nuclear research has discovered the additional particle, the *neutron,* which was postulated by Rutherford in 1920 and was first detected by his associate James Chadwick in 1932. It has a mass very similar to that of the proton but no electrical charge. This characteristic of being electrically neutral has made it one of the particles most utilized by physicists. These scientists continue to explore the atomic nucleus and to make new discoveries of nuclear particles. We will not consider these particles, as knowledge of them is not necessary for our understanding of the chemical nature of atoms.

We can now divide the atom into two parts—the *nucleus,* which contains neutrons and protons, and the *electron cloud* surrounding the nucleus that is involved in chemical reactions. It is not possible to determine the precise location of electrons in this cloud at any instant, but it is possible to estimate the most probable position of an electron in this cloud. We will picture these regions in which we would more likely find particular electrons as concentric spheres, or shells, that surround the nucleus (*see* fig. 2–8).

Chemical Bonds

In considering the chemical reactions in which atoms are involved, we will be concerned primarily with the

FIGURE 2–8 Bohr-Stoner Orbital Models for Atoms.

Each atom is composed of a positively charged nucleus with negatively charged electrons around it. The nucleus, which occupies very little space, contains most of the mass of the atom. The atomic number is equal to the number of protons (positively charged nuclear particles) in the nucleus of an atom of an element. Note that the first shell holds only two electrons. Other shells can hold no more than eight electrons when they are in the outer position.

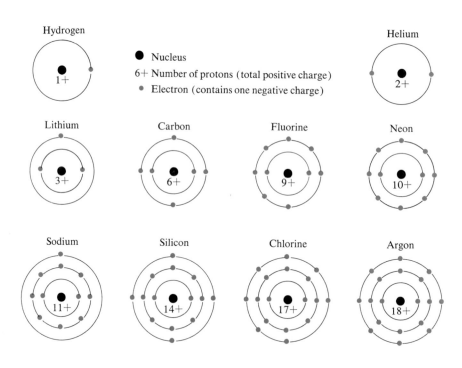

HYDROGEN

(Atomic $_1H^1$ (Mass units)
number)

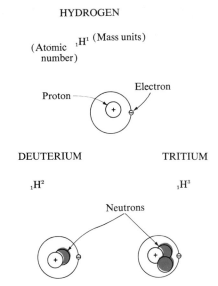

DEUTERIUM   TRITIUM

$_1H^2$ $_1H^3$

Neutrons

FIGURE 2–9 Neutrons as Nuclear Particles.

Isotopes are atoms of an element that have different atomic weights. The hydrogen atom ($_1H^1$) accounts for 99.98% of the hydrogen atoms on earth. It contains one nuclear particle, a proton. Deuterium ($_1H^2$) contains one neutron and one proton in its nucleus and combines with oxygen to form "heavy water" with a molecular weight of 20. Tritium ($_1H^3$) is a very rare radioactive isotope of hydrogen that has a nucleus containing one proton and two neutrons.

distribution of electrons in the outer shell. With some exceptions, when atoms combine to form compounds, they combine in one of two ways:

1. *covalent bonding*—sharing electrons in the outer shells of the atoms
2. *ionic bonding*—some of the atoms lose electrons from their outer shell, and others gain electrons in their outer shell

This latter process produces an *ion,* an electrically charged atom that no longer has the properties of a neutral atom of the element it represents. A positively charged ion, a *cation,* is produced by the loss of electrons from the outer shell, the positive charge being equal to the number of electrons lost. A negatively charged ion, an *anion,* is produced by the gain of electrons in the outer shell of an atom, and its charge is equal to the number of electrons gained.

In a statement of great oversimplification, let us say it is the "desire" of an individual atom to assume the outer shell electron content of the inert gases such as helium (2), neon (8), and argon (8) that makes it chemically reactive (*see* fig. 2–8). Normally, if an atom can assume this desired configuration by either sharing one or two electrons with an atom of another element or by losing or gaining one or two

electrons, the elements are highly reactive. By contrast, the elements are less reactive if three or four electrons must be shared, gained, or lost to achieve the desired configuration.

We can describe the combining power of atoms of different elements by the concept of *valence.* Valence can be defined as the number of hydrogen atoms with which an atom of a given element can combine. Valence can be either ionic or covalent. Elements with lower valences of one or two combine chemically in a more highly reactive manner than do those with higher valences of three or four, or in certain instances, more than four. Although those elements with higher valences do not react as violently as low-valence elements, they have a greater combining power and can gather about them larger numbers of atoms of other elements than can the atoms with lower valence values.

An example of a covalent bond is the sharing of electrons by hydrogen and oxygen atoms in the water molecule. In the formation of this molecule, both hydrogen atoms and the oxygen atom assume the inert gas configuration they seek by sharing electrons (fig. 2–10A).

An example of a compound formed by ionic bonding is seen in the metallic salt sodium chloride (fig. 2–10B). In the formation of this compound, the sodium atom, which has one electron in its outer shell,

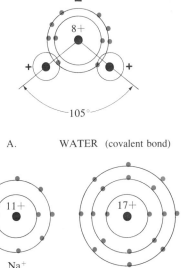

A.   WATER (covalent bond)

Na$^+$
11 protons = 11 +
10 electrons = 10 −

Cl$^-$
17 protons = 17 +
18 electrons = 18 −

B. SODIUM CHLORIDE (ionic bond)

FIGURE 2–10 Covalent and Ionic Bonds.

Notice the contrast in the outer shells of the ions of sodium (Na$^+$) and chlorine (Cl$^-$) with the atoms of these elements in figure 2–8.

loses this electron, forming a sodium ion with a positive electrical charge of one. The chlorine atom, which contains seven electrons in its outer shell, completes its shell by gaining an electron and becoming a chloride ion with a negative charge of one. These two ions are held in close proximity by an electrostatic attraction between the two ions of equal and opposite charge.

LIFE FORMS IN THE OCEANS

We did not include free oxygen in the excess volatiles that were produced at the earth's surface by volcanic activity. If such a gas were produced, it is certain that it could not remain free in the earth's atmosphere because we find iron that occurs in two forms in volcanic rock—ferrous iron (FeO) and ferric iron (Fe_2O_3).

The ferrous iron atom displays a valence of two, while the iron atom in a molecule of ferric iron has a valence of three. The ferric state is more highly oxidized (contains more oxygen per iron atom) than the ferrous. The iron in volcanic rock is generally observed to be ferrous. Any oxygen that surfaced during volcanic activity would be used up converting ferrous iron to the ferric state, thus preventing free oxygen from being found as an excess volatile. As a result of this condition, we may consider that the earth's early atmosphere did not contain this gas that makes up almost 21% of our present atmosphere.

$$8\ FeO + 2O_2 \longrightarrow 4\ Fe_2O_3$$
$$\text{Ferrous iron} + \text{Oxygen} \longrightarrow \text{Ferric iron}$$

Ozone Formation

We depend on the oxygen in our atmosphere for protection from the ultraviolet radiation that is constantly bombarding our planet. Much of the energy represented by this ultraviolet radiation is exhausted in our upper atmosphere, converting the free oxygen (O_2) into ozone (O_3), thus allowing very little of the ultraviolet energy to reach the surface of the earth.

Free oxygen molecules (O_2) absorb ultraviolet light, and the energy taken up exceeds the bond strength. The bond breaks to give two oxygen atoms.

$$O_2 + \text{Ultraviolet light} \longrightarrow O + O$$

The oxygen atoms produced are very reactive and quickly combine with an oxygen molecule to produce a molecule of ozone.

$$O_2 + O \longrightarrow O_3$$

The First "Organic" Substances

Of course, if oxygen was missing from the primitive atmosphere, the ultraviolet radiation would have readily penetrated the atmosphere and reached the surface of the young oceans. These oceans contained the gases methane (CH_4) and ammonia (NH_3), products of the chemical combination of hydrogen (H_2), nitrogen (N_2), and carbon (C) made available by the great outgassing of carbon dioxide (CO_2).

Laboratory experiments have shown that exposing a mixture of hydrogen, carbon dioxide, methane, ammonia, and water to ultraviolet light and electrical spark will produce a large assortment of organic molecules (fig. 2–11). The exposure of the mixture to ultraviolet radiation causes photosynthesis, carried on in the absence of chlorophyll. *Photosynthesis* is an endothermic reaction by which energy from the sun is stored in the products of the photosynthetic reaction—the organic molecules. The term organic molecules is used in reference to molecules of substances such as sugars, amino acids, and fats produced by organisms. The production of organic substance in the shallow oceans that developed early in the earth's history must have resulted in a vast amount of "organic" material, as do the laboratory experiments, but this material did not yet represent life on the planet earth.

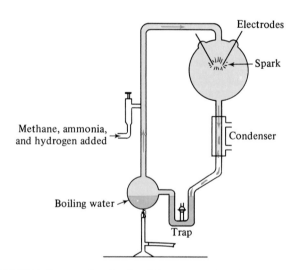

FIGURE 2–11   Synthesis of Organic Molecules.
The apparatus used by Stanley L. Miller in the 1952 experiment that resulted in the synthesis of the basic components of life, amino acids. A mixture of water vapor, methane, ammonia, and hydrogen were subjected to an electrical spark that provided the energy for synthesis. This mixture is thought to resemble the composition of the atmosphere and oceans that existed when "organic" molecules first formed on earth.

Carbon—the Organizer of Life

Let us now leave these general considerations for a moment and consider more specifically the element that seems to be the most important in allowing the build-up of the organic compounds we have just mentioned. This element is carbon. We even refer to the study of the chemistry of carbon compounds as *organic chemistry.* In studying the chemistry of carbon, we are constantly concerned with molecules which are several times larger than those with which the inorganic chemist is commonly concerned. The greater percentage of inorganic molecules contains fewer than 10 atoms, while in organic chemistry, molecules that exceed 100 atoms per molecule are not uncommon.

This fact arises out of the "combining power" of the carbon atom, which has four electrons in its outer electron shell. It has a capacity of accepting atoms of other elements with a total valence of four. It is tetravalent.

To demonstrate how this ability gives carbon an unusually high combining power, let us compare the carbon atom with other common atoms with valences less than four. We will consider valence to be the number of hydrogen atoms with which an atom can combine. If we use chlorine, which occurs as a gas, we can see that a single molecule of chlorine combines with a single molecule of hydrogen to produce two molecules of *hydrogen chloride,* HCl.

$$Cl_2 + H_2 \longrightarrow 2HCl \qquad\qquad Cl - H$$

Oxygen, which also occurs as a gas, has a valence of two. It combines with two molecules of hydrogen in the formation of *water,* H_2O.

$$O_2 + 2H_2 \longrightarrow 2H_2O$$

Another gas common in our atmosphere, nitrogen, is trivalent. One molecule of nitrogen will combine with three molecules of hydrogen to produce the gas *ammonia,* NH_3.

$$N_2 + 3H_2 \longrightarrow 2NH_3$$

We return to the carbon atom, with a valence of four,

and find that it will combine with two molecules of hydrogen to produce the gas *methane,* CH_4.

$$C + 2H_2 \longrightarrow CH_4$$

We can readily see that the higher the valence of an atom of an element, the greater the complexity of the molecules it can form. If we consider that carbon has a valence of four compared to the valence of one for chlorine, then carbon must have a much greater combining power than chlorine or any other element with a valence of one. Consider chlorine as it combines with the metal sodium, which also has a valence of one.

$$2Na + Cl_2 \longrightarrow 2NaCl$$

We can see that these two elements can combine to form only one compound, sodium chloride (common table salt).

By contrast, methane, an important constituent of the earth's early atmosphere, represents only the simplest of the group of carbon compounds called *hydrocarbons*—those containing only carbon and hydrogen atoms. Somewhat more complex are the *carbohydrates,* which are composed of carbon and water molecules. Sugars and starches belong to this group. Even more complex carbon compounds that are known as *fats* and *proteins* may or may not have been present in those early oceans as products of inorganic photosynthesis.

Acidity and Alkalinity

As water vapor and carbon dioxide were spewed into the earth's atmosphere by the volcanic activity on the earth, they combined to form *carbonic acid,* H_2CO_3.

$$H_2O + CO_2 \longrightarrow H_2CO_3$$

This acid is very weak, but it is important because it falls with water to the earth's surface and aids the water in dissolving the mineral substances locked within the rocks that make up the landmasses. These mineral substances were later incorporated into living organisms. Carbonic acid also played a very important role in keeping even the very earliest oceans slightly alkaline, as life evolved to fit this medium.

Before discussing this condition further we will describe what is meant by acid. An *acid* is a compound containing hydrogen that, when dissolved in water, sets hydrogen ions (H$^+$) free and thus increases the number of such ions per unit of water mass. A strong acid is one whose molecules readily release hydrogen ions when they are part of a dilute solution. The term *alkaline* refers to the presence of compounds in water that dissociate to increase the concentration of hydroxide ions (OH$^-$). A strong *base* is a compound whose molecules readily release hydroxide ions in dilute solutions.

Both hydrogen and hydroxide ions are present at all times in water as the water molecules dissociate and reform.

$$H_2O \longrightarrow H^+ + OH^-$$

If they are produced in pure water by the water molecules themselves breaking up, they are always found in equal concentrations. The water is neutral; there is no excess of either the hydrogen or the hydroxide ion.

If we add a given amount of hydrochloric acid (HCl) to a volume of water, the water will become acidic because it will receive a high concentration of hydrogen ions as a result of the dissociation of the HCl molecules. If we add a strong base (alkaline substance) such as sodium hydroxide (NaOH) instead, there will be an excess of hydroxide ions (OH$^-$) as the NaOH molecules break up.

The acidity or alkalinity of a solution is indicated by a pH scale ranging from 0 to 14. A solution with a pH of 7 is neutral, such as pure water. The lower the pH of the solution falls below 7, the stronger its acidity, indicating an increasing concentration of hydrogen ions. The higher the pH rises above 7, the stronger the alkalinity because of an increasing content of hydroxide ions.

pH scale

0	7	14
Acid	Neutral	Alkaline

The introduction of carbon dioxide into the oceans as carbonic acid serves to maintain the slightly alkaline conditions of the oceans (fig. 2–12). When the carbon dioxide molecule combines with a water molecule, two of the electrons in the oxygen atom of the water molecule are shared with two of the electrons of the carbon atom from the carbon dioxide molecule. This situation leaves the hydrogen nuclei weakly attached to the water molecule as it becomes part of the H$_2$CO$_3$ molecule. One of these hydrogen nuclei may be lost and become a hydrogen ion (H$^+$), transforming the carbonic acid molecule into a negatively charged bicarbonate ion, HCO$_3^-$.

Should the bicarbonate ion lose its hydrogen ion, it then is transformed into a double-charged negative carbonate ion, CO$_3^{--}$. The arrows in figure 2–12 indicate the reaction can go in both directions. If for any reason the pH of the ocean rises, the reaction proceeds to the right, releasing H$^+$ and lowering the pH. Should the pH of the ocean drop, the reaction proceeds to the left and raises the pH by removing H$^+$. At the normal pH of ocean water, about 8.1, most of the carbon dioxide in it occurs as the bicarbonate ion (HCO$_3^-$). This process is termed *buffering*.

A solution must be slightly alkaline before complex organic combinations can form within it. Solutions that are either highly acid or highly alkaline lack the necessary tolerance for the vast range of bonding arrangements needed to form organic compounds. Carbon dioxide combined with water as carbonic acid did much to prepare the environment in which life was to begin. Most important of all is the fact that the carbon dioxide molecule contains that atom which became the organizer of these complex organic building blocks—the carbon atom.

We do not know precisely how the organic material took on the characteristics that allowed it to become living substance, nor do we know when this occurred. Through some process, however, the organic material became chemically self-reproductive, developed the ability to actively seek light, and began to grow toward a characteristic size and shape dictated by forces from within. These living entities were composed mostly of a substance with a chemical composition not greatly unlike that of the water in which they originated. We call this material *protoplasm*. Once organisms formed in the ocean, they could be protected from the lethal ultraviolet radiation by sinking only a few millimeters beneath the ocean surface.

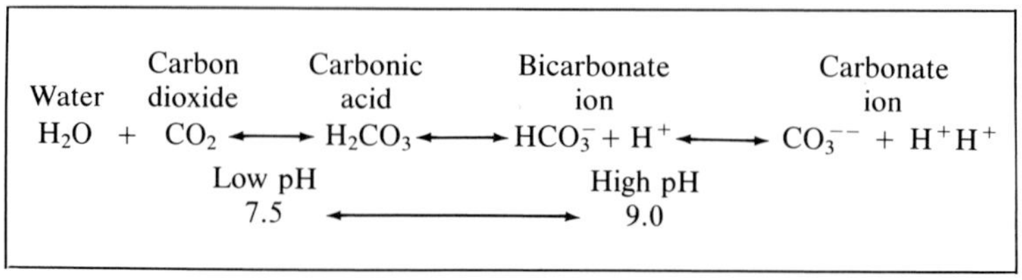

FIGURE 2–12 Maintenance of Ocean Alkalinity.

	Carbon dioxide	Carbonic acid	Bicarbonate ion	Carbonate ion
Water				
H$_2$O +	CO$_2$ ⟷	H$_2$CO$_3$ ⟷	HCO$_3^-$ + H$^+$ ⟷	CO$_3^{--}$ + H$^+$H$^+$
	Low pH 7.5	⟷	High pH 9.0	

DID LIFE BEGIN ON THE OCEAN FLOOR?

In our discussion of how life began on earth, we described a process by which it may have originated in the surface waters of the ocean. Critics of this hypothesis say the concentration of organic matter would have been too dilute to support the heterotrophic cells the theory postulates. Another problem with this theory is understanding the process by which such primitive cells developed the complex cell membrane needed by heterotrophic forms to transport organic compounds. Others believe that the first microorganisms came to earth from space.

John B. Corliss (fig. 2B) and associates propose an alternative hypothesis—life on earth began in the hydrothermal (hot-water) vents of the ocean floor. They believe the planetesimal bombardment of the inner solar system from about 4.2 to 3.9 billion years before the present (BYBP) would have provided up to one-fifth of the earth's mass and produced major volcanic activity. The earth's surface, made molten by this energy release, cooled as the bombardment diminished. Crustal plates formed and were carried across the earth's surface by the convecting magma beneath, initiating the process of global plate tectonics at about 3.9 BYBP. Soon the oceans filled the low-lying basins. Where crustal plates were moved apart by mantle convection, magma rose to form new ocean floor. Newly formed crust would crack as it was quickly cooled by contact with ocean water, and tensional stresses of the plates being moved apart produced linear fissures. As soon as the ocean water migrated down through the broken crust to be heated by the underlying magma, then rose again to the ocean floor as a hydrothermal spring (fig. 2C), the stage was set for the origin of life.

The process hypothesized to have produced the first protocells on earth—almost instantaneously—after the formation of the first hydrothermal spring follows:

1. Heat and carbon dioxide (CO_2) released by the magma interacted with the sea water near its magma contact.
2. The gases hydrogen (H_2), methane (CH_4), ammonia (NH_3), hydrogen sulfide (H_2S), and others, as well as many elements released by the crystallization of the magma, interacted with the heated water.
3. Heated to temperatures up to $900 + °C$ ($1652 + °F$) at the magma interface, the rising water would contain organic compounds of low molecular weight synthesized at high temperature.
4. The water continued to rise and cool rapidly through fractures coated with a catalyst, saponite, a magnesium clay formed by the alteration of basalt. In the presence of the catalyst, low-molecular-weight organic molecules polymerized into complex organic molecules that combined to form protocells.
5. Exiting vent water carried the protocells out onto the ocean floor at temperatures usually below 100°C (212°F), although some vents are known to spew 350°C (662°F) water into the 2°C (35.6°F) ambient water at the ocean floor.

FIGURE 2B John B. Corliss, Georgetown University.

Fossil Evidence   Fossils found in rocks from 3.5 to 3.8 BYBP lend support to the deposition of protocells on the ocean floor surrounding hydrothermal vents. The three ancient rock locations where such evidence is available are the *Isua* in southwest Greenland (3.8 BYBP), *Onverwacht* in South Africa (3.5 BYBP), and *Warrawoona* in western Australia (3.5 BYBP). There is a remarkable resemblance between filamentous "organiclike fossils" in the Onverwacht and organic structures in scanning electron micrographs of hydrothermal deposits recovered at 21°N on the East Pacific Rise in 1979 (fig. 2D). All of the ancient fossil locations contain rocks interpreted to have been deposited in a hydrothermal vent environment.

The Organisms   The chemosynthetic bacteria isolated from 21°N latitude and Galapagos vents may help answer the question of whether or not the first life

FIGURE 2C   Hydrothermal Spring.

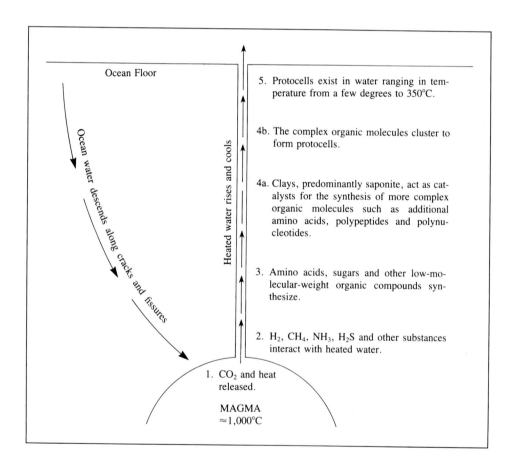

Ocean Floor

Ocean water descends along cracks and fissures

Heated water rises and cools

5. Protocells exist in water ranging in temperature from a few degrees to 350°C.

4b. The complex organic molecules cluster to form protocells.

4a. Clays, predominantly saponite, act as catalysts for the synthesis of more complex organic molecules such as additional amino acids, polypeptides and polynucleotides.

3. Amino acids, sugars and other low-molecular-weight organic compounds synthesize.

2. H_2, CH_4, NH_3, H_2S and other substances interact with heated water.

1. CO_2 and heat released.

MAGMA
$\approx 1,000°C$

forms were heterotrophs. Maybe they weren't. If life did originate at the hydrothermal vents as anaerobic chemosynthetic forms, it could have been sustained by the constant effluent of hydrothermal gases, CO_2, H_2, NH_3, and possibly H_2SO_4. From this beginning, evolution could have produced the more complex macromolecules that were heterotrophic and that reproduced by binary fission.

These early forms, "archaebacteria," may have reproduced by budding off chains of new individuals as is observed in protocells synthesized in the laboratory. They may have been methanogens that reduced CO_2 and SO_4 (sulfate) to CH_4 by using H_2. Present-day procaryotes, "eubacteria," may have evolved from methane-producing "archaebacteria."

The occurrence of petroleum condensate at the sea bed surrounding sediment-covered hydrothermal vents in the Guaymas Basin of the Gulf of California indicates the plume water must contain large amounts of methane and higher-molecular-weight organic matter. Although some of this hydrocarbon material is generated in the sediments, further investigation may show that such protocells as have been described above are still forming in the hydrothermal vents of today's oceans.

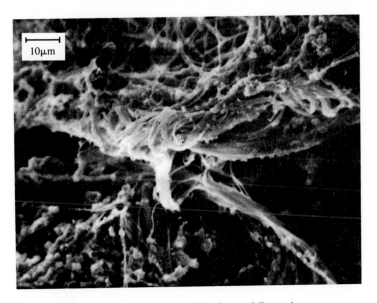

10μm

FIGURE 2D Organiclike Structures from Hydrothermal Deposits at 21°N on East Pacific Rise.

Plants and Animals Evolve

These very earliest forms of life must have been *heterotrophic* organisms, those that depend on an external food supply. That food supply was certainly abundant in the form of molecules of the same nonliving organic material from which the first living single-celled organisms originated. The *autotrophs,* those organisms that do not depend on an external food supply but manufacture their own, eventually evolved. These first autotrophic forms may well have been organisms similar to our present-day anaerobic bacteria, forms that can exist without free oxygen. They may have been able to oxidize various inorganic compounds that would release energy that could be used to produce their own food internally. This process is called *chemosynthesis*. At some later date, the more complex single-celled plants probably evolved. Having developed a green pigment called *chlorophyll,* plants could capture the sun's energy and produce through *photosynthesis* their own food supply from the carbon dioxide and water that surrounded them.

The autotrophs had developed security in the form of a built-in mechanism to assure them of a food supply. Other forms, not so endowed, were at a considerable disadvantage because they had to search for this necessity of life. To meet this need, animals eventually emerged with their inherent characteristic of mobility and began to develop an awareness or a consciousness that gave them a much higher place in the biological community than that possessed by the plants.

Plants and animals developed in a beautifully balanced environment where the waste products of one filled the vital needs of the other; we can see this expressed chemically by looking at the processes of photosynthesis and respiration. In the photosynthetic process, energy is being captured and stored in the form of organic compounds. This is an *endothermic* chemical reaction. Respiration is just the opposite—an *exothermic* reaction. It is the reaction by which the energy stored in the foods produced by photosynthesis is extracted by the plants and animals to carry on their life processes.

Photosynthesis $\longrightarrow$ *energy is stored (endothermic)*

$6H_2O + 6CO_2 + Energy \longleftrightarrow C_6H_{12}O_6 + 6O_2$

Water + Carbon + Energy   Sugar + Oxygen
      dioxide

Respiration $\longleftarrow$ *energy is released (exothermic)*

The first evidence we have in the fossil record of life on earth is in the form of algae and bacterialike forms that have been preserved in rocks more than three billion years old. Every living organism that inhabits the earth today is the result of a process of natural selection that has been going on since these early times and by which various life forms (species) have been able to inhabit increasingly numerous niches within the earth environment. As these diverse life forms developed in an attempt to exist and adapt to various environments, they also modified the environments in which they lived.

As the plants emerged from the oceans to inhabit the terrestrial environment, they changed the appearance of the landscape from the harsh, bleak panorama we may envision as being typical of a lunar surface to the soft green hues that cover much of the land surface of the earth today. Other changes were manifested in the ocean itself as vast quantities of the hard parts of dead organisms began to accumulate on the bottom. These accumulations of calcium carbonate ($CaCO_3$) and silica (SiO_2) formed the limestone and diatomaceous deposits that we now see exposed on the continents as evidence that these areas were once covered by the seas.

Probably the most important environmental modification was that which involved the atmosphere. It was this change that made it possible for animals to develop. A byproduct of photosynthesis in the early oceans was the free oxygen that makes up 21% of our present atmosphere. Carbon dioxide, which at one time must have made up a large portion of our atmosphere, was removed by photosynthesis to produce the present concentration of 0.03% of the earth's present atmosphere. These changes in the atmosphere must have had a very significant effect on the climates that have evolved throughout the earth's history.

We also see, in petroleum and coal deposits, the remains of plant and animal life that were not completely oxidized but were buried in a reducing (oxygen-free) environment that allowed their energy to remain stored. In some instances, especially in the formation of coal, the dead plant matter accumulated in such quantities that the reducing environment was produced by the decay of the organic matter using up the available oxygen. These deposits provide us with over 90% of the energy used for domestic and industrial purposes today. So we, as animals, are not only dependent upon the present productivity of plants to supply the energy required by our life processes, but we also depend very heavily on the energy stored by plants during the geologic past.

SUMMARY

Our **solar system**, consisting of the sun and nine planets, belongs to the **galaxy** of stars we call the Milky Way. All the stars we see with the unaided eye belong to this galaxy; yet there are some 100 million galaxies thought to exist in the universe. Most seem to be moving away from us as though all were set in motion by a single large explosion, and the farther away a galaxy is, the faster it seems to be moving away.

The galaxies are thought to be accumulations of the debris from this explosion in large eddies, inside of which are smaller eddies where concentration of matter has produced stars and their associated planets. **Stars** result from accumulation of masses large enough to produce sufficient internal temperatures and pressures to set off processes causing them to "shine" as they give off energy at high rates. The **planets** were smaller, so they did not reach this state.

Protoearth was composed mostly of **hydrogen** and **helium**, but as it condensed and heated up, these elements were driven off into space. There are indications the earth became molten and then developed an atmosphere rich in **water vapor** and **carbon diox**ide produced by volcanic activity. **Methane** and **ammonia** were also present in significant quantities. As the earth's surface cooled sufficiently, the water vapor condensed and accumulated in depressions on the surface to give the earth its first oceans.

Study of atomic bonding has revealed two predominant types. Water molecules typically display **covalent bonds**, while **ionic bonds** are characteristic of metallic salts.

Life is thought to have begun in these oceans. As **ultraviolet radiation** fell on the oceans with their dissolved carbon dioxide, methane, and ammonia, inorganic molecules may have combined to produce carbon-containing molecules that are now formed naturally on earth only by organisms. As these chance combinations became more complex, some of the combinations that sank deep enough into the ocean to be protected from the lethal ultraviolet rays evolved into **bacteria**. Eventually some included **chlorophyll** in their makeup, and **plants** were the result. The **photosynthesis** of plants extracted carbon dioxide from the atmosphere and produced the first free **oxygen** at the earth's surface. This produced an oxygen-rich atmosphere in which **animals** as we know them could survive. Eventually, both plants and animals evolved into forms that could survive on the stark environment of the continents, producing the lush continental environments we enjoy today.

QUESTIONS AND EXERCISES

1. Construct a graph with velocity in thousands of km/s plotted along the vertical axis from 0 to 100 and light-years of distance plotted along the horizontal axis from 0 to 2 billion. On this base, plot the values given for the constellations Virgo, Corona Borealis, and Hydra in figure 2–3. Establish a line by connecting the three points. How much farther from us is a galaxy for each increase in velocity of ten thousand km/s?

2. Why is it theorized that the observed motions of galaxies were initiated by an explosion?

3. Why is it logical to assume that, if argon, with a molecular weight of 40, was lost from the earth's atmosphere when the earth's surface was much hotter than it is today, the components of today's atmosphere (water vapor, nitrogen, and oxygen) would have also been lost at that time?

4. During its history, the earth may have possessed three distinctly different envelopes of gases, or atmospheres. Describe the atmosphere of protoearth, of the earth when the oceans first formed, and of the earth at present.

5. In what three components of the earth's makeup must we find the total mass of crystalline rock components that were eroded from the primary crystalline rocks of the earth's surface?

6. List the five most abundant excess volatiles. Why are they called excess volatiles?

7. Why does the fact that new water is continually being released to the atmosphere by volcanic activity not necessarily mean the oceans will progressively cover an increasing percentage of the earth's surface?

8. Locate sulfur (S) on the periodic table of the elements, appendix III. List its atomic number, atomic weight, and valence. Draw a diagram similar to those used for the atoms in figure 2–8 showing the number of protons, neutrons, and electrons in the most common isotype of the element.

9. List the two types of chemical bonds discussed in the text, describe the manner in which the outer shell of electrons is filled, and give an example of a compound containing each type of bond.

10. Why do we believe that free oxygen (O_2) was not released into the atmosphere by volcanic activity?

11. How does the presence of oxygen (O_2) in our atmosphere help reduce the amount of ultraviolet radiation that reaches the earth's surface?

12. How did the photosynthesis that produced the first organic molecules in the early oceans differ from plant photosynthesis? How does chemosynthesis differ from photosynthesis?

13. When we say carbon (C) has a greater combining power than oxygen (O), we are referring to chemical valence. How is valence related to combining power?

14. How are the pH scale, acidity-alkalinity, and relative H^+ ion and OH^- ion concentrations related?

15. How does the chain of reactions made possible by dissolving CO_2 in water help maintain a slightly alkaline condition in the oceans?

16. Describe some basic characteristics of living things.

17. Discuss photosynthesis and respiration, and explain their relationship to the chemical processes of endothermic and exothermic reactions.

18. As plants evolved on the earth, great changes in the earth's environment were produced. Describe some of these major changes caused by plants.

REFERENCES

Baker, R. H. 1955. *Astronomy.* Princeton, N.J.: Van Nostrand Reinhold.

Davis, K. S., and Day, J. A. 1961. *Water: The mirror of science.* Garden City, N.Y.: Doubleday.

Frank, L. A.; Sigwarth, J. B.; and Craven, J. D. 1986. On the influx of small comets into the earth's upper atmosphere—observations and interpretations. *Geophysical Research Letters.* 13:4, 303–310.

Glaessner, M. F. 1984. *The dawn of animal life: A biohistorical study.* Cambridge: Cambridge University Press.

Holland, H. D. 1984. *The chemical evolution of the atmosphere and oceans.* Princeton, N.J. Princeton University Press.

Lagemann, R. T. 1969. *Physical science: Origins and principles.* Boston: Little, Brown.

Rubey, W. W. 1951. Geologic history of sea water—an attempt to state the problem. *Geological Society of America Bulletin.* 62:1110–19.

Strom, K. M., and Strom, S. E. 1982. Galactic evolution: A survey of recent progress. *Science.* 216:4546, 571–80.

Weyl, P. K. 1970. *Oceanography: An introduction to the marine environment.* New York: Wiley.

SUGGESTED READING

Scientific American

Barghoorn, E. S. 1971. The oldest fossils. 224:5, 30–54.
   The evidence that bacteria and algae were present on earth over 3 billion years ago is discussed along with evolutionary considerations.

Barrow, J. D., and Silk, J. 1980. The structure of the early universe. 242:4, 118–28.
   Although the universe is inhomogeneous on the small scale of the solar system or a galaxy, it is very homogeneous on the scale of the universe as a whole.

Clarke, B. 1975. Causes of biological diversity. 223:2, 50–61.
   The diversity that is observed within species and its relationship to natural selection are pursued. The physiological and social implications of the data are discussed.

Frieden, E. 1972. Chemical elements of life. 227:1, 52–64.
   The roles of the twenty-four elements known to be essential to life are discussed, including some background on how they may have been selected from the physical environment in which life evolved.

Gott, J. R., and Gunn, J. E. 1976. Will the universe expand forever? 234:3, 62–79.
   Based on data regarding the recession of galaxies, average density of matter, and chemical elements, the authors suggest expansion will be reversed.

Levine, J. S., and MacNichol, E. F., Jr. 1982. Color vision in fishes. 246:2, 140–49.
   A discussion of the effect the blue-green environment of fishes may have had on the evolution of the fish eye.

Stebbins, G. L., and Ayala, F. J. 1985. The evolution of Darwinism. 253:1, 72–85. New advances in molecular biology and new interpretations of the fossil record add to the knowledge of evolution.

Vidal, G., 1984. The oldest eukaryotic cells. 250:2, 48–57.
   Cells with a nucleus appear to have evolved as marine plankters 1.4 billion years ago.

Wilson, A. C. 1985. The molecular basis of evolution. 253:4, 164–175.
   Mutations within the genes of organisms play an important role in evolution at the organismal level.

Woese, C. R. 1981. Archaebacteria. 244:6, 98–125.
   A discussion of two classes of bacteria, archaebacteria and eubacteria.

An understanding of the general bathymetric features of the ocean is important because these features will be related ultimately to the origin of the ocean basins as well as physical and biological phenomena to be discussed in following chapters. *Bathymetry* in oceanography is analogous to *topography* in geography. As investigations into the depth of the ocean have proceeded, it has become apparent that a broad shelflike feature that steepens and slopes off into deep basins has generally developed around the continents. Quite commonly, linear mountain ranges run through the basins. In all oceans, but especially around the margin of the Pacific Ocean, deep linear trenches up to 11 km (7 mi) in depth may separate the slopes from the deep-ocean basin (fig. 3–1). The underlying structures and their origin will be discussed in chapter 4.

The hypsographic curve showing the distribution of the earth's solid surface at elevations above and below sea level indicates that 71% of the earth's surface is to be found beneath the oceans (fig. 3–2 and plate 9). The mean elevation of the continents above the sea (840 m, or 2755 ft) and mean depth of the oceans (3865 m, or 12,677 ft) are the result of different densities of the continental and ocean crustal rocks. The ocean crust is denser.

CHAPTER THREE
MARINE PROVINCES

ISOSTASY

The crust can be divided into two distinct types. The basaltic oceanic crust has an average thickness of 7 to 8 km (4 to 5 mi) and a density of about 3.0 g/cm³. The continents themselves are composed of a lighter granitic crustal material ranging in density from 2.67 g/cm³ near the surface to 2.8 g/cm³ deep beneath the continental mountain ranges. The continental crust averages about 35 km (22 mi) in thickness but may reach a maximum thickness of 60 km (37 mi) beneath the highest mountain ranges.

The granitic continental blocks and basaltic oceanic crust "float" on the denser mantle beneath. This concept of continental flotation, which may be equated to the flotation of ice on water, is called *isostasy* (fig. 3–3).

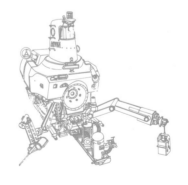

A toothpaste-like squeeze of lava observed at the FAMOUS sight in the North Atlantic Ocean. It is about 30 cm (1 ft) wide and 60 cm (2 ft) long. Photograph courtesy of the Scripps Institution of Oceanography, University of California/San Diego.

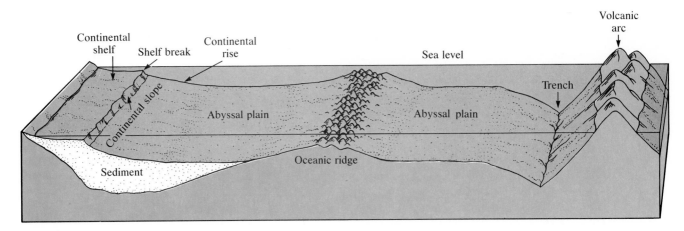

FIGURE 3–1 Marine Physiographic Provinces.

A longitudinal profile showing submarine physiographic provinces. To the left of the oceanic ridge is a continental margin typical of the Atlantic Ocean, with a well-developed continental shelf, slope, and rise at the surface of a thick sedimentary deposit. The features to the right are more typical of the Pacific Ocean, where trenches and associated volcanic arcs are common.

FIGURE 3–2 Hypsographic Curve.

This curve indicates the percentage of the earth's total area, land area, and ocean area at depths and elevations indicated along the left margin of the figure. After Sverdrup, Johnson, and Fleming, 1942.

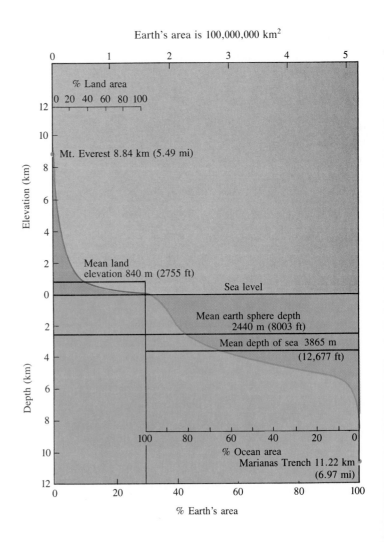

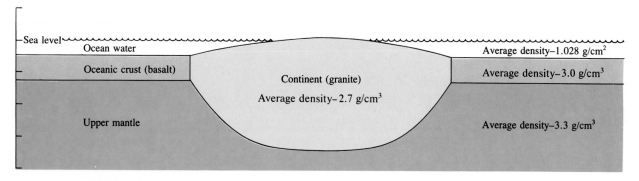

FIGURE 3-3 Isostasy.

Isostasy is a state of equilibrium reached by different components of the earth's crust and the mantle. So that the lithosphere (the rock sphere of the earth's structure), which contains the continental and oceanic crustal units as well as the upper mantle, may maintain a uniform average density, continental crustal units that extend up to 14 km (8.7 mi) above the floor of ocean basins must send deep roots into the mantle.

If we take a regularly shaped prism of ice with a density of 0.91 and float it on water with a density of 1.0, the ice will sink into the water until 91% of its mass is submerged. Therefore, 91% of the length of the prism of ice will be beneath the surface of the water. Similarly, about 55% of the mass of the continents is found submerged in the mantle. We would actually expect that about 91% of the continents would be submerged because they are, on the average, about 91% of the density of the upper mantle. However, where high heat flow in the mantle heats and reduces the density of the mantle beneath the continents, this low-density mantle provides *isostatic compensation*. In such situations as exist beneath the southwestern United States and eastern Africa, the low-density mantle material provides the buoyancy that would normally be provided by a deep root of continental crust.

TECHNIQUES OF DETERMINING BATHYMETRY

Needless to say, our greatest understanding of the ocean floor relates to the shallow shelf areas at the margins of the continents, over which the most important commercial fisheries have been located. They are presently being explored extensively because of their high potential for petroleum accumulation. Because of their economic importance and the fact that we have a greater knowledge of this region than any other ocean region, we will begin our description of the ocean floor with these marginal shelves.

However, before we discuss the various submarine provinces, we should discuss briefly the technique

that has been most used to obtain data regarding submarine geological processes, *seismic surveying.* Such surveys are made by emitting a sound signal that will travel through the water, bounce off an object, and return to be picked up by a receiver. If the velocity at which sound will travel in water has been determined, the distance to the object is equal to the velocity times one-half the time required for the sound signal to make the trip from the surface to the object or ocean floor and back to the receiver (fig. 3-4).

If one wishes to know only the distance to an object in the water or the ocean floor, relatively weak high-frequency sound signals of the *echo sounder* can be used. Marine geologists, however, need to know the thickness of various rock units beneath the ocean floor, and for this purpose they use stronger low-frequency signals generated by explosions. These sound signals can penetrate the rocks beneath the ocean floor, reflect off the contacts between rock units, and produce seismic reflection profiles such as those shown in chapters 3 and 5.

A variation of this technique is *side-scan sonar,* which allows a survey ship to map the topography of the bottom along a strip of ocean floor up to 5 km (3 mi) wide. The map is developed with the aid of computers and sound emitters directed away from both sides of the ship.

Owing to the gravitational effects on the ocean surface, sea level rises or falls 4 m (13 ft) with each decrease or increase in ocean depth of 1000 m (3280 ft). Altimeters aboard *SEASAT A* detected these differences and provided data to produce a map of the ocean surface (plate 14). The map reflects major bathymetric features of the ocean floor.

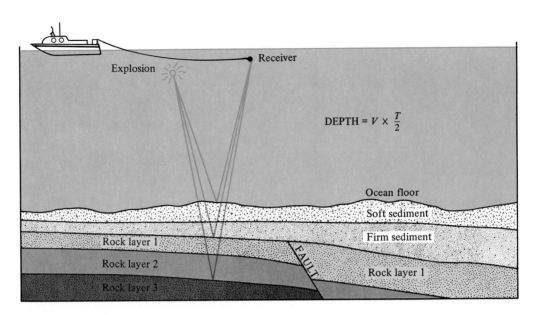

FIGURE 3–4 Seismic Profiling.
Low-frequency sound that can penetrate bottom sediments is emitted by an explosion. It reflects off the boundaries between rock layers and returns to the receiver. The depth of each reflecting layer is equal to the velocity of sound travel (V) times one-half the time (T) required for the sound to travel from the source to the reflecting layer and back to the receiver.

CONTINENTAL MARGIN

Continental Shelf

Extending from the shoreline is a shelflike feature called the *continental shelf,* which is geologically part of the continent. During the geologic past much of it was exposed above the shoreline. Because the shoreline has moved back and forth across the shelf, its general bathymetry usually can be predicted by looking closely at the topography of the adjacent coastal region. With few exceptions, this coastal topography can be expected to extend beyond the shore and onto the continental shelf. The continental shelf is defined as a shelflike zone extending from the shore to a point at which a marked increase in slope occurs. This point where an increase in the rate of slope occurs is referred to as the *shelf break,* and the steeper slope beyond the break is known as the *continental slope* (shown in fig. 3–1).

The width of continental shelves may vary from a few tens of meters to about 1300 km (800 mi). The broadest shelf developments occur off the northern coasts of Siberia and North America in the Arctic Ocean and in the North and West Pacific from Alaska to Australia. The average width of the continental shelf is about 70 km (43 mi) and the average depth at which the break occurs is about 135 m (443 ft). Around the continent of Antarctica, however, the shelf break occurs at a depth of 350 m (2200 ft). The mean slope of the continental shelf is 0°07′, or 1.9 m/km (10 ft/mi).

Sediment is transported across the continental shelf by current flow and mass wasting (submarine sediment slides), which are often generated by earthquakes. Evidence of earthquake-induced sediment failure producing a slide on the gently sloping (0.25°) Klamath River delta off the northern California coast was recorded in 1980. On November 8, 1980, an earthquake of magnitude 6.5 on the Richter scale occurred 60 km (37 mi) off the coast. During a December survey of the area, using high-resolution seismic reflection and deep-towed side-scan sonar, a slide zone 1 km (0.62 mi) wide and 20 km (12.4 mi) long was discovered. The failure seems to have occurred along the sediment boundary between fine to medium sand and a seaward deposit of mud (clayey silt), shown in figure 3–5.

Continental Slope

The continental slopes beyond the shelf break are features similar in relief to mountain ranges we find on the continent. The break at the top of the slope may be from 1 to 5 km (0.62 to 3 mi) above the deep-ocean basin at its base. In areas where the slope descends into submarine trenches, even greater vertical relief is measured. Off the west coast of South America the total relief from the top of the Andes Moun-

FIGURE 3–5 Location of Klamath River Delta Failure Zone Resulting from November 8, 1980, Earthquake.

The slide zone is 1 km (0.62 mi) wide and about 20 km (12.4 mi) long at an average depth of 60 m (197 ft). It seems to be associated with a sediment contact where near-shore sandy sediment changes seaward to mud. Courtesy of Michael Field, U.S.G.S.

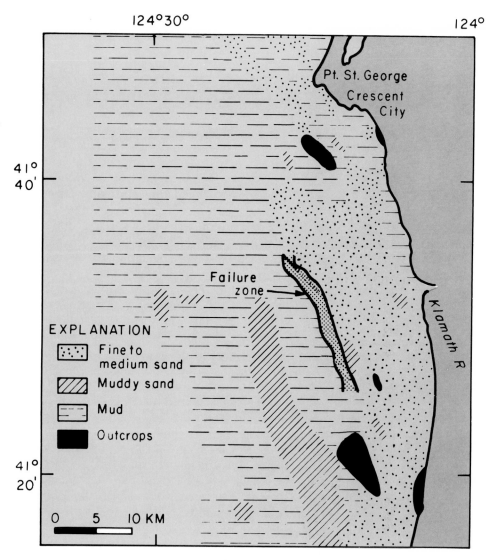

tains to the bottom of the Peru-Chile Trench is about 15 km (9.3 mi).

Continental slopes vary in steepness from 1° to 25° and average about 4°. Around the margin of the Pacific Ocean, where the slope is associated with the processes forming coastal mountain ranges and submarine trenches, the continental slopes are steeper than in the Atlantic and Indian oceans. Slopes in the Pacific Ocean average more than 5°, while those in the Atlantic and Indian oceans are about 3°.

Submarine Canyons The continental slope and, less commonly, the continental shelf are cut by large submarine canyons that resemble the largest of canyons cut on land by rivers. Like the canyons cut by rivers, the submarine canyons have tributaries and steep V-shaped walls that expose a wide range of rock types varying greatly in geological age. Exposed in

the walls of the canyons are rocks ranging from soft shales to quartzite and granite.

The canyons are obviously erosional features; the problem is to explain how the erosion occurred. If the first person to observe a submarine canyon had observed those off the west coast of the French island of Corsica in the Mediterranean, he would have concluded that, without doubt, submarine canyons were related to land river systems since the canyons lead right into the mouths of the rivers that drain the western flank of the island. Sediments recovered during Deep Sea Drilling Project cruises in the Mediterranean Sea indicate this body of water dried up at least once in its history. Therefore, the submarine canyons off the island of Corsica may well have been created primarily or entirely by stream erosion. The majority of submarine canyons, however, do not tie so nicely with land drainage systems, and many are confined

SEASAT MAPS THE OCEAN FLOOR

The shape of the ocean floor (the study of which is called bathymetry) has long eluded human knowledge. The first clues to the ocean's depths were provided through individual soundings obtained by lowering a weighted line over the side of a ship. This method was used on the voyage of the *Challenger* (1872–1876). Continuous determinations of depth along a ship's course were made possible by the development of the echo sounder, first used in deep-sea oceanographic research aboard the *Meteor* (1925–1927). Until recently, maps of ocean-floor bathymetry have been constructed by tedious hand compilation of individual soundings and echo sounding data. Many gaps in data exist, especially in the southern oceans, where there can be 300-km (186-mi) distances between ship paths. The use of satellites may, however, provide the solution to this need for data in the southern oceans.

Exciting new maps of world ocean bathymetry have recently been developed using data obtained from a microwave altimeter flown aboard the *SEASAT* satellite from July 7 to October 10, 1978. The altimeter measured the distance from the satellite to the ocean surface. This surface is significantly affected by gravitational attraction and tends to conform to the earth's geoid, a theoretical surface that is a close approximation of the ocean surface. The ocean surface may vary by as much as 1 m (3.28 ft) from the geoid because of currents and atmospheric pressure variations, but is usually within 20 cm (8 in) of the geoid (fig. 3A). The geoid may vary by as much as 200 m (656 ft) relative to another theoretical surface, the reference ellipsoid. Conforming to these variations, the ocean surface stands 80 m (262 ft) above the ellipsoid just north of Australia and drops 100 m (328 ft) below south of the tip of India. These large-scale surface features (greater than 5000 km, or 3100 mi, across) probably result from mass anomalies deep within the mantle that are not yet well understood. Smaller-scale (less than 500 km, or 310 mi, across) changes in the level of the ocean surface are well correlated with ocean bathymetry.

The construction of these new *SEASAT* maps required computer analysis of 50 million physical measurements that were corrected for atmospheric and other interferences. The maps are designed to show small-scale ocean-floor features ranging in size from 50 to 500 km (31 to 310 mi) in length. Plate 14 shows slopes of up to 5 m/degree (approx. 111 km, or 69 mi) along northwest-to-southeast lines. The flat ocean floor is medium blue. It darkens with increased rate of downslope and lightens with increased rate of upslope.

The relationship of the level of the ocean surface to changes in depth is demonstrated by a lowering of sea level by approximately 20 m (66 ft) over a trench that drops 5000 m beneath the depth of the surrounding deep-ocean floor and a rise of 4 m (13 ft) above a seamount that stands 1000 m (3280 ft) above the deep-ocean floor.

The major ocean trenches and seamount alignments are clearly visible on the maps. One major topographic feature of the ocean floor that does not show up well in the maps is the East Pacific Rise. This discrepancy results because it is a broad region of rapidly forming, relatively plastic young lithosphere. Features that represent younger volcanic structures forming on older, more rigid lithosphere show up more distinctly. A good example of this condition is the

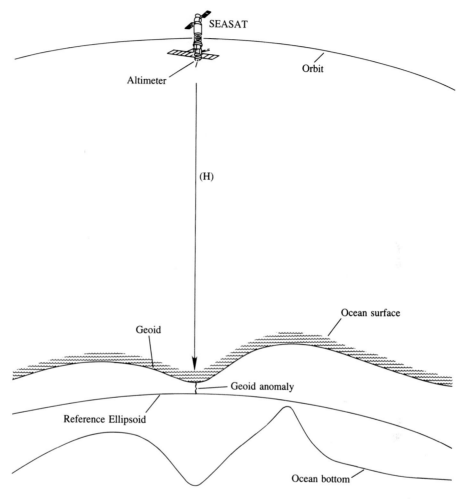

FIGURE 3A *SEASAT* Maps the Ocean Floor.

The *SEASAT* altimeter measures the distance between the satellite and the ocean surface (H). The ocean surface is usually less than 20 cm (8 in) above or below the geoid, a surface determined by the earth's gravitational field. Very rarely, it may vary from the geoid by as much as 1 m (3.3 ft). The variation between another theoretical surface, the reference ellipsoid, and the geoid (α the ocean surface) is caused by density anomalies within the mantle and ocean-floor topography. It is called the geoid anomaly, which can be mapped to show the effect of topography and density anomalies on the ocean surface.

alignment of the Emperor Seamount–Hawaiian Island chain, where the hot spot volcanism created volcanoes on lithosphere that was as much as 75 million years old.

Plate 14 shows a map of the ocean surface that correlates with sea-floor features that are less than 600 km (372 mi) in scale. It is contoured to show the elevation of the sea surface from 4 m (13 ft) above to 4 m below the reference ellipsoid. Noted on the map are significant features on the ocean floor that were not known to exist before the *SEASAT* data became available or were significantly modified by it. They all lie in the southern Pacific and Indian oceans, where ship tracks are widely spaced.

In the southwest Pacific Ocean, new seamounts were identified on the Louisville Ridge, and a large rise which was poorly known shows up east of the Louisville Ridge.

Features identified in the Indian Ocean were a large trough east of the Kerguelen Plateau, a rise and seamount south of Elan Bank, and a small plateau east of Crozet Island. The *SEASAT* maps also established that the Conrad Rise is larger and of different shape than previously indicated, the Marion Dufresne Seamount is 100 km (62 mi) to the east of where it was thought to stand, and a previously unknown seamount lies north of the seamount.

The data obtained from the *SEASAT* altimeter, in addition to confirming the bathymetry of the oceans, will aid in determining the tectonic history of the ocean basins. For instance, strong correlations have been established between the increased magnitude of the geoid anomaly and a decreasing rate of spreading at oceanic spreading centers as well as the increasing difference in the age of ocean crust and the age of volcanic features supported by the crust.

Reference: Dixon, T. H., and Parke, M. E. 1983. Bathymetry estimates in the southern oceans from *SEASAT* altimetry. *Nature* 304:5925, 406–411.

exclusively to the continental slope. The most obvious objection to explaining them as drowned river valleys is the fact that they continue to the base of the continental slope, which averages some 3500 m (11,500 ft) below sea level. Since rivers lose their ability to erode shortly after reaching the ocean, it seems impossible that rivers could have cut canyons this far below sea level.

Submarine canyons were one of the primary objects of study of the "father of marine geology," the late Francis P. Shepard, who began his informal studies of the ocean floor in the 1920s (fig. 3–6). Dr. Shepard continued to investigate coastal changes and the effect of volcanic activity on climate until his death at age 87 in 1985.

Recent work done with side-scan sonar indicates that the continental slope is dominated by submarine canyon topography along the Atlantic coast from Hudson Canyon to Baltimore Canyon (fig. 3–7). Smooth intercanyon areas are particularly rare on the middle and upper surfaces of the slope. There is also a correlation between the spacing of canyons and the continental slope gradient. No canyons are present where the slope is less than 3°. Canyons are from 2 to 10 km (1.2 to 6 mi) apart where the slope is between 3° and 5°, and they are 1.5–4 km (0.9–2.5 mi) apart where the slope gradient exceeds 6°.

Canyons confined to the continental slope are straighter and have steeper canyon floor gradients than those that extend into the continental shelf. This condition suggests that the canyons are initiated on the continental slope by some marine process and cut their way headward into the continental shelf as they age.

Turbidity Currents
Probably the most widely supported theory concerning the origin of submarine canyons is erosion by *turbidity currents,* flows of sediment-laden water that periodically move down the canyon. Turbidity flows resulting from river input near the heads of some Pacific-coast submarine canyons that extend well onto the continental shelf have been observed. However, turbidity currents in many canyons confined to the continental slope apparently are initiated after sediment moves across the continental shelf into the head of the canyon and accumulates there. The actual initiation of the current may be produced by an earth tremor or other disturbance that makes the mass collapse and start to move down the canyon under the force of gravity. As it moves, the sediment will mix with the water, producing a high-density mass that moves down the canyon, spreads out, and is deposited at the base of the

FIGURE 3–6 Francis P. Shepard (1898–85), father of marine geology.

continental slope as part of a *deep-sea fan* structure commonly found at the foot of a submarine canyon. The flow moves out across the fans somewhat as a river flow moves across the surface of a delta. A pattern of distributaries develops. They are bordered by levees that are deposited as the turbid mass overflows the banks of the relatively shallow distributary channels (fig. 3–8).

It is not surprising that such events have never been directly observed. However, in 1971, measurement of water properties at a depth of 5200 m (17,000 ft) immediately above the Sohm Abyssal Plain in the North Atlantic (fig. 3–9) indicated a turbidity current may have occurred just prior to the time the samples were taken. In these samples, the values for temperature, salinity, and dissolved oxygen were all higher than normally found near the deep-ocean floor. These properties resembled those typical of water at 4200 m (13,800 ft). In addition to these anomalies, the water was many times more turbid than is normal for this location.

A similar condition was recorded during a continuous seismic survey southwest of the Canary Islands

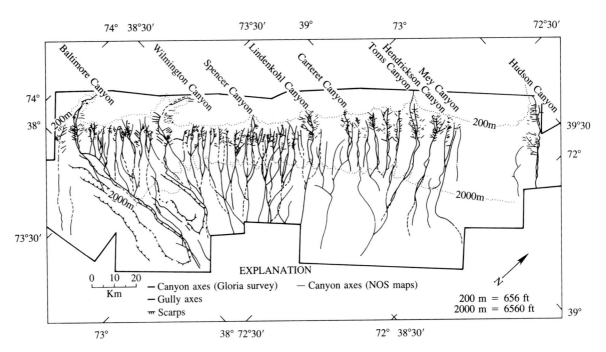

FIGURE 3–7 Submarine Canyon Topography along the Atlantic Coast.

There is a correlation between the formation of submarine canyons and the steepness of the continental slope. Between Hudson Canyon and Mey Canyon the angle of slope is less than 3° and no canyons have formed. In the area of Spencer Canyon, the slope steepens to over 6°, and canyons are very closely spaced. Throughout most of the area mapped, the slope steepness ranges between 3° and 5°. Within this region of intermediate slopes, the distance between canyons decreases as the slope steepness increases.

FIGURE 3–8 Submarine Canyons and Turbidity Currents.

Erosional and depositional features associated with the formation of submarine canyons.

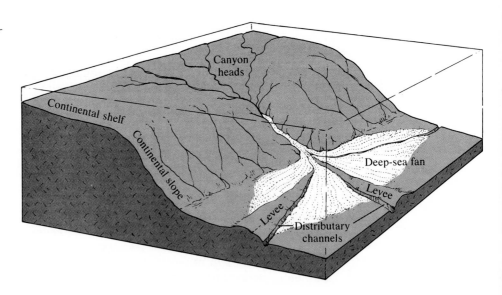

on June 26, 1981. A dense cloud 2 km (1.2 mi) wide and extending 215 m (700 ft) above the 4900 m (16,000 ft) deep Canary Abyssal Plain floor was recorded (fig. 3–9). Upslope from the cloud, which shrank in a few hours, the water was clear enough to record bottom features with precision. A channel down which the turbidity current could have traveled was clearly present.

The remote sonar recording of a turbidity current generated on August 26, 1976, by the discharge of mine tailings into Rupert Inlet, British Columbia (fig. 3–10), may be the only real-time observation of a ma-

FIGURE 3—9 Floor of the North Atlantic Ocean.

Circle indicates location of sites where turbidity currents may have been accidently recorded, as described on pages 55, 56, and 58. Arrow indicates the part of North Atlantic Deep Water or western boundary undercurrent. Hexagon indicates cable breaks associated with the 1929 Grand Banks earthquake. North Atlantic panaroma excerpted from *The World Ocean Floor* by Bruce C. Heezen and Marie Tharp, published by the Lamont-Doherty Geological Observatory of Columbia University and sponsored by the U.S. Navy through the Office of Naval Research. Copyright © 1977 by Marie Tharp.

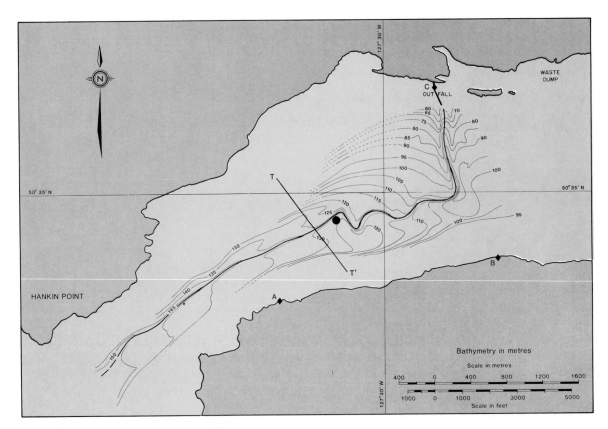

FIGURE 3–10   Rupert Inlet, British Columbia.

Bathymetry of submarine channel. Line T-T' is path followed by the Canadian Survey Ship *Vector* on August 26, 1976, when it recorded the occurrence of a turbidity current. The black circle denotes the approximate location of a bottom-moored current meter, and the diamonds labeled A, B, and C are locations of positioning transponders. Reprinted by permission of American Association for the Advancement of Science from *Science,* © 1982. Photo courtesy of Alex E. Hay.

rine turbidity current moving through its channel. Figure 3–11 shows acoustic sounding profiles across the submarine channel in Rupert Inlet before and during the event that lasted for about one and one-half hours.

Another line of evidence that indicates that turbidity currents move across the ocean bottom is associated with trans-Atlantic cable breaks that occurred on the continental shelf and slope south of Newfoundland after an earthquake on November 18, 1929 (fig. 3–9). These cables were broken in a pattern that indicated that the cables closest to the earthquake were broken first and those that crossed the shelf at greater depths and distances from the epicenter (point on ocean floor directly above earthquake) were broken later. This phenomenon could be explained by a dense flowing mass moving away from the epicenter down the slope and snapping the cables as it passed. The velocity required in this particular case would be approximately 27 km/h (16.8 mi/h), which seems quite high. The possibility that such a mass may move with this velocity makes it easier to understand

how the submarine canyons may have been formed on the continental slope and continental shelf.

Continental Rise

Deep-sea fans accumulate as deposits at the mouths of submarine canyons (fig. 3–8). The merging of these deep-sea fans along the base of the continental slope is partly responsible for the development of the continental rise.

The Amazon Cone is one of the largest deep-sea fans. It extends 700 km (434 mi) northeast of Brazil (fig. 3–12). Examination of the fan's distributary system using GLORIA long-range side-scan sonar revealed the following features. The main channel extends seaward from the Amazon Submarine Canyon. At a depth of about 2200 m (7200 ft), the channel splits into western and eastern systems that extend across the middle of the fan. Additional branching and development of meanders occur across the middle of the fan.

A.   B.   C.

FIGURE 3–11   Acoustic Sounding Profiles along Line T–T'.

In *A*, the ship crossed the channel at about 12:53 P.M. and no turbidity current was flowing in the channel. Note the levees on either side of the main channel. The discrete spots in the water column above the bottom are thought to be fish. In *B*, the ship was over the channel at 1:03 P.M. and the turbidity current had begun to flow in the main channel. By 1:28 P.M. in *C*, the turbidity current had overspilled the levees. The turbidity current was not detectable 1.5 h after its onset. The steep wall is on the north side of the channel. The inclined vertical lines are the result of transmitted pulses from another sounder. Reprinted by permission of American Association for the Advancement of Science from *Science,* vol. 217, 4562, 27 Aug., ©1982. Photo courtesy of Alex E. Hay.

FIGURE 3–12   The Amazon Cone.

The Amazon Canyon branches into two major distributary systems, the Western Levee Complex and the Eastern Levee Complex. They are composed of meandering channels bounded by levees. Note the large debris flows originating on the steep upper slope of the fan cone. Map courtesy of John E. Damuth and Roger D. Flood. Permission granted by the Geological Society of America.

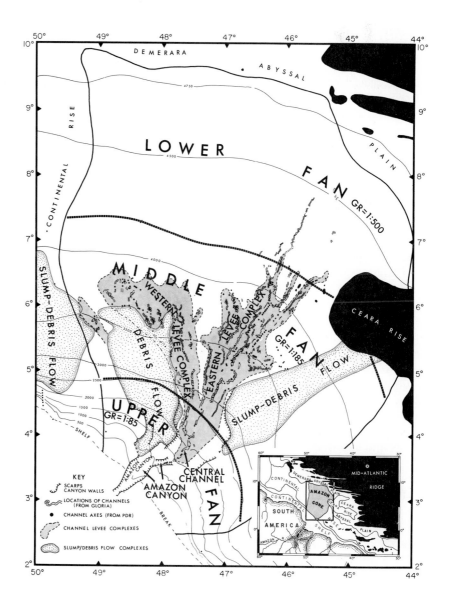

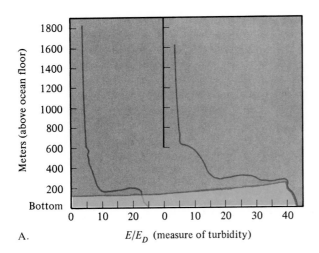

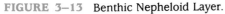

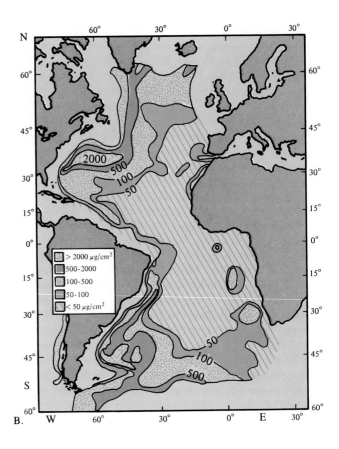

FIGURE 3–13 Benthic Nepheloid Layer.

A, two light-scattering curves showing the nepheloid layer on the Blake-Bahama Outer Ridge. *B,* distribution of suspended material in the nepheloid layer in the Atlantic Ocean. Note the high concentrations found only in the western portions of the ocean overlying the continental rise. Compare the concentration pattern in the North Atlantic with the location of the western boundary undercurrent in figure 3–9. After Biscaye and Eittreim, 1977.

The onset of meandering of the distributary channels across the middle fan corresponds well to the decreased gradient of the middle fan surface as compared to that of the steeper upper fan. The meander patterns, as well as other features—abandoned meander loops, cutoffs, etc.—closely resemble in form and scale those found on the floodplains of mature river systems on land such as the lower Mississippi River.

A major factor in shaping the continental rise is the strong *western boundary undercurrent* (WBUC) or *slope current* that flows toward the equator at the base of the continental slope along the western boundaries of most ocean basins. This deep boundary current and the continental rise over which it flows have been extensively studied along the east coast of North America using submersibles, surface ships, and side-scan sonar.

The Coriolis Effect (described in chapter 7) forces the WBUC, which originates in the Norwegian Sea and flows into the North Atlantic through deep troughs between Greenland and Scotland, to veer to the right and flow snugly against the base of the continental slope at all times. Picking up volcanic debris from Iceland and sediment from periodic turbidity flows, and scouring sediment from some portions of the continental slope and rise, the western boundary undercurrent, flowing at velocities that reach 40 cm/s (1 mi/h), creates a *benthic nepheloid layer* (BNL) of suspended sediment above the ocean floor (fig. 3–13). As the current is forced around bends in the continental slope, its velocity decreases and the rate of deposition increases, producing the deposits called *drifts* or *ridges,* shown in figure 3–9.

The surfaces of large drifts and ridges consist of *mud waves* that have an average distance between crests of 2–3 km (1.2–1.9 mi). These mud waves, shown in figure 3–14, are also found along much of the lower continental rise off the central Atlantic states and are known as lower continental rise hills. Cutting across the mud waves are *furrows* that are a few meters in width and run parallel to the direction of current flow. The presence of the furrows suggests that sediment has been removed to produce them. The furrows, in turn, have *ripples* cutting across them that are about 10–15 cm (4–6 in) from crest to crest. The ripples are undoubtedly created by the current flow along the furrows.

Increased knowledge of continental-margin features and the processes that create them may be important in selecting ocean sites for toxic waste disposal and effective exploitation of oil, gas, and minerals.

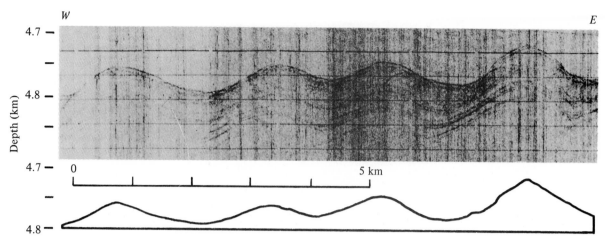

FIGURE 3–14 Mud Waves.
Echo sounding of mud waves taken along an east-west line across the Blake-Bahama Outer Ridge, a large depositional feature on the continental rise north of the Bahamas. The near-bottom bathymetric profile beneath the echo sounding gives a more accurate shape of the waves. Courtesy of Roger Flood, Lamont-Doherty Geological Observatory, Columbia University.

DEEP-OCEAN BASIN

Abyssal Plains

Extending from the base of the continental rise are the deep-ocean basins, which include flat surfaces with slopes of less than 1:1000 that cover extensive portions of the basins (fig. 3–15). These *abyssal plains* are particularly extensive and flat in the Atlantic Ocean, although they are occasionally interrupted by volcanic peaks. The peaks are called *seamounts* if they rise more than 1 km (0.62 mi) above the abyssal plain. Smaller peaks are called *abyssal hills*. Most abyssal plains lie at depths of between 4500 and 6000 m (14,760 and 19,700 ft).

The benthic nepheloid layer that is so well developed over the continental slope is also found (although not so well developed) throughout the deep ocean, where bottom current speeds average at least 8 cm/s (approx. 0.16 mi/h). Such currents may play an important role in distributing fine sediment on the deep abyssal plains. Just north of the equator in the eastern Pacific, average bottom-current velocities (measured 30 m, or 98 ft, above the ocean floor) of up to 18.5 cm/s (0.46 mi/h) have been recorded.

Trenches

Although the continental rise is a feature commonly found at the base of the continental slope where it meets the abyssal plain, the slope sometimes descends into long, narrow, steep-sided *trenches*. The deepest portions of the world's oceans are found in these trenches. Such features are characteristic of the margins of the Pacific Ocean along the coasts of South America, Central America, and the Aleutian Islands and are also characteristic of the western margin of the Pacific Ocean (fig. 3–16). A depth of 11,022 m (36,152 ft) has been recorded in the Challenger Deep of the Marianas Trench. Table 3–1 presents some of the dimensional characteristics of trenches. The landward side of the trench rises as a *volcanic island arc* (Japan) or a *volcanic mountain range* at the margin of a continent (the Andes).

Oceanic Ridges and Rises

Extending across some 65,000 km (40,365 mi) of the deep-ocean basin are continuous mountainous features. Those with steeper and irregular slopes are usually called ridges, and those with gentler slopes

TABLE 3–1 Dimensions of Trenches

Trench	Depth m (ft)	Mean Width km (mi)	Length km (mi)
Peru-Chile	8,055 (26,420)	100 (62)	5,900 (3,664)
Aleutian	7,697 (25,246)	50 (31)	3,700 (2,298)
Marianas	11,022 (36,152)	70 (43)	2,550 (1,584)
Tonga	10,882 (35,693)	55 (34)	1,400 (869)

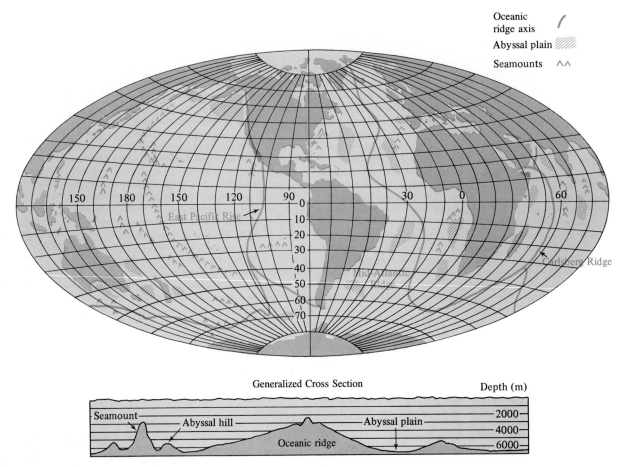

FIGURE 3–15   Oceanic Ridges, Abyssal Plains, and Volcanic Peaks.
65,000 km (40,300 mi) of oceanic mountain ranges extend across the floors of the world ocean.
Flat-floored basins with sediment cover produce abyssal plains at the base of many oceanic mountain ranges. Volcanic peaks rising less than and more than 1 km (0.62 mi) above the ocean floor
are named *abyssal hills* and *seamounts,* respectively. Base map courtesy of National Ocean Survey.

are rises. The Mid-Atlantic Ridge divides the Atlantic
Ocean in half. This ridge rises 2.5 km (1.5 mi) above
the abyssal plains on either side. Other major ridges
are the Carlsberg Ridge, which extends north across
the Indian Ocean to the Red Sea, and the East Pacific
Rise, which runs north across the southeastern Pacific Ocean to the North American continent in the
region of the Gulf of California (fig. 3–15).

   This system of mountains is entirely volcanic and
formed from lavas with a basaltic composition characteristic of the ocean crust. Some stretches of this
mountain system have a rift valley (down-dropped
block) near the crest. The rift is particularly well-
developed on the Mid-Atlantic Ridge. Rugged fault
scars called *fracture zones* cut across the oceanic
ridge and rise systems and may extend for hundreds
of kilometers into the ocean basins on either side.

SUMMARY

Much of the knowledge we have gained about the nature of the ocean floor has been obtained using **echo
sounders, seismic surveys,** and **side-scan sonar.** Seismic surveys can reveal the topography of the ocean
floor and the location of rock unit boundaries beneath it. Side-scan sonar is used to make strip maps
of ocean-floor topography along the course of the survey ship.

   Much as ice floats on water, continents and the
ocean crust follow the same principle of buoyancy
and float on the earth's mantle. This phenomenon as
it applies to the crust and mantle is called **isostasy.**

   Extending from the shoreline of continents are
gently sloping **continental shelves** that extend out to
sea for an average distance of 70 km (43 mi). Sedi-

FIGURE 3–16 Trenches.

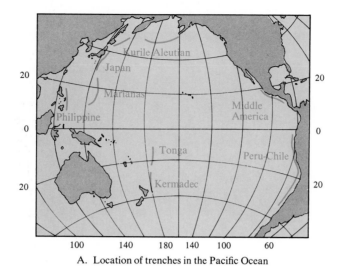

A. Location of trenches in the Pacific Ocean

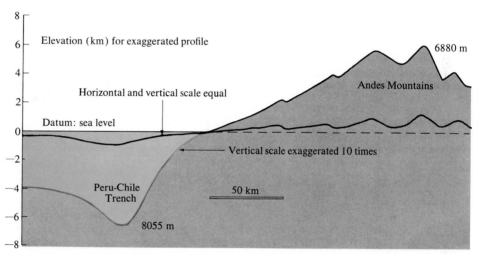

B. Profile across Peru-Chile Trench and Andes Mountains off the coast of Chile

ment is transported across the continental shelves by currents and mass wasting (submarine sediment slides). When the shelves reach an average depth of 135 m (443 ft), they steepen into a feature called the **continental slope**. The slopes continue to the deep-ocean floor with steepness ranging from 1° to 25° and averaging 4°. Cutting deep into the slopes and reaching well onto the continental shelves, in some cases, are **submarine canyons**. The origin of the canyons is unknown, but they may be caused in part by periodic flows that are a mixture of sediment and water called **turbidity currents**. Turbidity currents deposit their sediment loads at the base of the continental slope to produce **deep-sea fans** or **cones** through a series of meandering **distributary channels**, which branch out from the mouth of the submarine canyon. The merging of these fans produces a gently sloping **continen-**

tal rise at the base of some continental slopes. A major factor in distributing sediment along the continental slope is the **western boundary undercurrent (WBUC)** or slope current that flows toward the equator along the western boundary of ocean basins. The sediment suspended and carried above the bottom by these currents creates a **benthic nepheloid layer**. When the current slows to make turns along the base of the continental slope, it deposits sediment onto features called **drifts** or **ridges** that become prominent features of the continental rise.

The continental rises gradually become flat, extensive, depositional features of the deep-ocean basin called **abyssal plains**. These plains are often penetrated by volcanic peaks. In some places, particularly around the margin of the Pacific ocean, the continental slope does not merge into a continental rise but continues down into deep linear **trenches**. Trenches are usually bounded on their landward side by **volcanic island arcs** or **continental volcanic mountain ranges** if the trench lies near the continent. **Oceanic ridges** and **rises** are volcanic mountain ranges running through the deep-ocean basins and rising about 2.5 km (1.5 mi) above the abyssal plains on either side. The volcanic peaks protruding above the abyssal plains are called **seamounts** if they extend over 1 km (0.62 mi) above the deep-ocean floor and **abyssal hills** if they are smaller. Rugged fault scars called **fracture zones** cut across oceanic ridges and rises.

QUESTIONS AND EXERCISES

1. Why do the continents float on the mantle with most of their mass embedded in it? What is the name of the concept that is used to explain this phenomenon?

2. Discuss the slope and width of the continental shelf as well as the means by which sediment is transported across it.

3. Describe the most widely accepted theory explaining the development of submarine canyons on the continental slope. Include a discussion of the physical evidence that lends support to the theory.

4. Why is the distributary pattern observed on the Amazon Cone of such great interest to geologists studying processes that produce the depositional features of the continental margins?

5. What are the sources of sediment that become suspended in the nepheloid layer?

6. Why is the nepheloid layer best developed along the western margins of the ocean basins?

7. Special features of the continental rise are mud waves, furrows, and ripples found on ridges and drifts. Make a diagram showing how these features are related to one another and to the deep boundary current that produces them.

8. In which ocean basin are the most ocean trenches found?

REFERENCES

Biscaye, P. E., and Eittreim, S. L. 1977. Suspended particulate loads and transports in the nepheloid layers of the abyssal Atlantic Ocean. *Marine Geology* 23:155–72.

Damuth, J. E.; Kolla, V.; Flood, R. D.; Kowsmann, R. O.; Monteiro, M. C.; Gorini, M. A.; Palma, J. J.; and Belderson, R. H. 1983. Distributary channel meandering and bifurcation patterns on the Amazon deep-sea fan as revealed by long-range side-scan sonar (GLORIA). *Geology* 11:2, 94–98.

Fairbridge, R. W., ed. 1966. *Encyclopedia of oceanography.* New York: Van Nostrand Reinhold.

Field, M. E.; Gardner, J. V.; Jennings, A. E.; and Edwards, D. E. 1982. Earthquake-induced sediment failures on a 0.25° slope, Klamath River delta, California. *Geology* 10:10, 542–545.

Hay, A. E.; Burling, R. W.; and Murray, J. W. 1982. Remote acoustic detection of a turbidity current surge. *Science* 217:4562, 833–835.

Hollister, C. D.; Flood, R. D.; and McCave, I. N. 1978. Plastering and decorating the North Atlantic. *Oceanus* 21:1, 5–13.

Keen, M. J. 1968. *An introduction to marine geology.* New York: Pergamon Press.

Menard, H. W. 1964. *Marine geology of the Pacific.* New York: McGraw-Hill.

Shepard. 1973. *Submarine geology.* New York: Harper and Row.

Sverdrup, H. U.; Johnson, M. W.; and Fleming, R. H. 1942. Renewal 1970. *The oceans: Their physics, chemistry, and general biology.* Englewood Cliffs, N.J.: Prentice-Hall.

Twichell, D. C., and Roberts, D. G. 1982. Morphology, distribution and development of submarine canyons on the United States Atlantic continental slope between Hudson and Baltimore canyons. *Geology* 10:8, 408–412.

Weirich, F. H. 1984. Turbidity currents: Monitoring their occurrence and movement with a three-dimensional sensor network. *Science* 224:4647, 384–387.

SUGGESTED READING

Sea Frontiers

Mark, K. 1976. Coral reefs, seamounts, and guyots. 22:3, 143–49.
A discussion of the role of global plate tectonics in explaining the distribution of seamounts, guyots, and the evolution of coral reefs.

Schafer, C., and Carter, L. 1986. Ocean-bottom mapping in the 1980s. 32:2, 122–30. The use of SeaMARK (Seafloor Mapping and Remote Characterization), a side-scan sonar device, in mapping the continental margin off the coast of Labrador is described.

Shepard, F. P. 1975. Submarine canyons of the Pacific. 21:1, 2–13.
A description of canyons around the Pacific margin and discussion of their possible origin.

Scientific American

Emery, K. O. 1969. The continental shelves. 221:3, 106–25.
The nature of the continental shelves and the effect of the advance and retreat of the shoreline across them as a result of glaciation are discussed.

Heezen, B. C. 1956. The origin of submarine canyons. 195:2, 36–41.
Theories explaining the origin of submarine canyons are presented along with data on the location and nature of such features.

Menard, H. W. 1969. The deep-ocean floor. 221:3, 126–45.
A summary of the dynamic effects of sea-floor spreading is presented with a description of related sea-floor features.

CHAPTER FOUR
THE ORIGIN OF OCEAN BASINS: THEORY OF GLOBAL PLATE TECTONICS

The fact that we live on a dynamic earth on which movement is the rule rather than the exception has long been accepted by geologists. Geologic processes responsible for the ever-changing landscape of the continents are believed to require long periods of geologic time to build such features as mountains. However, many geologists were not prepared until recent years to accept a dynamic definition of the earth that was of a much broader scope—the movement of continents. When geologists first spoke of the movement of continental masses across the earth's surface, they referred to this phenomenon as *continental drift*.

Marine geologists who studied the ocean floor and in the 1960s developed theories to explain why the continents are moving relative to one another used the term *sea-floor spreading*. The name for this theory is derived from evidence indicating that new oceanic crust and rigid upper mantle material is being added along a series of mountain ranges on the ocean floor and moving at right angles away from these mountainous ridges. The continents, which are thought to float on this denser material, are carried along on a moving layer, or plate, of denser rock.

A third term used to describe this entire process is *global plate tectonics*. Study of the process has indicated that there are a number of major plates into which this rock sphere, or *lithosphere*, can be divided. The interaction of these plates as they move builds the structural features of the earth's crust that we observe. Thus, *tectonics* refers to the building of the earth's crustal structure and is derived from Greek *tektonikos*, which means *to construct*.

The idea that the continents may be moving across the surface of the earth is not new. Such a possibility was suggested early in the nineteenth century, and the first major theory attempting to explain the movement was presented in 1912. With such a long history of awareness of the possibility of such movement, why has the acceptance of theories about the process been so long in coming?

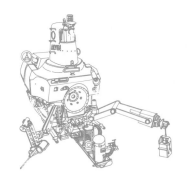

The deep-sea research submersible *Alvin* during a sequence of dives to explore hydrothermal vent biological communities on the East Pacific Rise. Photo courtesy of Woods Hole Oceanographic Institution.

DEVELOPMENT OF THE THEORY

The first scientific observations related to the possible breakup of the continents that moved to either side of the Atlantic Ocean were made by Antonio Snider-Pelligrini in 1858. He presented the first reconstruction of the continents as they may have existed prior to their separation. Snider-Pelligrini believed that the 300-million-year-old coal deposits so widespread in western Europe and eastern North America must have been deposited when these landmasses were united (fig. 4–1).

Also using geological evidence, particularly the origin and alignment of mountain ranges on opposite sides of the Atlantic, were two Americans, Frank Taylor and Howard Baker. Taylor, who included in his discussion the possible origin of these mountain ranges, presented in 1908 a well-developed argument for large-scale continental movements.

Alfred Wegener, however, is considered by most scientists to be the pioneer of the modern continental drift theory. This German scientist originally was drawn to the concept in an attempt to explain the ancient climates that were recorded in the rocks deposited in ocean basins and on landmasses of the past. Wegener's theory was published in 1912 and met a great deal of resistance from the scientific community.

Wegener considered that about 200 million years ago all of the continental mass of the earth was one large continent, *Pangaea*. About 150 million years ago the continent began to break up, and the various continental masses we know today started to drift toward their present positions. Because it was impossible to present evidence that would support a mechanism for such transportation, the theory did not receive wide acceptance.

It is interesting that John Joly, an Irish physicist, suggested that the heat generated in the interior of the earth by radioactive decay found its way to the surface through slowly moving convection cells that could carry the continents laterally across the earth's surface. Although we call on a poorly understood convection process today to account for the movement of continental masses, Joly's theory ran into great difficulty because there were no observational data that would indicate the presence of convection cells in the earth's interior. Such a theory was further developed by Arthur Holmes in 1927. Holmes, a strong proponent of continental drift, derived much of his supporting data from the radiometric methods of dating rocks, a field in which he was a pioneer.

Although many Southern Hemisphere geologists had accepted continental drift as a geological reality,

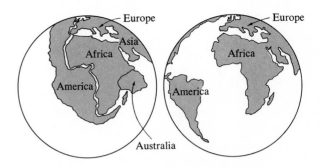

FIGURE 4–1   Snider-Pelligrini's Reconstruction.
Snider-Pelligrini's diagram was the first to explain the similarities of fossils in coal deposits in Europe and North America, suggesting the continents were once united. From *Continental drift* by Don and Maureen Tarling, © 1971 by G. Bell & Sons, Ltd. Reprinted by permission of Doubleday & Co., Inc.

it was not until the 1950s that geologists of the Northern Hemisphere began to give it serious attention. The impetus for the renewed attention arose from the study of the earth's ancient magnetism. The British geophysicist S. Keith Runcorn explained his observations of the magnetic properties of the rocks of Europe and North America in terms of continental movements. In 1963 Frederick Vine, a British oceanographer, suggested that parallel bands of magnetic anomalies on the ocean floor resulted from the combined effects of the generation of new ocean floor along the axes of oceanic ridges and reversals of polarity of the earth's magnetic field. The details of these considerations will be discussed later in this chapter.

On the basis of this study of the earth's ancient magnetic fields, convincing arguments could be made for the fact that the continents had moved relative to one another. As study continued, more evidence was gathered to support this movement, and additional data suggested the mechanism by which the movement might have taken place. The confirmation of the spreading process was achieved during the early stages of the Deep Sea Drilling Project in 1968.

SUPPORTING EVIDENCE

Continental Jigsaw

Before we look at the bulk of scientific evidence that has been gathered in relation to plate tectonics, we should consider how well the continents do fit together. Attempts were made by many of the early investigators to arrange the continents in a manner that would achieve a reasonable fit and support their data. Most of these attempts used the existing shoreline as

FIGURE 4–2 Relative Intensity of the Earth's Magnetic Field over Continents and Oceans.

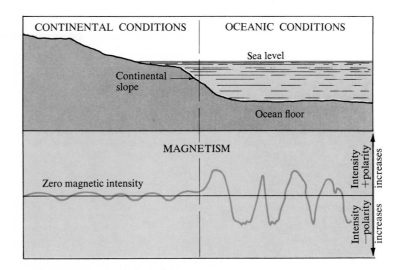

the margin of the continent. Based on our previous discussion of the geomorphic provinces of the ocean basin, we considered the continental shelf and continental slope to be part of the continental mass. Studies of the magnetic intensity of the earth's crust support such a contention. A presentation of the general results of such investigations is shown in figure 4–2.

There is a distinct increase in the magnetic intensity of oceanic crustal rocks over that of the continental crustal rocks. The increased intensity of the magnetism of these rocks is caused by higher iron content of oceanic crustal rocks.

Sir Edward Bullard, an English geophysicist, constructed a computer fit of all the continents in 1965. The arrangement that produced minimum overlaps and gaps was found to be formed by continents outlined by the 2000-m (6560-ft) depth contour. This contour represents a depth that is approximately halfway down the continental slope (fig. 4–3).

CONTINENTAL GEOLOGY

To test the fit of the continents, we may compare the rocks along the margins of the two continents that were at one time united. We need to identify rocks of the same type and age on the continents along their common margin. Identification is not easy in some areas, since during the millions of years since their separation, younger rocks have been deposited and cover those rocks that might hold the key to the past history of the continents. However, there are many areas where such rocks are available for observation, and in these areas we can compare the ages by the use of (1) *fossils,* the remains of ancient organisms preserved in rocks, and (2) radiometric dating of the rocks.

FIGURE 4–3 Computer Fit of Continents by Sir Edward Bullard.

Sir Edward Bullard's fit of the continents was attempted in 1965 after a convincing fit pattern had been achieved in 1958 by the Australian geologist S. Warren Carey without the aid of computers. Bullard's fit was in complete agreement with that of Carey. From *Continental drift* by Don and Maureen Tarling, © 1971 by G. Bell & Sons, Ltd. Reprinted by permission of Doubleday & Co., Inc.

The Fossil Record

The use of fossils became an important dating device in the early nineteenth century when it was realized that a particular layer of sedimentary rock could contain the remains of organisms that, as a group, were unique in time. Rocks laid down earlier and those laid down later would have a distinctly different assemblage of plant and animal fossils. Once these assemblages are recognized, a geologist can tell if rocks exposed in one area are younger or older than in another. The unique character of these assemblages is the result of evolution throughout geologic time. Appendix IV shows the general pattern of this evolution and the geologic time scale.

Relative Dating   Dating by assemblages of organisms is referred to as *relative dating.* One can tell only whether or not an assemblage is the same age or younger or older than another, but not its actual age

in years (fig. 4–4A). Such assemblages have been found abundantly in sedimentary rocks of the last 600 million years. Note that this is but a little over one-eighth of the time of earth's existence, which is thought to exceed 4.5 billion years. Therefore, this method is not effective in comparing the ages of the continental margins in those rare instances where they are composed of rocks older than 600 million years.

Absolute Dating   Most of the rocks we find on the continents contain small amounts of radioactive elements such as uranium, thorium, and potassium, which break down into atoms of other elements. Each radioactive *isotope* (atom of an element with an atomic weight different than that of other atoms of the element) has a specific *half-life,* the time it takes for one-half of the atoms present in a sample to decay to atoms of some other element. By comparing the quantities of the radioactive isotope with the quanti-

FIGURE 4–4   Relative and Absolute Dating.

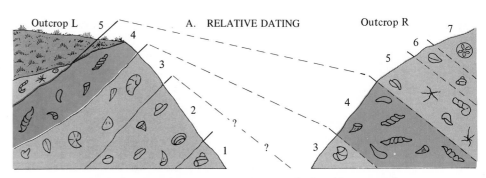

A.   RELATIVE DATING

Fossil assemblages used in relative dating can be found in outcrops L and R. Assemblages 3, 4, and 5 may be matched, telling us the rocks in segments of the outcrops formed at the same time. We cannot, however, tell how many years ago this formation occurred.

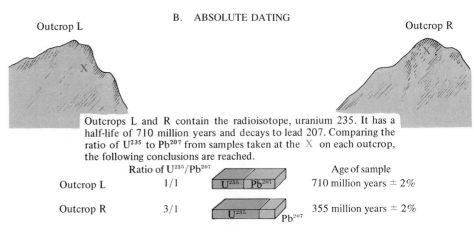

B.   ABSOLUTE DATING

Outcrops L and R contain the radioisotope, uranium 235. It has a half-life of 710 million years and decays to lead 207. Comparing the ratio of U^{235} to Pb^{207} from samples taken at the X on each outcrop, the following conclusions are reached.

	Ratio of U^{235}/Pb^{207}		Age of sample
Outcrop L	1/1	U^{235} | Pb^{207}	710 million years ± 2%
Outcrop R	3/1	U^{235} | Pb^{207}	355 million years ± 2%

Dating with radioisotopes is possible for many types of rocks but is more broadly possible with igneous rocks like the basalt and granite that make up most of the oceanic and continental components, respectively, of the earth's crust.

ties of decay products in given rocks and by knowing the rate of decay (half-life), the age of the rock may be determined within 2–3% of its actual age. Although in very old rocks this error can be quite significant, this method of age determination is a very powerful tool and is the first method that we have had of giving the actual age of rocks in years. We refer to such dating as *absolute dating.* As compared to the relative dating possible with the use of fossils, absolute dating not only tells us which rocks are younger or older, but also their actual age in years (fig. 4–4B).

Ancient Life and Climates

The fossil record of plants and animals in the sedimentary rocks can tell us much about the environments of the past. For instance, it is relatively simple to determine whether an organism lived in the ocean or on land. This distinction can normally be made also by the characteristics of the sedimentary rock itself. However, finer differences in environmental conditions may also be determined.

Upon examining the distribution of some fossil assemblages and other characteristics of rocks that give clues to the climate under which they were formed, we find some that could not have formed under the present climatic conditions. These anomalous locations of fossil assemblages and rock types may be explained by assuming that either (1) the factors that control climatic belts today are different than the factors that controlled climates in the past or (2) the factors controlling climate throughout geologic time have not changed. The second assumption, which seems more logical, lets us conclude that these assemblages and rock types representing sediments of the past must have been carried to their present positions by the movement of the continental masses.

The dominant factor controlling climatic distribution is latitudinal position on the rotating earth. Assuming that the earth's axis of rotation has not changed significantly throughout its history, we may conclude that latitudinal belts have possessed climatic characteristics that have not changed greatly during the evolution of life on earth.

Laurasia* and *Gondwanaland Modern reef-building corals are known to exist in an environment where the water is clear and shallow and its temperature does not fall below 18°C (64.2°F). Although we cannot be sure such conditions have been required throughout the long history of coral evolution, we can assume that the conditions required were similar.

Exactly the same species of coral are found in rocks 350 million years old in western Europe and eastern North America, as well as throughout the Alps and Himalayas, implying that they were in close proximity at that time. Other evidence indicates that from 350 to 250 million years ago a major ocean, the *Tethys Sea,* separated the large continent *Laurasia,* composed of North America, Europe, and Asia, to the north, from the large continent *Gondwanaland,* composed of South America, Africa, India, Australia, and Antarctica, to the south (fig. 4–5). The record also indicates that the two supercontinents periodically came into direct or close contact across the Tethys Sea. Both continents were joined in forming one landmass, *Pangaea,* about 200 million years ago (fig. 4–6). Pangaea was surrounded by the ocean *Panthalassa*, the ancient precursor of the Pacific Ocean.

Fossils from sediments that were laid down within a land environment aid in determining the latitudinal positions of the two large continents. From 350 to 285 million years ago there existed two distinct floral assemblages on the two supercontinents. The Laurasian floral assemblage included many species of tropical plants that were incorporated into the sediment to form the extensive coal beds mined throughout the eastern United States and Europe. These tropical flora imply a low-latitude position for Laurasia during that time. Throughout Gondwanaland, there existed an assemblage represented by a few species of plants that are thought to have grown in a cold climate, presumably at a high southern latitude. Supporting this belief are indications of glaciation in South America, Africa, India, and Australia, which at that time must have been very near the southern polar region (figs. 4–5 and 4–7).

Continental Magnetism

Although we cannot reconstruct the relative positions of continents prior to 200 million years ago with as much accuracy, there is enough evidence from that time to the present to give us some idea of the paths followed by the continents after the breakup of Pangaea. The first clues to such movements came from the study of the magnetism of continental rocks.

Most igneous rocks contain some particles of magnetite (an iron oxide, Fe_3O_4) that become magnetized by and align themselves with the earth's magnetic field at the time of the rocks' formation. Volcanic lavas such as basalt are high in magnetite content and solidify from molten material that possesses temperatures in excess of 1000°C (1832°F). As they cool below 600°C (1112°F) the magnetite particles become oriented in the direction of the earth's magnetic field,

POLAR GONDWANALAND FLORA

TROPICAL LAURASIAN FLORA   A   TETHYS CORAL ASSEMBLAGE   C

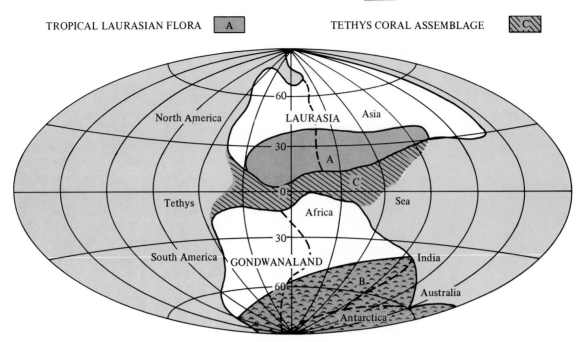

FIGURE 4–5   Laurasia and Gondwanaland.
Laurasia and Gondwanaland may have been in the relative positions shown here during the interval 350–250 million years ago. The Tethys coral assemblage could have been deposited in the narrow Tethys Sea while the tropical Laurasian floral assemblage and the polar Gondwanaland floral assemblage were developing on the continents.

FIGURE 4–6   Pangaea.
About 200 million years ago, near the beginning of the Mesozoic Era, all the present continents were joined together into one continent—Pangaea. The ocean that surrounded Pangaea was called Panthalassa.

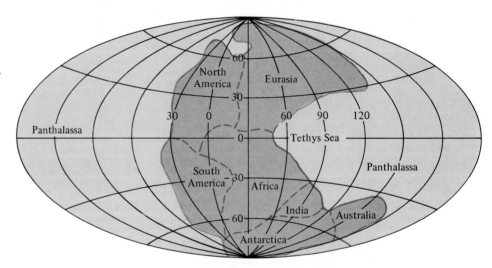

recording that field relative to the rock location permanently. If the earth's magnetic field changes subsequent to the formation of the igneous rock, the alignment of these particles will not be affected.

Magnetite is also deposited in sediments. While the deposit is in the form of a sediment surrounded by water, magnetite particles have an opportunity to align themselves once more with the earth's magnetic field. This alignment is preserved when the sediment is buried.

Although a number of rock types may be used for the study of the earth's paleomagnetism, the basaltic lavas and other igneous rocks high in magnetite content are best for such studies. The magnetite particles act as small compass needles, as shown in figure 4–8. They not only point in a north-south direction but also point into the earth at an angle relative to the earth's surface called the *magnetic dip,* or *inclination,* which is related to latitude. At the equator, the "needle" will not dip at all, but will lie horizontally. It will

FIGURE 4–7 Fossil and Glacial Evidence Supporting Continental Drift.

A fossil coral assemblage that flourished in the Tethys Sea and continental floral assemblages A and B that developed to the north and south of the Tethys suggest Pangaea was divided into Laurasia to the north and Gondwanaland to the south during the time span covering 350 to 250 million years ago. The B flora are thought to be cold-climate forms. This belief is supported by evidence of glaciation.

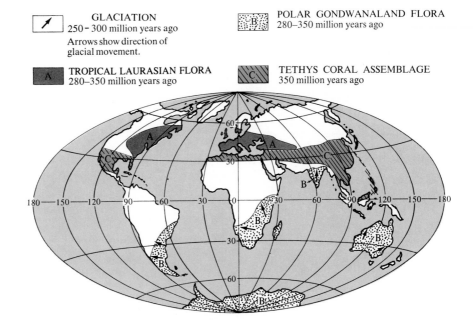

GLACIATION
250 – 300 million years ago
Arrows show direction of glacial movement.

B POLAR GONDWANALAND FLORA
280–350 million years ago

A TROPICAL LAURASIAN FLORA
280–350 million years ago

C TETHYS CORAL ASSEMBLAGE
350 million years ago

point straight into the earth at the magnetic north pole and straight out of the earth at the magnetic south pole. At points between the equator and the pole the angle of dip increases with increased latitude. It is this dip that is retained in magnetically polarized rocks. By measuring the dip angle, we can estimate the latitude at which the rock formed.

Apparent Polar Wandering Because the present magnetic poles do not coincide with geographical poles related to latitude, it might be felt that determining latitude by magnetic dip would give some incorrect determinations. However, for the last few thousand years the average positions of the magnetic poles have coincided with those of the geographic poles. If we can assume that this is true for the past, we can determine average positions of the magnetic poles during a time interval and consider them to represent the geographic poles.

As these average positions for the magnetic poles are determined for rocks on the continents, it is found that their positions changed with time. It appears the magnetic poles were wandering. The wandering curve of the pole for North America shows an interesting relationship with that determined for Europe. Both curves had a similar shape, but for all rocks older than 70 million years the pole determined from North American rocks lies to the west of that determined by the study of European rocks. This difference implies that North America and Europe have changed position relative to the pole and relative to each other (fig. 4–9).

If there can only be one north magnetic pole at any given time and its position must be at or near the

north geographic pole, a problem exists that can be solved only by moving the continents. The two wandering curves can be made to coincide by moving the two continents together as we go back in time. Now we have a single wandering curve, which shows the magnetic north pole to be much too far south during the interval of time from 200 million to 300 million years ago. Rotating the merged continents can bring the pole into the proper position. The fact that moving the continents was the only solution to this problem

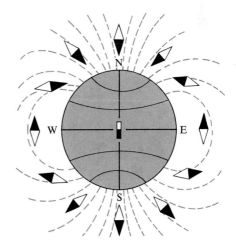

FIGURE 4–8 Nature of Earth's Magnetic Field.

Over long periods of time, the earth's average magnetic field can be considered to be that of a large magnet with its poles aligned along the earth's axis of rotation. As are the compass needles in the illustration, magnetite particles in newly forming rocks of the earth's crust are aligned with this field. Once the rocks solidify, the magnetite particles are frozen into position and become fossil compass needles that tell today's investigators about the strength and alignment of the earth's magnetic field when the rocks of which they are a part formed.

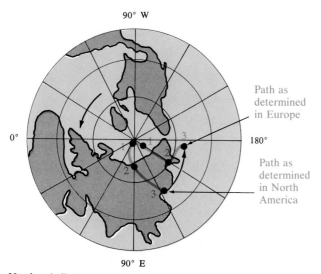

Numbers indicate
position of poles
hundreds of millions
of years ago

FIGURE 4–9 Apparent Wandering Curves of North Magnetic Pole Determined from Rocks of North America and Europe.

Apparent wandering curves of the north magnetic pole as determined from North America and Europe, going back in time from the present to 300 million years ago, follow divergent paths. If one moves the apparent position of the pole as determined from North America to a position that makes it coincide with the pole as determined from Europe, the direction and distance equal that required to close the Atlantic Ocean between the two continents. This still leaves the apparent north magnetic pole out in the North Pacific Ocean about 40° south of the geographic north pole. Again, movement of the continent is required to solve this problem. Rotation of Laurasia, which is formed by closing the Atlantic Ocean and joining Europe and North America, can bring the wayward apparent north magnetic pole to a position coincident with the north geographic pole.

was very strong evidence in support of the movement of continents throughout geologic time.

Estimating the latitude of formation for rocks from many areas studied throughout the continents on the basis of paleoclimatic and paleomagnetic evidence produces similar results. The most logical explanation for the changes that have occurred throughout the past in both climate and magnetic dip of the rocks at a given location is that the continents have moved.

Magnetic Reversals Not only have the magnetic poles as determined from continental rocks seemed to wander throughout geologic time, but also the *polarity,* the direction of the magnetic field, seems to have reversed itself periodically. Reversals in polarity can be described in the following way. A compass needle that would point to the magnetic north today would, during a period of reversal, point south. It is

not known why these reversals occur, but they have for the last 76 million years occurred once or twice each million years. The period of time during which a change in polarity occurs lasts for a few thousand years. It is identified on the basis of magnetic properties of rocks by a gradual decrease in the intensity of the magnetic field of one polarity until it disappears, followed by the gradual increase in the intensity of the magnetic field with the opposite polarity. The time during which a particular paleomagnetic condition existed can be determined by the radiometric dating of the igneous rock from which the paleomagnetic measurements were taken (fig. 4–10).

The earth's present magnetic field has been weakening for the last 150 years, and some investigators believe that the present polarity will disappear in another 2000 years.

MARINE GEOLOGY

We have so far considered only data obtained from the continents in considering the theory of moving continents. This continental evidence did not convince many geologists that the continents had moved. For many years no other data were available since extensive sampling of the deep ocean bottom did not become technologically feasible until the 1960s.

The Deep Sea Drilling Project, which began in 1968, and the Ocean Drilling Program, which followed it in 1983, have added greatly to our knowledge of the ocean floor. The impetus for such programs was provided in great part by the need for observations of oceanic sediment and crust to check the theory of sea-floor spreading. Extensive geophysical studies, including those related to the paleomagnetic characteristics of the earth's crust, were first carried out. The results of these studies gave the planners of the Deep Sea Drilling Project clues as to where the drilling should be concentrated in order to gain the greatest amount of new knowledge.

Paleomagnetism

A detailed study of the magnetism of the Pacific Ocean floor by Scripps Institution of Oceanography identified narrow strips where magnetic properties of oceanic crust differed from those currently forming (magnetic anomalies). The anomalies ran parallel to the East Pacific Rise. Each anomaly represents a period when the polarization of the earth's magnetic field was reversed compared to today's. The anomalies were separated by bands of ocean crust that displayed present-day polarity.

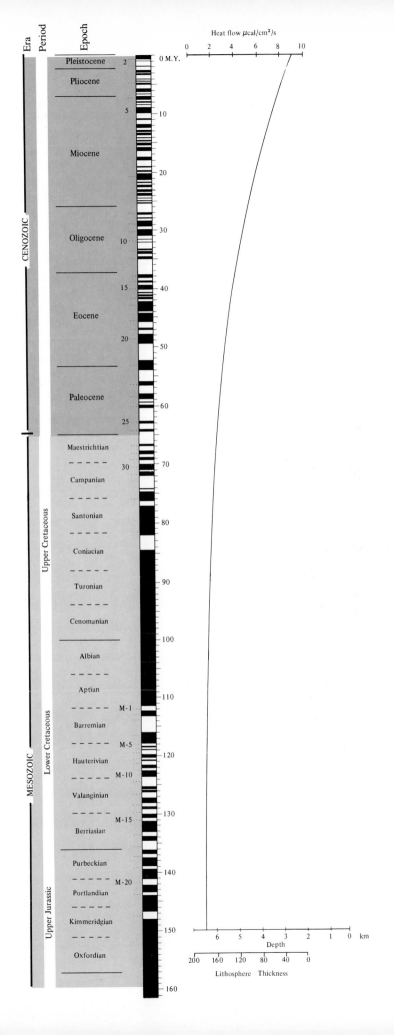

FIGURE 4–10 Magnetic Time Scale, Heat Flow, Ocean Depth, and Lithosphere Thickness.

Black bands represent periods when the magnetic polarity of the earth was the same as at present, and white bands represent periods when it was reversed. The curve shows that the ocean gets deeper and the lithosphere thicker with increasing age while the rate of heat flow to the earth's surface through the lithosphere decreases as the lithosphere ages. Although this curve could not be accurately applied to any particular region of the ocean basins, it indicates that, in general, heat flow decreases while ocean depth, thickness of the lithosphere, and age of the ocean floor increase with increasing distance from the spreading centers. The geologic time scale is shown to the left (*see* Appendix IV).

75

Additional study showed the sequence of polarity changes on one side of an oceanic ridge is identical to the sequence of reversals on the opposite side of the ridge. During dating of the reversal points on both flanks of the ridge, it became apparent that the rocks became older with increased distance from the ridge axis. The evidence indicated new oceanic crust was being formed at the oceanic ridges and moving away from them on opposite sides of the ridges. Having determined from continental studies the dates at which many of the more recent reversals occurred, it was possible to determine the age of the ocean floor at each strip boundary. Dividing the width by the number of years that polarity lasted produced the rate at which the ocean floor appeared to be moving away from the ridge (fig. 4–11).

Confirmation of the spreading of the oceanic crust away from the oceanic ridges required obtaining actual samples of the crust at various locations for radiometric dating. In addition, since sediment could not be laid down on crustal material until it formed at the oceanic ridges, it would be expected that the fossil assemblages observed in sediments immediately overlying the oceanic crust should be representative of organisms that existed at the time of the crustal formation. Also, it should be expected that the sediment thickness would be greater on older sea floor than on younger sea floor. A significant task of the Deep Sea Drilling Project was to check the age of the ocean bottom by drilling through sedimentary sections into the oceanic crust. Although all attempts to do this did not meet with success, enough data based on fossil assemblages immediately overlying the crust and radiometric age determinations of the crust itself confirmed that the oceanic floor is moving in the manner suggested by paleomagnetic data (fig. 4–12). The youngest oceanic crust was found at the axes of the oceanic ridges, with the age of the crust increasing at greater distances from the ridge crests.

Sea-Floor Spreading

New ocean floor appears to be forming at the oceanic ridges and rises. It then is carried away from the axes

FIGURE 4–11 Geomagnetic Record Contained in the Ocean Floor.

The polarity of the earth's magnetic field is recorded at the time of the ocean floor's formation. Like a tape recorder, the ocean floor stores a record of changes in polarity. The pattern recorded on one side of an oceanic ridge is the mirror image of that recorded on the opposite side.

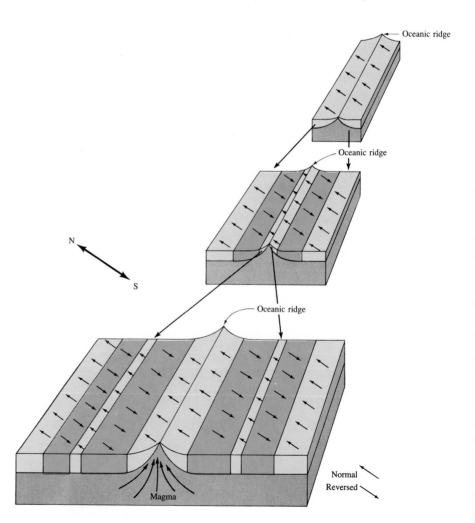

of such features, and volcanic processes fill the void with yet a younger strip of ocean floor. Thus, the axes of oceanic ridges and rises are referred to as *spreading centers*. The process is often referred to as sea-floor spreading.

It is, however, more than the sea floor that is moving. Although the details are far from clear, the mechanism of spreading involves two fundamental units of the earth's structure confined to the uppermost 650 km (400 mi) (fig. 4–13). The *lithosphere* (rock sphere) is a relatively cool, rigid shell that includes the crust and upper mantle. The lithosphere is broken into a dozen or so lithospheric plates (fig. 4–14 and plate 10) that we observe moving across the earth's surface. Underlying the lithosphere is a high-temperature plastic layer within the mantle. This layer, called the *asthenosphere* (weak sphere), can flow slowly, allowing the rigid lithospheric plates resting at its upper surface to move.

The ultimate fate of a lithospheric plate is its *subduction*—a process by which it descends beneath another plate and is ultimately resorbed into the asthenosphere (fig. 4–15).

A Possible Mechanism for Spreading   An absolute answer to the mechanism that might cause the formation of lithosphere at the oceanic ridges and its movement away from those ridges at right angles has not been determined. However, the lithosphere thickens from a few kilometers near the ridge to over 200 km (124 mi) beneath some continental regions. The increasing thickness of the lithosphere correlates well with increasing depth of the ocean and decreasing heat flow (*see* fig. 4–10).

Within the earth, radioactive atoms are breaking down and releasing energy that must find its way to the surface as heat. It is conceivable that this heat moves to the surface through convection cells that carry the heat up in the regions of the oceanic ridges. If this is the case, there must be some regions on the earth where the cooler portions of the mantle descend to complete the convection cell.

Heat-flow measurements taken throughout the earth's crust show that the quantity of heat flowing to the surface along the oceanic ridges is as much as eight times in excess of the average for the earth's crust (1.25 μcal/cm^2/s). In addition, areas where

FIGURE 4–12   Confirmation of Sea-Floor Spreading.

If new oceanic crust is being added at oceanic ridges and spreads away from these structures, the age of fossil assemblages in sediments immediately above the crust and the radiometrically determined ages of the basal sediments and crust should increase as one moves away from the oceanic ridges. Data of this type recovered from the ocean floor support the theory of sea-floor spreading originating at the ridges.

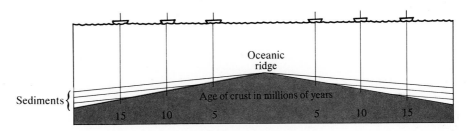

FIGURE 4–13   Lithosphere and Asthenosphere.

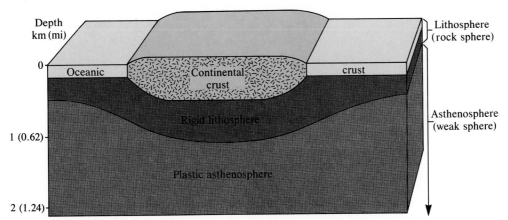

The lithosphere is the rigid outer shell of the earth's structure that includes the crust and cooler upper mantle. It is supported and allowed to move across the earth's surface by flow within the warmer, plastic underlying portion of the mantle called the asthenosphere. This weaker, possibly partially molten layer is thought to extend to a depth of approximately 650 km (400 mi).

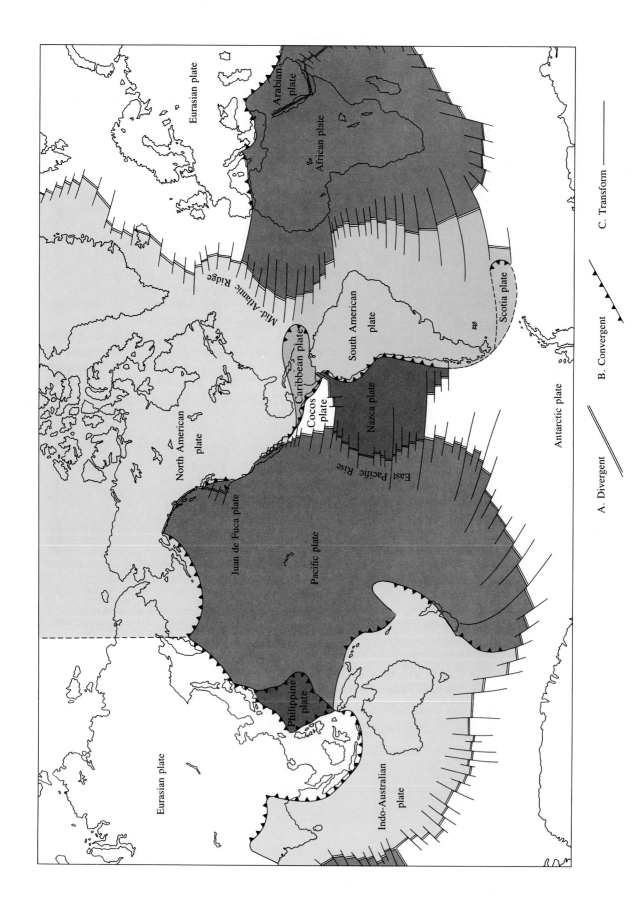

FIGURE 4–14 Crustal Plates and Associated Features.
After W. B. Hamilton, U.S. Geological Survey.

A. Divergent   B. Convergent   C. Transform

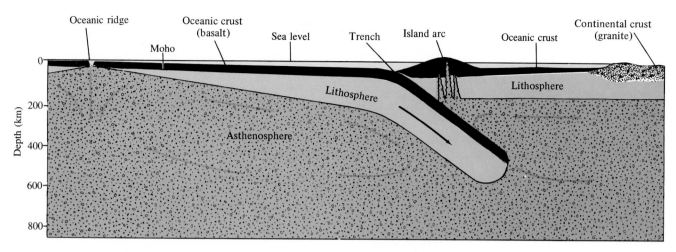

FIGURE 4–15 Movement in the Asthenosphere.
Lateral flow within the asthenosphere is a possible mechanism for the lateral movement of the lithosphere, including the crust. The asthenosphere is a low-strength plastic zone within the mantle where pressure is low enough and temperature sufficiently near the melting point of the material to allow it to flow slowly. Some form of thermal convection within the asthenosphere appears to create new lithosphere at the oceanic ridges and rises and may carry old lithosphere back into the mantle to be subducted beneath oceanic trench systems.

ocean trenches are known to exist have heat-flow characteristics as little as one-tenth the average.

The lateral movement that occurs between the regions of creation (ocean ridges and rises) and destruction (trenches) of the lithosphere may originate within the asthenosphere. These movements in the asthenosphere could carry the more rigid lithosphere, which contains the crust and upper mantle (fig. 4–15). It has been observed that the trenches do not fit the isostasy patterns characteristic of the continents, and it appears that there may be some force other than gravity pulling the lithosphere down in these regions. The descending convection cells might well provide that force.

The observed relationship of increasing lithospheric plate thickness, age of ocean floor, and ocean depth, along with a decrease in the rate of heat flow through the ocean floor with increasing distance from the ridge and rise spreading centers can be explained by the convection process (see fig. 4–10). The lithosphere at the spreading centers would be thin because of the proximity of high temperatures immediately beneath the ocean floor. This heat would also cause expansion of the upper mantle rocks, raising the oceanic ridges and rises to elevations well above those of the ocean floor on either side of the zones of upwelling heat. As the lithosphere moves away from the spreading centers toward zones of convection downwelling beneath the trenches, it will encounter decreasing rates of heat flow through the ocean floor from the underlying mantle. This would result in in-

creasing thickness of the lithosphere as cooling asthenosphere at the base of the lithosphere is converted to rigid lithosphere. The thermal contraction of hot, fluffy asthenosphere into cold, brittle lithosphere accounts for the increasing depth of the ocean away from the spreading centers.

The relationship of plate characteristics and ocean depth to convection cell motion represents only one possibility. Many investigators do not believe convection is an important factor in plate motion. Some believe the plates are pulled down into the mantle by the weight of their cooler dense edges. At present there seems to be no clearly defined process that can be identified as driving the lithospheric plates.

PLATE BOUNDARIES

The process of sea-floor spreading sets up stresses within the lithosphere as it moves across the spherical surface of the earth. These stresses have broken the present lithosphere into about a half-dozen major and another half-dozen or more minor plates, some of which are shown in figure 4–14 and plates 10, 12, and 13. The plate boundaries where new lithosphere is being added along oceanic ridges are called *constructive* or *divergent plate boundaries;* boundaries where plates collide and one subducts beneath the other are called *destructive* or *convergent boundaries.* Lithospheric plates also move past one another along *slip-slide* or *shear boundaries.*

Constructive Boundaries

Oceanic Ridges and Rises Project FAMOUS, the French-American Mid-Ocean Undersea Study, conducted in 1974, made observations in the rift valley running along the axis of the Mid-Atlantic Ridge. These observations provide a clue to the spreading process (fig. 4–16 and plate 10).

Fissures oriented parallel to the ridge axis range from hairline size near the center of the rift valley to more than 10 m (32.8 ft) in width near both margins of the valley. This may indicate the plates are being continuously pulled apart rather than being pushed apart by upwelling of material beneath the ridges.

Possibly, the upwelling of magma beneath the oceanic ridges is more that of filling in the void left by the separating plates of lithosphere. Whatever the mechanism of emplacement, a large mass of magma or viscous igneous material must exist near the surface beneath the ridges. This is manifested in the oceanic ridges being elevated in mountainlike proportions above the deep-ocean basins and the lithosphere thinning significantly beneath the ridges.

Destructive Boundaries

Island Arc Trench System Oceanic ridges and trench systems both represent significant discontinu-

FIGURE 4–16   Rift Valley Fissures.
A, a fissure in the rift valley of the Mid-Atlantic Ridge photographed from the submersible *Alvin* during Project FAMOUS, 1974. Courtesy of Woods Hole Oceanographic Institution. *B,* a more readily observable fissure above sea level in the rift valley of Iceland. Here the Mid-Atlantic Ridge rises above the ocean surface as an island because of the high rate of volcanic activity in the Iceland hot spot. Courtesy of Br. Robert McDermott, S.J.

A.

B.

ALVIN EXPLORES SPREADING-CENTER MEDIAN VALLEYS

In the summer of 1974, a series of explorations of the spreading centers of oceanic ridges and rises began, using the submersible *Alvin* and other underwater exploration vehicles and devices. This project led to a surprising and significant increase in our knowledge of the rift valleys, where new ocean floor is being created (fig. 4A). Project FAMOUS (French-American Mid-Ocean Undersea Study) investigated the median valley of the Mid-Atlantic Ridge 448 km (278 mi) southwest of the Azores. In addition to *Alvin* (fig. 4B), the French submersible *Cyana* and bathyscaphe *Archimede* examined the median valley to observe directly the process by which new ocean crust is being created along the axis of the Mid-Atlantic Ridge.

Sources of volcanism found near the center of the median valley and called *haystacks* were about 5 m (16 ft) high and 7 m (23 ft) wide. Radiating from the haystacks were *pillow tubes* formed by lava flowing beneath a solidifying crust that periodically ruptured, extruding lava that eventually cooled into pillow-shaped masses called *pillow lava* (fig. 4C). *Fissures* (fig. 4D) created by tectonic plates moving away from the median valley on either side were found running parallel to the valley and increased in width toward the sides of the valley. *Master lava tubes* (fig. 4E), much larger than the pillow tubes, allowed for the formation of *sheet flow lava* (fig. 4F) without the pillow character of lava that cooled in direct contact with the ocean water. No evidence of active volcanism was found, but some of the volcanic features observed were believed to be no more than a few hundred years old.

Cayman Trough *Alvin*'s next spreading-center dive in February 1976 investigated the Cayman Trough of the Caribbean Sea. Extending to depths in excess of 4700 m (15,416 ft), the Cayman Trough is created by a transform fault running west from Haiti to south of Grand Cayman Island. Here it angles south for 110 km (68 mi) as a north–south-trending spreading center. At the southern end of the spreading-center rift, the trough continues west as a transform fault to the Gulf of Honduras. Because *Alvin* could dive to only 3660 m (13,005 ft), the main objective of the dives was to examine the steep east wall of the spreading center's rift valley. Pilot Dudley Foster and diving geologist Dr. Robert D. Ballard from Woods Hole Oceanographic Institution collected some very significant rock samples from the rift valley wall.

Scientists have identified four distinct layers of materials beneath the surface of the ocean floor—layer 1 (sediments), layer 2 (basalt), layer 3 (gabbro), and layer 4 (mantle) (fig. 4G). Until this dive, layers 3 and 4 had not been observed in place beneath the ocean. All four layers were sampled in the Cayman Trough Rift Valley.

Galapagos Rift In February 1977 the submersibles dove again in the Galapagos rift about halfway between the Galapagos Islands and the South American mainland. Although this expedition was undertaken for geological research, it soon became of great interest to biologists. Thirteen of the first 3000 pictures taken by the camera sled ANGUS (Acoustically Navigated

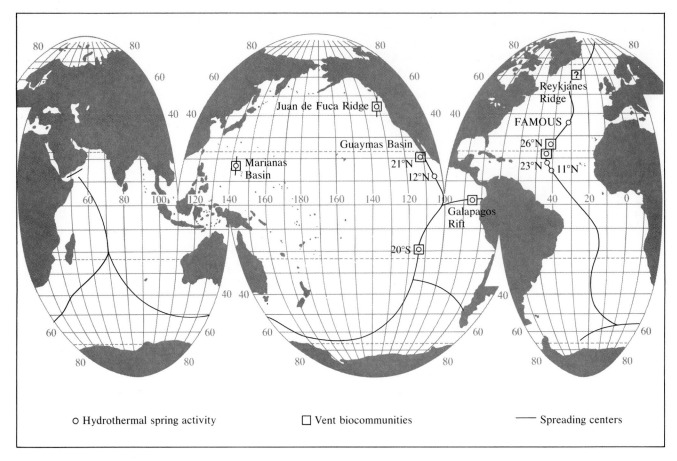

FIGURE 4A Hydrothermal Vents and Biocommunities

The box containing a question mark indicates the location of an early sighting of large clams that has not yet been confirmed (*see* page 86).

Underwater Survey System, fig. 4H) in the region of a temperature anomaly showed hundreds of large clams and mussels at depths below 2500 m (8200 ft) (plates 2 and 3).

The immediate question was "How do they survive?" The answer was not long in coming. Flowing out of small cracks in the lava was warm water. Samples of the warm spring water, analyzed aboard *Knorr,* a research vessel for Woods Hole Oceanographic Institution, showed it to be rich in hydrogen sulfide. It appears that bacteria use the energy released by their oxidation of hydrogen sulfide to sulfur and sulfate to fix carbon dioxide into organic matter in a process called *chemosynthesis.* This process can be carried out in the absence of sunlight and allows these bacteria to replace photosynthetic phytoplankton in the deep ocean as the primary producers of organic matter.

The ambient temperature (2°C, or 35.6°F) was found to support populations of spaghettilike acorn worms (fig. 4I) and benthic siphonophores (fig. 4J) on the outer edges of the biotic community surrounding the warm-water vents. The siphonophores are relatives of the Portuguese man-of-war and resemble dandelions gone to seed.

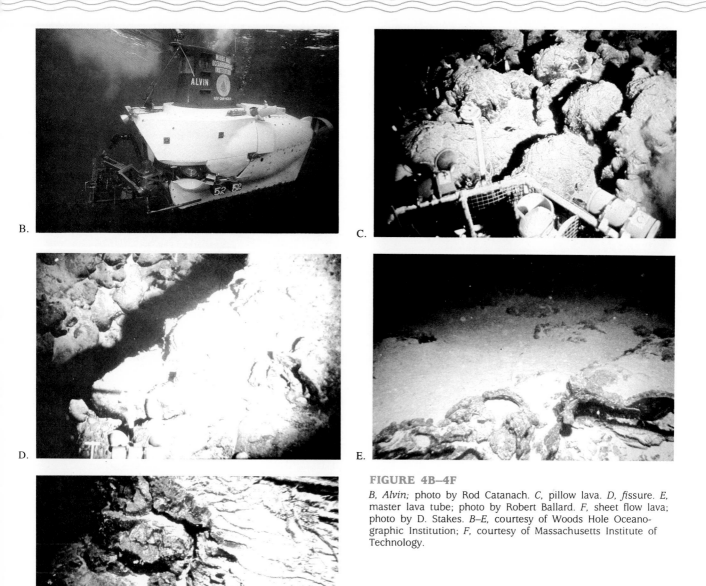

FIGURE 4B–4F

B, Alvin; photo by Rod Catanach. *C,* pillow lava. *D, fissure. E,* master lava tube; photo by Robert Ballard. *F,* sheet flow lava; photo by D. Stakes. *B–E,* courtesy of Woods Hole Oceanographic Institution; *F,* courtesy of Massachusetts Institute of Technology.

FIGURE 4G Cayman Trough.

The Cayman Trough is composed of two transform faults which separate the westward-moving North American Plate from the eastward-moving Caribbean Plate and are connected by a 110-km (68-mi) spreading-center rift valley. It is this north–south trending rift valley that *Alvin* explored to find the upper mantle exposed.

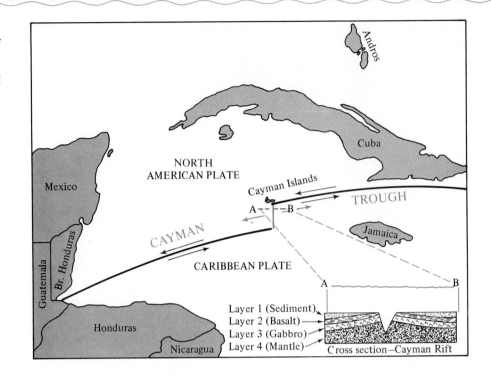

At intermediate distances, where temperatures were between 2° and 15°C (35.6° and 59°F), clams, mussels, small anemones, and serpulid worms were found. Thriving in the warm 10°–15°C (50°–59°F) water near the vent openings were pogonaphoran worms *(Riftia pachyptila, fig* 4K) living in tubes up to 2 m (6.6 ft) long and filter-feeding limpets. Additionally, a large number of galatheid (fig. 4L) and brachyuran crabs (fig. 4K) and grenadier fish roamed the area. A relict vent area was identified by the shells of dead clams. When the vent became inactive it ceased to provide the hydrogen sulfide needed by the bacteria, which served as a food supply for the clams (fig. 4J). Dives to the East Pacific Rise at 21°N latitude off the tip of Baja California revealed another biotic community similar to that in the Galapagos Rift. The hot spring activity, however, was much more intense. Black smokers (fig. 4M) spewed out 350°C (662°F) water rich in metallic sulfides. Much of the bottom in the vicinity of the vents was covered with a yellow to reddish-brown coating of metallic sulfides (fig. 4N) that precipitated out of the 350°C water as it came into contact with the 2°C ambient water. The mineral-rich water supported a concentration of suspended organic matter up to 500 times greater than normal for this bottom environment and up to 4 times greater than that found in productive surface waters.

The large tube worms *(R. pachyptila),* with their furry blood-red tentacles protruding from plasticlike tubes, were common in water with temperatures of 3° to 23°C (37.4° to 73.4°F) (fig. 4K). Beyond them were the large, 30-cm (12-in)-long clams concentrated in the crevices between the pillows of lava. Although the mussels found in the Galapagos Rift were

H.

I.

J.

K.

L.

FIGURE 4H–4L

H, ANGUS; photo by Al Gunderson. *I,* acorn worms—Galapagos; photo by James J. Childress. *J,* siphonophores and dead clams; photo by Robert Hessler. *K,* pogonophoran worms and brachyuran crabs; photo by Fred Grassle. *L,* Galatheid crabs; photo by Robert Hessler. *H, K,* courtesy of Woods Hole Oceanographic Institution; *I,* courtesy of University of California, Santa Barbara; *J, L,* courtesy of Scripps Institution of Oceanography.

absent at 21°N, the rest of the biological community was present. Additionally, a previously undescribed white fish, "the 21°N vent fish" (fig. 4O), was found. It has not been seen elsewhere.

Continued investigation has resulted in the discovery of similar occurrences on the East Pacific Rise at 12°N (February 1982) and in the Guaymas Basin of the Gulf of California (January 1982), as well as on the Juan de Fuca Ridge (September 1981) off the coast of Oregon. The biological community of the Juan de Fuca vents appears to be composed of less robust forms than observed on the East Pacific Rise (fig. 4P).

In the Guaymas Basin, where the spreading center is covered with sediment carried to the Gulf of California by rivers, large mounds of sediment were found to overlie the spreading axis. Sulfide samples recovered were saturated with petroleum, the source of which is still under investigation.

Dives to examine the mounds showed they were tens of meters wide and hundreds of meters long. Atop them were pagodalike structures (fig. 4Q) venting petroleum-bearing water with temperatures as high as 315°C (599°F). The "pagodas" are covered with groves of tube worms. The sediment is covered with thick mats of bacteria. Also found were large patches of black coral, large white octopuses, and large red spider crabs not found at other sites. Apparently the hydrocarbons contribute to the food chain in this environment, where the abundance and diversity of life exceeds that observed at bare ocean crust vent sites.

A site similar to the Guaymas Basin occurrence was observed in April 1987 in the sediment-filled Marianas Basin of the Western Pacific (fig. 4A). Methane gas plumes indicated a possible vent on the east flank of the Marianas Spreading Center, where a biocommunity was discovered.

The first confirmed biotic community associated with hydrothermal vents in the rift valley of the Mid-Atlantic Ridge was discovered in July 1985 by scientists working in the NOAA Vents program. The vent biota was found at the Trans-Atlantic Geotraverse (TAG) site located at approximately 26°N latitude. A major component of the TAG site biota are two species of a new genus of shrimp, *Rimicaris.* About 3 to 5 cm (1.2 to 2 in) in length, they are shown swarming around a 2-m (6.6 ft) black smoker chimney in figure 16–29B. This photograph was taken from *Alvin* during the summer of 1986. Although many believed hydrothermal vent communities would be found only on spreading centers with spreading rates exceeding 6 cm (2.4 in)/yr, the TAG site is located where the spreading rate is no more than 2 cm (0.8 in)/yr. Before the discovery of the East Pacific vent biota, Bruce Heezen observed large clams on the crest of Reykjanes Ridge south of Iceland while diving in the U.S. Navy's nuclear powered research submarine, *NR-1.* A 1984 attempt to confirm this sighting failed.

As deeper diving submersibles are made available and exploration of the spreading centers and transform faults that offset them continues, we can expect the discovery of many new hydrothermal vent biotic communities. Active spreading centers may possess hydrothermal vents regardless of their spreading rates. If so, we can expect new and exciting finds in all of the ocean basins.

M.

N.

O.

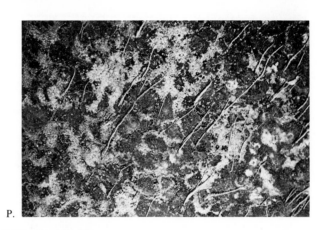

P.

Q.

FIGURE 4M–4Q

M, black smoker; photo by Fred Spiess. *N,* coating of metallic sulfides; photo by Robert Ballard. *O,* 21°N vent fish; photo by Horst Felbeck. *P,* Juan de Fuca pogonophoran worms. *Q,* Guaymas pagoda structures; photo by Robert Brown. *M, O, Q,* courtesy of Scripps Institution of Oceanography; *N,* courtesy of Woods Hole Oceanographic Institution; *P,* courtesy of University of Washington.

ities in the earth's crustal structure and are characterized by earthquakes and volcanic activity. There is a significant difference in the earthquake activity of the two regions, however. Shallow quakes, usually less than 10 km (6 mi) in depth, are associated with the constructive ridge boundaries where the lithosphere is thin, and movements that cause earthquakes in the trench regions may occur at depths down to 650 km (405 mi).

The earthquake *focus,* the point at which the movement that causes the quake occurs, is usually shallow in the immediate area of the trench. As earth-

quake activity moves toward the continent, the depth of the focus becomes greater and reaches a maximum value of 650 km. The *Benioff seismic zone* is a band 20 km (12.5 mi) thick that dips from the trench region under the continents within which all of the earthquake foci are located. The angle of dip is approximately 45°, becoming steeper at greater depths (fig. 4–17A).

This band within which the earthquakes originate must lie within the slab of the lithosphere that is descending. The fact that the oceanic crust generated at the oceanic rises and ridges is subducted at the

FIGURE 4–17 Effects of Plate Collisions.

A, classic oceanic trench systems develop where oceanic lithosphere subducts beneath a plate with oceanic crust at its leading edge and a continent trailing some distance behind. A basaltic island arc-trench system develops. The sloping belt in which earthquake foci are located is the Benioff seismic zone.

B, when an oceanic lithospheric plate subducts beneath a plate with continental crust at its leading edge, an andesitic continental arc develops. The volcanic rock andesite is named for the Andes Mountains, where a continental arc exists. Andesite has a composition between that of granite and basalt. It is thought to form as melted basalt from the descending oceanic plate rises through the continental granite to produce the volcanic continental arc.

C, as two plates carrying continents near their leading edges converge, trenches do not develop because sediments that accumulated in the sea between the continents are too light to be subducted. The shortening of the earth's crust in such regions is accommodated by the folding of the sediments into mountain ranges such as the Appalachians, Alps, Himalayas, and Urals. The lower lithosphere and asthenosphere may be involved in low rates of subduction beneath the mountains.

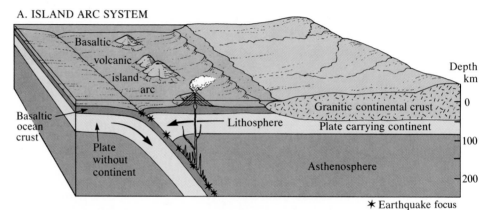

A. ISLAND ARC SYSTEM

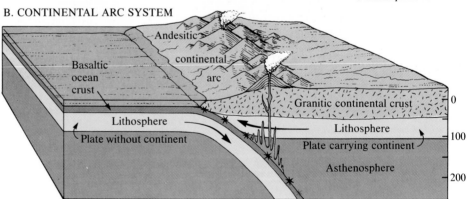

B. CONTINENTAL ARC SYSTEM

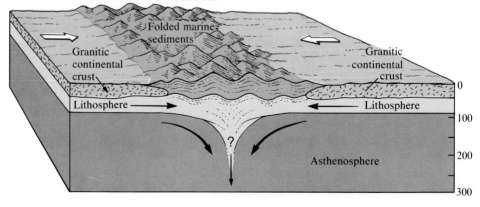

C. CONTINENTAL MOUNTAIN SYSTEM

trenches is supported by the fact that no oceanic crustal material more than 200 million years old has been recovered from the ocean bottom.

As the lithosphere is carried into the asthenosphere, it is heated by being subducted to greater depths. Water and other volatiles are freed, producing a low-density mixture that rises to the surface on the continental side of the trenches to produce basaltic volcanoes that may become island arcs.

Continental Arc Trench System Should an oceanic plate subduct beneath a plate with a continent at its leading edge, the melting of the subducting oceanic plate occurs beneath the continent. Consequently, the rising basalt melt passes through and mixes with the granite of the continental crust. This results in a continental arc of volcanoes along the edge of the continent (fig. 4–17B). The volcanoes are composed of a rock called *andesite,* which is of a composition intermediate between those of basalt and granite and is thought to form from the mixing of the two basic crustal rock types.

If the spreading center producing the subducting plate is far enough from the subduction zone, an oceanic trench is well developed along the margin of the continent. The Peru-Chile Trench is an example. It is associated with the Andes Mountains continental arc of volcanoes from which the rock, andesite, takes its name. The Cascade Mountains continental arc extending from northern California into Canada has no visible trench associated with it. This is probably primarily the result of the nearness of the Juan de Fuca spreading center to the edge of the North American Plate.

Continental Mountain System If two lithospheric plates collide and both contain continental crust near their leading edges, the surface expression will be a mountain range composed of the folded sedimentary rocks derived from sediments deposited in the sea that previously separated the continental blocks. These sedimentary rocks do not subside in a zone of subduction because of their relatively low density, having been derived from the continents. The oceanic crust itself may, however, subside beneath such mountains. The exact nature of the subsidence of oceanic crustal material in these regions has not been determined (fig. 4–17C).

Some degree of folded mountain formation is always associated with plate collisions, but ocean trenches are absent in some areas—probably for a number of reasons. One possible explanation for their absence has to do with the rate of plate convergence. Where the rate of convergence is less than 6 cm/yr

(2.4 in/yr), the crust may be able to absorb the compression by folding itself into mountain ranges. Higher rates of convergence cause one plate to break free and move past the other while sinking back into the mantle, thus producing the ocean trench.

Shear Boundaries

Transform Faults As newly formed lithosphere moves away from the axis of an oceanic ridge, it travels perpendicular to the axis. On a spherical surface like that of the earth, all of this new lithosphere will converge toward two points. For instance, if the equator were an oceanic ridge, the points of convergence would be the North and South poles. Obviously, all this new matter cannot be accommodated at a single point, and stresses are set up within the lithosphere. The stresses cause the lithosphere to break along lines perpendicular to the ridge axis. Movement along the breaks offsets the ridge axis, producing a *transform fault* (fig. 4–18). Along such faults one plate slides past another, producing shallow earthquakes in the thin lithosphere. An example of such a fault that has come ashore is the San Andreas Fault, which runs across California from the head of the Gulf of California to the San Francisco area.

Fracture Zones Close observation of the axes of oceanic ridges and rises shows they do not move in a continuous meandering path through the world's oceans, but are offset periodically by transform faults that run perpendicular to the axes. Continuing along the alignment of the transform fault beyond the offset axes are fracture zones. The transform faults represent active displacements of the axes; the fracture zones are evidence of past transform-fault activity. On opposite sides of the transform fault, the lithospheric plates are moving in opposite directions, whereas there is no relative plate motion occurring along a fracture zone. In other words, the transform faults serve as plate boundaries, while the fracture zones are embedded in a single plate (fig. 4–18). Earthquakes with foci above a depth of 10 km are common along the transform faults, whereas the fracture zones are aseismic.

Because the age of the ocean floor on one side of a fracture zone is greater than that of the ocean floor on the opposite side, the older ocean floor will have subsided more from thermal contraction. This results in a vertical escarpment that faces the older ocean floor and increases in height with increased difference in the age of ocean floor on opposite sides of the fracture zone.

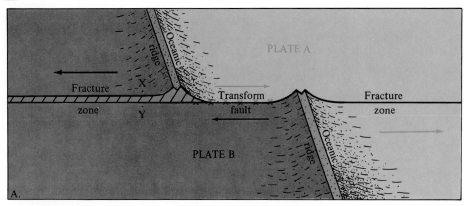

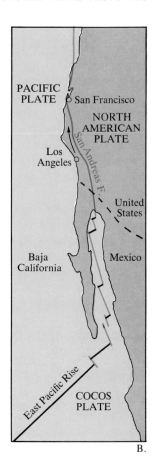

x – Shallow focus
earthquakes

➡ Arrows indicate
the direction of
plate movement.

FIGURE 4–18 Transform Faults—Fracture Zones.

A, the axes of oceanic ridges are offset by transform faults due to the stresses resulting from the motion of rigid plates moving acros the spherical surface of the earth. Earthquakes are common and of relatively great magnitude along the transform faults. They are called transform faults because they cut across the larger physiographic features of the ocean floor, oceanic ridges and rises. Extending beyond the offset ridges are fracture zones, where earthquakes are rare because there is no relative movement on the opposite sides of the fracture zone. Fracture zones are scars of old transform fault activity and are maintained as significant topographic features because the age of the ocean floor on one side of the fracture zone is older than the ocean floor immediately opposite it on the other side of the fracture zone. Point X represents younger ocean floor than point Y. Therefore point Y has subsided more because of cooling and thermal contraction than the ocean floor at point X. Thus there is an escarpment along the fracture zone in Plate B that faces the older ocean floor on the side of the fracture zone where point Y is located. There is also an escarpment on Plate A, but it faces the opposite direction. *B,* the San Andreas Fault is a classic example of a transform fault that offsets an oceanic ridge or rise and forms a boundary between plates that are sliding past each other. Along this contact, the Pacific Plate is moving north relative to the North American Plate. Base map courtesy of Open University Press.

GROWTH OF OCEAN BASINS

Having discussed the evidence concerning the general nature of the plate tectonics process, we will now take a closer look at the three major ocean basins, considering specific features of the plate tectonics process and the rate at which these basins appear to have developed. Since we have the greatest knowledge of the Atlantic Ocean, we will consider it before the lesser-known Pacific and Indian oceans. As we look more closely at these ocean basins, we will become aware that the plate tectonics theory is providing the long-sought answers to some major geological questions. New questions are raised, however, as certain features of the ocean floor do not appear to fit the pattern as well as might have been expected. The changing positions of continents during the last 150 million years as they moved to their present positions can be observed in figure 4–19. In these paleoceanic reconstructions, the African Plate has been chosen as the reference. Present-day coastlines have been maintained in the projections throughout the 150-million-year span. As a result, continents that are actually in contact along some line currently found along their continental slope will appear separated on the projections.

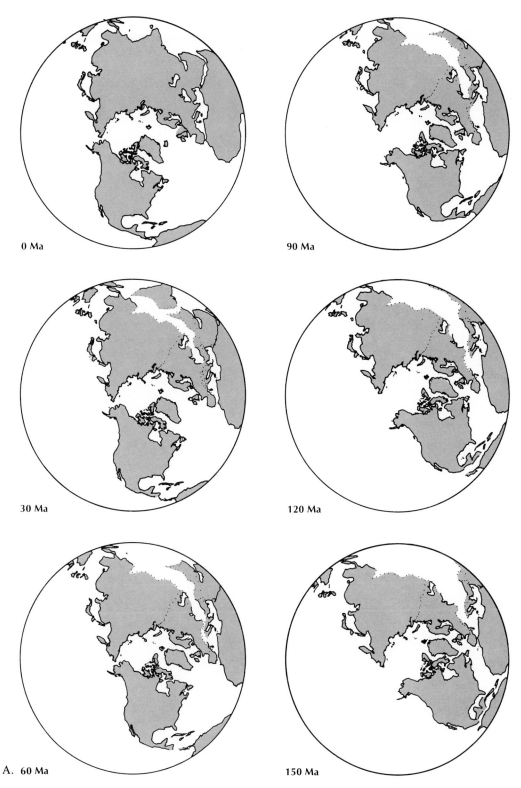

0 Ma

90 Ma

30 Ma

120 Ma

A. 60 Ma

150 Ma

FIGURE 4–19 Paleoceanic Reconstructions.

These paleoceanic projections represent an attempt to determine the relative positions of the continents over the last 150 million years. Ages are quoted in Ma (millions of years ago) at intervals of 30 million years. *A*, Northern Hemisphere; *B*, Southern Hemisphere (p. 92).

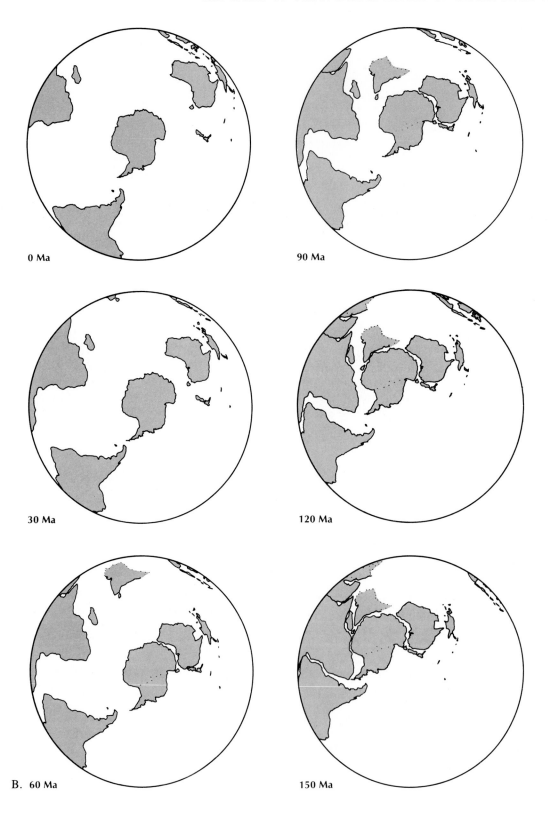

FIGURE 4–19 Paleoceanic Reconstructions (continued).

The projections are based on data taken from the first 60 legs of the Deep Sea Drilling Project, conducted during a decade of deep-ocean drilling from August 1968 to August 1978. Courtesy of P. L. Firstbrook, B. M. Funnell, A. M. Hurley, and A. G. Smith in association with Scripps Institution of Oceanography, University of California, San Diego.

ATLANTIC OCEAN

We can generally consider that the Atlantic Ocean is the only new ocean to have formed in the 200 million years since Pangaea was surrounded by the ancestral Pacific Ocean, Panthalassa. Since that time, the Pacific Ocean has been decreasing in size as the Atlantic Ocean increases in dimensions. The Indian Ocean simply represents a portion of Panthalassa that has been more or less separated from the Pacific Ocean as a result of the rearrangement of the continental masses of Australia, India, and eastern Asia.

Before the last breakup of Pangaea into Laurasia and Gondwanaland, the continental mass of the earth was distributed rather evenly between the Northern and Southern hemispheres. The subsequent movements have shifted two-thirds of the continental mass into the Northern Hemisphere. According to the reconstruction of the continents by Robert Dietz and John Holden, North America, Europe, and Africa were in contact at a point near the present position of Ascension Island (fig. 4–20).

About 180 million years ago, rifting of the Pangaean continent began with the separation of North America from South America and Africa, after which it joined with Eurasia in the formation of Laurasia. By 135 million years ago, the North Atlantic rift had produced a considerable separation between North America and South America while extending itself to the north to begin the separation of Greenland from the North American mainland. By this time, South America had begun to separate from Africa along a rift zone.

By 65 million years ago the separation of South America from Africa was completed, and the South Atlantic had reached a width in excess of 3000 km (1860 mi). The North Atlantic had not been growing so rapidly, and the separation from Europe had not yet been totally achieved. Shortly thereafter, the separation of Greenland from Europe occurred, and the Atlantic Ocean opened throughout its present length. During the last 65 million years, North and South America were joined by the Isthmus of Panama, and the ocean widened to its present dimension.

PACIFIC OCEAN

As North America and South America move in from the east and Asia and Australia from the west, the Pacific Ocean becomes smaller. It is in this ocean that we see the greatest evidence for the subduction of oceanic crust in the form of the many trenches marking the ocean margin. The sources of new crustal material in the Pacific Ocean are the South Pacific and East Pacific rises.

Additional features associated with plate tectonics of interest in the Pacific Ocean are the region where North America has been pushed onto the East Pacific Rise, the extremely long fracture zones in the North Pacific, the relatively numerous northwest–southeast-trending chains of volcanic islands, and the trench

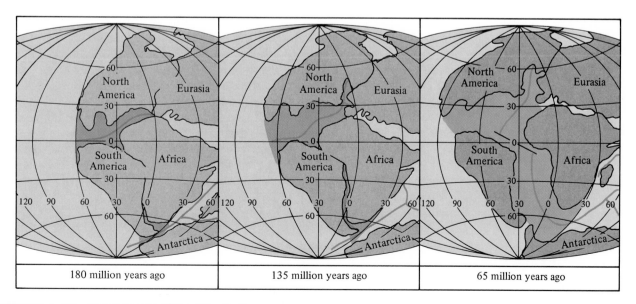

| 180 million years ago | 135 million years ago | 65 million years ago |

FIGURE 4–20 Development of the Atlantic Ocean.

OPHIOLITES

It is a well-accepted tenet of the global plate tectonic theory that new oceanic lithosphere created at spreading centers ultimately is *subducted* and returns to the underlying mantle. However, it seems that about one one-thousandth (0.001) percent of the oceanic lithosphere does not return. Instead, it is *obducted* onto the edge of a continental mass (fig. 4R). These fragments of oceanic lithosphere embedded into the continents are called *ophiolites*. They characteristically consist of a thin layer of sedimentary rock that undoubtedly formed in the ocean, underlain by a sequence of crystalline rocks typical of ocean crust and the underlying mantle. Many of the ophiolites contain rich metallic ores with patterns of enrichment similar to those known to occur in the sediments and oceanic crust near hydrothermal vents on oceanic spreading centers.

With our present understanding of the sea-floor spreading process and its role in global plate tectonics, we have developed a reasonably clear view of how the units of an ophiolite assemblage may have formed. How they become obducted onto the continents is not so well understood.

In the idealized ophiolite sequence of rock types, there is a layer of marine sediments and an underlying series of crystalline rocks. Immediately beneath the sediments, one finds *pillow basalts* that flowed out on the ocean floor and show the pattern characteristic of submarine cooling. Descending through the lava flow, one encounters an increasing concentration of basaltic dikes that replace the flows completely about 1 km (0.62 mi) below the pillow lava surface (fig. 4S). It is believed that these *sheet dikes* form as vertical sheets of lava fill injected into narrow gaps formed as lithospheric plates are pulled away from the axis of spreading. At a distance of approximately 3 km (1.9 mi) beneath the pillow lava surface, the dikes give way to *massive gabbro,* a rock similar to basalt in composition but of a coarser texture owing to slower cooling. The gabbro may represent the cooling of magma near the roof of the magma chamber that was the source of the overlying basalts. The massive gabbro may be up to 300 m (984 ft) thick. It is underlain by *layered gabbro* that may have crystallized within the upper region of the magma chamber. Layered *peridotite,* a mantle rock composed mostly of the iron- and magnesium-rich minerals pyroxene and olivine, is found beneath the layered gabbro. The boundary between the gabbro and peridotite is thought to represent the *Moho,* or the contact between the ocean crust and the underlying mantle. The Moho is usually found about 4.5 km (2.8 mi) below the top of the pillow lavas.

The main features of the idealized ophiolite sequence observed on the margins of continents were partially encountered in place beneath the ocean floor. The Deep Sea Drilling Project drill hole No. 504B, located on the flank of the eastern Galapagos Spreading Center, was drilled to a depth of 1.35 km (0.84 mi) beneath the ocean floor in 3.46 km (2.15 mi) of water. The crustal rocks penetrated are thought to be just over 6 million years old and the sequence of rocks penetrated in the hole are as follows:

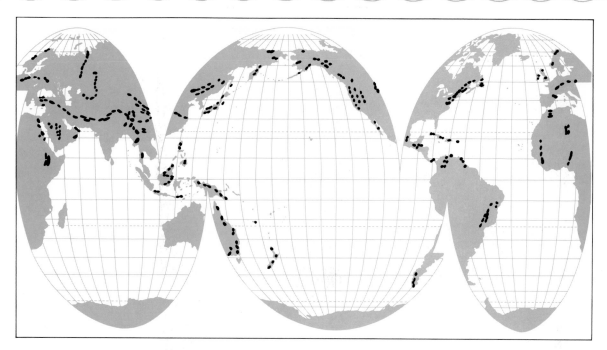

FIGURE 4R Ophiolite Distribution.

Distribution of ophiolites throughout the world. Ophiolites near the margins of the present continents where subduction is occurring are generally younger (less than 200 million years old) than those embedded deep within the continents. Some ophiolites are thought to be as much as 1.2 billion years old.

FIGURE 4S Ophiolite Rock Types.

The column on the left shows the sequence of rock types found in ophiolite complexes such as the Troodos Massif of Cyprus. On the right is the sequence of rock types encountered in the Deep Sea Drilling Project drill hole No. 504B, drilled to a total depth of 1350 m (4428 ft) beneath the ocean floor. It closely resembles that of the upper ophiolite complex and adds support to the belief that ophiolites are pieces of oceanic lithosphere that have been obducted onto the margins of continents.

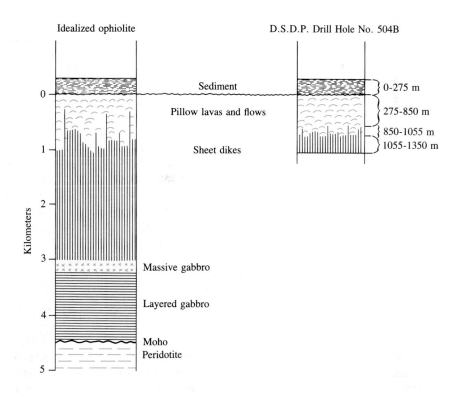

1. Sediment   0–275 m (0–902 ft)
2. Pillow lavas and flows   275–850 m (902–2788 ft)
3. Pillow lavas, flows, and sheet dikes   850–1055 m (2788–3460 ft)
4. Sheet dikes   1055–1350 m (3460–4428 ft)

Figure 4S shows how well this sequence resembles that observed in the ophiolite complexes.

One of the oldest known ore deposits is the Troodos Massif of the island of Cyprus (fig. 4T). It has been mined for copper since the time of the Phoenicians. The main sulfide ore bodies are found in the volcanics near the top of the ophiolite sequence, and the sediment section above the volcanics is rich in

FIGURE 4T Ophiolite Mining. *A,* location of the Troodos Massif ophiolite on the island of Cyprus in the eastern Mediterranean Sea. *B,* mining the Troodos Massif. The sulfide ore is the black material in the bottom of the pit. Photo courtesy of R. Koski, U.S.G.S.

A.

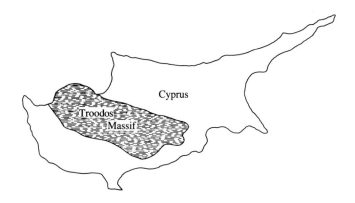

B.

iron and manganese. Such a sequence of metal enrichment is known to be occurring now at oceanic spreading centers as a result of hydrothermal activity.

The difficult question of how the ophiolites become obducted onto the continents may have numerous answers. One of the more straightforward explanations (fig. 4U) is as follows:

1. A plate containing a continent has a section of oceanic lithosphere between the continent and a subducting plate of oceanic lithosphere.
2. The subducting oceanic plate contains water-bearing minerals such as zeolites and amphiboles. Subduction will heat these minerals and release water. The heated water converts the mantle rock, peridotite, to serpentine, which is much less dense than the surrounding peridotite. The serpentine rises and pushes up the overlying crust and mantle so that the oceanic crust is raised above sea level.
3. Continued compressional stresses may cause the oceanic crust and associated serpentine to break into layers that are obducted onto continental rock.

Continued study of ophiolites is needed to answer many questions that remain about the nature of all the processes involved in forming and obducting these ore-rich bodies along the margins of continents.

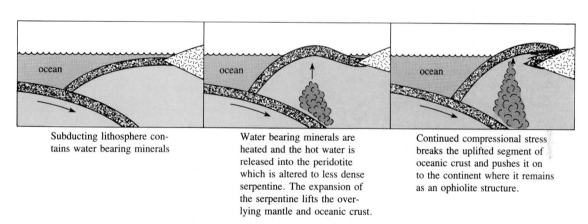

Subducting lithosphere contains water bearing minerals

Water bearing minerals are heated and the hot water is released into the peridotite which is altered to less dense serpentine. The expansion of the serpentine lifts the overlying mantle and oceanic crust.

Continued compressional stress breaks the uplifted segment of oceanic crust and pushes it on to the continent where it remains as an ophiolite structure.

FIGURE 4U   The Obduction of an Ophiolite onto a Continent.

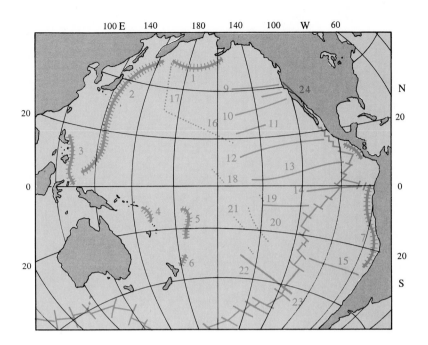

1. Aleutian Trench	9. Mendocino Fracture Zone	17. Emperor Seamounts
2. Kuril-Japan-Mariana Trench System	10. Murray Fracture Zone	18. Line Islands
3. Mindanao Trench	11. Molokai Fracture Zone	19. Marquesas Islands
4. New Hebrides Trench	12. Clarion Fracture Zone	20. Tuamotu Islands
5. Tonga Trench	13. Clipperton Fracture Zone	21. Society Islands
6. Kermadec Trench	14. Galapagos Fracture Zone	22. Eltanin Fracture Zone
7. Peru-Chile Trench	15. Easter Fracture Zone	23. East Pacific Rise
8. Mid-America Trench	16. Hawaiian Islands	24. San Andreas Fault

FIGURE 4–21   Physiographic Features of Pacific Ocean Basin.
See plate 12 for a detailed view of these features. Base map courtesy of Open University Press.

systems along the North American and South American coasts that have no island arcs along them (fig. 4–21).

Forming at the South Pacific and East Pacific rises and moving off in a northwesterly direction is the large Pacific Plate, which underlies most of the Pacific Ocean. Moving away from the rises to the east and south are the Antarctic Plate; the Nazca Plate, which subducts beneath the South American continent at the Peru-Chile Trench; and the smaller Cocos Plate, which produces the Mid-America Trench as it descends directly beneath Central America.

Seamounts and Tablemounts

Although the features describd below are abundant in all oceans, the evidence of volcanic activity on the bottom of the Pacific Ocean is particularly wide-

spread—more than 20,000 volcanic features are found there. Some marine volcanic activity produces widespread gently sloping surfaces where large volumes of lava flowed out in broad sheets, solidified, and became *archipelagic aprons.* Such features commonly surround volcanic islands.

Volcanic peaks that extend more than 1 km (0.62 mi) above the abyssal floor are termed *seamounts,* unless they have tops flattened by erosion, in which case they are classified as *tablemounts,* or *guyots.* Tablemounts are less common than seamounts and are found in linear trends in the Pacific Ocean. The tops of most tablemounts in the Pacific Ocean are between 1800 and 3000 m (5904 and 9840 ft) below the present ocean surface.

Volcanic features on the ocean bottom that are less than 1 km (0.62 mi) above the ocean floor are called *abyssal hills.* They cover a large percentage of the en-

FIGURE 4–22 Formation of Seamounts and Tablemounts. Volcanic activity develops near the crest of the ridge because of tensional fractures that develop parallel to the ridge. These fractures serve as conduits for lava but usually seal within 30 million years.

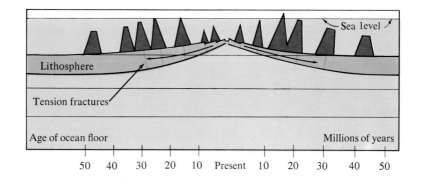

tire ocean basin floor and have an average relief of about 200 m (656 ft), while many such features are found buried beneath the sediments of abyssal plains.

A theory related to the formation of such features in association with the oceanic ridge systems could account for the existence of many seamounts and tablemounts (fig. 4–22). Considering oceanic ridge systems on a worldwide basis, active oceanic volcanoes are characteristic of the axes of the ridges. Moving away from the ridge axis, the volcanic activity gradually decreases as the cracks in the earth's crust serving as conduits for rising lava near the axis of these ridges are sealed off as they move away. Active volcanoes over 30 million years of age are rare.

Let us consider the life history of a theoretical volcano that progressively moves away from an oceanic ridge, being carried by the lithospheric plate. A volcano may originate near the axis of the ridge and gradually increase in size as it moves downslope. By the time it is 10 million years old, if it has been a very active volcano, it may develop into an island that may exist for another 10 to 15 million years if the volcano remains active. The volcano will probably become dormant within 30 million years of its origin, and as it continues to move down the flank of the oceanic ridge, the top will be flattened by wave action and be ultimately submerged. The flat-topped structure will sink deeper and deeper beneath the ocean surface as it descends the slope of the ridge. It has now become a guyot or tablemount, most of which are at least 30 million years of age.

Coral Reef Development

Although the development of a comprehensive theory of global plate tectonics was not to be developed until over 100 years later, Charles Darwin developed a theory on the origin of coral reefs that depended on the subsidence of volcanic islands. Darwin published his theory in the book *The Structure and Distribution of Coral Reefs* in 1842. Figure 4–23 illustrates Darwin's theory of reef development. Drilling into reefs before and after World War II provided the evidence to prove the validity of Darwin's theory.

There are three basic types of reefs—fringing, barrier, and atoll (fig. 4–23). *Fringing reefs* can develop at the margin of any landmass where the temperature, salinity, turbidity, and depth of the water are suitable for reef-building corals. Because of the close association between the landmass and the reef, there will be inevitable periods of time when runoff from the landmass will carry too much sediment for the coral in a particular area to survive. Very commonly, the amount of living coral in a fringing reef at any given time will be relatively small. The greatest concentration of live reef will be on the seaward margin, which is more protected from sediment and salinity changes due to runoff. Without some change in sea level relative to the landmass during the existence of the fringing reef, it is not likely that the other two types of reefs we have mentioned, the barrier reef or the atoll, will form.

Barrier reefs are linear or circular developments separated by a lagoon from the continent or island on whose margin they are growing. It is thought that barrier reefs form when the substratum on which a fringing reef is built begins to subside. As the island or margin of the continent subsides, the reef maintains its position at the optimum water depth by growing upward. Studies of reef growth rates indicate most have grown 3 to 5 m/1000 yr (9.8 to 16.4 ft/1000 yr) during the recent geologic past. There is some evidence, however, that reefs in the Caribbean have grown at rates of over 10 m/1000 yr (32.8 ft/1000 yr).

The most outstanding example of a barrier reef is the Great Barrier Reef along the northeast coast of Australia. It is 150 km (93 mi) wide and more than 2000 km (1242 mi) long. Smaller barrier reefs are found around the island of Tahiti and other islands in the western Atlantic and Pacific oceans. Coral growth is much more active on the portions of the reef that

FIGURE 4–23 Darwin's Theory of Development of Reef Types.

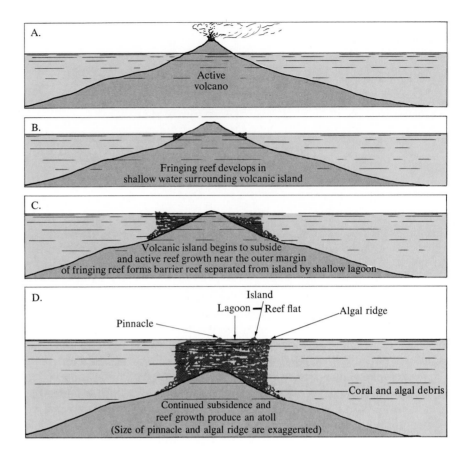

are better protected from land runoff. While the barrier is actively growing, the lagoon floor may be covered with sediment derived both from the reef and the adjacent landmass.

Atolls may form either on a subsiding continental shelf or around the margin of a sinking volcanic island in the open ocean. Most atolls are on sinking volcanic peaks, and the greatest number occurs in the equatorial Pacific. Atolls may take a circular or irregular shape and surround a lagoon that is usually 30 to 50 m (98 to 164 ft) deep at its deepest part. The fringing reef will generally have many channels that allow circulation between the lagoon and the open ocean. The reef flat is commonly broader on the windward side of the atoll, and channels are more numerous on the leeward side (fig. 4–24).

Islands large enough to be inhabited may develop as deposits of coral debris from the windward reef flat, which may be characterized by a well-developed algal ridge on the windward edge of the reef. The algal ridge, built by calcium-secreting algae, can extend above the low-tide level, but beyond it on the reef front will be found a profuse growth of coral below the low-water line.

The floor of the lagoon will be filled with coral debris. Oval-shaped reefs called *pinnacles* or *knolls* maintain active reef growth in regions of the lagoon in good communication with the open ocean through the channels.

Pacific Ocean Hot Spots

Many northwest–southeast-trending island chains are located on the Pacific Plate. The most intensely studied of these is the chain of islands between Midway Island and Hawaii. There are no active volcanoes on Midway Island; volcanic activity appears to have ceased 20 million years ago. The ages of the islands decrease toward Hawaii, where the only active volcanism in the chain exists. It is manifested in the volcano Kilauea, which is less than 1 million years old. An active submarine volcano, Loihi, currently exists southeast of Hawaii and may either become part of that island or create a new island (fig. 4–25 and plate 11).

The explanation for the existence of this chain of islands and seamounts with a progressive increase in age toward the northwest is that a hot spot, a station-

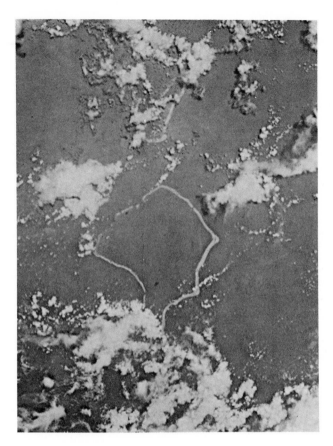

FIGURE 4–24 Marshall Island Atoll.
Courtesy of NASA.

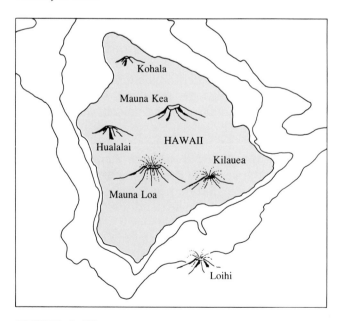

FIGURE 4–25 Hawaiian Hot Spot.
The submarine volcano Loihi is located 40 km (25 mi) off the southeast coast of Hawaii. This island, which is the youngest island in the Hawaiian chain, has two recently active volcanoes, Mauna Loa and Kilauea, along its southern margin. Loihi, Kilauea, and Mauna Loa overlie the hot spot that has been creating islands for the last 70 million years.

ary upwelling (plume) of magma, exists beneath the Pacific Plate in the region of the island of Hawaii. As the plate moves across this hot spot, volcanic eruptions produce islands that subsequently become dormant, erode, and subside as they move off to the northwest (*see* plates 12 and 13).

Continuing northwest from Midway Island, submerged seamounts have the same alignment as the island chain from Midway to Hawaii. At about 188°W longitude and 30°N latitude, an abrupt change in the alignment of the seamounts occurs. From this point on, the alignment is slightly west of true north. The age of the seamounts at this juncture of the two trends is thought to be between 30 and 40 million years, whereas the oldest seamounts in the northward-trending Emperor Seamount Chain formed some 70 million years ago. If the Emperor Seamount Chain and the Hawaii seamounts and islands were generated by the same hot spot, there must have been an abrupt change in the direction of the movement of the Pacific Plate between 30 and 40 million years ago (fig. 4–21).

The Line Islands to the south of the Hawaiian Islands do not show such a distinct pattern of progressive formation as can be measured for the Hawaiian Islands and the Emperor Seamount Chain. Beginning with Christmas Island near the southern end of the Line Islands, just north of the equator, an age of approximately 70 million years seems to be the minimum that can be determined. At the northern end of the chain, the age appears to be no more than 85 million years. Although the islands at the southern end are younger, the difference in age is not as great as might be expected had the hot spot theory accounted for this alignment.

South of the equator are other northwest–southeast-trending volcanic island chains—the Marquesas Islands, Tuamotu Islands, Society Islands, and Cook Islands. It appears that groups of islands in a given chain may have formed more or less at the same time. Although there are indications these chains of islands may have a hot spot origin, the evidence is not as clearly developed as for the Hawaiian Islands.

Fracture Zones

Major fracture zones extend into the Pacific Plate for thousands of kilometers between the Mendocino Fracture Zone (which intersects the San Andreas Fault in northern California) to the Eltanin Fracture Zone in the South Pacific. The alignment of these zones is shown in figure 4–21. The physiographic relief near the fracture zones may extend through a width of up to 200 km (124 mi). The ocean floor on either side of a fracture zone may have depths differ-ing by as much as 1500 m (4920 ft). For example, the ocean bottom on the southern side of the Mendocino Fracture Zone is more than 1000 m (3280 ft) deeper than that to the north. Therefore, this feature is some-times referred to as the Mendocino Escarpment. Much of this difference in elevation can be attributed to the fact the ocean floor on the south side of the escarpment is as much as 40 million years older than that to the north and has undergone much more thermal contrac-tion and deepening than the younger ocean floor to the north (*see* plates 12 and 13).

Oceanic Trenches

Around the margin of the Pacific Ocean we see ocean trenches as dominant physiographic features of the ocean bottom. It is in these regions that the older lithospheric material is being subducted as new litho-sphere is produced along the ridges and rises. The Pacific trenches range in depth from 8 km (5 mi) to just over 11 km (6.8 mi) in the Marianas Trench.

By comparing three marginal regions throughout the Pacific, we may develop a picture of conditions that can develop as plates carrying continental crustal blocks converge with plates of purely oceanic crustal material. As was previously mentioned, the oceanic plate will subduct beneath the plate containing the continent because of its higher average density. But the nature of the landforms resulting from the con-vergence may depend on the position of the continen-tal mass relative to the trench. They may change as the trench and the continent converge (fig. 4–26).

Tonga Trench   The convergence of the Indo-Austra-lian Plate, which is moving in a northerly direction, with the Pacific Plate, which contains no continents and is moving in a northwesterly direction, produces an ocean trench called the Tonga-Kermadec Trench System. A cross section of the northerly portion of this structure, the Tonga Trench, reveals that the litho-sphere of the Pacific Plate, containing only oceanic crust, subducts predictably beneath the lithosphere of the Indo-Australian Plate, which contains both con-tinental and oceanic crustal material.

As the oceanic crust descends into the mantle, it is heated to its melting point, which is several hundred degrees below that of the mantle material. Water and other volatiles contained within the melting litho-sphere are driven off, and the mixture of molten rock and gas seeks the surface because of its relatively low density. Only a small portion of the melted mass finds its way to the surface. This surfacing occurs through the tensional fractures that develop on the

continental side of the trench. Friction between the plate carrying the continent and the descending plate pulls down the edge of the plate containing the con-tinental crust. An andesitic volcanic chain, which eventually grew into the Tonga Islands, developed as a result of this activity. The island arc is separated from the mainland by a shallow sea.

Peru-Chile Trench   In the case of this trench on the opposite side of the Pacific Ocean, the Nazca Plate is descending beneath the South American Plate. There is a head-on collision between the two plates as they move in opposite directions, the Nazca Plate moving east and the South American Plate mov-ing west. During the geologic past, possibly 200 mil-lion years ago, a trench similar to the Tonga Trench may have existed off the western coast of South America. Sediments that were being eroded from the continents were being deposited between the conti-nental shoreline and the arc of islands to the east of the trench. Some sediment undoubtedly found its way into the trench itself. As the distance between the trench and the continental mass decreased owing to the westward movement of South America, the sedi-ments that had accumulated between the islands and the mainland were folded into a high-rising belt along the western margin of the continent—the Andes Mountains.

It is no longer possible for the island arc to develop seaward of the continent. By the time the oceanic crustal plate reaches a depth at which it melts, it has moved laterally to a position beneath the margin of the continent. Thus the volcanic activity of the Peru-Chile Trench is found not in the form of an island arc some distance from the mainland, but is an active part of the mountain-building process at the margin of the continent.

Western Margin of the North American Conti-nent   To the north we see that the encroachment of the North American continent has resulted not only in the disappearance of the island arc, but also in the disappearance of the trench itself. The East Pacific Rise begins to be overridden by the continent at the southern end of Baja California and reappears off northern California as the Gorda Ridge.

North America, for the most part, is being carried in a west-northwesterly direction by the North Amer-ican Plate that is being generated along the Mid-At-lantic Ridge, and the Pacific Plate is moving at a more rapid rate in a north-northwest direction. Because of this relative movement, the San Andreas Fault has developed at the margin of these two plates. Baja Cal-ifornia and the portion of southern California on the

FIGURE 4—26 Three Stages of Plate Convergence: Pacific Ocean.

A. TONGA TRENCH

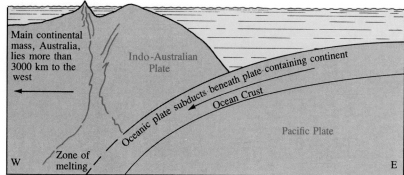

Stage 1—An oceanic plate collides with a plate with oceanic crust at its leading edge. The result is an ocean trench-island arc system. A volcanic island arc develops on the continental side of the trench but is separated from the continent by a marginal sea.

B. PERU-CHILE TRENCH

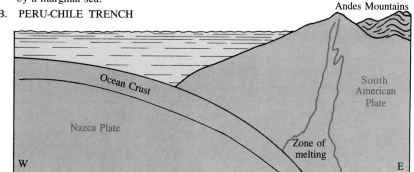

Stage 2—An oceanic plate collides with a plate with a continent at its leading edge. The result is an ocean trench-continental arc system. As the continent and the rise producing oceanic crust come closer together, the island arc becomes the margin of the continent and sediments of the marginal sea that once separated islands from the continent are compressed into a mountain range—the Andes.

C. NORTH AMERICAN CONTINENTAL MARGIN

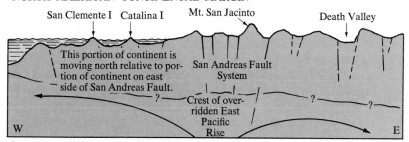

Stage 3—Tensional faults (dashed lines) may result from stress created by convection cells associated with an overridden segment of the East Pacific Rise, tending to carry portions of the continent on either side of the buried feature in opposite directions.

west side of this transform fault move in a northerly direction about 5 cm/yr (2 in/yr) relative to the rest of the North American mainland to the east. That portion of the continent on the west side of the fault has actually become part of the Pacific Plate and is now being carried toward the Aleutian Trench, which it may reach in some 60 million years.

The ripping away of Baja California from the mainland in association with the movement along the San

Andreas Fault and other transform faults is only one aspect of the continental features resulting from the overriding of the oceanic ridge. The basin-range province that runs through southern California, eastern Arizona, and most of Nevada is composed of mountain ranges that have been produced by the same kind and magnitude of faulting that produces the mountains of the oceanic ridge systems. These are quite unlike the Andes Mountains to the south, which

FIGURE 4–27 Mt. St. Helens.
A, tectonic features of the Mt. St. Helens region. The mountains shown are volcanoes within the Cascade Mountain Range.
B, the eruption of Mt. St. Helens resulted from the subduction of the Juan de Fuca Plate beneath the North American Plate.

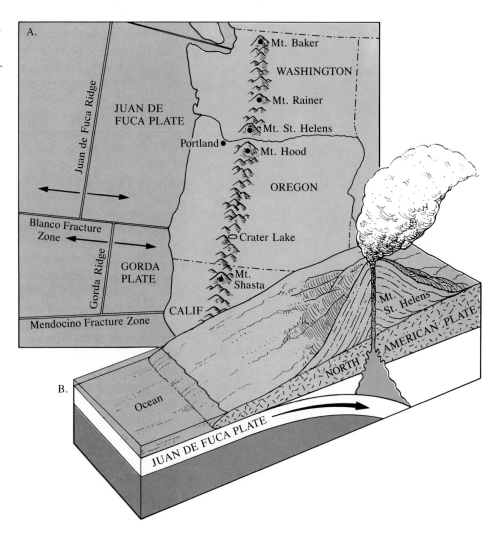

were produced by the folding and compression of sediments near the margins of the continents. The mountains of the basin-range province are the result of tension as the earth's crust is attempting to stretch in this region. Because the lithosphere is very rigid, instead of stretching it breaks into blocks separated by faults that run in a generally north-south direction. The blocks move vertically relative to one another and thus produce fault block mountains.

North of the Mendocino Fracture Zone, off the coast of northern California, the spreading center appears again as the Gorda Ridge. The ridge is offset to the west by the Blanco Fracture Zone and is called the Juan de Fuca Ridge off the coast of Oregon and Washington. The Gorda and Juan de Fuca plates subduct beneath the northwestern United States and produce the Cascade Mountains. The Cascades are topped by a series of volcanic peaks extending from Mt. Baker in northern Washington to Mt. Lassen in northern California. Many of the Cascade volcanoes have been active within the last 100 years, with the most recent being Mt. St. Helens. When Mt. St. Helens erupted in May of 1980, 62 people were killed (fig. 4–27).

INDIAN OCEAN

The Indian Ocean is characterized by extensive abyssal plains bounded by well-defined ridges that are associated with past and present events related to the origin of the ocean basin. Two significant features related to the movement of the Indian continental mass from Antarctica to its collision with the Asian continent are the presently inactive Chagos-Laccadive Ridge and the Ninety East Ridge. The presently active oceanic ridge system includes the Southwest Indian Ridge, which extends into the Indian Ocean from between Africa and Antarctica to merge with the Southeast Indian Ridge at 25°S latitude and 65°E longitude. The Southeast Indian Ridge enters the Indian Ocean from between Australia and Antarctica. This northwest-trending ridge continues between the Seychelles Islands and the Chagos Islands as the Carlsberg Ridge, which runs to the southern tip of Arabia. As shown in figures 4–28 and 4–29 and plate 12, the ridge continues through the Gulf of Aden, where it splits into a northwest-trending branch, responsible for the formation of the Red Sea, and a southerly branch, which produces the East Africa Rift Zone.

Microcontinents

Unusual features of the Indian Ocean are the microcontinents, or linear segments of continental crust that can be found widely separated from the larger continental masses. The Seychelles-Mascarene Plateau in the western Indian Ocean and Broken Ridge east of Australia are such features. The precise answer to how these continental fragments have reached their present positions has not been determined.

Movement of the Indian Plate

Although it is difficult to establish the early history of movement of crustal plates within the Indian Ocean, it is likely that the Indian subcontinent was at one time a portion of Gondwanaland and attached to those continents we now refer to as Africa and Antarctica. Although the exact timing of the breakaway cannot be determined, it is possible that by 140 million years ago India had separated itself from Gondwanaland and moved along a northerly course toward Laurasia.

About 75 million years ago India was moving north-northeast on a plate that was being separated from Africa and Antarctica along the ridge shown in figure 4–30A. The ridge was offset in many places by major transform faults.

The presently inactive ridge known as the Owen Fracture Zone marked the northwestern boundary of the Indian Plate, and the Ninety East Ridge was an active transform fault marking the southeastern boundary of the plate. The northern boundary of the plate was probably marked by a trench system that developed as the Indian Plate subducted beneath the Asian continent. The Chagos-Laccadive Ridge is a remnant resulting from a transform fault that lay between the two boundary faults previously mentioned and served as a partial boundary between the Indian and the African plates.

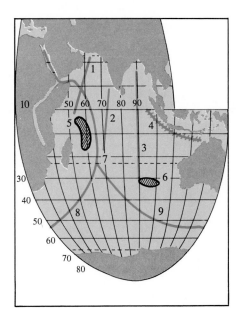

1. Owen Fracture Zone
2. Chagos-Laccadive Ridge
3. Ninety East Ridge
4. Java Trench
5. Mascarene Plateau
6. Broken Ridge
7. Carlsberg Ridge
8. Southwest Indian Ridge
9. Southeast Indian Ridge
10. East Africa Rift Zone

FIGURE 4–28 Indian Ocean Physiographic Features. See plate 12 for more detailed view of these features. Base map courtesy of National Ocean Survey.

Australia Begins to Move

The ridge that was providing the new crustal material between India and Antarctica extended to the east to separate Australia from Antarctica about 50 million years ago (fig. 4–30B). This separation had a very significant effect on the character of the present Indian Ocean floor. With the breakaway of Australia, the Ninety East Ridge Fracture Zone ceased to be active. India and Australia were now moving on the same plate. At about this same time, India had converged sufficiently with the Asian mainland to cause the folding of sediments that had been deposited in the sea separating the two continental masses. As the width of this sea was decreased by the encroachment of India, the sediments were folded and thrust into the Himalaya Mountains.

Until about 35 million years ago, the movement across the Indian Ocean floor was generally north. But after the collision of India and the Asian mainland the east-west trending ridge systems changed their orientation drastically to produce the complicated pattern that presently exists on the Indian Ocean floor

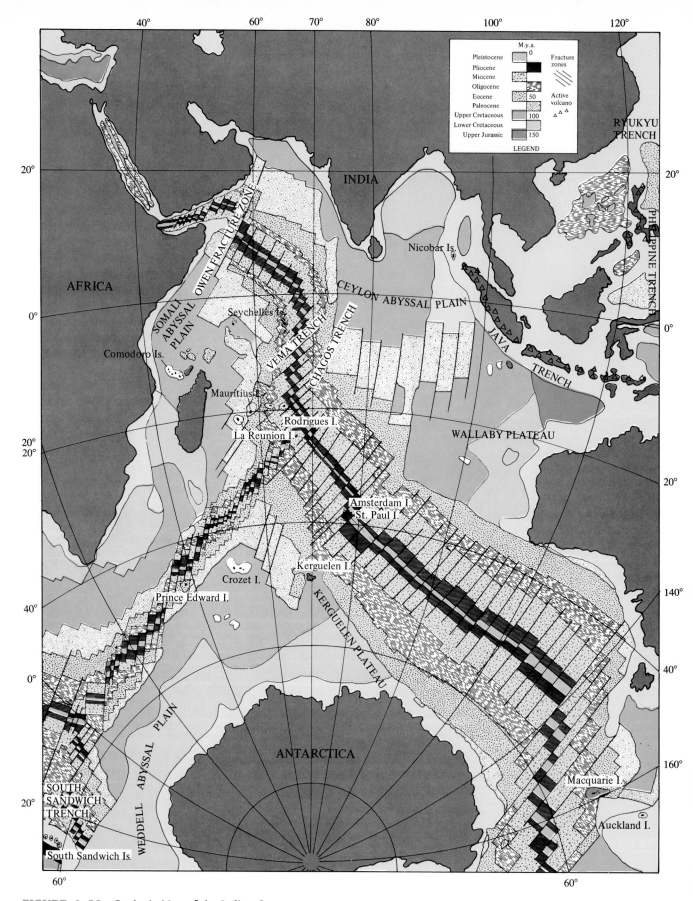

FIGURE 4–29 Geologic Map of the Indian Ocean.

Adapted from *Geologic Map of the Indian Ocean* by Bruce C. Heezen; Raymond P. Lynde, Jr.; and Daniel J. Fornari; original cartographic work by Suzanne B. MacDonald; Lamont-Doherty Geological Observatory of Columbia University. Courtesy of UNESCO.

FIGURE 4–30 Development of Indian Ocean.

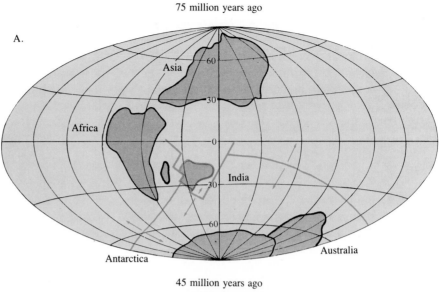

75 million years ago

A.

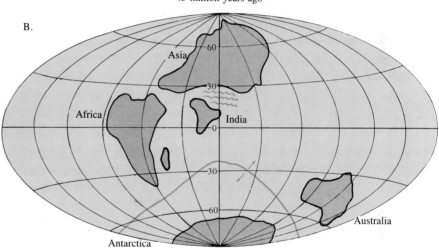

45 million years ago

B.

(fig. 4–29). The Carlsberg Ridge developed as a northeasterly trending system offset by many transform faults, and it has become one of the most active ridge systems in all of the world's oceans. The separation of Arabia from Africa and the formation of the East African Rift System commenced about 20 million years ago.

SUMMARY

Scientific investigation of evidence suggesting continents are moving relative to one another began with **Alfred Wegener** at the start of the twentieth century. The evidence gathered by the late 1960s indicated all the present continents were combined into one large continent, **Pangaea**, about 200 million years ago. Just before the formation and possibly also after the breakup of Pangaea, two large continents appear to have existed, **Laurasia** in the north and **Gondwana**-land to the south.

The discovery that the earth's **magnetic field** has changed **polarity** throughout the earth's history and that the record of these changes was permanently recorded in the rocks of the oceanic crust made it possible to develop a hypothesis of **sea-floor spreading** as the process by which the continents could be moved. According to the hypothesis, new rock was added to rigid **lithospheric plates** by intrusion and extrusion of molten material along submarine mountain ranges. These plates are composed primarily of dense mantle rocks with a thin crust of basaltic lava beneath the ocean basins. The continents "float" on these plates because they are composed of a less dense igneous rock, granite. As material is added to the plates near the axis of the oceanic ridges and rises, it moves away to make room for new material surfacing at the axis. Thus the crustal rocks beneath the oceans increase in age as one moves from the axis of the mountain ranges down their slopes into the deep-ocean basins. Continents located on a lithospheric plate would be carried passively along as the plate moved away from the oceanic ridge or rise. The sea-floor–spreading hypothesis was confirmed when the **Glomar Challenger** began the Deep Sea Drilling Project to sample the sediments and crust beneath the oceans in 1968. The movement of the lithosphere appears to be made possible by the existence of a plastic region of mantle material with temperatures near the melting point of the mantle, the **asthenosphere**, immediately beneath the lithosphere.

As new mass is added to the lithosphere at the oceanic rises and ridges, the leading edge of the same plates is **subducted** into the mantle at ocean trench systems. Additionally, there are plate boundaries called **transform faults** along which oceanic ridges and rises are offset and plates slide past one another. Although the earth is over 4.5 billion years old, no rocks have been found on the ocean floor that are as much as 200 million years old. It is not known exactly how the convectionlike movement is driven, but it is obvious that the process produces recycling of material between the lithosphere and the deeper mantle.

Since the breakup of Pangaea began about 180 million years ago, the entire Atlantic Ocean has been created and the ancient ocean **Panthalassa** has been divided by continental displacement into the present Indian and Pacific oceans.

The Atlantic Ocean began to form with the separation of North America from Africa and South America. By 135 million years ago, the **Mid-Atlantic Ridge** had extended northward, separating North America from southern Europe. The separation of Greenland from North America had also begun. South America had just started its separation from Africa. Sixty-five million years ago the separation of Africa and South America was well advanced, and the north end of the Mid-Atlantic Ridge shifted to the east of Greenland, which was still connected to Europe. During the last 65 million years, North America and South America were joined by the formation of the Isthmus of Panama, and North America and Europe completed their separation.

The lithospheric plates originating along the South Pacific Rise and East Pacific Rise contained no conti-

nents. Because of their greater average density, they are subducted beneath the encroaching plates containing continental material. Trenches with associated volcanic mountains and island arcs developed around most of the margin of the Pacific, which has become an ocean characterized by subduction of lithospheric plates.

Additional features common to the Pacific Ocean floor are **seamounts** and guyots **(tablemounts)** that formed as volcanic peaks and islands along the spreading centers and at hot spots. Another feature that appears to have its classic development in the Pacific Ocean is the "hot spot," a region of local upwelling of molten material. Very long **fracture zones** such as the **Mendocino Fracture Zone** display extreme horizontal and vertical displacements.

Three major types of **tectonic plate margins** are found around the Pacific Ocean. The **Tonga Trench** is typical of the trench-island arc subduction segments and the **Peru-Chile Trench** represents a continental arc along the margin of a continent. Along the southern California coast, the trench that once existed there has been overrun by the North American continent. The **San Andreas Fault** separates the part of the continent that has been pushed onto the **Pacific Plate** from that which remains on the **North American Plate**. North of the Mendocino Fracture Zone, the spreading center produces the **Gorda** and **Juan de Fuca plates** that subduct beneath North America to produce the volcanic **Cascade Mountain Range**.

The Indian Ocean floor shows the remnants of transform faults that separated the **Indian Plate** from the **Pacific Plate** and **African Plate** during the northward movement of India from about 140 to 45 million years ago. These are the **Owen Fracture Zone, Chagos-Laccadive Ridge**, and the **Ninety East Ridge**. The present active spreading centers are the **Southwest Indian**, the **Southeast Indian**, and the **Carlsberg Ridges**. Other features characteristic of the Indian Ocean are the **microcontinents**, such as the Seychelles-Mascarene Plateau and Broken Ridge. About 50 million years ago, when India began its collision with Asia, the present ridge system began to form. Australia began to rotate north as part of the **Indo-Australian Plate** while India served as the pivot point, pushing hard against Asia to create the Himalaya Mountains. The separation of Africa and Arabia began about 20 million years ago.

QUESTIONS AND EXERCISES

1. Why couldn't Alfred Wegener convince most scientists that the continents were indeed moving across the earth's surface?

2. How do the magnetic properties of the continents and ocean basins differ? Why?

3. What discovery made absolute dating of rock units possible? Why was this technique a significant improvement over relative dating?

4. What is the age of a rock that has a ratio of U^{235} to Pb^{207} of 1 to 3?

5. Describe the evidence found on the present continents that suggests Pangaea had broken into two continents, Laurasia and Gondwanaland, between 250 and 350 million years ago.

6. How does magnetic dip of magnetite particles found in igneous rocks tell us at what latitude they formed?

7. How does continental movement account for the two apparent wandering curves of the north magnetic pole as determined from Europe and North America?

8. What property of the magnetic record found in oceanic crustal rocks gave rise to the idea of sea-floor spreading?

9. Describe a possible mechanism responsible for moving the continents over the earth's surface. Include a discussion of the lithosphere and asthenosphere.

10. Describe the general relationships that exist among distance from the spreading centers of the oceanic ridges, heat flow, age of the oceanic crustal rock, and ocean depth.

11. Discuss the three types of plate boundaries, spreading centers of the oceanic ridges, trenches, and transform faults. Explain why earthquake activity is usually confined to depths less than 10 km (6 mi) along the spreading centers and transform faults while it may occur as deep as 650 km (405 mi) near the trenches. Construct a plan view and cross section showing each of the types of boundary and direction of plate movement.

12. Describe the relationship between fracture zones and transform faults, including a comparison of the earthquake activity along each.

13. Why is an oceanic ridge centered in the Atlantic Ocean while ridges and rises in the Indian and Pacific oceans are not?

14. Explain why seamounts and guyots increase in age and depth with increased distance from the oceanic ridges. Discuss the length of time they probably were active volcanoes.

15. Discuss the possible relationship of coral reef development to sea-floor spreading.

16. Describe how the "hot spot" theory explains the age, alignment, and degree of submergence characteristic of volcanoes found along the Hawaiian Island and Emperor Seamount chains.

17. Compare the horizontal displacements indicated on plate 13 for the Paleocene-Cretaceous contacts along the Mendocino Fracture Zone and the Murray Fracture Zone. Describe the displacements in terms of relative magnitude and left or right lateral motion.

18. What major difference exists in the features of the Tonga Trench as compared to those of the Peru-Chile Trench?

19. What difference in the stresses that produced them is responsible for the folded nature of the Andes Mountains as compared to the fault-block structure of mountain ranges of southern California and Nevada?

20. Describe the events that affected India-Asia, India-Australia, and Australia-Antarctica about 50 million years ago.

REFERENCES

Anderson, D. L. 1975. Accelerated plate tectonics. *Science* 187:1077–79.

Bullard, Sir Edward. 1969. The origin of the oceans. *Scientific American* 221:66–75.

Carrigan, C. R., and Gubbins, D. 1979. The source of the earth's magnetic field. *Scientific American* 240–2:118–33.

Dietz, R. S., and Holden, J. C. 1970. The breakup of Pangaea. *Scientific American* 223:30–41.

Haggerty, J. A.; Schlanger, S. O.; and Silva, I. P. 1982. Late Cretaceous and Eocene volcanism in the southern Line Islands and implications for hotspot theory. *Geology* 10:8, 433–36.

Heezen, B. C., and MacGregor, I. D. 1973. The evolution of the Pacific. *Scientific American* 229:102–12.

Heirtzler, J. R. 1968. Sea floor spreading. *Scientific American* 219:60–70.

Jordan, T. H. 1979. The deep structure of the continents. *Scientific American* 240–1:92–107.

McKenzie, D. P., and Sclater, J. G. 1973. Evolution of the Indian Ocean. *Scientific American* 228:62–72.

Menard, H. W. 1969. The deep ocean floor. *Scientific American* 221:53–63.

Schilling, J. G.; Thompson, G.; Kingsley, R.; and Humphris, S. 1985. Hotspot—migrating ridge interaction in the South Atlantic. *Nature* 313:5999; 187–191.

Shepard, F. P. 1977. *Geological oceanography.* New York: Crane, Russak.

Tarling, D., and Tarling, M. 1971. *Continental drift: A study of the earth's moving surface.* Garden City, N.Y.: Doubleday.

SUGGESTED READING

Sea Frontier

Burton, R. 1974. Instant islands. 19:3, 130–36.
A description of the 1973 eruption on the island of Heimaey south of Iceland and its relationship to plate tectonics processes.

Dietz, R. S. 1976. Iceland—where the mid-ocean ridge bares its back. 22:1, 9–15.
A description of the rift zone of Iceland and its relationship to the Mid-Atlantic Ridge and sea-floor spreading.

————. 1977. San Andreas: An oceanic fault that came ashore. 23:5, 258–66.
A discussion of the San Andreas Fault and its relationship to global plate tectonics.

————. 1971. Those shifty continents. 17:4, 204–12.
A very readable and informative presentation of the crustal features and the possible mechanism of plate tectonics.

Emiliani, C. 1972. A magnificent revolution. 18:6, 357–72.
A discussion of advances in studying earth science from the sea, including climate cycles, plate tectonics, and the economic potential of resources lying within marine sediments.

Mark, K. 1974. Earthquakes in Alaska. 20:5, 274–83.
A discussion of the origin, nature, and effects of earthquakes in Alaska.

————. 1972. Ocean fossils on land. 18:2, 95–106.
A discussion of the significance of marine fossils found in rocks now many miles inland is centered on the work of an early American paleontologist, James Hall.

Rona, P. 1984. Perpetual seafloor metal factory. 30:3, 132–141.
A discussion of how metallic mineral deposits form in association with hydrothermal vents located on oceanic ridges and rises.

Scientific American

Bonatti, E., and Crane, K. 1984. Oceanic fracture zones. 250:5, 40–51
The role of oceanic fracture zones in the plate tectonics process is related to the spherical shape of the earth.

Courtillot, V., and Vink, G. E. 1983. How continents break up. 249:1, 43–49.
The process of continental rifting and deformation is considered.

Dietz, R. S., and Holden, J. C. 1970. The breakup of Pangaea. 229:4, 30–41.
A description of the breakup of the hypothetical supercontinent and the movement of continents to their present positions. The possible mechanism of continental drift is explored.

Edmond, J., and Von Damm, K. 1983. Hot springs on the ocean floor. 248:4, 78–93.
The roles of spreading-center hydrothermal springs in depositing metallic ores and sustaining life in the absence of light are discussed.

Francis, P., and Self, S. 1983. The eruption of Krakatau. 249:5, 172–187.
An anlysis of the 1883 event based on the volcanic deposits and the timing of air and sea waves created by the eruption.

Gass, I. G. 1982. Ophiolites. 247:2, 122–31.
The relationship between ophiolite emplacements and global plate tectonic processes is discussed.

Hallam, A. 1975. Alfred Wegener and continental drift. 232:2, 88–97.
This article presents a history of Wegener's development of his theory of continental drift presented in 1912 and the reasons for its general rejection until the 1960s.

Heirtzer, J. R., and Bryan, W. B. 1975. The floor of the Mid-Atlantic Rift. 233:2, 78–91.
The data gathered by Project FAMOUS (French-American Mid-Ocean Undersea Study) are summarized. This investigation of the Mid-Atlantic Rift south of the Azores involved the use of the submersibles *Alvin, Archimede,* and *Cyana.*

Jones, D. L.; Cox, A.; Coney, P.; and Beck, M. 1982. The growth of western North America. 247:5, 70–84.

Menard, H. W. 1969. The deep-ocean floor. 221:3, 126–45.
A summary of the dynamic effects of sea-floor spreading is presented with a description of related sea-floor features.

Molnar, P. 1986. The structure of mountain ranges. 255:1, 70–79. Folded mountain ranges and their relationship to tectonic plates are discussed.

Pollack, H. N., and Chapman, D. S. 1977. The flow of heat from the earth's interior. 237:2, 60–76.
Based on data determined from thousands of heat-flow measurements on continents and the ocean floor, this article describes the relationship between lithospheric thickness and the heat flow through it.

Sclater, J. G., and Tapscott, C. 1979. The history of the Atlantic. 240:6, 156–75.
The history of the formation of the Atlantic Ocean basin is described on the basis of data gathered from measurements of heat flow and magnetism as well as rock samples recovered by the Deep Sea Drilling Project.

Tokosöz, M. N. 1975. The subduction of the lithosphere. 233:5, 88–101.
A discussion of how the subduction of the lithosphere is related to ocean trenches, island arcs, volcanism, and earthquakes.

Over half of the rocks exposed at the surface of the continents above the shoreline were formed by the lithification of sediment laid down in past ocean environments. This relationship between the continents and oceans had puzzled geologists until global plate tectonics theory explained the mechanism for incorporating ocean sediments into the continents: folded mountain ranges formed when two plates collided and compressed the sediments against the margin of the existing continent. Geologists once believed they could examine the sediments in the deep-ocean basins, which they assumed had existed since the earth's oceans first formed, and find an undisturbed record of the earth's history. Knowledge of the plate tectonics process has removed that dream, but much of that ancient record is still available in ancient marine sediments exposed in continental mountain ranges.

Although it seems unlikely that marine sediments more than 180 million years old will be found in the present ocean basins, the enthusiasm of marine geologists for examining them has not dimmed. There is still much to learn about this shorter period of history. Clues to past climates, movements of the ocean floor, changes in ocean circulation patterns, and nutrient supply for marine plant populations are embedded in the sedimentary deposits throughout the ocean basins.

SEDIMENT TEXTURE

Sediment texture, which is determined primarily by grain size, is indicative of the energy condition under which a deposit is laid down. The Wentworth scale presented in table 5–1 classifies particles in categories ranging from boulders to colloids. Between these extremes are cobbles, pebbles, granules, sand, silt, and clay-sized particles. Deposits that are laid down in areas where wave action is strong (areas of high energy) may be composed primarily of the larger particles, cobbles, and boulders. The deposition of clay-

CHAPTER FIVE
MARINE SEDIMENTS

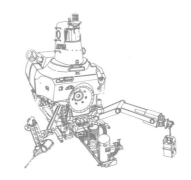

Silica skeletons of radiolarians, single-celled marine animals, are shown here magnified 280 times life size. Geologists study the evolution of these microfossils as a basis for assigning ages to ocean sediments. Picture was taken using camera attached to Cambridge S4 scanning electron microscope at Scripps's Analytical Facility. Photo courtesy of Scripps Institution of Oceanography, University of California, San Diego.

TABLE 5–1   Wentworth Scale of Grain Size for Sediments.

	Particle	Minimum Size (mm)
	Boulder	256
	Cobble	64
	Pebble	4
	Granule	2
SAND	Very coarse sand	1
	Coarse sand	1/2
	Medium sand	1/4
	Fine sand	1/8
	Very fine sand	1/16
SILT	Coarse silt	1/32
	Medium silt	1/64
	Fine silt	1/128
	Very fine silt	1/256
CLAY	Coarse clay	1/640
	Medium clay	1/1024
	Fine clay	1/2360
	Very fine clay	1/4096
	Colloid	1/4096

Wentworth (1922), after Udden (1898)

sized particles occurs in areas where the energy level is low and the current speed is minimal.

Sediments composed of particles primarily in the same size classification are well sorted. A beach sand is usually a well-sorted deposit. Glacial deposits sometimes found on the continental shelf are an example of a poorly sorted deposit because they contain particles ranging in size from the colloid to boulders that have been carried by glaciers and dropped out as the glaciers melted.

As particles are carried from the source to their point of deposition, they increase in *maturity*. This change results from their association with moving water, which has the capacity to carry particles of a

certain size away from a deposit and leave behind other larger particles. With continued reworking by wave and current action, clay content of the sediment is reduced. As the duration of transport increases, particles of sand size or larger become more rounded through chemical weathering and abrasion. Increasing sediment maturity is indicated by (1) decreasing clay content, (2) increasing degree of size sorting, and (3) increasing rounding of the grains within the deposit.

The poorly sorted glacial till, which contains relatively large quantities of clay-sized and larger particles that have not been well rounded, is an example of an immature sedimentary deposit. The beach sand we used as an example of a well-sorted sediment contains no clay-sized material and is usually composed of well-rounded particles that have undergone considerable transportation. The beach sand is an example of a mature sedimentary deposit. Figure 5–1 illustrates the nature of sediments of varying degrees of maturity based on the criteria.

SEDIMENT TRANSPORT

The sediment carried to the margins of the ocean by the rivers and glaciers flowing from the continents either settles out to fill estuaries, becomes a part of a delta, is spread across the continental shelf by currents, or is carried beyond the shelf to the deep-ocean basin. The greatest volume of sediment transport in the ocean is achieved by the longshore currents caused by breaking waves at the margins of the continent (*see* chapter 9). Lower-energy currents distribute the finer components of the sediment along the margin of the continental shelf and even into the deep-ocean basin.

Figure 5–2 shows the relationship between average horizontal current velocity and the erosion, transportation, and deposition of particles ranging in size from 0.001 to 100 mm (0.0000394 to 3.94 in). As might be expected, the curve that separates transpor-

FIGURE 5–1   Sediment Maturity.

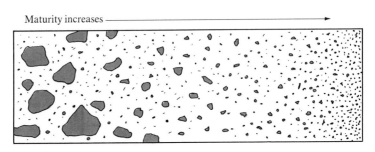

Maturity increases ⟶

Degree of sorting increases ⟶
Clay content decreases ⟶
Rounding of particles increases ⟶

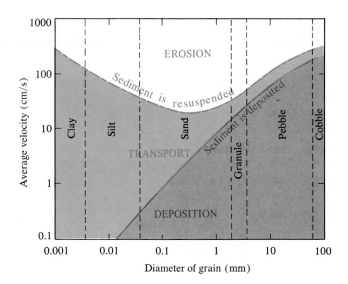

FIGURE 5–2 Hjulstrom's Curve.

Horizontal current velocity vs. erosion—deposition of sediment by particle size.

tation of particles from their deposition shows that larger particles will settle out at higher velocity than smaller ones.

Observing the curve that separates the process of erosion from transportation, we can see that it takes higher-velocity currents to *erode* (pick up and carry away) larger particles from the sand-sized particles up. The surprising fact that tiny clay-sized particles require a higher-velocity current than that needed to erode the larger sand-sized particles results from the fact that clay-sized particles are flat rather than round like typical sand-sized particles. Thus they have a greater surface area per unit of mass in contact with one another when they are laid down in a deposit. This causes a great cohesive force in clay sediments. Overcoming this cohesive attraction and picking up these particles once they are deposited requires a high-velocity current.

Particles less than 20 μm (0.008 in) in diameter can be carried far out over the open ocean by prevailing winds. Such particles either settle out as the velocity of the wind decreases or find their way into the ocean during precipitation. They commonly serve as nuclei around which raindrops and snowflakes form. Some particles that reach high altitudes can be carried very rapidly by the jet stream that exists in the midlatitudes.

CLASSIFICATION OF MARINE SEDIMENT

The fundamental classification of marine sediment particles is based on the origin of the particle. There

are four basic sources for marine sediment—rock (lithogenous), organic material (biogenous), compounds dissolved in water (hydrogenous), and outer space (cosmogenous).

Lithogenous means *derived from rocks.* Most of the rock mass that supplies lithogenous sediment is that which makes up the continents. The volcanic islands found in the open ocean are also important sources of lithogenous sediment. All parts of the earth's crust originally formed from the solidification of molten material into igneous rocks (*ignis* is Latin for *fire*). Igneous rocks solidify at temperatures and pressures well above those of atmospheric conditions and in an environment in which free oxygen and water are scarce. They are unstable under the conditions that prevail at the earth's surface and begin to undergo chemical and physical breakdown known as *weathering.* As they break up, the particles are carried to the ocean, where they are deposited. By far the greatest quantity of lithogenous material is found around the margins of the continents, but there are no parts of the ocean basins where traces of this sediment cannot be found.

Biogenous means *derived from organisms.* The remains of organisms, such as bones and teeth of animals and the shells of minute plants and animals, are deposited on the ocean bottom. The most common chemical compounds found in biogenous sediment are calcium carbonate ($CaCO_3$) and silica (SiO_2). The most common particles found in calcareous sediments are the shells of foraminifers, coccolithophores, and pteropods. Diatom and radiolarian shells are the most important contributors of the siliceous particles. These organisms will be discussed in the following chapters on marine biology.

Hydrogenous means *derived from water.* These deposits are formed by a chemical reaction that occurs within seawater. Manganese deposits, phosphorite, and glauconite are minerals that form by chemical precipitation from water. The rate of accumulation of this type of sediment is the slowest of all except the following classification.

Cosmogenous means *derived from beyond the earth's atmosphere.* Typical particles of cosmic origin found in marine sediment are magnetic spherules rich in iron that range in size from 10 to 640 μm (0.004 to 0.25 in).

Lithogenous Sediment

Common Rock-forming Minerals We will first direct our attention to the composition of sediments derived from rock, either from the continental masses or from volcanic activity in the open ocean. Since the sources of lithogenous sediment are the rocks of the earth's crust, we will be interested in identifying the most abundant compounds in these rocks. We call the elements and compounds that make up the rocks *minerals.* Although over 2000 minerals exist, we need to recognize fewer than two dozen of these to be able to identify those that make up most of the earth's crust. As we discuss these minerals, we will refer to their chemical composition. Table 5–2 lists the eight elements that are the most abundant constituents of the common rock-forming minerals and therefore the most abundant elements in the earth's crust. These are the only ones we will need to consider as we discuss the chemical composition of the common rock-forming minerals. In the common rock-forming minerals, ions of metals combine with a negatively charged ion composed of one atom of silicon and four atoms of oxygen, the *silicate tetrahedron* (fig. 5–3A).

Those minerals composed of silicate tetrahedra and ions of iron and magnesium are called *ferromagnesian* minerals. They are dark or black and have higher densities than other rock-forming minerals. The most important ferromagnesian minerals are olivine, augite, hornblende, and biotite. Because of different crystalline structures of these minerals, they have different physical appearances (fig. 5–3B). Olivine usually occurs in greenish granular masses, while augite and hornblende are usually black and cleave, or break cleanly, into blocky forms. Biotite has excellent cleavage in one direction along a plane where sheets of molecules are poorly bonded together. As a result, the sheets of biotite, black mica, can be separated along very smooth, shiny surfaces. The density of these iron- and magnesium-containing minerals ranges from 2.8 to 3.4 g/c^3.

Three other common rock-forming minerals that do not contain iron and magnesium and are therefore known as *nonferromagnesian* minerals, are varieties of feldspar, muscovite, and quartz (fig. 5–3C). All the

TABLE 5–2   Most Abundant Elements in Earth's Crust.

Element		% by Weight
Iron	(Fe)	35
Oxygen	(O)	30
Silicon	(Si)	15
Magnesium	(Mg)	13
Nickel	(Ni)	2.4
Sulfur	(S)	1.7
Calcium	(Ca)	1.1
Aluminum	(Al)	1.1
Others		0.7

A. Silicate tetrahedron

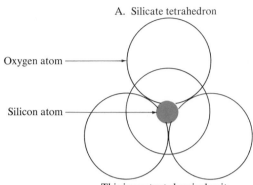

This important chemical unit is the basic component of the common rock-forming minerals. It contains one atom of silicon surrounded by four atoms of oxygen.

B.

C.

FIGURE 5–3   Silicate Tetrahedron, Common Rock-Forming Minerals.

A, silicate tetrahedron. *B,* ferromagnesian minerals: A) hornblende, B) augite, C) olivine, D) biotite. *C,* nonferromagnesian minerals: A) muscovite, B) feldspar, C) quartz. *B* and *C,* photos from Gary Cawthon.

varieties of feldspar are aluminum silicates that contain varying amounts of potassium, sodium, or calcium. Muscovite is similar to biotite in composition, except that iron and magnesium are excluded, and quartz is composed exclusively of the silicate tetrahedra without metallic ions. The density of these light-colored nonferromagnesian minerals ranges from 2.6 to 3.1 g/c³.

Classification of Igneous Rocks *Igneous rocks,* those that solidify from a molten mass, are classified on the basis of texture (mainly the size of the crystals of which the rock is composed) and mineral composition. The texture of igneous rocks is determined by how rapidly they cool. Rocks extruded at the earth's surface during volcanic activity cool rather rapidly and have a fine-grained or even glassy texture if the cooling occurs so rapidly that crystals do not have a chance to form. They are *extrusive rocks.* A typical example of fine-grained igneous rock is *basalt,* which is characteristic of the oceanic crust and is the most common igneous rock in volcanic islands rising above the deep ocean basins.

Coarse-grained igneous rocks form as a result of slow cooling at great depth. They are intruded masses that do not reach the surface of the earth's crust before they solidify and therefore cool more slowly because of the layer of insulating rock between them and the earth's atmosphere. The most common such rock is *granite,* which is the rock type characteristic of the continents. These are *intrusive rocks.*

Figure 5–4 shows the most common varieties of igneous rocks, both of intrusive and extrusive origin. Granite and rhyolite, its volcanic equivalent, are light-colored rocks on one end of the classification scheme that are composed primarily of nonferromagnesian minerals. This composition accounts for the fact that these rocks are light in color and of relatively low density. Such rocks are sometimes referred to as *sialic* because they characteristically contain high percentages of the elements silicon and aluminum.

Basalt and its coarse-grained equivalent, gabbro, are near the opposite end of the scale and are characterized by dark color and higher density. Because they have a relatively large ferromagnesian mineral

FIGURE 5–4 Igneous Rock Classification.

A, igneous rocks. This table shows the most common types of igneous rocks. The coarse-grained ones cool slowly deep beneath the earth's surface, and granite, the typical intrusive rock of the continents, is the most common variety. The fine-grained rocks, which are associated with volcanic activity, cool rapidly at or near the earth's surface. The most common variety is basalt, which is characteristic of the oceanic crust. The rocks on the left of the table are light colored and contain mostly nonferromagnesian minerals. Moving to the right, ferromagnesian minerals replace nonferromagnesian minerals, and the rocks become darker in color. *B,* examples of igneous rocks: A) granite, B) andesite, C) basalt. Photo from Gary Cawthon.

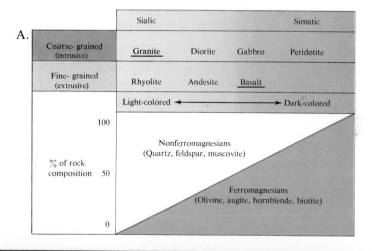

content, the rocks at this end of the scale are referred to as *simatic*. This term is derived from the elements silicon and magnesium. The rock peridotite is composed almost entirely of ferromagnesian minerals; it is relatively rare at the earth's surface but is representative of the composition of the mantle.

The common rock-forming minerals that make up igneous rocks form at temperatures and pressures well above those of atmospheric conditions at the earth's surface. They are therefore chemically unstable and begin to *weather* (break down) into more stable compounds when exposed to the lower temperatures and pressures of atmospheric conditions. Physical and chemical processes of weathering free the individual grains of minerals from the parent rock mass, after which they are transported and deposited by whatever means of transportation might be available. Most of this material ends up in major river systems that carry it to the ocean basins.

The various rock-forming minerals have different degrees of stability under atmospheric conditions, quartz being the most stable and undergoing the slowest alteration. Where granitic masses are available as a source, quartz will commonly provide the majority of the coarse sand-sized particles in beach deposits. Feldspars and ferromagnesian minerals alter to clays and oxides of iron.

The weathering of basalt and other ferromagnesian rocks will produce primarily clays and iron oxides. In addition to the particles that result from the decomposition of igneous rocks, other substances that will be dissolved in the ocean water are also produced. These dissolved substances include ions of calcium, carbonate, iron, magnesium, silicon, potassium, sodium, and oxygen.

Biogenous Sediment

The chemical composition of organically produced particles in marine sediment will usually be either calcium carbonate ($CaCO_3$) or silica (SiO_2). These are the compounds most used by organisms to secrete protective external coverings. Contributing most of the silica component of biogenous particles, which rarely exceed 100 μm in diameter, are microscopic plants *(diatoms)* and animals *(radiolarians)*. *Foraminifera,* close relatives of radiolarians, are responsible for most of the calcium carbonate component, but other significant calcareous contributors are the larger pteropods and microscopic algae, *coccolithophores* (fig. 5–5).

Coral reef material is also in the biogenous classification. Composed of the calcareous ($CaCO_3$) skeletal structure secreted by corals and algae, coral reefs are massive sedimentary rock structures.

Hydrogenous Sediment

Manganese Nodules Potentially of economic importance are the manganese nodules (polymetallic nodules) found on the deep-ocean floor. Since the voyage of the *Challenger,* we know them to be relatively abundant in all of the major ocean basins. They consist mainly of MnO_2 and Fe_2O_3, with the average manganese dioxide content ranging around 30% and the average iron oxide content around 20% by weight. Also occurring in economically significant concentrations in the nodules are copper, cobalt, and nickel. Although these concentrations are usually less than 1% they can exceed 2% by weight (fig. 5–6).

Since iron and manganese both occur in hydrogenous deposits in concentrations greater than those they reach in igneous rocks, there exists a problem of explaining the concentration of these elements in the marine deposits. This problem has not yet been successfully solved, but it has been concluded that there are three possible sources for manganese and iron. It is thought that they are derived from (1) the weathering of volcanic material produced by volcanic activity on the ocean floor, (2) concentration in hydrothermal spring-water found at the axes of oceanic ridges and rises (*see* page 77), and (3) runoff that carries the iron and manganese minerals as soluble compounds from the continents. There seems to be little variation in the chemistry of deep-ocean water throughout the world ocean. Studies on the relationship between manganese modules and organisms suggest that the presence of microorganisms, primarily bacteria, may be a determining factor in the formation of nodules. The Manganese Nodule Project (MANOP), which began in 1977, uses sediment traps to catch particles falling to the ocean floor and a Bottom Lander, shown in figure 5–6C, to carry out seabed experiments designed to explain how nodules form and why they have the observed range of composition.

Phosphorite Phosphorus in the form of P_2O_5 is found abundantly as a precipitate in nodules and in the form of a thin crust on the continental shelf and banks at depths above 1000 m (3280 ft). Concentrations of phosphates in such deposits commonly reach 30% by weight. The deposits seem to be associated with areas of upwelling water rich in nutrients. A study of phosphate nodules on the Peruvian continental margin, an area of intense upwelling, found

A.

B.

FIGURE 5-5 Microscopic Skeletons from the Deep Sea. *A,* magnified 160 times by scanning electron microscopy, the skeletons above are foraminifers and radiolarians found in sediment cores taken at Site 55 in the Caroline Ridge area of the west Pacific Ocean by the Deep Sea Drilling Project. They were recovered in water 2843 m (9315 ft) deep, with further penetration of 130 m (426 ft) into the ocean bottom. Rounded globose forms are forminifers, whose skeletons are calcareous. Forms with reticulate skeletons are radiolarians and are siliceous. In life, these tiny organisms lived near the ocean surface, but upon death their skeletons fell to the sea bed to become entombed in the sediments. By carefully studying them, scientists can determine how long ago they lived and can learn much about the history of the ocean. Courtesy of the Deep Sea Drilling Project, Scripps Institution of Oceanography, University of California, San Diego. *B,* coccoliths magnified about 10,000 times. Photo courtesy of Deep Sea Drilling Project, Scripps Institute of Oceanography. *C,* foraminifer, *Orbulina universa.* Photo by James M. King.

C.

nodules forming preferentially where the oxygen minimum zone of the water column intersects the ocean floor at depths from 800 to 1000 m (2624 to 3280 ft). The nodules are growing slowly (1 to 10 mm/1000 yrs, or 0.0394 to 0.394 in/1000 yrs). Instead of growing in all directions, as do manganese nodules, the phosphorite nodules grow down into the sediment. Their tops are older than their bottoms. The source of phosphorus and the other elements found in the nodules is the water trapped in the pores between sedimentary particles.

Glauconite This generally greenish hydrous silicate has a complex and variable composition which includes ions of potassium, magnesium, and iron. Glauconite may form by submarine weathering of biotite, although the exact mode of formation is not known. Glauconite, which is found to depths of 2500 m (8200 ft) but more commonly on topographically high areas near the coastline, is considered to form only in the marine environment. Forming as casts in organic remains and as encrustations, glauconite is more commonly found in grains of silt or sand size

A.

B.

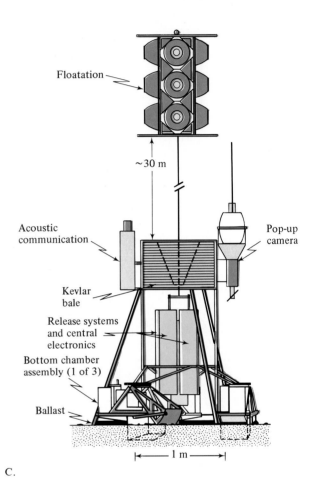

C.

FIGURE 5–6 Hydrogenous Sediment.

A, photograph of the surface of a manganese nodule, magnified 156 times by a scanning electron microscope. The small tube and dome structures were built by organisms for shelter, according to Dr. Jimmy Greenslate at Scripps Institution of Oceanography. He has studied 71 nodules from the Pacific Ocean and found biologically derived structures both buried inside them and covering large portions of their surfaces. Nodules may owe their existence to organisms that participate in their formation. *B,* manganese nodules on the Pacific Ocean sea floor are formed as chemical precipitates. They contain quantities of manganese, iron, copper, nickel, and cobalt. Although the manganese is inferior to manganese ores mined on land, it is still a valuable commodity, making harvesting of the lumps a probability in the future. Courtesy of Scripps Institution of Oceanography, University of California, San Diego. *C,* MANOP Bottom Lander designed for geochemical experiments on the sea floor to learn how nodules form and why they have the observed range of composition. Courtesy R. F. Weiss, O. H. Kirsten, and R. Ackermann, Scripps Institution of Oceanography, University of California, San Diego.

HOW FAST ARE THE CONTINENTS GROWING?

A study conducted by the U.S. Geological Survey has created a possible budget for the growth and denudation of continents. By calculating the yearly flux of various components of marine sediment, seamounts, and island arc volcanoes, a somewhat tentative rate of growth for continents since the breakup of Pangaea has been determined.

It was determined that

1. Ocean crust is recycled every 110,000,000 years and its average age is 55,000,000 years.
2. Lithogenous particles accumulate in the deep ocean at a rate of 1.27 km^3/yr. Biogenous particles accumulate at the rate of 0.38 km^3/yr. This totals to a *rate of continental denudation of 1.65 km^3/yr.*
3. Adding to this the 0.05 km^3/yr of volcanic sediment that accumulates gives a total yearly rate of sediment accumulation of *1.7 km^3/yr.*
4. Seamounts add *0.2 km^3/yr* of mass to the ocean crust, and volcanic island arcs add *1.1 km^3/yr.*
5. *Total material available to accrete to continents during subduction is 1.7 + 0.2 + 1.1, or 3 km^3/yr.*
6. Subtracting the rate of continental denudation (1.65 km^3/yr) from this total leaves *1.35 km^3/yr of possible continental growth.*
7. Whatever amount of deep-sea sediment or seamounts is subducted will have to be subtracted from this growth rate. The numbers for these variables are not yet known.

Reference: Howell, D. G., and Murray, R. W. 1986. A budget for continental growth and denudation. *Science* 233:4762, 446–449.

in mud and sand deposits. Deposits in which the glauconite is sufficiently abundant are frequently referred to as *green sands* or *green muds* owing to the coloration that results from the presence of glauconite.

Carbonates   The two most important carbonates in marine sediment are aragonite ($CaCO_3$) and calcite ($CaCO_3$). Aragonite and calcite have the same chemical composition but different crystalline structures. Aragonite is the least stable form and over a period of time changes to calcite.

During the geologic past, deposits of limestone ($CaCO_3$) in the marine environment appear to have been widespread. Some contain fossil evidence indicative of a biogenous origin, while others contain no fossils or characteristics that give any indication they formed organically. There appear to be very few places where nonbiogenous carbonates are presently forming. The Bahamas Banks is the largest region where significant carbonate precipitation is presently occurring, although deposits are also forming on the Great Barrier Reef, in the Persian Gulf, and in other local areas in the low latitudes. Carbonate precipitation may occur directly from the water through physiochemical precipitation when water conditions are right. The most common forms of precipitated carbon are aragonite crystals less than 2 μm (0.0008 in) in length and *oolites,* onionlike spheres that develop around a nucleus and reach diameters of less than 2 mm (0.08 in).

Since ocean water is essentially saturated with calcium (Ca^{++}), the only factor preventing precipitation of calcium carbonate (the most common carbonate) is the absence of the carbonate ions (CO_3^{--}). Carbon dioxide (CO_2) is important in carbonate chemistry. To see how it influences the precipitation of carbonates, we need to look at the various forms in which CO_2 is stored in ocean water. The reactions presented in chapter 2 and below tell us that when carbon dioxide combines with water in the ocean, carbonic acid (H_2CO_3) is formed. Carbonic acid may dissociate and produce a hydrogen ion (H^+) and a bicarbonate ion (HCO_3^-). On further dissociation of the bicarbonate ion, an additional hydrogen ion may be formed along with carbonate ion (CO_3^{--}).

$$H_2O + CO_2 \rightleftharpoons H_2CO_3 \rightleftharpoons H^+ + HCO_3^- \rightleftharpoons H^+ + H^+ + CO_3^-$$

| Water | Carbon dioxide | Carbonic acid | Hydrogen ion | Bicarbonate ion | Hydrogen ions | Carbonate ion |

Low pH ⟵―――――――――⟶ High pH

7.5   9.0

The presence of carbon dioxide and the carbonate ion in ocean water is mutually exclusive. Carbon dioxide is present in free form only at pH values be-

low that of the average ocean pH, 8.1. The concentration of carbonate ions increases with increased pH, so it is necessary to raise the pH of ocean water to make CO_3^{--} available for combination with Ca^{++}.

Removal of carbon dioxide raises the pH, so we might look to areas where photosynthesis is occurring at rates that remove significant amounts of carbon dioxide from the water for areas of carbonate precipitation. Heating water will also reduce its ability to dissolve carbon dioxide and decrease carbon dioxide concentration. Both factors are at work in areas where precipitation is known to occur. For instance, in the Bahama Banks, water that is moving north from the deep flow of the Florida Straits moves up over the shallow bank, where it is heated and exposed to high plant productivity. Both changes, along with lower pressure in near-surface waters, tend to reduce carbon dioxide content of the water and enhance precipitation of carbonates.

Other Hydrogenous Minerals   Other minerals found on the ocean bottom are secondary zeolites and clays commonly associated with deposits of the deep-ocean basin. The most likely process by which these minerals—zeolite (phillipsite) and montmorillonite (clay)—form is through the chemical alteration of volcanic material on the ocean bottom.

Zeolites in general are hydrated silicates of aluminum, calcium, sodium, and potassium. They have the peculiar property of gradually losing their water of hydration and possess the ability to substitute various metals for the alkaline metal originally present in their structures. Phillipsite is a white to reddish mineral that helps give the abyssal red clay deposits, particularly in the Pacific Ocean, their color. Although it may contain no sodium, phillipsite always contains considerable potassium. The total zeolite concentration in deep ocean sediment frequently exceeds 50%.

Cosmogenous Sediment

Particles that originate in space occur in marine sediment as either nickel-iron spherules that are magnetic or as silicate chondrules composed of olivine and augite and some of their ferromagnesian relatives called *pyroxenes* (fig. 5–7). Such particles had been thought to be fragments remaining from meteorites. However, recent studies indicate they form in the asteroid belt as sparks produced when asteroids collide. Iron spherules are generally about 30 μm (0.012 in) in diameter, while silicate chondrules group in diameters of 30 μm or 125 μm (0.012 or 0.05 in). The overall size range of these cosmic particles falls between 10 and 640 μm (0.004 and 0.25 in).

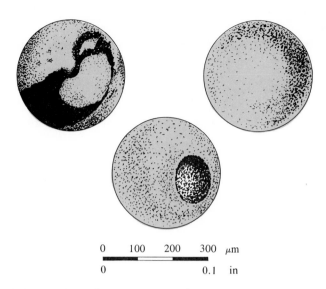

0   100   200   300   μm

0   0.1   in

FIGURE 5–7   Cosmogenous Sediment.

DISTRIBUTION OF SEDIMENTS

As a result of marine sediment deposition, the relative amounts of lithogenous, biogenous, hydrogenous, and cosmogenous particles found in marine sedimentary deposits will vary considerably. The two largest categories into which marine sedimentary deposits can be divided are neritic deposits found near the continents and oceanic deposits characteristic of the deep-ocean basins. These two broad categories, the subgroups into which they are divided, and the processes determining their formation are discussed in the following paragraphs.

Neritic sediment refers to material with a wide range of particle size that is composed primarily of lithogenous particles derived from the continents. For the most part, such material enters the ocean at or near the shore. It accumulates rapidly on the continental margin. Neritic sediment also contains biogenous, hydrogenous, and cosmogenous particles. These constitute only a minor percentage of the total sediment mass because of the rapid deposition rate of lithogenous particles near the continents. Neritic sediment is distributed along the continental shelf by waves, currents, and mass wasting. It is transported down the continental slope to be deposited on the continental rise by sediment slumps and turbidity currents. Here the sediment is further distributed by deep boundary currents.

Oceanic sediment is the very fine-grained material that accumulates at a slower rate on the deep ocean basin floor. It is possible for a greater variety of oceanic sediment types to accumulate because the rate of deposition of lithogenous particles is greatly reduced from that observed on the continental mar-

gins. Although the lithogenous component dominates the sediment found in most of the deeper basins of the ocean floor, biogenous and hydrogenous components are abundant in many areas of the ocean bottom.

SEDIMENTS OF THE CONTINENTAL MARGIN (NERITIC SEDIMENTS)

Owing to the rise in sea level that occurred with the melting of glaciers at the end of the last ice advance, many rivers of the world today deposit their sediment in estuaries rather than carry it onto the continental shelf as they did at various times during the geological past (fig. 5–8). The continental margin of the Atlantic coast of North America is an area where most river sedimentation is presently taking place in estuaries. In such regions the sediments that cover the continental shelf, *relict* sediments, may range from 3000 to 7000 years in age. Relict sediments currently cover about 70% of the world's continental shelf.

Of the 20 billion tn of sediment carried to the earth's continental margins by rivers each year, almost 80% is provided by runoff from Asia. Figure 5–8 shows that of the ten most important rivers contributing to this total, seven drain the Asian continent. The Hwang Ho, Yangtze, and Mekong empty their sediment into the marginal seas of the western Pacific Ocean. The Ganges, Brahmaputra, and Irrawaddy flow into the Bay of Bengal between India and Burma at the north end of the Indian Ocean. Also flowing into the Indian Ocean via the Arabian Sea is the Indus River.

The sediment of the Colorado River remains near the northern end of the Gulf of California. The Mississippi River is building a major delta on the continental shelf of the Gulf of Mexico, while the Amazon flows into the equatorial Atlantic.

Sediment that reaches the shore is sorted by current and wave action that generally leaves the sand to be transported along the shore to form beach deposits. The finer particles, silt- and clay-sized, are carried farther out on the shelf and deposited in areas where wave energy has decreased sufficiently. Where sediments are being deposited on the continental shelf, the rate at which they accumulate may be greater than 10 cm/1000 yr (4 in/1000 yr). Relative to the deposition of other types of sediments in the marine environment, this is a very high rate of accumulation that allows burial of sediment before it has time to be acted upon chemically by the ocean water. Therefore, these deposits usually contain materials that have not been significantly altered since they en-

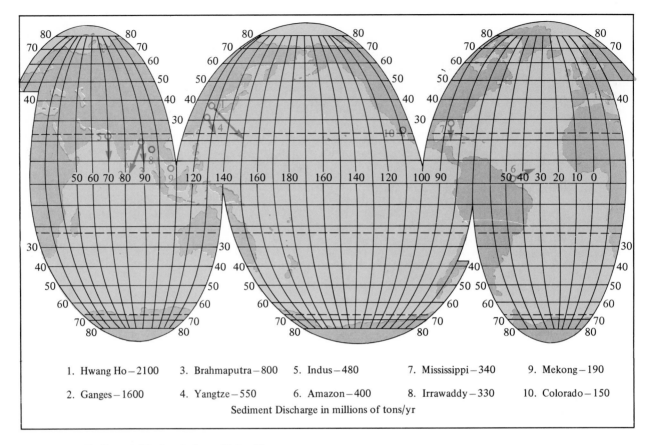

| 1. Hwang Ho – 2100 | 3. Brahmaputra – 800 | 5. Indus – 480 | 7. Mississippi – 340 | 9. Mekong – 190 |
| 2. Ganges – 1600 | 4. Yangtze – 550 | 6. Amazon – 400 | 8. Irrawaddy – 330 | 10. Colorado – 150 |

Sediment Discharge in millions of tons/yr

FIGURE 5–8   Sediment Discharge from Major Rivers.
Base map courtesy of National Ocean Survey.

tered the marine environment. They are commonly gray to greenish in color.

Turbidites

Although it seems unlikely that wave action and ocean current systems could carry coarse material beyond the continental shelf into the deep-ocean basin, there is evidence that much neritic sediment has been deposited at the base of the continental slope forming the continental rise. These accumulations thin gradually toward the abyssal plains. Such deposits are called *turbidites* and are thought to have been deposited by turbidity currents that periodically move down the continental slopes through the submarine canyons, carrying loads of neritic material that spread out across the continental rise (fig. 5–9).

Turbidites are characterized by graded bedding. This means that each deposit that is laid down has coarser material at its base and decreasing particle size toward the top of the layer. This pattern results from particles of different sizes settling out of a moving water mass as the velocity decreases. The coarser material settles out first; the finer material stays in suspension longer.

During the millions of years that sediments accumulated at the margins of continents, the continental shelf, slope, and rise developed into a sedimentary wedge more than 10 km (6.2 mi) thick that contains over 75% of the sediment to be found on the ocean floor. Less than 25% of marine sediment will be found in the deep-ocean basins that cover in excess of 80% of the ocean floor.

Although the greatest percentage of sediment making up the continental margin was provided by streams, other types of sediment are characteristic of certain regions of the continental shelf and will be discussed in the following paragraphs.

Glacial Deposits

Poorly sorted deposits containing particles of all sizes, from boulders to clay, may be found in the high-latitude portions of the continental shelf. These glacial deposits were laid down by melting glaciers that covered the continental shelf area during the Pleistocene Epoch, when glaciers were more widespread than they are today and sea level was lower. Glacial deposits are still forming around the continent of Antarctica

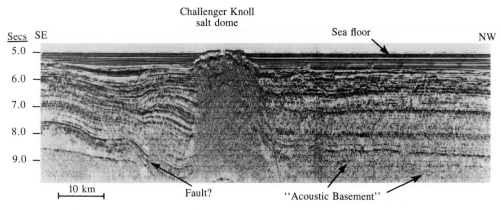

FIGURE 5–9 Seismic Profile Showing Turbidites.
This multifold seismic section was made in the area of the Sigsbee Knolls in the Gulf of Mexico. It shows a spectacular salt dome (Challenger Knoll) and Pleistocene turbidites. The latter are closely spaced, uniform horizontal reflectors immediately below the sea floor. The Challenger Knoll was cored during the first voyage of the *Glomar Challenger* in the Deep Sea Drilling Project. Courtesy J. Lamar Worzel, Geophysical Laboratory of Marine Science Institute, University of Texas.

and around the island of Greenland by *ice rafting,* a process by which rock particles trapped in glaciers are carried out to sea by icebergs that break away from glaciers as they push into the coastal ocean. As the icebergs melt, the lithogenous particles are released and settle to the ocean floor.

Stromatolites

Stromatolites are interesting calcareous sedimentary structures that are often dome-shaped (fig. 5–10). To produce these deposits, algal mat mucous secretions trap fine oolitic calcium carbonate sediment. Other forms of algae produce long filaments that are responsible for binding the carbonate particles to-

FIGURE 5–10 Stromatolites in the Bahamas.
Subtidal oolitic stromatolites on the crest of an oolitic tidal bar on Eleuthera Bank, Bahamas. Most are 0.5 to 1.0 m (1.6 to 3.3 ft) high. They are colonized by forms of algae that produce a mucous scum which traps sediment. Photo courtesy of Jeff Dravis. Permission granted by Exxon Production Research Company and American Association for the Advancement of Science.

gether. When the filamentous algae die, the filaments become calcified and cement the oolitic particles into the stromatolite structure.

Commonly observed in rocks derived from sediments deposited in ancient oceans, stromatolites have long been known to be forming in Hamelin Pool of Shark Bay, Western Australia. In this location, the hypersalinity of the water is believed to protect the algae from grazing gastropods. It had been thought that stromatolites could form only in such an environment. However, they have also been observed forming on Eleuthera Bank, Bahamas (fig. 5–11), where normal marine salinity exists.

It has been suggested that migrating bars of oolitic sand that periodically cover and uncover the stromatolites, as well as a veneer of unconsolidated oolitic sand shifting in response to tidal currents, sufficiently limit gastropod grazing to allow the stromatolites to form in the Eleuthera Bank occurrence.

Blake Plateau

Just beyond the narrow continental shelf off the coast of Florida and extending northward to the Cape Hatteras area is a special feature of the continental margin that is receiving very little sediment at present, except for the hydrogenous accumulations of phosphorite and manganese nodules. The floor of the Blake Plateau lies at the base of the Florida-Hatteras Slope at a depth of about 700 m (2296 ft). The plateau is actually a terrace that begins to form south of Cape Hatteras and increases to a width in excess of 300 km (186 mi) off the Florida coast. This gently sloping surface, which slopes at a rate similar to that of the average continental shelf, reaches a depth of

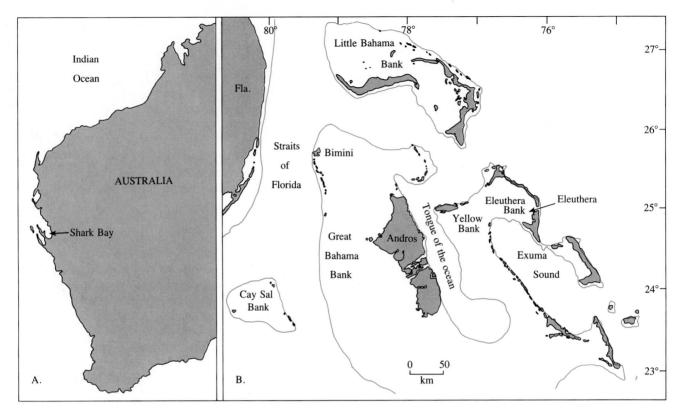

FIGURE 5–11   Locations of Two Regions of Present-Day Stromatolite Development.
A, map of western Australia showing Shark Bay. Arrow denotes approximate location of Hamelin
Pool stromatolites. Restricted circulation with the open ocean produces hypersaline conditions that
prevent algal grazing by gastropods. *B,* map of the Bahamas showing the locations of Eleuthra
Bank and southern end of Tongue of the Ocean locations of stromatolite development (arrows).
The physical movement of oolitic sediment helps to protect the algal mats from mollusk grazing in
this environment where water salinity is normal. Map courtesy of Jeff Dravis. Permission granted by
Exxon Production Research Company and American Association for the Advancement of Science.

about 1000 m (3280 ft). The ocean floor then descends at slopes of up to 30° along the Blake Escarpment into the deep ocean basin. The presence of this
terrace is usually ascribed to the fact that the Gulf
Stream flowing northward against the Florida-Hatteras Slope has sufficient velocity to prevent the deposition of sediment, therefore preventing the growth
of a broad continental shelf. Extensive accumulations
of manganese nodules are found on the outer Blake
Plateau, and phosphorite nodules occur in a belt
along its inner margin (fig. 5–12).

SEDIMENTS OF THE DEEP OCEAN
BASIN (OCEANIC SEDIMENTS)

As was previously discussed, most of the neritic sediment is deposited on the continental margin as part
of the continental shelf, continental slope, or continental rise. Since the continental rise is a transitional
feature between the continental margin and the
deep-ocean basin, we may well conclude that neritic

material is deposited in significant quantities in the
deep-ocean basin as part of the deep-sea fan deposits
making up the continental rise. This is true, but when
we consider the total surface area of the deep-ocean
basin, we find that more commonly no neritic material will be found in the sedimentary section. Covering most of the deep-ocean floor are two types of
oceanic sediments, abyssal clay and oozes.

Abyssal Clay

Accumulating at a rate of approximately 1 mm/1000
yr (0.04 in/1000 yr), *abyssal clays* cover most of the
deeper ocean floor. They are commonly red-brown or
buff in color due to phillipsite and oxidized iron.
Abyssal clay consists predominantly of clay-sized particles derived from the continents. The particles are
carried by winds or ocean currents to the open-ocean
regions and, along with some cosmic and volcanic
dust, settle slowly to the ocean floor.

Significant components of the abyssal clay are the
authigenic minerals montmorillonite and phillipsite,

F rom the peak of Mt. Everest to the bottom of
the Challenger Deep in the Marianas Trench,
the maximum relief of the earth's surface is
over 20 km (12.5 mi).

Because the continental crust is less dense than the
ocean crust, the continents rise above the ocean surface with an average elevation above sea level of 840
m, or just over .5 mi. The average depth of the oceans
is 3865 m (2.5 mi).

Seventy-one percent of the earth's surface is covered by ocean water.

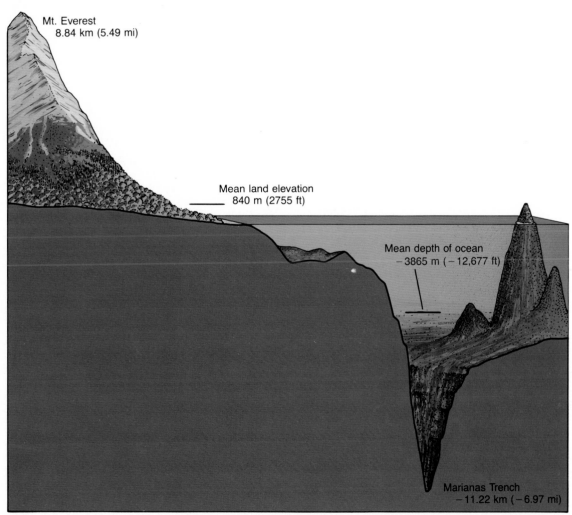

Mt. Everest
8.84 km (5.49 mi)

Mean land elevation
840 m (2755 ft)

Mean depth of ocean
−3865 m (−12,677 ft)

Marianas Trench
−11.22 km (−6.97 mi)

PLATE 9

PLATE 10

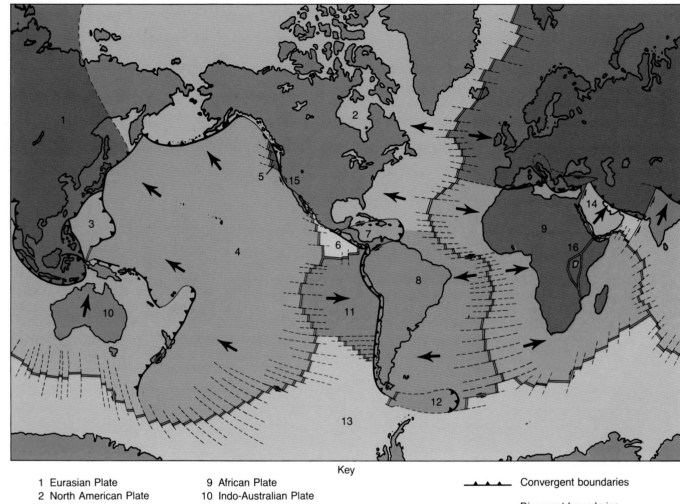

Key

1 Eurasian Plate	9 African Plate
2 North American Plate	10 Indo-Australian Plate
3 Philippine Plate	11 Nazca Plate
4 Pacific Plate	12 Scotia Plate
5 Juan de Fuca Plate	13 Antarctic Plate
6 Cocos Plate	14 Arabian Plate
7 Caribbean Plate	15 San Andreas Fault
8 South American Plate	16 East African Rift Valley

▲▲▲ Convergent boundaries

═══ Divergent boundaries

═══ Transform fault boundaries

← Direction of plate movement

This relief is directly related to the interactions of a dozen or so lithospheric plates that form at the axes of submarine mountain ranges (divergent boundaries) and dive back into the earth's interior beneath deep-ocean trenches (convergent boundaries). The divergent plate boundaries are offset by transform faults, along which the plates slide past one another in opposite directions.

The plates move at rates ranging from 1.5 cm (0.8 in) per year for the American plates to more than 10 cm (5.4 in) per year for the Pacific Plate.

Sixty-five thousand kilometers (40,000 mi) of oceanic mountain ranges called oceanic ridges and rises mark the divergent plate boundaries and transform fault plate boundaries and meander through the oceans. In places, they invade the land, producing such features as the San Andreas Fault in California and the East African Rift valley.

Convergent plate boundaries are associated with ocean trench–volcanic island arc systems like the Aleutian Islands. The Peru-Chile Trench and adjacent volcanic belt of the western Andes Mountains are elements of an ocean trench–continental volcanic arc system that also marks a convergent plate boundary. Folded mountain ranges such as the Himalayas and Alps are also manifestations of plate convergence.

New lithosphere is added to the plates by volcanic processes operating at the divergent plate boundaries. This volcanism is fed from chambers of molten rock or magma located only a few kilometers beneath the ocean floor. Another source of volcanism, represented by a stationary plume of magma that may rise from depths of over 2000 km (1200 mi), is called a hot spot.

Hot spots may be responsible for such continental volcanic features as Yellowstone National Park's hot springs, and there is much evidence of marine hot spots.

The island of Iceland is a large volcanic feature located on a divergent plate boundary or spreading center called the Mid-Atlantic Ridge. While most of the Mid-Atlantic Ridge is more than 2 km (1.2 mi) beneath the ocean surface, the additional lava produced by the hot spot has helped Iceland rise to the surface as a volcanic island.

The Hawaiian Islands–Emperor Seamount chain in the North Pacific Ocean was produced by a hot spot that is far from a divergent plate boundary. This hot spot has been in existence for more than 70 million years. Its present location is beneath the southeastern shore of the Island of Hawaii—the only volcanically active island in the chain. As the Pacific Plate moved in a west-northwest direction over the hot spot, it burned through the lithosphere to produce a series of volcanic islands and seamounts, which are volcanoes that did not become large enough to become islands. The oldest seamount in the chain is located just south of the western end of the Aleutian Islands, and it is over 70 million years old. Moving south along the chain toward the Hawaiian Islands, the volcanic peaks become progressively younger. The island of Kauai is the oldest major Hawaiian island, and it began to form 5.5 million years ago. The island Hawaii began to form only .75 million years ago.

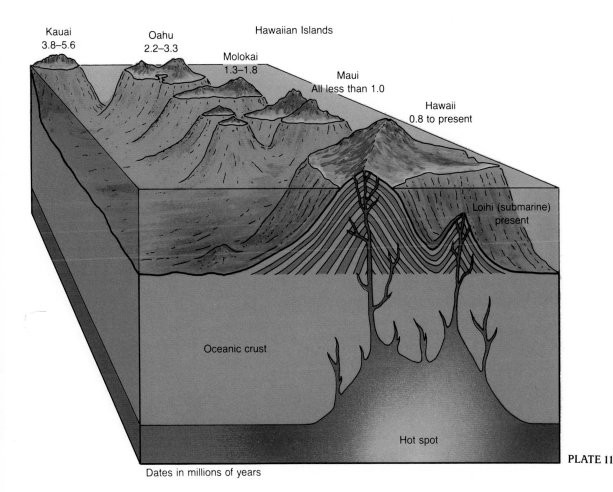

PLATE 11

Dates in millions of years

Plate boundaries are readily visible on a bathymetric map of the ocean floor. The youngest ocean floor is found at the axes of the long chains of mountains that run through the ocean basins. The ocean floor gradually deepens toward the trenches, where the oldest ocean floor is subducting back into the mantle.

Painted by Tanguy de Rémur.

THE FLOOR OF THE OCEANS

PLATE 12

Based on Bathymétric studies by
Bruce C. Heezen and Marie Tharp
of the Lamont Doherty Geological Observatory
Columbia University Palisades, New York, 1964

SUPPORTED BY THE UNITED STATES NAVY
OFFICE OF NAVAL RESEARCH

American Geographical Society, Broadway at 56th St. New York, N.Y. 10032.

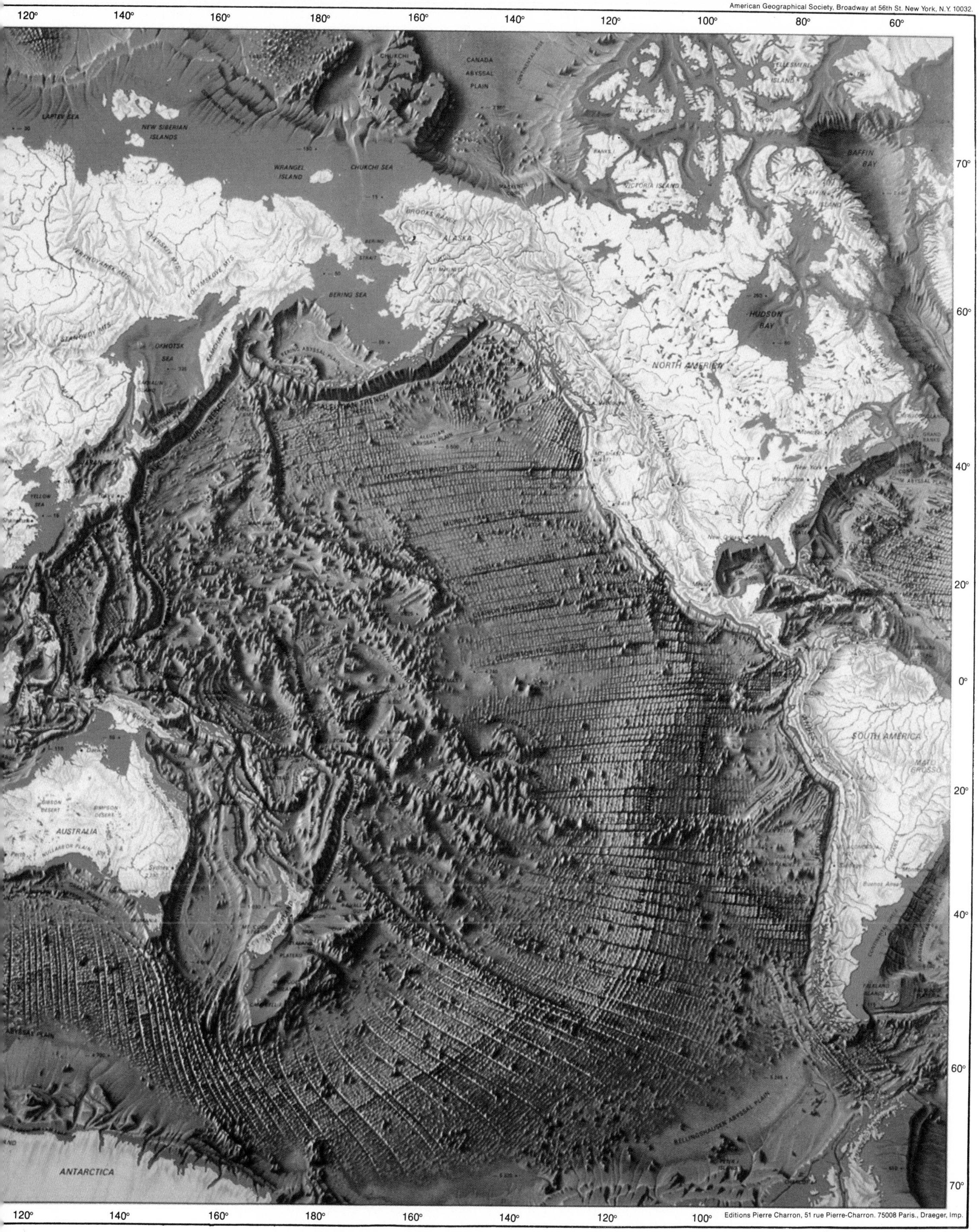

Editions Pierre Charron, 51 rue Pierre-Charron. 75008 Paris., Draeger, Imp.

Mercator Projection 1 46,000,000 at the Equator.
Depth and Elevations in Meters.

PLATE 13

COMMISSION DE LA CARTE GÉOLOGIQUE DU MONDE

COMMISSION FOR THE GEOLOGICAL MAP OF THE WORLD

Atlas géologique du monde
Geological World Atlas

Coordonnateurs géneraux/*General Coordinators*: G. CHOUBERT et A. FAUSE MURET

Océan Pacifique/Pacific Ocean   1/36 000 000

Coordinateurs/*Coordinators* Bruce C. HEEZEN and Daniel J. FORNARI
Travaux Cartographiques/*Cartographic work:* Ann KEBABIAN
Lamont—Doherty Geological Observatory–COLUMBIA UNIVERSITY

Légende/Legend

DEEP SEA DRILLING AND CORING
Carottage et forage en mer profonde

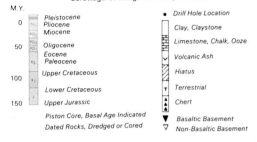

M.Y.

0	Pleistocene / Pliocene / Miocene
50	Oligocene / Eocene / Paleocene
100	Upper Cretaceous
	Lower Cretaceous
150	Upper Jurassic

Piston Core, Basal Age Indicated
Dated Rocks, Dredged or Cored

- Drill Hole Location
- Clay, Claystone
- Limestone, Chalk, Ooze
- Volcanic Ash
- Hiatus
- Terrestrial
- Chert
- ▼ Basaltic Basement
- ▽ Non-Basaltic Basement

THICKNESS OF UNMETAMORPHOSED SEDIMENTS ON THE OCEAN FLOOR
Épaisseur des sédiments non métamorphisés sur le fond de l'océan

Isopachs in Hundreds of Meters — 5

Presence of Opaque Layer

NEOTECTONICS
Néotectonique

Faults

 Active Volcano

Earthquakes

Shallow Focus

Intermediate Focus

Deep Focus

→ Direction of First Motion of Recent Earthquakes

BATHYMETRY
Bathymétrie

Contours in Thousands of Meters

— 4000 —

CONTINENTAL BASEMENT AGE
Age du socle continental

Age of Basement x 10^6

- • <.44
- • .44–.8
- • .8–1.3
- • 1.3–1.7
- ○ 1.7–2.35
- ○ 2.35–2.7
- ○ >2.7

THICKNESS OF UNMETAMORPHOSED SEDIMENTS ON THE CONTINENTAL BASEMENT
Épaisseur des sédiments non métamorphisés sur le socle continental

Isopachs in Hundreds of Meters — 10

Recorder = Recorder Seamount
Ocean I. = Ocean Island

/////// Magnetic Anomalies / Anomalies Magnétiques

CONTINENTAL SHELF AND SLOPE
Plateau et pente continentaux

Continent

Printed in Australia by Mercury Walch Pty Ltd. Hobart
©C.G.M.W.—Unesco

Publié par l'Unesco, Paris
Published by Unesco, Paris

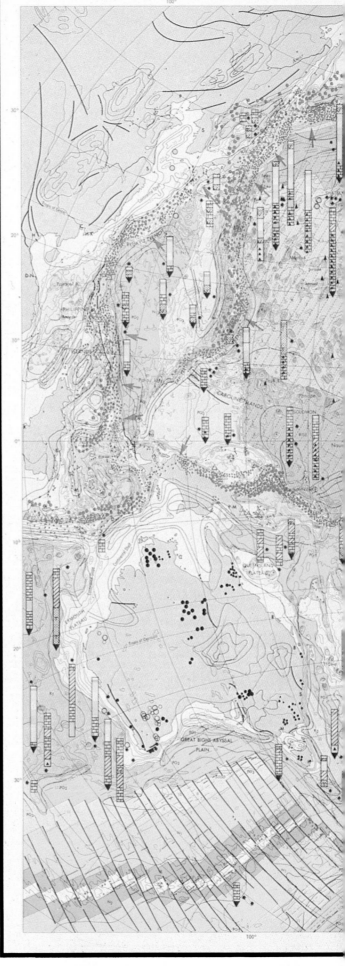

When the sediment is stripped from the ocean floor, as it has been in the geologic map of the Pacific Ocean, the age of the ocean floor can be seen to exhibit the trend just de-

scribed. The abrupt change in the age of ocean floor along lines running perpendicular to the oceanic mountain range in the eastern and southern Pacific is the result of trans-form faults offsetting the axis of the mountain range during the early stages of plate formation.

PLATE 14

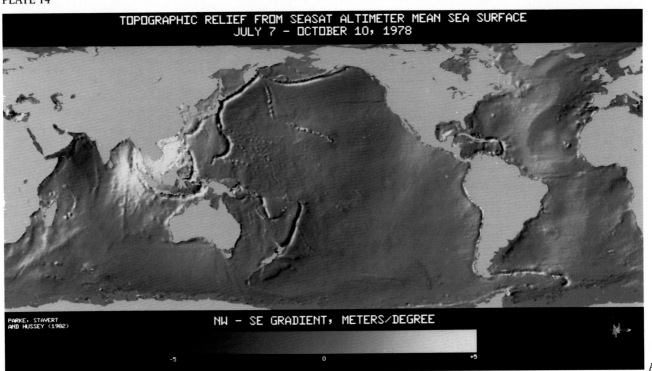

TOPOGRAPHIC RELIEF FROM SEASAT ALTIMETER MEAN SEA SURFACE
JULY 7 – OCTOBER 10, 1978

PARKE, STAVERT
AND HUSSEY (1982)

NW – SE GRADIENT, METERS/DEGREE

-5   0   +5

A

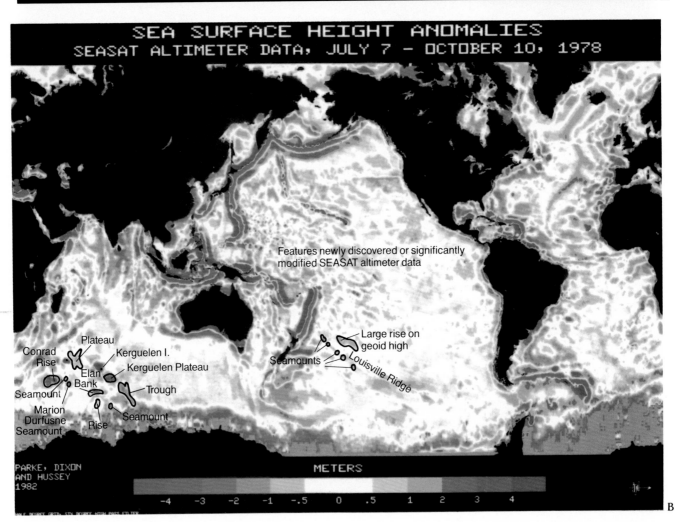

SEA SURFACE HEIGHT ANOMALIES
SEASAT ALTIMETER DATA, JULY 7 – OCTOBER 10, 1978

Features newly discovered or significantly
modified SEASAT altimeter data

Plateau
Conrad
Rise
Kerguelen I.
Elan
Bank
Kerguelen Plateau
Seamount
Trough
Marion
Durfusne
Seamount
Rise
Seamount

Large rise on
geoid high
Seamounts
Louisville Ridge

PARKE, DIXON
AND HUSSEY
1982

METERS

-4   -3   -2   -1   -.5   0   .5   1   2   3   4

B

Due to gravitational effects, the ocean surface rises over high volcanic peaks and drops over depressions such as trenches on the ocean floor. An altimeter on the SEASAT satellite that operated in 1978 was able to measure these changes in the ocean surface with sufficient precision to map major bathymetric features on the ocean floor. In the southern Indian Ocean and Pacific Ocean, this remote sensing of the ocean surface was able to identify some new bottom features and provide information that significantly altered the shape or location of previously identified features (circled).

which are thought to form from the interaction of ocean water with volcanic and other lithogenous particles.

Oozes

At somewhat shallower depths, biogenous material makes up a significant portion of the sediment that reaches the ocean bottom. It consists primarily of the minute shells of microscopic plants and animals that accumulate on the ocean floor. If sediment contains 30% or more skeletal material by weight, that deposit is called an *ooze.* Oozes may be classified as *calcareous* ($CaCO_3$) or *siliceous* (SiO_2), depending on the chemical composition of the dominant type of skeletal material making up the deposit. Oozes can form only where the deposition of material other than the remains of plants and animals occurs at a very low rate.

More specific names may be given to oozes to indicate the types of organic remains most abundant in the deposits. *Diatom oozes* are siliceous oozes that contain mostly the remains of diatoms. Other descriptive names commonly used on this same basis are *radiolarian ooze* (siliceous), *foraminiferan ooze* (calcareous), and *pteropod ooze* (calcareous). The predominant type of ooze is *Globigerina ooze,* which is especially widespread in the Atlantic and Pacific oceans. *Globigerina* is an abundant and widespread genus of foraminifer. Pteropods are snails with thin shells and a foot modified for swimming.

The rate of accumulation of biogenous material on the ocean floor depends upon three fundamental processes—productivity, destruction, and dilution. *Productivity* of the planktonic* forms that contribute most of the biogenous material in oceanic sediments is greater in the areas of upwelling along the margins of continents and associated with the diverging water masses in equatorial regions.

Destruction of skeletal remains occurs primarily through solution. The ocean is undersaturated with silica at all depths. So for siliceous shells to last until they are incorporated in the sediment, they must be thick, as thin fragments will be dissolved into the ocean water before they can settle out and be covered by the sediment. About 80% of the silica reach-

---

*In this instance, primarily floating, microscopic plants and animals.

FIGURE 5-12 Blake Plateau.
The relationship of the Blake Plateau to the Gulf Stream can be seen as the mighty current sweeps along the Florida-Hatteras Slope.

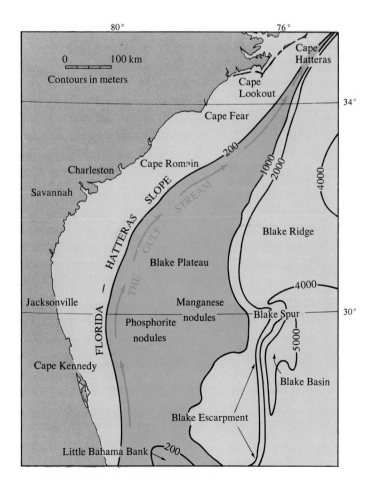

ing the ocean floor is thought to be dissolved back into the water before it is protected by being covered.

The solubility of calcium carbonate increases with increased CO_2 content of the water and increased pressure. Near the surface, where temperatures are high, there is little CO_2 in the water and the water does not dissolve much $CaCO_3$. At greater depths, the temperature decreases and pressure increases. This lower temperature and higher pressure increase the capacity of the water for dissolving carbon dioxide; this, in turn, increases the water's ability to dissolve calcium carbonate. Below 4500 m (14,760 ft) the dissolved carbon dioxide increases to the point that calcium carbonate dissolves quite readily. Carbonate oozes are generally rare below a depth of 5000 m (16,400 ft). The depth at which calcium carbonate is dissolved at the same rate it is arriving from above is the *calcium carbonate compensation depth.* It may be as deep as 6000 m (19,680 ft) in parts of the Atlantic Ocean, but above 3500 m (11,480 ft) in regions of low biological productivity in the Pacific Ocean. About 50% of the total calcium carbonate reaching the deep-ocean floor is dissolved into the ocean water before it is protected by being covered with sediment. Essentially all of this dissolving of deposited calcium carbonate occurs below a depth of 3500 m. Therefore, even though calcium carbonate particles may reach the deep-ocean floor, they may not survive to be incorporated into the sediment.

Dilution is illustrated by the fact that calcareous oozes are not found on the continental margins, where rapid rates of deposition of lithogenous sediment prevail. Even in areas of upwelling where productivity is high, oozes do not occur on the continental shelf, because the concentration of biologically derived silica or calcium carbonate remains below 30% by weight as a result of the high rate of deposition for lithogenous materials in these areas. The rates of deposition for biogenous sediment range from 1 to 15 mm/1000 yr (0.04 to 0.06 in/1000 yr), depending on the net effect of productivity and destruction. Figure 5–13 shows the relationships among calcium carbonate compensation depth, sea-floor spreading, and productivity in determining what type of sediment will accumulate on the ocean floor.

Manganese Nodules

Although what is known about the formation and distribution of manganese nodules is greatly exceeded by what is not known, we do know they are found only in deep-ocean environments where other sediments accumulate at rates of less than 7 mm/1000 yr (0.27 in/1000 yr). The nodules grow very slowly, adding layers from 1 to 200 mm (0.04 to 8 in) thick every million years.

There appears to be a correlation between the biological productivity of overlying waters and the me-

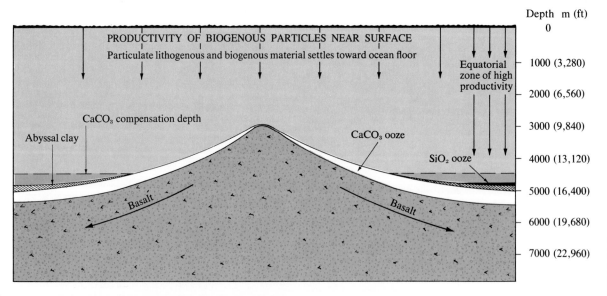

FIGURE 5–13 Sea-Floor Spreading and Sediment Accumulation.

New basaltic oceanic crust is forming at ridges that run across the ocean floor and rise to depths of about 3000 m (9840 ft). As the crust moves away from the ridges and sinks to greater depths, the type of sediment that accumulates on it changes. Until it sinks below the $CaCO_3$ compensation depth, the typical sediment that forms is $CaCO_3$ ooze. Below this depth $CaCO_3$ is dissolved, and abyssal clays dominate. If the crust passes beneath a region of high biological productivity, such as the Pacific equatorial region, oozes could again accumulate.

COBALT-RICH OCEAN CRUSTS

Cobalt is an important strategic metal that is not mined in the United States. It is required to produce dense, strong alloys with other metals for use in high-speed cutting tools, powerful permanent magnets, and jet engine parts.

Major enrichments of cobalt in the earth's crust are confined to central southern Africa and deep-ocean nodules and crusts. Considering the unstable political situation in Africa, the United States has been looking to the ocean floor as a more dependable source. However, the political and economic uncertainties associated with the United Nations Law of the Sea Convention have made mining interests back away from the mining of deep-sea nodules beyond the 200-mile-wide Exclusive Economic Zone (EEZ). Fortunately, in 1981 cobalt-rich manganese crusts were found on the upper slopes of islands and seamounts that lie within the EEZs of the United States and its allies. The cobalt concentrations in these crusts are on the average half again as rich as in the richest African ores and at least twice as rich as in deep-sea manganese nodules.

The greatest concentrations seem to be associated with depths between 1000 and 2500 m (3280 and 8200 ft) on the flanks of islands and seamounts in the central Pacific Ocean. Additional sampling needs to be done to assure that crust deposits are extensive enough to provide a dependable source of cobalt. At least one international consortium, International Hard Minerals Company, is moving into the study of the feasibility of crust mining. If the results of these investigations are positive, the first deep-sea mining operations may be for cobalt and other metals found in these crusts that lie within the EEZs of the United States or one of our allies.

tallic composition and growth rates of manganese nodules. Beneath unproductive areas near the center of the subtropical oceans, the nodules are rich in iron and cobalt extracted from the bottom waters, and they grow slowly (less than 5 mm/million yr, or 0.2 in/million yr). Beneath waters of moderate levels of biological productivity, such as those north and south of the equatorial Pacific Ocean, nodules have increased concentrations of copper, nickel, and manganese. It is believed that these metals are transported to the bottom in the remains of plants and animals that settle from above. They are transferred to the nodules through a process that is not yet fully understood. In this environment nodules grow at rates of from 5 to 10 mm/million yr (0.2 to 0.4 in/million yr). Beneath very productive waters, another process that is not fully understood produces nodules so rich in manganese that the economically important copper, nickel, and cobalt are diluted to very low levels of concentration. These nodules grow very rapidly. From 10 to 200 mm/million yr (0.4 to 8 in/million yr) may be added. Figure 5–14 shows the pattern of copper concentration in Pacific Ocean nodules. The nod-

ules with the greatest economic potential are those that form under conditions of moderate levels of biological productivity in the overlying waters. They are rich in copper and nickel.

An interesting finding resulting from the study of small nodules in the high-copper-nickel belt north of the equator at 140° W longitude and 11° N latitude was that the core of these nodules was composed of molds of biogenic material: coccoliths and radiolarians. All of the $CaCO_3$ and SiO_2 of which these forms were originally composed has been replaced by a form of manganese oxide. It must be emphasized that the fundamental reason for these composition and distribution patterns is unknown. Future research conducted as part of the Manganese Nodule Program (MANOP) may help identify the processes that affect the composition and growth rates of manganese nodules.

The Future of Deep-Sea Mining on pages 136–138 is an interview with Dr. G. Ross Heath, Dean of the College of Ocean and Fishery Sciences, University of Washington, concerning the factors that influence decisions regarding the mining of manganese nodules.

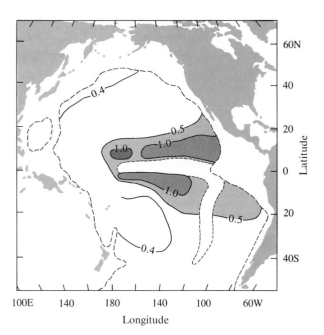

FIGURE 5–14 Copper Content of Nodules—Percent of Copper by Weight.
The areas of high copper content (greater than 1%) form under conditions where biological productivity in overlying waters reaches moderate levels. Nodules found here are also enriched in nickel and manganese. Blank areas are nodule-free because of the deposition of lithogenous particles near the continents and calcareous ooze along the equator and on the East Pacific Rise exceeds 7 millimeters (0.28 in) per thousand years. Courtesy of G. Ross Heath, University of Washington.

DISTRIBUTION OF OCEANIC SEDIMENTS

Oceanic sediments cover about 75% of the ocean bottom. Table 5–3 shows that calcareous oozes make up almost 48% of oceanic sediments, abyssal clay accounts for 38%, and siliceous oozes 14% of the total area. Calcareous oozes are the dominant oceanic sediment in the Indian and Atlantic oceans, while abyssal clay is the principal oceanic sediment in the Pacific Ocean. This difference is undoubtedly related to the fact that the Pacific Ocean is deeper, so that more of its bottom lies beneath the calcium carbonate compensation depth. Siliceous oozes cover the smallest percentage of the ocean bottom in all of the oceans because regions of high productivity of diatoms and radiolarians, the major components of these deposits, are restricted.

Fecal Pellets

A major problem that once puzzled marine geologists studying oceanic sediment is that sediments on the deep-ocean floor very closely reflect the particle composition of the surface water directly above. This was difficult to understand because it would typically take these very tiny particles from 10 to 50 years to sink from the ocean surface to abyssal depths. During this interval a horizontal ocean current of only 1 cm/s

TABLE 5–3 Distribution of Oceanic Sediment.
This graph shows that the percentage of an ocean-basin floor covered by calcareous ooze decreases with increasing mean depth of the basin. This shift probably occurs because the deeper an ocean basin is the greater the percentage of its floor that lies beneath the calcium carbonate compensation depth. The mean depths calculated for this table exclude the shallow adjacent seas, where little oceanic sediment accumulates. Notice that the dominant oceanic sediment found in the deepest ocean basin, the Pacific Ocean, is abyssal clay, while calcareous ooze is the most widely deposited abyssal sediment in the shallower Atlantic and Indian oceans.
After Sverdrup, Johnson, and Fleming, 1942.

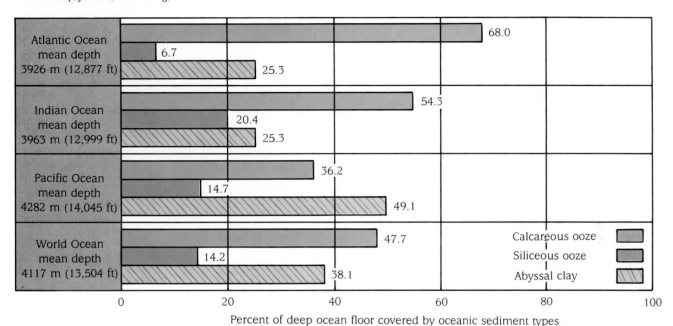

Percent of deep ocean floor covered by oceanic sediment types

(0.03 ft/s) (about 0.035 km/h, or 0.02 mi/h) would carry them from 3000 to 15,000 km (1860 to 9300 mi) laterally before they reached the deep-ocean floor. To further complicate the picture, the particle assemblage at middepth is not similar to that found in the surface water and sediment below.

Such findings could be understood only if there was some mechanism by which the small surface particles could be aggregated into larger particles to sink faster. Continued study revealed such a mechanism—*fecal pellets*. Most particles in the surface water, organic and inorganic, are eaten at least once. Recovery of samples during the GEOSECS study of the ocean's chemical makeup and circulation confirmed the presence of fecal pellets among the flux of material falling from the ocean's surface to the ocean floor. The GEOSECS samples revealed an interesting pattern of concentration of suspended particulate matter in the ocean. They show that the highest concentrations of particulate matter in the water column at high latitudes are found at the surface and near the bottom. The surface concentration results from high biological productivity of microscopic plants and animals, and the deep increase results from the re-

suspension of bottom sediment into the nepheloid layer.

Further study of these samples shows that 99% of particles that fall to the ocean floor do so as a part of fecal pellets produced by small animals and fish. These pellets (fig. 5–15), though still small (50 to 100 μm , or 0.02 to 0.04 in) in their smallest dimension), are large enough to allow the particulate matter to reach the abyssal ocean floor in 10 to 15 days. If this process is important in transporting particles from surface water to abyssal depths, it could readily explain the similarity between particle composition of surface waters and bottom sediments immediately beneath them. Figure 5–16 shows a sediment trap set on the deep-ocean floor to intercept the flow of particles from the surface during particle flux studies.

Another process that may contribute significantly to increasing the settling rates of particles is the aggregation that results from particles being trapped in mucus networks produced by organisms such as salps and coccolithophores.

There also appears to be a seasonal pattern of flux of material from the surface zone to the deep sea. The highest levels of flux have been observed to fol-

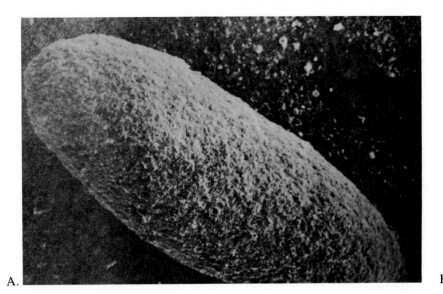

A.   B.

FIGURE 5–15   Fecal Pellets.

A, fecal pellet produced by zooplankton. The pellet is 200 μm (0.08 in) long. *B,* view of the surface of the fecal pellet showing the remains of small phytoplankton and detritus. Photos courtesy of Susumu Honjo, WHOI.

FIGURE 5–16   Sediment Traps.

Sediment traps that can be set on the ocean floor to catch particles falling from the surface. Photo courtesy of Susumu Honjo, WHOI.

low periods of high biological productivity in the surface waters of tropical to temperate oceans.

Calcareous Ooze

Atlantic Ocean   Since more than two-thirds of the deep Atlantic Ocean basin is covered with calcareous oozes, we will first discuss their distribution and general composition (fig. 5–17). The most widespread calcareous oozes in the Atlantic Ocean are composed of foraminifers, many of which are planktonic (floa-

ters). However, up to 10% of the carbonate in some Atlantic oozes is provided by benthic (bottom-dwelling) forms of foraminifers. Also contributing to the calcareous oozes are the *Coccolithophoridae,* algae with minute calcareous plates that form a protective covering. The aragonite shells of pteropods are also found in calcareous oozes to a restricted degree. Aragonite is more readily dissolved than the calcite form of calcium carbonate, so pteropod remains are found only in restricted shallow warm-water areas where the water is essentially saturated with calcium carbonate. Such conditions exist on the shallow plateau around the Azores in the North Atlantic, between Ascension Island and Tristan da Cunha Island on the Mid-Atlantic Ridge in the South Atlantic, and on the Rio Grande Rise between the South American mainland and the Mid-Atlantic Ridge.

Pacific Ocean   Since the North Pacific Ocean floor is for the most part below 4000 m (13,120 ft) and productivity is low in regions far from land, very little carbonate is found in the sediment there. Carbonate oozes are confined to an area south of 10°N latitude, with the exception of local accumulations in relatively shallow ridges in the North Pacific. Carbonate oozes in the east equatorial region contain up to 75% calcium carbonate. Even higher percentages, up to 90%, may be found in the region of the South Pacific Rise, which is actually less productive than the equatorial zone. In the South Pacific, this high concentration can be accounted for by a low rate of dilution, since the area is far from any sources of lithogenous material and the accumulation of siliceous remains is minimal.

Indian Ocean Calcareous oozes are the dominant oceanic sediment in the Indian Ocean, which is intermediate in depth between the Atlantic Ocean and the Pacific Ocean. About 54% of the total surface area of the Indian Ocean is covered by calcareous oozes, while clays cover a little over 25%, and siliceous oozes cover just over 20% of the total area of oceanic sediment.

Foraminiferan oozes are the dominant oceanic sediment on the bottom where the depth is less than 5000 m (16,400 ft). This area includes that portion of the Indian Ocean west of the 5000-m depth contour on the eastern flank of the oceanic ridge. Only in local deep basins do abyssal clays become the oceanic sediment type within this region. A thin belt of calcareous ooze occurs at the base of the continental rise along the western margin of Australia and the East Indian Archipelago.

Figure 5–18 shows the percent by weight of CaCO₃ accumulation in the ocean basins. It shows that high concentrations (in excess of 80%) are found high on the oceanic ridges, whereas little is found in deep-ocean basins like the North Pacific, where the bottom lies beneath the calcium carbonate compensation depth.

Abyssal Clay

Atlantic Ocean An interesting pattern of abyssal clay distribution is found in the Atlantic Ocean. The greatest portion of the ocean bottom covered by abyssal clays is on the west side of the Mid-Atlantic Ridge, which divides the ocean basin in half. This distribution pattern is a result of the different calcium carbonate compensation depths on the two sides of the ridge. Calcium carbonate is found in the sediments to depths in excess of 6000 m (19,680 ft) on the east side of the ridge, while deposition below 5000 m (16,400 ft) is rare in the basins to the west. This difference results from the fact that the Antarctic Bottom Water originating in the Weddell Sea to the south and the North Atlantic Deep Water originating off Greenland in the north carry low-temperature water with

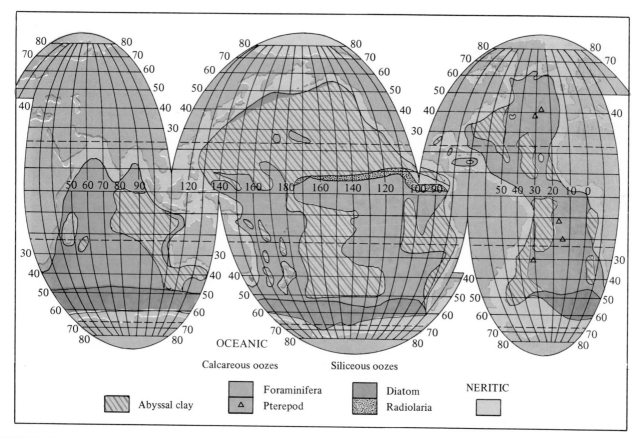

FIGURE 5–17 World Distribution of Neritic and Oceanic Sediments.
Base map courtesy of National Ocean Survey.

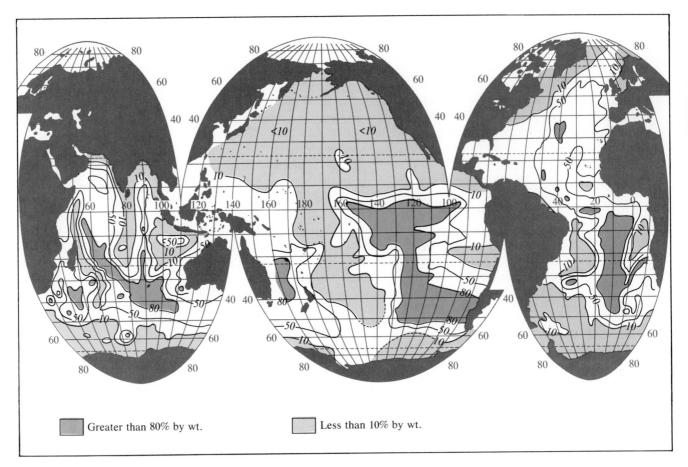

FIGURE 5–18   Calcium Carbonate in Surface Sediments of the World's Oceans.
Concentrations of calcium carbonate (CaCO₃) in ocean sediments is negatively correlated with
ocean depth. The deeper the ocean floor, the less CaCO₃ found in the sediment. It is also low in
sediments accumulating beneath cold, high-latitude waters. Extensive areas of concentrations of
CaCO₃ in the sediment in excess of 80% are associated with relatively shallow ocean floor on the
Carlsberg Ridge in the Indian Ocean, the East Pacific Rise, and the Mid-Atlantic Ridge. Note that
the deepest ocean basin, the North Pacific, lies for the most part beneath the CaCO₃ compensation
depth and has very little CaCO₃ in its accumulating sediment. After Biscaye et al., 1976; Berger et
al., 1976; and Kolla et al., 1976.

high carbon dioxide content in the deep basins on the west side of the ridge. The ridge protects the basins to the east from the invasion of this low-temperature water with its high capacity for dissolving calcium carbonate. Although abyssal clays are deposited in the deepest parts of the basins on the east side of the ridge, they replace the calcareous oozes over a much greater area and to shallower depths in the western basins.

Pacific Ocean   The Pacific Ocean is so vast that the existing sampling of bottom sediments has left wide gaps unobserved. On the basis of what sampling is available, it appears that the oceanic sediment covering almost half of the Pacific Ocean floor is abyssal clay. Abyssal clays are dominant in the North Pacific,

where they mantle fully 80% of the deep-ocean basin. Calcareous oozes cover about 36% and siliceous oozes about 15% of the Pacific Ocean floor. The reason the Pacific Ocean floor is covered more extensively by clays than the other ocean basins is its greater depth. Generally, calcium carbonate is dissolved below 4500 m (14,760 ft), leaving only minor concentrations of biologically derived silica among the terrestrial and extraterrestrial components.

Transportation of land-derived clays from the continents into the deep-ocean basin is severely restricted by the system of trenches that separates the deep Pacific from all land but a segment of the North American coast. Transportation of lithogenous material to the deep-ocean basin must be achieved in great part by the prevailing westerly winds that blow

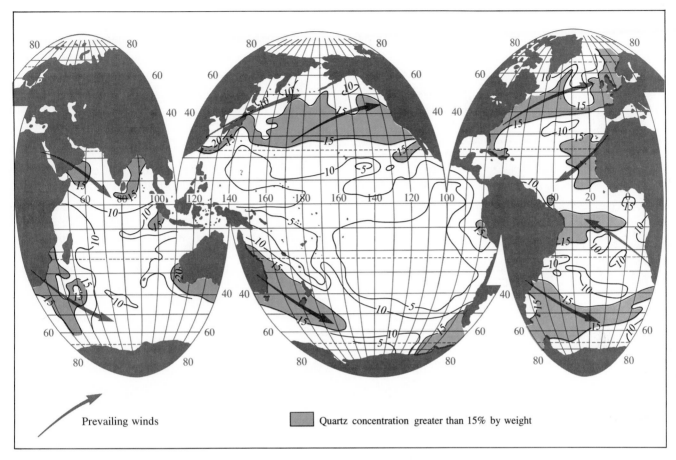

FIGURE 5–19 Lithogenous Quartz (SiO_2) in Surface Sediments of the World's Oceans.

Occurring as chips and shards mostly within the 5–10 μm (0.002–0.004 in) range, most quartz found in oceanic sediments is believed to have been transported by winds. Desert areas of Africa, Asia, and Australia are important sources of eolian (wind-blown) quartz. Turbidity flows have been significant sources of quartz in the equatorial, northwestern, and northern Atlantic Ocean and the Bay of Bengal, where deep-bottom currents have also modified sediment distribution. The distribution of lithogenous clays transported by winds may show a similar pattern. After Leinen et al., 1986.

across the Asian mainland and the continent of Australia (fig. 5–19). Ice rafting in the high latitudes has periodically played an important role in this transportation. Lithogenous sediment is provided as a land-derived material from the North American continent, which is not separated from the North Pacific basin by any trench structure. This condition, which allows a greater amount of lithogenous material to be carried by currents into the North Pacific basin, has resulted in a more rapid rate of accumulation for abyssal clays in the North Pacific than in the South Pacific, where all of the continental masses are separated from the deep-ocean basin by trench systems.

Indian Ocean East of the oceanic ridge at depths below 4500 m (14,760 ft) in some cases and at all depths below 5000 m (16,400 ft), calcareous content of the sediment is sharply reduced, and the dominant type becomes abyssal clay.

Siliceous Ooze

Atlantic Ocean Siliceous oozes are less abundant in the ocean basins because ocean water is never saturated with silica (SiO_2). Diatoms and radiolarians provide most of the skeletal material for siliceous oozes, but few of the diatom or radiolarian skeletons ever reach the bottom or become incorporated in the sediment before they are dissolved. Diatoms are plentiful in all areas of high productivity, but diatom oozes can be found only in areas below the calcium carbonate compensation depth. Above this depth they are usually diluted to low concentrations by higher

Interview with Dr. G. Ross Heath (fig. 5A), dean of the College of Ocean and Fishery Sciences, University of Washington.

Q. Dr. Heath, it seems that manganese nodules must represent great potential wealth. Otherwise, how can one explain the fact that this one issue has been such a problem for the Third United Nations Conference on the Law of the Sea? Do these deposits, in fact, deserve so much consideration?

A. I don't think anyone can confidently give you a yes or no answer to this question. One source of possible overoptimism in this area may be the 1965 estimate of manganese nodule reserves in the Pacific Ocean of 1600 billion metric tons. Although only a tiny fraction of that amount would be mineable, this large figure may have stuck in the minds of the representatives of the developing nations. More recently J. L. Mero, who made that estimate, has suggested that the reserves of mineable high-grade nodules between the Clarion and Clipperton fracture zones are about 38 billion tons (*see* plates 12 and 13).

FIGURE 5A Dr. G. Ross Heath.

As is the case with many of the leading oceanographers, Dr. Heath must assume more than one role to make the fullest use of his abilities. To the left, we see him in his role as Dean of the College of Ocean and Fishery Sciences at the University of Washington; at the right he is dressed less formally while participating in the geological research conducted aboard the *Glomar Challenger*. Photos courtesy of Oregon State University.

This region may be very unusual in that high-grade nodules are found in relatively great abundance—over 5 kg (11 lb)/m^2. No other areas have been found with the economic potential of this area. In contrast, at least one study has indicated that grade and abundance of nodules are negatively correlated on a global scale. Recent data indicate that a 20% recovery of the nodules by individual operations of 3 million tons annual production each would recover from 0.6 to 2 billion tons. This is a much smaller number than the 1600 billion ton number that may have given rise to so much optimism. The number of potential mining sites in the Clarion-Clipperton area ranges from 8 to 26, and the life of an operation would be from 20 to 25 years. The profitability of the industry is still very much open to question.

Q. Then we should not expect to see operations under way in the next decade because of this question of profitability?

A. Commercial exploitation will very likely be technically feasible within the next ten years. Environmentally and scientifically, mining the deep ocean seems quite feasible. Although additional research is needed to reduce technological risk, there is confidence this can be achieved in a few years. The main risks at present lie in uncertainties in the economic, political, and legal areas. The prices of the metals enriched in Clarion-Clipperton nodules—especially nickel and copper—are currently depressed. Cobalt-enriched nodules are of more economic appeal at the moment and are found on submarine mountains. They will be much harder to mine.

Q. If the developed nations with mining interests do not sign the U.N. treaty, what are the prospects for the commencement of mining operations?

A. Four nations, West Germany, the Soviet Union, Britain, and the U.S., that have legislation to protect the right of their licensed deep-sea mining operators did not vote for the adoption of the L.O.S. Convention. The U.S., Britain, West Germany, and France are negotiating to harmonize their license recognition and eliminate mining site overlaps. Complicating this picture is the fact that France and Japan, nations with mining legislation, signed the convention but have not ratified the treaty. Two other nations, Belgium and Italy, also have mining interests and abstained on the vote to adopt the L.O.S. Convention in 1982 but signed in 1984. Whether or not they decide to ratify the treaty could have a significant effect on the political and legal future of deep-sea mining. To the extent that mining consortia are multinational, they are likely to operate from the member country that faces the fewest legal and political obstacles. Thus, failure to sign the treaty may not preclude participation in mining ventures under its jurisdiction. The level of effort expended to resolve the issue will undoubtedly be correlated to the price of the metals in the nodules.

Q. What is known of the environmental effects of deep-sea mining of manganese nodules?

A. Since 1968, the ocean-mining industry, government, and academia have been cooperating to assess the effect deep-sea mining will have on the environment. These efforts have been focused in the Deep Ocean Mining Environment Studies (DOMES I & DOMES II). Obviously, the benthic ecosystem of the mined area will be totally disrupted. The plume of sediment stirred up on the ocean floor may also affect benthos in adjacent areas. However, in the Clarion-Clipperton area natural erosion by deep current does occur, so the environment is not pristine or stable. Most benthos of the deep ocean are deposit feeders and would be less affected by transient turbid water than the suspension feeders more common on the rocky shallow bottoms.

The DOMES data also indicate that particulate plumes discharged at the surface sink or disperse rapidly. Evidence of the discharge is undetectable a day or two later. Therefore, the negative effect on plankton may not be too serious. Longer-term assessments are needed, but it appears that a small number of mining operations will not cause irrevocable environmental damage. It is, however, essential that well-designed monitoring programs accompany mining activities.

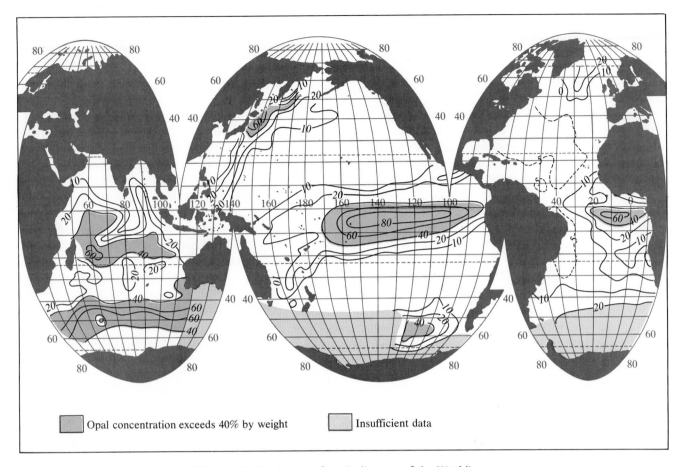

FIGURE 5–20 Biogenic Silica (SiO$_2$·nH$_2$O–Opal) in Surface Sediments of the World's Oceans.

The distribution pattern of opal in surface sediments of the ocean shows maximum concentrations in areas of highest biological productivity. These particles are produced by diatoms and radiolarians in the surface waters. In the equatorial and northwest Pacific, the opal is produced predominantly by radiolarians. Elsewhere, diatom remains are more abundant. The highly productive waters south of the Antarctic Convergence show up well in the south Indian Ocean and southeast Pacific Ocean. In the southern Atlantic Ocean and southwest Pacific Ocean there are insufficient data to determine the relationship between highly productive surface waters and opal concentrations in the surface sediment. After Leinen et al., 1986.

rates of deposition of CaCO$_3$ biogenous particles. Except for local concentrations of diatom remains in the sediment that are great enough to produce oozes in the equatorial region at about 20°W longitude, there are no highly productive areas in the North Atlantic (fig. 5–20).

Although based on limited sampling, there is presumed to be a significant belt of diatom ooze near Antarctica. This accumulation would be made possible by the high diatom productivity of the region. Conditions for the formation of diatom ooze in the Antarctic are enhanced by the fact that calcareous organisms do not occur in high concentrations where water temperatures are extremely low. The concentration of foraminifers decreases greatly south of the

Antarctic Convergence at about 50°S latitude (fig. 5–18).

Radiolarian oozes are very restricted in the Atlantic Ocean. The only sediments identified as such are found in the equatorial region near the previously discussed area of diatom oozes.

Pacific Ocean Diatom ooze is found in the Antarctic under conditions similar to those described for the Atlantic Ocean. Diatom ooze is also present in the extreme North Pacific because of the low-temperature waters that prohibit high productivity of carbonate forms. The diatom ooze of the North Pacific does not have the high silica concentration of the Antarctic ooze because of dilution by the greater rate of deposition of lithogenous material in the North Pacific.

Along a band north of the equator, radiolarian ooze can be found because calcareous remains are dissolved, leaving the content of radiolarian remains high enough to produce an ooze deposit. The thin belt of radiolarian ooze is bounded on the north by a line determined by decreased radiolarian productivity and on the south by a line determined by decreased dissolution of carbonate remains. Other localized deposits of radiolarian ooze are found in the southwest Pacific.

Indian Ocean   The continuation of the diatom ooze development around the margin of the continent of Antarctica can be seen in the southern Indian Ocean. In addition to this siliceous accumulation, there is a development of radiolarian ooze surrounded by calcareous ooze deposits in an area of high productivity south of the equator.

Biogenous silica can be differentiated from lithogenous silica because it contains molecular water. The distribution of biogenous silica, *opal* ($SiO_2 \cdot nH_2O$), shown in figure 5–20, is greatest in areas of high biological productivity.

SUMMARY

Sediments that accumulate on the ocean floor are classified by origin as **lithogenous** (derived from rock), **biogenous** (derived from organisms), **hydrogenous** (derived from water), and **cosmogenous** (derived from beyond the earth's atmosphere). **Sediment texture**, determined in part by the size and sorting of the particles, is affected greatly by the type of transportation (water, wind, or ice) that brought it to the deposit and the energy conditions under which it was deposited.

Lithogenous sediment is composed of fragments of inorganic compounds called *minerals,* which make up rocks. **Ferromagnesian minerals**, rich in iron and magnesium, are abundant in the dark igneous rock **basalt**, characteristic of ocean basins. Minerals low in iron and magnesium, **nonferromagnesian minerals**, are characteristic of the igneous rock **granite**, which makes up the continents. **Biogenous** sediment is composed primarily of the compounds **calcium carbonate** ($CaCO_3$) from the remains of foraminifers, pteropods, and coccolithophores and **silica** (SiO_2) from the remains of diatoms and radiolarians. **Hydrogenous** sediment includes a wide variety of materials that precipitate from the water or form from interaction of substances dissolved in the water with materials on the ocean floor. **Manganese nodules,**

phosphorite, glauconite, carbonates, and zeolites are examples. **Cosmogenous** sediment is composed of nickel-iron spherules and silicate chondrules rich in pyroxenes that probably represent sparks created as asteroids collided in the asteroid belt.

Neritic sediment accumulates rapidly along the margins of continents. It is dominated by particles of lithogenous origin. Owing to the recent rise in sea level from melting glaciers, many rivers throughout the world, including those flowing into the Atlantic along the east coast of the United States, are not now depositing sediment on the continental shelves, but rather in their estuaries. About 70% of the sediment covering the continental shelves is relict sediment from 3000 to 7000 years old. Of the 20 billion tn of sediment carried each year to the oceans by rivers, almost 80% is provided by the rivers of Asia.

Turbidity currents are thought to transport shelf sediment down submarine canyons and deposit it as **turbidite** on the deep-ocean floor. More than 75% of the sediment mass found on the ocean floor is part of the thick sediment wedge underlying the continental shelves, slopes, and rises. In high latitudes this accumulation includes poorly sorted **glacial deposits**.

Stromatolites are dome-shaped, $CaCO_3$-depositional structures that result from the entrapment of $CaCO_3$ particles in algal mats in environments that inhibit the development of algae-grazing gastropods.

The **Blake Plateau**, which extends from Cape Hatteras south off the Florida coast and ranges in depth from 700 to 1000 m (2296 to 3280 ft), does not receive sediment because of the high velocity of the northward-flowing Gulf Stream.

Oceanic sediment accumulates at low rates on the floor of the deep-ocean basins far from the continents. In the deeper basins where most biogenous sediment is dissolved before reaching bottom, **abyssal clay** deposits predominate. **Oozes** composed of over 30% by weight of biogenous sediment are found in shallower basins and on ridges and rises. The rate of biological productivity measured against the rates of destruction and dilution of biogenous sediment determines whether abyssal clay or oozes will form on the ocean floor. Much of the destruction of $CaCO_3$ occurs because the CO_2 content of the ocean water increases with depth. At the **calcium carbonate compensation depth**, ocean water dissolves $CaCO_3$ at a rate equal to the rate it settles from above, thus preventing accumulation on the ocean floor below that depth.

Manganese nodules are found on the deep-ocean floor where other sediments are accumulating at rates of less than 7 mm (0.3 in)/1000 yr. There is a relationship between the biological productivity of overlying waters and metallic content of the nodules.

Although it would take from 10 to 50 years for individual sediment particles in the ocean surface waters to settle to the bottom, most appear to be combined into larger aggregates as **fecal pellets** of small marine animals and reach the ocean floor in 10 to 15 days. This process of particle transport explains how the particle composition of the surface waters is similar to that of the sediment on the ocean floor immediately below.

Because the bottom water in the western Atlantic Ocean is colder, the calcium carbonate compensation depth is shallower on the west side of the Mid-Atlantic Ridge and abyssal clay in the western basins has a broader distribution. In the Pacific Ocean, abyssal clay covers more of the ocean floor than in other oceans because it is deeper (especially in the North Pacific) and lies beneath the calcium carbonate compensation depth. Silicous oozes are well developed in the North Pacific Ocean and around the southern margins of all oceans because little lithogenous sediment reaches these areas and foraminifers that produce $CaCO_3$ do not thrive in these cold waters.

QUESTIONS AND EXERCISES

1. List the four basic sources of marine sediment.

2. What characteristics of marine sediment indicate increased maturity?

3. Why is a higher-velocity current required to erode clay-sized particles than larger, sand-sized particles?

4. List the four ferromagnesian and the three nonferromagnesian minerals designated as common rock-forming minerals. Why are the nonferromagnesian minerals of lower density?

5. What igneous rocks are most characteristic of the oceanic and continental crusts?

6. List the two major chemical compounds of which most biogenous sediment is composed and the organisms that produce them.

7. What is the chemical composition of hydrogenous deposits of manganese nodules, phosphorite, glauconite, and carbonate? In which environments do they form? What is the relationship between a deposit's formation and its enviroment?

8. What are the two most common forms in which calcium carbonate precipitates from ocean water? Describe the environment in which such precipitation is most likely to occur. Include a discussion of the pH, biological processes, and water temperature conditions that enhance precipitation.

9. Describe the most common types of cosmogenous sediment and give the probable source of these particles.

10. Describe the basic differences between neritic sediment and ocean sediment.

11. Discuss the processes by which sediments are carried to and distributed across the continental margin.

12. In the two locations of stromatolite occurrence discussed in the text, what processes inhibit the development of algae-grazing gastropods?

13. How do oozes differ from abyssal clay? Discuss how productivity, destruction, and dilution combine to determine whether an ooze or abyssal clay will form on the deep-ocean floor.

14. Refer to figure 5–13 and explain why abyssal clay deposits on the floor of deep-ocean basins are commonly underlain by calcareous ooze.

15. Describe the apparent relationships that exist between biological productivity of overlying waters and metallic content of manganese nodules.

16. How do fecal pellets help explain that the particles found in the ocean surface waters are closely reflected in the particle composition of the sediment directly beneath? Explain why you would not expect this.

17. Explain why the abyssal clay deposition on the east side of the Mid-Atlantic Ridge is more restricted than on the west side.

REFERENCES

Berger, W. H.; Adelseck, C. G., Jr.; and Mayer, L. A. 1976. Distribution of carbonate in surface sediments of the Pacific Ocean. *Journal of Geophysical Research* 81:15, 2617–2629.

Billett, D. S.; Lampitt, R. S.; Rice, A. L.; and Mantoura, R. F. C. 1983. Seasonal sedimentation of phytoplankton to the deep-sea benthos. *Nature* 302:5908, 520–522.

Biscaye, P. E.; Kolla, V.; and Turekian, K. K. 1976. Distribution of calcium carbonate in surface sediments of the Atlantic Ocean. *Journal of Geophysical Research* 81:15, 2595–2602.

Deuser, W. G.; Brewer, P. G.; Jickells, T. D.; and Commeau, R. F. 1982. Biological control of the removal of abiogenic particles from the surface ocean. *Science* 219:4583, 385–86.

Dravis, J. J. 1983. Hardened subtidal stromatolites, Bahamas. *Science* 219:4583, 385–86.

Heath, G. R. 1982. Manganese nodules: Unanswered questions. *Oceanus* 25:3, 37–41.

Kolla, V., and Biscaye, P. E. 1976. Distribution of calcium carbonate in surface sediments of the Atlantic Ocean. *Journal of Geophysical Research* 81:15, 2602–2616.

Leinen, M.; Cwienk, D.; Heath, G. R.; Biscaye, P. E.; Kolla, V.; Thiede, J.; and Dauphin, J. P. 1986. Distribution of biogenic silica and quartz in Recent deep-sea sediments. *Geology* 14:3, 199–203.

Shepard, F. P. 1973. *Submarine geology.* New York: Harper and Row.

Smith, R. V., and Kinsey, D. W. 1976. Calcium carbonate production, coral reef growth, and sea level change. *Science* 194:937–38.

Spencer, D. W.; Honjo, S.; and Brewer, P. G. 1978. Particles and particle fluxes in the ocean. *Oceanus* 21:1, 20–26.

Sverdrup, H. U.; Johnson, M. W.; and Fleming, R. H. 1942. Renewal 1970. *The oceans: Their physics, chemistry, and general biology.* Englewood Cliffs, N.J.: Prentice-Hall.

Udden, J. A. 1898. Mechanical composition of wind deposits. *Augustana Library Pub.* 1.

Wentworth, C. K. 1922. A scale of grade and class terms for clastic sediments. *Journal of Geology* 30, 377–392.

SUGGESTED READING

Sea Frontiers

Dietz, R. S. 1978. IFOs (Identified Flying Objects). 24:6, 341–46.
The source of Australasian tektites and microtektites is discussed.

Dudley, W. 1982. The secret of the chalk. 28:6, 344–49.
An informative discussion of the formation of marine chalk deposits.

Dugolinsky, B. K. 1979. Mystery of manganese nodules. 25:6, 364–69.
The problems of origin, growth, and the environmental implications of manganese nodule mining.

Feazel, C. T. 1986. Asteroid impacts, seafloor sediments, and extinction of the dinosaurs. 32:3, 169–78.
A discussion of the possibility that high concentrations of iridium and osmium in a marine clay deposited at the time dinosaurs and many other species died out 65,000,000 years ago may have resulted from the collision of the earth with a meteor 6 mi in diameter.

La Que, F. L. 1979. Nickel from nodules? 25:1, 15–21.
The economic and political problems related to nodule mining are discussed.

Shepard, F. P. 1975. Submarine canyons of the Pacific. 21:1, 2–13.
A description of canyons around the Pacific margin and discussion of their possible origin.

Shinn, E. 1981. Time capsules in the sea. 27:6, 364–74.
    A discussion of how stress caused by climatic and tectonic changes are recorded
    in the banding patterns of coral reefs.

———. 1985. Mystery muds of Great Bahama Bank. 31:6, 337–46.
    Possible explanations of the white patches of milky water west of Andros and
    Abaco islands are considered.

Scientific American

Emery, K. O. 1969. The continental shelves. 221:3, 106–25.
    The nature of the continental shelves and the effect of the advance and retreat of
    the shoreline across them as a result of glaciation.

Heezen, B. C. 1956. The origin of submarine canyons. 195:2, 36–41.
    Theories explaining the origin of submarine canyons along with data on the loca-
    tion and nature of such features.

Menard, H. W. 1969. The deep-ocean floor. 221:3, 126–45.
    A summary of the dynamic effects of sea-floor spreading with a description of re-
    lated sea-floor features.

Rona, P. A. 1986. Mineral deposits from sea-floor hot springs. 254:1, 84–93.
    A discussion of the creation and composition of metal-rich deposits associated
    with oceanic hot springs.

CHAPTER SIX
PROPERTIES OF WATER

Before we can go further in our investigation of the ocean, we must establish a background that will allow us to understand the various properties of the substance that makes up 96.5% of the mass of the oceans—water. We at first may not consider water to be an unusual substance, inasmuch as it is probably the one we would mention if we were asked to name the most common or prevalent substance that exists on the earth. Water most certainly is abundant, but it does have unique properties; we find that it is a very uncommon substance.

It is the presence of water on the planet earth that makes life possible. Water controls the distribution of heat over the earth's surface as well as other conditions for life.

PROPERTIES OF WATER

The Water Molecule

When two hydrogen atoms combine with one oxygen atom to form one water molecule, they join in such a manner that the hydrogen atoms are separated by an angle of 105° rather than being at opposite ends of a linear molecule and separated by 180° (*see* fig. 6–1A). As the electrons move around within the structure, this arrangement of atoms within a molecule produces a greater concentration of electrons around the nucleus of the oxygen atom than around the hydrogen nuclei. As a result, a positive charge associated with the unshielded proton in the nucleus of each hydrogen atom is concentrated at the location of the hydrogen atoms. This situation produces polarity of electrical charge, in which the end of the molecule represented by the oxygen atom is slightly more negatively charged and the end with the hydrogen atoms is slightly more positively charged: the water molecule is *dipolar*, and water is known as a polar substance.

Solvent Properties of Water

The polarity in the distribution of electrical charge does not make the water molecule behave as an ion.

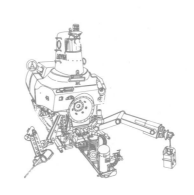

Pure water is actually a very poor conductor of electricity, since the molecule will not move toward either the negatively charged pole or the positively charged

A. WATER MOLECULE

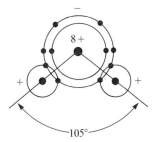

With the angle of 105° between the two hydrogen atoms in the water molecule, a dipolar molecule results. The oxygen end of the molecule is negatively charged, and the hydrogen regions exhibit a positive charge.

B. SOLID STATE

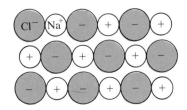

Ionic bonds hold sodium ions (Na^+) and chloride ions (Cl^-) together in the compound sodium chloride (NaCl).

C. IN SOLUTION

Water molecules

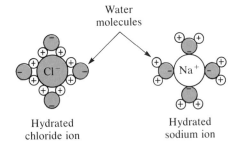

Hydrated   Hydrated
chloride ion   sodium ion

The positively charged hydrogen end of the dipolar water molecule is attracted to the negatively charged Cl^- ion; the negatively charged oxygen end is attracted to the positively charged Na^+ ion. The ions become hydrated when surrounded by water molecules that are attracted to them. In this state they have become solute ions in a water solution.

FIGURE 6–1   Water as a Solvent.

pole in an electrical system. Instead, the water molecule simply becomes oriented in an electrical field with its positively charged hydrogen nuclei toward the negative plate and its negatively charged oxygen end toward the positively charged plate. This orientation of water molecules in an electrical field tends to neutralize the field.

Owing to their polar nature, water molecules form bonds between one another. These bonds are referred to as *hydrogen bonds.* The positively charged hydrogen areas of the water molecule attract the negatively charged oxygen portion of the neighboring water molecules and bond together by electrostatic forces. When polar molecules are bound together in complexes, as are the water molecules, their ability to reduce the intensity of an electrical field acting on the water and, therefore, the electrostatic attraction between ions of opposite charges, is greatly increased. In fact, these forces can be reduced to 1/80 their value out of water.

If we consider sodium chloride, a compound containing *ionic bonds*, we can see that simply by placing that substance in water, we have reduced the electrostatic attraction between the sodium and chloride ions by 80 times. This reduced attraction makes it much easier for the ions to be dissociated. Once they do become dissociated, the sodium ions are attracted to the negative pole of the water molecule and the chloride ions are attracted to the positive pole of the water molecule (fig. 6–1).

As more and more ions of sodium and chlorine are freed by the weakening of the electrostatic attraction that is holding them together, they become surrounded by the polar molecules of water, or *hydrated.* Sodium ions are surrounded by water molecules with their negative poles toward the ion, and chloride ions are surrounded by water molecules oriented so that their positively charged poles are directed toward the ion.

Thermal Properties of Water

Freezing and Boiling Points of Water   The Celsius (centigrade) temperature scale was constructed on the basis of the characteristics of water at one atmosphere of pressure, with the freezing point of water, the temperature at which it changes from a liquid to a solid, representing 0° and the boiling point of water, the temperature at which it changes from a liquid to a gas, representing 100°. If we compare these values to those of some compounds similar to water in makeup, we will see that the freezing and boiling points of water are uniquely high. The other compounds occur in nature as gases, while water occurs in the gaseous, liquid, and solid states of matter within the narrow range of atmospheric conditions.

We should now consider the nature of the forces that must be dealt with to change the state of a compound. There exists between all molecules of any compound a relatively weak attraction that results from the electrostatic attraction between the nuclear parts of one molecule and the electrons of another. This attraction slightly exceeds the electrostatic repulsion that exists between the nuclear parts of the same molecules and the electrons of the same molecules.

This intermolecular force is known as *van der Waals force,* and it becomes significant only when molecules are very close together, such as in the solid or liquid states. Generally, the heavier the molecule, the greater the van der Waals attraction between individual molecules of the compound. Therefore, with increasing molecular weight, a greater amount of energy is needed to overcome that attraction and allow a change of state, say from a solid to a liquid or a liquid to a gas, to occur. Consequently, the melting and boiling points of compounds generally increase as the molecular weight increases.

What form of energy is used to change the state of matter? We should first define a few terms before we go further in our discussion. The first is *heat,* which is a form of energy associated with and proportional to the energy level of molecules. It may be generated by combustion, chemical reaction, friction, or radioactivity. So that we can measure the amount of energy that is being added to or removed from the molecules, we must introduce the term *calorie* as a unit for measuring quantity of heat. A calorie is defined as the amount of heat required to raise the temperature of one gram of water one degree Celsius. So that we can measure the amount of energy that the molecules within the substance have, we will need to understand the meaning of the word *temperature.* Temperature is defined as the direct measure of the *kinetic energy* of the molecules that make up a substance. Kinetic energy is energy of motion, so the higher the temperature, the greater the velocity of the molecules of the substance for which the temperature is being measured.

Let us return now to the van der Waals force that causes molecules of a given substance to be attracted to one another. We can see that if this attraction is to be broken, energy must be given to the molecules so that they can move more rapidly, in the ways described in figure 6–2, to overcome this force. It is this attraction that must be broken if we are going to pro-

duce a change of state in any substance from solid to liquid or liquid to gas.

We find that in a crystalline solid state, although bonds are constantly being broken and reformed, the predominant relationship between molecules is that of a rather firm attachment produced by the nearness of the molecules and the great effect the van der Waals force has upon molecules that are close together. Vibration is the dominant type of molecular movement in the solid state as the molecules vibrate but remain in relatively fixed positions.

In the liquid state the molecules have gained enough energy to overcome many of the van der Waals forces that bound them together in the solid state. The molecules have enough freedom to move relative to one another. In this state we see all forms of molecular movement—vibration, rotation, and translation. The molecules are free to move relative to one another but are still attracted by one another. Bonds are being formed and broken at a much greater rate than in the solid state.

In the gaseous state, translation has become the dominent type of motion. Molecules are now moving at random, and there exists no attraction between individual molecules. The only effect that one molecule will have on another is that which is produced by collision during random movement.

Returning to our comparison of the properties of water and compounds of similar composition containing two hydrogen atoms and an atom of another element, we will look at H_2S, H_2Se, and H_2Te. The

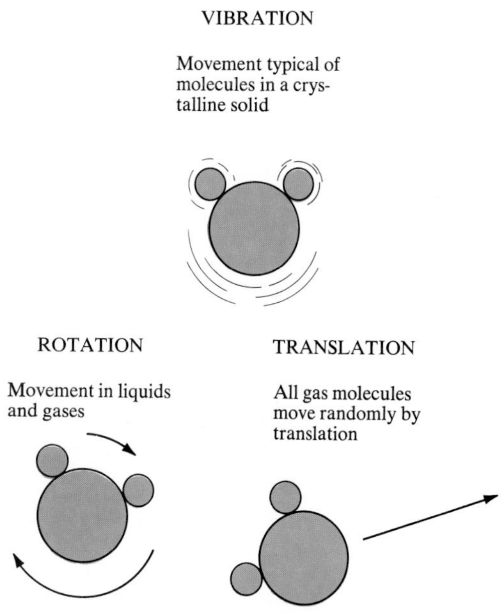

VIBRATION

Movement typical of molecules in a crystalline solid

ROTATION

Movement in liquids and gases

TRANSLATION

All gas molecules move randomly by translation

FIGURE 6–2   Motion of Molecules.

molecular weights of the four compounds are H_2O (18), H_2S (34), H_2Se (80), and H_2Te (129). Figure 6–3 shows that, as was predicted in the discussion of the van der Waals force, the freezing points and boiling points for H_2S, H_2Se, and H_2Te increase with increased molecular weight. When we insert H_2O into the scale, the freezing point and boiling point based on its molecular weight should be about $-90°C$ ($-130°F$) and $-68°C$ ($-108.4°F$), respectively. The fact that water freezes at 0°C (32°F) and boils at 100°C (212°F), seems to be a violation of natural order. Here we see again the great significance of the polarity of the water molecule and the hydrogen bond that this structure produces. The high freezing and boiling points of water are the manifestations of the additional kinetic energy required to overcome not only van der Waals force but also the hydrogen bonds to achieve a change of state.

Heat Capacity   Another direct result of the hydrogen bond is the high heat capacity of water. As noted, a calorie is the amount of heat required to raise the temperature of 1 g (0.035 oz) of water 1°C (1.8°F). The heat capacity of water compared to that of most other substances is great, and we use the capacity of water to absorb heat as a standard against which we compare the heat capacities of other substances. The amount of heat that is required to raise the temperature of 1 g of any substance 1°C is the *heat capacity* of the substance. It is 1 cal for water and less than 1 cal for most other substances. It is this property that produces the mild climates in coastal regions. Great amounts of heat can be gained or lost by the coastal water without causing extreme changes in temperature.

Latent Heats of Melting and Vaporization
Closely related to water's unusually high heat capacity are its high latent heat of melting and latent heat of vaporization. The continuous addition of heat to a substance in the solid or liquid state will bring about a change of state in the substance. A solid converts to a liquid at a temperature called its *freezing* (melting) *point* and a liquid is changed to a gas at a temperature defined as its *boiling* (condensation) *point.*

When changing the state of any substance, there may be no increase of temperature at that point where a change of state occurs even though heat is continuously being added. The heat energy is being used entirely to break all the bonds required to complete the change of state. While the bonds are being broken, a mixture of the substance in both states is in equilibrium. Only upon completion of the change of state will the temperature again rise. The heat that is added to 1 g of a substance at the melting point to break the required bonds to complete the change of state from solid to liquid is the *latent heat of melting.* The heat applied to effect a change of state at the boiling point is the *latent heat of vaporization.*

The use of the word *latent* in describing the heats of vaporization and melting is necessary because the heat that must be added to a given mass of ice or water to convert it to a higher energy state, water or water vapor, is held in reserve or "hidden" by that mass of water or water vapor. When water vapor returns to a liquid, it *condenses* and heat is released into the surrounding air. Release of heat also occurs with the freezing of water, which is the change of state from liquid to solid. During condensation and freezing, the same amount of heat that is added to change the state of that water from a liquid to a gas or from a solid to a liquid is released.

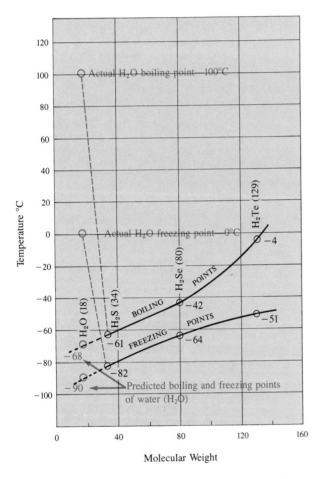

FIGURE 6–3   Molecular Weight—Freezing and Boiling Points.

Because of the hydrogen bonds that must be broken in addition to overcoming the van der Waals force in achieving changes of state, the freezing and boiling points of water are higher than would be expected for a compound of its molecular makeup.

A practical application of these principles of heat transfer can be seen in the use of ice for the purpose of refrigeration. A block of ice set in a closed container among food articles will lower their temperature because heat energy is drawn from the food articles and added to the molecules of ice to change them to the liquid state as the ice melts. The principle of cooling air with water is similar: in arid climates, the hot dry air that passes through a surface coated with liquid water will lose heat to the water. The water is converted to a vapor. Thus, after passing through or across the water-covered surface, the air is considerably cooler.

To help clarify this phenomenon, we will discuss the heat transfer and the change of state of H_2O from a solid at $-40°C$ ($-40°F$). Observe the heat-temperature graph, which has a vertical temperature scale ranging from $-40°$ to $120°C$ ($248°F$) and a horizontal scale which begins at 0 and continues through 800 cal (fig. 6–4).

For 1 g of ice, we find that the addition of 20 cal of heat raises the temperature 40°, from $-40°C$ to $0°C$. Thus, the heat capacity of ice is 0.5 cal/g, half that of liquid water.

Once the temperature has been raised to 0°C (32°F), the continuous addition of heat does not result in an increse in temperature until 80 cal have been added. We do not observe an increase in temperature during the addition of these 80 cal of heat because all the heat energy added was used to break the bonds that were holding the water molecules in the solid state. The temperature will remain unchanged until all of the necessary bonds are broken and the mixture of ice and water has been changed to 1 g of water. The amount of heat required to convert 1 g of ice to 1 g of water, 80 cal, is termed the

latent heat of melting, and it is higher for water than for any other commonly occurring substance.

The bonds that are broken in converting most substances from a solid to a liquid are the van der Waals bonds. In water, however, not only the van der Waals bonds but some of the hydrogen bonds must also be overcome. It is necessary only to break the ice structure down into numerous small clusters surrounded by individual water molecules to the extent that these remaining ice clusters can move relative to one another and allow the mass to assume a liquid state. Liquid water, particularly at low temperature near the freezing point, could well be described as a pseudo-crystalline liquid, as there are still many small clusters of ice crystals contained within it. Once enough hydrogen bonds have been broken to allow freedom of movement among the clusters and individual water molecules, the temperature of the water will again rise with the addition of heat.

After the phase change from ice to liquid water has occurred at 0°C, the addition of heat to the water causes the temperature to rise again. As it does, note that it requires 1 cal of heat to raise the temperature of water 1°C (1.8°F). We must therefore add a total of 200 cal before one gram of water reaches the boiling point of 100°C (212°F). At this temperature we note the development of another plateau, far more prominent on the graph than that which represents the latent heat of melting, 80 cal/g. This plateau at 100°C represents an addition of 540 cal, the latent heat of vaporization, to the gram of water before complete conversion to the vapor state.

Why is so much more heat energy required to convert 1 g of water to water vapor than was required to convert 1 g of ice to water? We defined a gas as a substance in which the molecules were moving at

FIGURE 6–4   Change of State Graph—Water.

The latent heat of melting (80 cal) is much less than the latent heat of vaporization (540 cal) because only a few hydrogen bonds must be broken to convert 1 g of ice to a liquid, while all remaining hydrogen bonds must be broken to convert 1 g of liquid water to a gas.

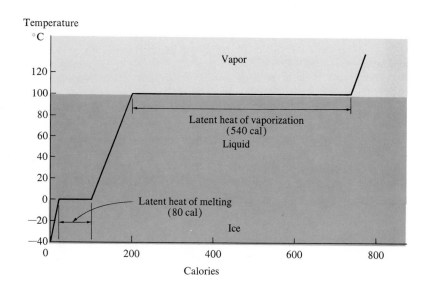

random, free from the influence of other molecules, except when they collide. To make the conversion from ice to water, not all the hydrogen bonds had to be broken, but only enough to allow freedom of movement along the various ice clusters that remained and the individual molecules that were also present in the system. To convert water to water vapor, every molecule must be freed from the attraction of other water molecules. Therefore, every hydrogen bond must be broken (fig. 6–5). To do this requires a great amount of heat energy.

We can now see that large amounts of heat energy are absorbed or released by water as it changes state, particularly if the change of state is from a liquid to a vapor or vice versa. Of course, we do not have temperatures of 100°C, at the surface of the ocean where this phenomenon of conversion of water to vapor occurs in nature. Sea surface temperatures are about 20°C (68°F) or lower.

The conversion of a liquid to a gas below the boiling point is called *evaporation*. At ocean surface temperatures, individual molecules that are being converted from the liquid to the gaseous state have a lower amount of energy than do the molecules of water at 100°C. Therefore, to gain the additional energy necessary to break free of the surrounding water molecules, an individual molecule must capture energy from its neighbors. This phenomenon explains the cooling effect of evaporation. The molecules left behind have lost heat energy to those that escape. To produce 1 g of water vapor from the ocean surface at temperatures less than 100°C requires more than the 540 cal of heat that are required to make this conversion at the boiling point. For instance, the *latent heat of evaporation* is 585 cal/g at 20°C and 595 cal/g at 0°C. This higher value is due to the fact that more hydrogen bonds must be broken at this lower temperature for a gram of water molecules to enter the gaseous state.

We can readily see the significance of the evaporation-condensation cycle in the earth's surface temperatures. Evaporation removes heat energy provided by the sun and stored in the oceans. This energy is carried high into the atmosphere as the water vapor rises and is released there when the vapor condenses and falls as precipitation. It is water's latent heat of evaporation that accounts for the removal of great quantities of heat from the low-latitude oceans by evaporation. It is later released in the heat-deficient higher latitudes after the vapor is transported through the atmosphere and condenses there as rain and snow (see *Hurricanes—Nature's Safety Valves* on pp. 152–156).

Surface Tension

Next to mercury, water has the highest surface tension of all commonly occurring liquids. We observe a surface tension phenomenon in filling a container with water to the brim and even beyond (fig. 6–6A). You will note that the water can be piled up above the rim of the container, forming a convex surface that is the water interface with the atmosphere. Water drops are also a very common manifestation of surface tension. These phenomena result from the tendency for water molecules to attract one another or to cohere at the surface of any accumulation of water. Because of this cohesive tendency, it is possible to float on water objects that are much heavier than water. A razor blade carefully laid on the surface will float, although it is normally five times as dense as water. Many insects use the surface of water as if it were a solid surface, moving around it at will.

FIGURE 6–5 Hydrogen Bonds in H₂O.

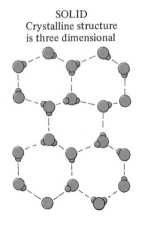

SOLID
Crystalline structure
is three dimensional

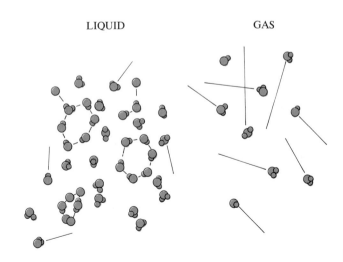

LIQUID

GAS

A. A container can be filled above its rim due to the strength of the surface layer.

C. Capillarity-water climbs the walls of a tubular container. The smaller the diameter, the higher it will climb.

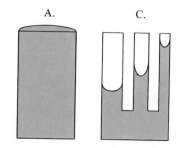

B. Hydrogen bonds

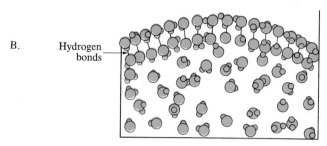

FIGURE 6–6 Surface Tension Phenomena.

A surface tension layer forms as hydrogen bonds form a strong attraction between the top layer of water molecules and the underlying molecules.

Surface tension is a manifestation of the presence of the hydrogen bond. Those molecules of water that are at the surface are strongly attracted to the molecules of water below them by their hydrogen bonds. The air above the surface has a very low density of molecules as compared to the water itself, and although water molecules are attracted to molecules of other substances, the attraction of the hydrogen bond holds the water molecules in the surface layer to those below (fig. 6–6B).

Water clings to the surface of many substances; we refer to this phenomenon as *wetting*. Water will adhere strongly to glass, organic substances, and inorganic material such as rocks and soil. When water is poured into a container made of a substance to which it strongly adheres, the adhesive force, or the force of attraction between water molecules and molecules of another substance, will cause the surface tension layer to take on a form unlike the convex one described previously. In this case the surface tension layer is concave, being drawn up on the container's sides (*see* fig. 6–6C).

If the container is glass, the positively charged portions of the water molecules are attracted to unbonded electrons in the oxygen atoms that are part of the makeup of the glass container. Being strongly attracted by these oxygen atoms, the water molecules will "climb" up the side of the container and will be held back only by the hydrogen bond attraction that exists between individual water molecules. In fact, if the diameter of the container is decreased to a very

fine bore, the combination of cohesion, which holds the water molecules together, and the adhesive attraction between the water molecules and the glass container will pull the column of water to great heights. This phenomenon is known as *capillarity*.

Salinity of Ocean Water

Though the proportion of water to dissolved salts may vary within the ocean, the major component ions are distributed in ocean water in relatively constant proportions. This *constancy of composition* was first established by Dittmar in his analysis of the *Challenger* expedition water samples discussed in chapter 1. Water makes up, on the average, 96.5% of the ocean's mass, so it determines most of the physical properties that we observe in ocean water.

It seems probable that when the techniques are developed for making such measurements, all of the known elements will be found to be dissolved in ocean water. Merely six elements, however, account for over 99% of the dissolved solids in ocean water. These major components are chlorine, sodium, magnesium, sulfur (as SO_4), calcium, and potassium (table 6–1).

TABLE 6–1 Ocean Salinity.

Considering the average salinity of the world ocean, 34.7 by weight, a 1000 g sample of this water would contain 34.7 g of dissolved solids. Six ions account for 99.28% of the dissolved solids.

Major Constituents (over 100 parts per million)	
Ion	Percentage
Chloride, Cl^-	55.04
Sodium, Na^+	30.61
Sulfate SO_4^{-2}	7.68
Magnesium, Mg^{+2}	3.69
Calcium, Ca^{+2}	1.16
Potassium, K^+	1.10
	99.28

Minor Constituents (1–100 parts per million)	
Bromine	65.0
Carbon	28.0
Strontium	8.0
Boron	4.6
Silicon	3.0
Fluorine	1.0

Trace Elements (less than 1 part per million)	
Nitrogen	Iodine
Lithium	Iron
Rubidium	Zinc
Phosphorus	Molybdenum

HURRICANES—NATURE'S SAFETY VALVES

Ten to 60-km (6- to 37-mi) holes in the earth's atmosphere serve to release excess heat energy accumulated at latitudes below the tropics. Such storms usually develop near the end of the summer season in both hemispheres. These tropical storms begin as low-pressure cells that break away from the equatorial low-pressure belt and grow as they pick up heat energy from the warm ocean. By means of the large heat capacity and latent heat of evaporation of the ocean's water, they transport great quantities of heat energy into heat-deficient higher-latitude regions by moving in curved paths around the western sides of the oceanic subtropical high-pressure cells (fig. 6A and plates 16 and 17). When wind velocity within the storms reaches 120 km/h (74 mi/h), tropical storms are classified as hurricanes in the Western Hemisphere and typhoons in the Eastern Hemisphere.

Hurricanes can have diameters exceeding 800 km (496 mi). More typically, they will have a diameter measuring less than 200 km (124 mi). Although hurricanes are usually smaller in areal extent than storms occurring in temperate latitudes, the tropical variety is much more violent. As air moves across the ocean surface toward the low-pressure eye of the hurricane and is drawn up to the outer reaches of the atmosphere, its velocity increases to as much as 300 km/h (186 mi/h). Once near the eye, the air begins to spiral upward and horizontal wind velocities may be negligible. Because of the Coriolis effect, the air flowing toward the eye of a storm veers to the right in the North-

FIGURE 6A Hurricanes and Typhoons.

A, air rising in a counterclockwise spiral around the eye of a hurricane. *B*, plan view of air at the ocean surface moving toward the eye of a hurricane. *C*, hurricanes begin as low-pressure cells moving out of the equatorial low-pressure belt toward the higher latitudes in curved paths around the oceanic subtropical highs. They are called typhoons in the western Pacific and Indian oceans and hurricanes in the western Atlantic and eastern Pacific oceans.

A.

B.

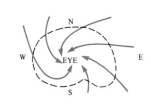

C.

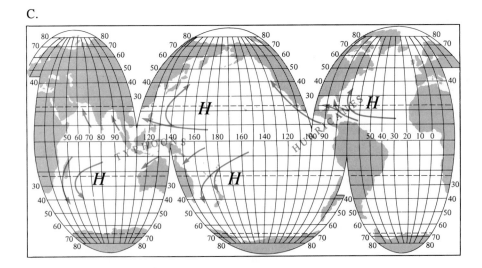

ern Hemisphere and to the left in the Southern Hemisphere, producing counterclockwise and clockwise rotation, respectively.

The energy contained in one hurricane is greater than that generated by all energy sources over the last twenty years in the United States. The energy released by an atomic bomb detonation in the center of a hurricane would have no effect on the storm. It would quickly be sucked up through the eye and distributed throughout the system.

The worst natural disaster in U.S. history occurred on September 8, 1900, when a storm surge associated with a 135-km/h (84-mi/h) hurricane inundated the barrier island city of Galveston, Texas, killing 6000 people.

More recently, at 1200 Greenwich Meridian Time (GMT) on August 25, 1979, a tropical depression began to develop near 12°N latitude and 36°W longitude in the North Atlantic Ocean. During the next fourteen days, Hurricane David grew into one of the strongest and most destructive hurricanes since 1900. It devastated the Caribbean islands, battered the Atlantic coast of the United States and Canada from southern Florida to Newfoundland, and then moved off into the North Atlantic Ocean (fig. 6B).

FIGURE 6B Path of Destruction.

The path followed by Hurricane David from its beginning on August 25, 1979.

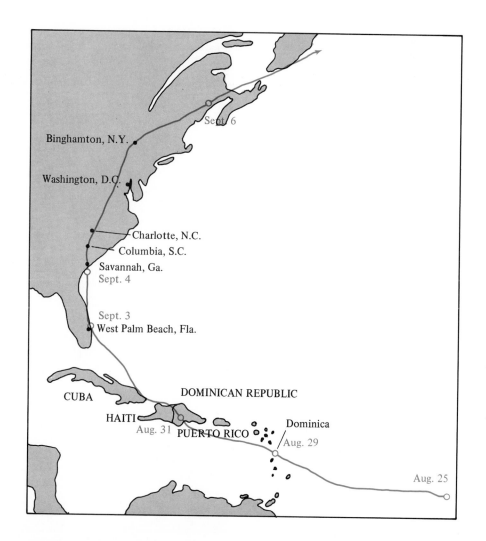

Measuring almost 500 km (310 mi) across with a 50-km (30-mi) eye, David made its first major landfall on August 29 after developing winds exceeding 200 km/h (fig. 6CA). Striking the island of Dominica, it left 37 dead and 60,000 homeless. Two days later, after killing 9 in Puerto Rico, David slashed across the Dominican Republic, killing 1100, injuring 3000, and leaving 225,000 homeless (fig. 6CB). After coming ashore briefly north of West Palm Beach, Florida, where six were left dead, David made its final landfall at Savannah, Georgia, on September 4 and took two more lives (fig. 6CC). Moving north

FIGURE 6C Hurricane David. *A*, 1901 GMT, August 29. Eye of hurricane has just passed over the island of Dominica, leaving 37 dead. *B*, 1901 GMT, August 31. Eye of hurricane reaches shore of Dominican Republic, killing 1100 and injuring 3000. *C*, 2001 GMT, September 4. Eye of hurricane makes landfall at Savannah, Georgia, taking its last two lives before moving north through the coastal states. Courtesy of NOAA.

A.

B.

C.

through coastal states, the storm spawned flooding, tornadoes, and destruction totaling over one-half billion dollars in the United States and Canada before heading back into the Atlantic on September 8.

Before the destruction of Hurricane David could be assessed, another hurricane, Frederic, zeroed in on Mobile, Alabama, on September 12 packing 210-km/h (120-mi/h) winds (fig. 6D). Frederic cut Dauphine Island, a barrier island at the mouth of Mobile Bay, in two and battered the area between New Orleans, Lousiana, and Panama City, Florida.

FIGURE 6D Hurricane Frederic.

A, huge piles of sand cover the roadway on Dauphine Island, a barrier island off Mobile Bay, Alabama. Hurricane Frederic, with 200-km/h (124-mi/h) winds, cut across the island on September 12, 1979. Courtesy of Wide World Photos. *B*, Frederic coming ashore at Mobile, Alabama. David is moving southeast in the North Atlantic as an extratropical storm as it continues its clockwise path around the North Atlantic subtropical high or Bermuda high. Courtesy of NOAA.

A.

B.

Even islands near the centers of ocean basins may be struck by hurricanes. The Hawaiian Islands were hit hard by Hurricane Dot in August 1959 and Hurricane Iwa on November 23, 1982. Hurricane Iwa hit very late in the season with winds of up to 130 km/h (81 mi/h). Well in excess of $100 million in damage was done to the islands of Kauai and Oahu. Niihau, a small island occupied by 230 native Hawaiians, was directly in the path of the storm and suffered severe damage, but none of the population received serious injuries (fig. 6E).

Hurricanes such as David, Frederic, and Iwa will always be a threat to life and property. There is little hope that they will ever be controlled. Inhabitants of areas subject to a hurricane's destructive force need to be cautious of this important natural phenomenon. With present-day warning capabilities, prudent responses to hurricane warnings should help prevent much property damage and eliminate any great loss of life. By ignoring safety precautions, however, many lives will needlessly be lost in future hurricanes.

Hurricanes may well be thought of as natural safety valves that help to release the heat energy stored up during summer months in the low latitudes. The property of water that makes possible the removal of great quantities of excess heat from the low latitudes is its high latent heat of evaporation. For every gram of water evaporated from the ocean's surface, the hurricane gains 585 cal of heat energy. As destructive as they are, they play a major role in transferring the excess heat energy accumulated in the lower latitudes to the heat-deficient higher latitudes. Hurricanes are an integral part of the earth's "heat engine" system.

FIGURE 6E *GOES* Satellite Infrared Image of Hurricane Iwa as Observed at 2:15 A.M., November 24, 1982.

The three most northerly inhabited islands, Niihau, Kauai, and Oahu, suffered the most damage.

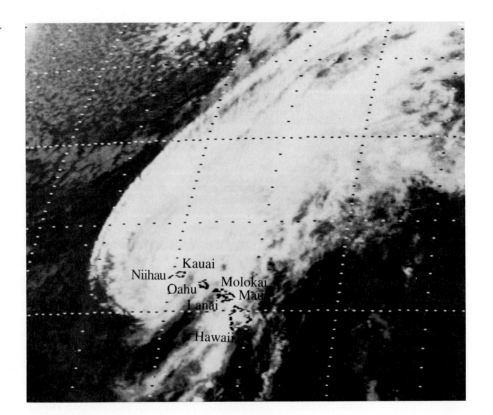

Salinity is defined as the total amount of solid material dissolved in a kilogram of seawater when all the carbonate has been converted to oxide, all bromine and iodine replaced by chlorine, and all organic matter completely oxidized. Given this definition, it may appear that salinity is a very complex property and very difficult to measure. However, since the oceans are so very well mixed and the relative abundance of the major constituents is essentially constant, we have a condition that makes the measurement of salinity relatively simple.

Because of constancy of composition, it is necessary to measure the concentration of only one of the major constituents in order to determine the salinity of a given water sample. The constituent that occurs in the greatest abundance and is the easiest to measure accurately is the chloride ion, Cl^-. The portion of the weight of a given sample of water that is the direct result of the presence of this ion is called *chlorinity* and is usually expressed in the terms grams per kilogram of ocean water (g/kg) or parts per thousand (‰). The chloride ion always accounts for 55.04% of the dissolved solids in any sample of ocean water. Therefore, by measuring its concentration, we can determine the total salinity in parts per thousand by the following relationship:

Salinity (‰) = 1.80655 × Chlorinity (‰)

For example, given the average chlorinity of the ocean, 19.2 ‰, the ocean's average salinity can be calculated as 1.80655 × 19.2 ‰, or 34.7 ‰. Although 1/0.5504 equals 1.81686 (not 1.80655), ocean-ographers have agreed on 1.80655 because the constancy of composition has been found to be more of an approximation than an absolute condition.

From 1902 to 1975, the Hydrographic Laboratory at Copenhagen, Denmark, provided Standard Seawater samples to assure that salinity determinations made throughout the world would be based on the same reference. In 1975, this duty was taken over by the Institute of Oceanographic Services located in Wormly, England. Standard Seawater consists of ocean water analyzed for chloride ion content to the nearest ten-thousandth of a part per thousand. It is then sealed in ampules and labeled to be sent to laboratories throughout the world. The chlorinity of water samples taken and measured in other parts of the world can then be compared if the titration solution or conductivity instrumentation is calibrated with the standard water sample (fig. 6–7).

The chemical determination of the chloride content in seawater can be made with a high degree of accuracy but requires a great deal of time and care. It is tedious to carry out the chemical titration with silver nitrate aboard ship. With the advancements in instrumentation in the field of oceanography, this task has been greatly simplified. Since electrical conductivity of ocean water is known to increase with increased temperature and salinity, salinity is now most commonly determined by measuring the electrical conductivity of ocean water (at a constant temperature of 15°C, or 59°F) with a modern conductance-measuring instrument, the *salinometer*. Salinity can be determined to better than 0.003 ‰ by this method.

FIGURE 6–7 An Ampule of Standard Seawater.

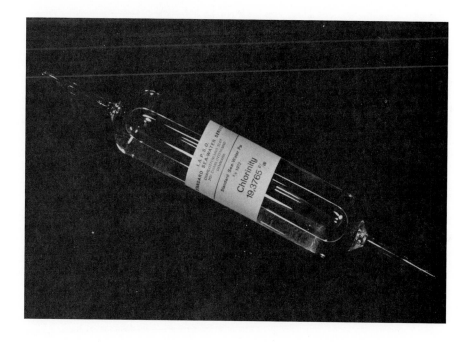

Water Density

The density of water in its various states and at different temperatures is of great importance in considering the movement of water in the ocean. Anything that is more dense than water will sink into it, and a substance that is less dense will float on the surface. We define *density* as mass per unit of volume, usually grams per cubic centimeter (g/cm^3).

Density is affected by temperature, salinity, and pressure. With most substances we observe that a decrease in temperature produces an increase in the density of the substance. The increase in density is the result of the same number of molecules occupying less space as they lose energy. This condition is also found in water. As the temperature of water decreases, the density increases as long as this temperature decrease occurs above 4°C (39.2°F). If we consider the change in the density of water below 4°C, we encounter considerations that must include the hydrogen bond. As the temperature of water is lowered from 4° to 0°C (32°F), we observe that its density decreases.

This anomalous behavior can be explained only by considering the molecular structure of water and the hydrogen bond. As we lower the temperature of water from 20°C (68°F) and reduce the amount of thermal agitation of the water molecules, the unbonded water molecules occupy less volume because of their decreased energy. But, as we approach the freezing point below 4°C, this reduction in volume owing to the decreased energy of unbonded molecules is not sufficient to compensate for another phenomenon that is occurring. Ice crystals, open six-sided structures in which water molecules are widely spaced, are becoming more abundant. The greater rate of in-

FIGURE 6–8 Formation of Ice Clusters in Water.

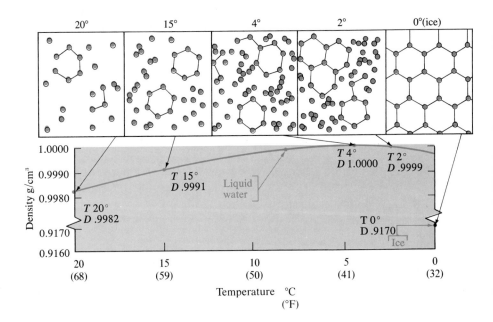

FIGURE 6–9 Salinity, Freezing Point, and Temperature of Maximum Density.

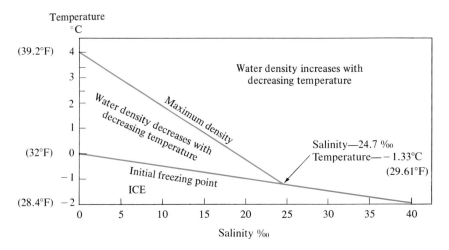

crease of ice crystals as the temperature approaches the freezing point accounts for the decreased density of water below 4°C (fig. 6–8).

The temperature of maximum density for fresh water, 4°C, is lowered by increasing pressure or adding solid particles such as salt, both of which will inhibit the formation of the ice clusters. Thus, more energy must be removed to produce crystals equal in volume to those that could be produced at 4°C in fresh water, causing a greater reduction in temperature of maximum density. This interference with the formation of ice crystals also produces a progressively lower freezing point for water as more solids are added. The freezing point and temperature of maximum density converge as they drop and coincide when the salinity reaches 24.7 parts per thousand (‰) at a temperature of −1.33°C (29.61° F)(fig. 6–9). When salinity is higher than 24.7 ‰, the density of water increases with decreasing temperature until the water freezes. Thus, average seawater (34.7‰) has no maximum density anomaly.

Because the components of salt have a density greater than the density of water, density of ocean water increases with increasing salinity. Physical oceanographers are greatly concerned with the relationships among water density, salinity, and temperature since density distribution is an important determinant of ocean circulation. Density cannot be measured directly so to determine density distribution within a region of study, oceanographers measure temperature and salinity and then use these values to calculate density.

RESIDENCE TIME

The *residence time,* or the average length of time an atom of an element will spend in the ocean, can be calculated by the following equation:

$$\text{Residence Time} = \frac{\text{Amount of element in the oceans}}{\text{Rate at which element is added to or removed from the oceans}}$$

The oceans are considered to be in a steady state, that is, the rate at which an element is removed from the oceans must be equal to the rate at which it is being added. For that reason, we could use either of these values as the denominator in the equation.

The residence time of an element in the ocean depends on how reactive it is with the marine environment. Those elements that are more reactive have a shorter residence time. The reactive elements aluminum and iron have residence times of 100 and 140

years, respectively, while less reactive sodium has a residence time of 260 million years (*see* table 6–2).

Elements are removed from ocean water through biological reactions, or they are incorporated into sediment. Each year a 1-m (3.28-ft) layer of the ocean's surface is evaporated. This water is returned directly by precipitation or indirectly by runoff from the continents. The ocean, which has an average depth of 4000 m (13,120 ft), would completely evaporate in 4000 years, but because the water is returned, this is the time required to completely recirculate the ocean. Thus, the residence time of water in the oceans is 4000 years.

The short residence time for aluminum shown in table 6–2 (100 years) is the average for particulate and dissolved aluminum. Recent investigation has shown that dissolved aluminum has a residence time of 1400 years, indicating that particulate aluminum must be removed from ocean water at a very rapid rate. It is believed that comparisons of the residence times of soluble fractions of elements in the ocean must be known to understand their chemical reactivity in the oceans since the residence time of particles is affected primarily by physical mixing processes.

GEOSECS (GEOCHEMICAL OCEAN SECTIONS)

The GEOSECS project mentioned in chapter 1 is a huge systematic program designed to increase understanding of circulation patterns and mixing processes in the oceans. All of the major ocean basins have been sampled throughout the depth of the water column. The sampling system includes electronic sensors that telemeter temperature, salinity, dissolved oxygen, pressure, and particulate content to a console aboard ship. As rosettes of 30-liter (7.9 gal) sampling bottles are lowered into the ocean, oceanographers have the option of deciding to sample at

TABLE 6–2 Residence Time of Some Elements in the Oceans.

Element	Amount in Ocean (g)	Residence Time in Years
Sodium, Na	147×10^{20}	2.6×10^8
Potassium, K	5.3×10^{20}	1.1×10^7
Calcium, Ca	5.6×10^{20}	8.0×10^6
Silicon, Si	5.2×10^{18}	1.0×10^4
Manganese, Mn	1.4×10^{15}	7.0×10^3
Iron, Fe	1.4×10^{16}	1.4×10^2
Aluminum, Al	1.4×10^{16}	1.0×10^2

Data from Goldberg and Arrhenius, 1958

FIGURE 6–10 GEOSECS Sampling Bottles.
One of the two rosettes carried aboard Scripps Institution of Oceanography's research vessel *Melville* during GEOSECS Pacific operations shows nonmetallic bottles used in collecting sea water samples. The ends of each bottle can be triggered to close by the scientist operating the shipboard console. Photo from Scripps Institution of Oceanography, University of California, San Diego.

any depth where an interesting value for any of the properties shows on the console (fig. 6–10).

The collected water is analyzed for some 23 chemical species, 15 isotope species, and the amount and types of particulate matter. From the results thus far obtained, oceanographers think the GEOSECS program will not only provide a much improved understanding of physical processes but also will supply much information about biological cycles in the ocean.

SUMMARY

The **hydrogen bond,** the bond between water molecules resulting from the dipolar nature of the molecule, plays the major role in giving water its many unusual properties.

Water is a great **solvent** because its dipolar molecules can attach themselves to charged particles, **ions,** that make up many substances such as sodium

chloride, hydrate them, and put them into solution.

The presence of the hydrogen bond also accounts for the unusual **thermal properties** of water such as its high **freezing point** and **boiling point,** high **heat capacity,** and high **latent heat of melting** and **latent heat of vaporization.**

The **surface tension** phenomenon that makes water form drops and allows it to be poured into a container until it stands well above the sides of the container results from the water molecules being strongly attracted by the hydrogen bond to those water molecules beside and beneath them. This **cohesive attraction** combined with the **adhesive attraction** of water to glass, organic substances, rock, and soil produces the **capillarity** that pulls water to great heights in small tubular openings.

Salinity is the amount of dissolved solids in ocean water, averaging about 34.7 g of dissolved solids per kg of ocean water. Salinity is usually expressed in **parts per thousand** (‰). Over 99% of the dissolved solids in ocean water are accounted for by six ions, in order of decreasing abundance: **chloride, sodium, sulfate, magnesium, calcium,** and **potassium.** In any sample of ocean water these ions will be in the same relative proportions, making it possible to determine salinity by measuring the concentration of only one of them, usually the chloride ion.

The **density** of pure water increases with decreased temperature, as does the density of most substances, to a temperature of 4°C (39.2°F). Below 4°C, its density decreases with decreased temperature. This change is due to the fact that the rate of formation of open ice crystals increases dramatically below 4°C, the **temperature of maximum density** for fresh water. By the time water freezes, the density of the ice is only about 0.9 that of water at 4°C. Density increases with increased salinity and pressure. Due to the fact that the components of salinity have a greater density than water, increased salinity increases the density of the ocean water. The **residence time** of elements in the ocean water ranges from 100 yr for aluminum to 260,000,000 yr for sodium. The time required to totally recirculate the water in the oceans, its residence time, is 4000 yr. To understand the relative chemical reactivity of elements, the residence times of their soluble fraction must be determined because the residence time of particles is affected primarily by physical mixing.

The GEOSECS program has helped give oceanographers an increased understanding of physical and biological processes of the oceans. It features an advanced system of sampling and chemically analyzing the properties and content of ocean water.

QUESTIONS AND EXERCISES

1. Describe the condition that exists in the water molecule to make it dipolar.

2. Discuss how the dipolar nature of the water molecule makes it such an effective solvent for ionic compounds.

3. Define freezing point and boiling point.

4. There is a fundamental difference between the intermolecular bonds that result from the dipolar nature of the water molecule (hydrogen bond) and the van der Waals force as compared to the chemical bonds (covalent and ionic). What is it?

5. Why are the freezing and boiling points of water higher than would be expected for a compound of its molecular makeup?

6. How does the heat capacity of water compare with that of other substances? Describe the effect this characteristic of water produces in climate.

7. Why does the heat energy added as latent heats of melting and vaporization for water not produce an increase in temperature? Why is the latent heat of vaporization so much greater than the latent heat of melting?

8. Describe how excess heat energy absorbed by the earth's low-latitude regions could be transferred to the heat-deficient higher latitudes through a process that makes use of water's latent heat of evaporation.

9. How does hydrogen bonding produce the surface tension phenomenon of water?

10. As water cools, two distinct changes in the behavior of molecules take place. One tends to increase density while the other decreases density. Describe how the relative rates of their occurrence cause pure water to have a temperature of maximum density at 4°C (39.2°F) and makes ice less dense than liquid water.

11. As water becomes more saline, the temperature of maximum density and the freezing temperature of water decrease and converge. At what salinity does water cease to have a temperature of maximum density above its freezing temperature?

12. What condition of salinity makes it possible to chemically determine the total salinity of ocean water by measuring the concentration of only one constituent, Cl^-?

13. If there is an estimated 18×10^{20} g of magnesium (Mg) in the ocean, and it is being added (or removed) at the rate of 56.5×10^{12} g per year, what is the residence time of magnesium in the ocean? (*See* appendix I.)

REFERENCES

Davis, K. S., and Day, J. S. 1961. *Water: The mirror of science*. Garden City, N. Y.: Doubleday.

Goldberg, E. D., and Arrhenius, G. O. 1958. Chemistry of Pacific pelagic sediments. *Geochimica et Cosmochimica Acta* 13:152–212.

Hammond, A. L. 1977. Oceanography: Geochemical tracers offer new insight. *Science* 195:164–66.

Harvey, H. W. 1960. *The chemistry and fertility of sea waters*. New York: Cambridge University Press.

Kuenen, P. H. 1963. *Realms of water*. New York: Science Editions.

MacIntyre, F. 1970. Why the sea is salt. *Scientific American* 223:104–15.

Pickard, G. L. 1975. Descriptive physical oceanography. 2nd ed. New York: Pergamon Press.

Revelle, R. 1963. Water. *Scientific American* 209:93–108.

Yokoyama, Y.; Guichard, F.; Reyss, J-L; and Van, N. H. 1978. Oceanic residence times of dissolved beryllium and aluminum deduced from cosmogenic tracers [10]Be and [26]Al. *Science* 201:1016–17.

SUGGESTED READING

Sea Frontiers

Gabianelli, V. J. 1970. Water—the fluid of life. 16:5, 258–70.
   The unique properties of water are lucidly described and explained. Topics covered include hydrogen bond, capillarity, heat capacity, ice, and solvent properties.

Smith, F., and Charlier, R. 1981. Saltwater fuel. 27:6, 342–49.
   The potential for using the salinity difference between river water and coastal marine water to generate electricity is considered.

Scientific American

Baker, J. A., and Henderson, D. 1981. The fluid phase of matter. 245:5, 130–39.
   The structure of gases and liquids is modeled using hard spheres.

MacIntyre, F. 1970. Why the sea is salt. 223:5, 104–15.
   A summary of what is known of the processes that add and remove the elements dissolved in the ocean.

CHAPTER SEVEN
AIR-SEA INTERACTION

In this first chapter on physical oceanography we will consider some properties of ocean water, the Coriolis effect, the heat budget of the ocean, and the process by which solar energy is converted into kinetic and heat energy of moving masses of ocean water.

PHYSICAL PROPERTIES OF OCEAN WATER

Light Transmission

The color of ocean water ranges from a deep indigo blue in tropical and equatorial regions of little biological productivity to a yellow green in the coastal waters of high-latitude areas where biological productivity is carried on seasonally at a very high rate. The blue of the tropical waters where there is little particulate matter is due to scattering by water molecules of wavelengths of solar radiation, which we perceive as the color blue. This process is also responsible for the blue color of the sky. Higher concentrations of dissolved organic matter, especially where microscopic plants are abundant, result in a greater scattering of longer wavelengths. This produces the greenish color characteristic of coastal waters.

The many forms of electromagnetic energy radiated by the sun may be seen in figure 7–1—the *electromagnetic spectrum*. The very narrow segment of this spectrum designated as visible light may be broken down by wavelength into violet, blue, green, yellow, orange, and red. Combined, these different wavelengths of light produce white light. The shorter-wavelength forms of energy to the left of visible light represent highly dangerous forms of electromagnetic energy, including X rays and gamma rays. To the right of the visible segment of the electromagnetic spectrum, we find longer-wavelength forms of energy that are the basis for the technological fields dealing with heat transfer and communication. We will consider further the electromagnetic spectrum when we consider the role of solar energy in setting the ocean masses in motion. Now only that portion of the spectrum that is manifested as visible light will be discussed.

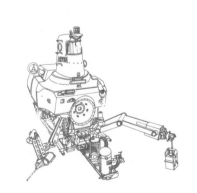

FIGURE 7–1 The Electromagnetic Spectrum and Transmission of Visible Light in Water.

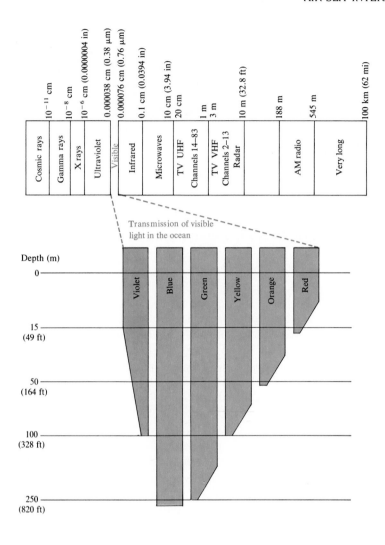

The reason we see things in color is because objects reflect wavelengths of light that represent the colors of the visible spectrum. If those wavelengths of light are not present in the light that falls upon an object, colors cannot be seen. In the ocean the absorption of visible light is greater for the longer-wavelength colors, thus the shorter-wavelength portion of the visible spectrum is transmitted to greater depths. As a result of this condition, the red wavelengths are absorbed within the uppermost 15 to 20 m (49 to 66 ft), the yellow will disappear before a depth of 100 m (328 ft) has been reached, but green light can still be perceived down to 250 m (820 ft). Only the blue and some green wavelengths extend beyond these depths, and their intensity becomes low. It is because of this pattern of absorption that objects in the ocean usually appear to be blue green. It is only in the surface waters that the true colors of objects can be observed in natural light, since it is only in the surface waters that all of the wavelengths of the visible spectrum can be found.

To measure the transmission of visible light in the ocean, a device that measures the degree to which light is scattered is used. Increased scattering of light indicates increased *turbidity*—the amount of suspended material in the water. Increased turbidity increases the rate of absorption, thus decreasing the transmission of visible light.

Density

Density is expressed in grams per cubic centimeter and ranges from 1.02200 to 1.03000 g/cm^3 in the open ocean. Density of ocean water is increased by an increase in salinity or pressure or by a decrease in temperature.

Oceanographers convert density to a more convenient term, σ_t (sigma tee). Normally the density of ocean water is expressed to five decimal places. We will frequently use, throughout our following discussions, σ_t in place of density. To derive it, ocean water density is first converted to specific gravity by dividing it by the density of pure water at 4°C (39.2°F):

$$\text{Specific gravity} = \frac{\text{Density of ocean water}}{\text{Density of pure water}}$$
$$= \frac{1.02567 \text{ g/cm}^3}{1.00000 \text{ g/cm}^3} = 1.02567$$

This eliminates the density units, g/cm^3. The specific gravity of ocean water is then inserted into the following equation to convert it to σ_t.

$$\sigma_t = (\text{Specific gravity} - 1) \times 1000$$

$$25.67 = (1.02567 - 1) \times 1000$$

Not a function of pressure, σ_t is useful in comparing the density of surface water and deep water.

It can be determined by observing the temperature-salinity-density relationships in figures 7–2 and 7–3 that temperature change can be expected to have a much greater effect on density in the high-temperature, low-latitude areas than in polar regions. It can be noted that the σ_t *isopleths* (lines of constant density) are more nearly parallel to the temperature axis for low temperatures than for high temperatures, showing a greater density change per unit of temperature change in the high-temperature range.

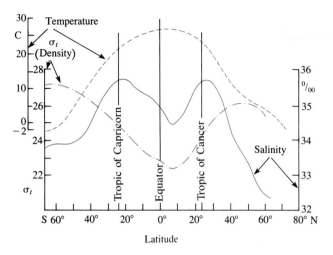

FIGURE 7–3 Average Temperature, Salinity, and Density of World Ocean by Latitude.

Sigma tee increases from about 22 near the equator to maxima of 26–27 near 50°–60° latitude in the surface water of the ocean. At higher latitudes it decreases slightly. Salinity maxima in the tropics do not seem to affect density, pointing out the importance of temperature in controlling density in low latitudes. After G. L. Pickard, *Descriptive physical oceanography* © 1963. By permission of Pergamon Press Ltd., Oxford, England.

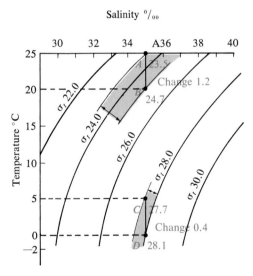

FIGURE 7–2 Temperature-Salinity-Density Relationship.

In high-latitude areas characterized by low-temperature water, temperature has less effect on density (sigma tee, σ_t) than in the high-temperature, low-latitude areas. Measured at a salinity of 35‰, σ_t values determined for points A, B, C, and D show the density change is greater over a 5° temperature span at a higher temperature range. The change in σ_t across a temperature range of 20–25°C (68°–77°F) is 1.2 compared to a change of only 0.4 across an equal range at the lower temperatures of 0–5°C (32–41°F). Thus, a change in the temperature of warm, low-latitude water has about three times the effect on density that an equal change in temperature occurring in colder, high-latitude waters has. After G. L. Pickard, *Descriptive physical oceanography* © 1963. By permission of Pergamon Ltd., Oxford, England.

Sigma tee and the variables that most affect it, temperature and salinity, are important water characteristics, defined as *conservative properties,* used in the identification of water masses. Conservative properties of water are those that are not significantly altered by any processes other than mixing and diffusion once the water sinks beneath the surface.

A *nonconservative property* of water is changed by some process other than mixing and diffusion after ocean water sinks beneath the surface. An example is dissolved oxygen, which will be changed by biological processes.

Since the gain or loss of heat energy at the ocean surface is of primary importance in determining density characteristics of the ocean water, we find, as with most water properties, that the rate of change in density is greater in a vertical direction than in a horizontal direction.

The density of surface water is affected primarily by temperature changes in the open ocean. In the extreme high-latitude areas of the ocean, however, where temperatures remain relatively constant, salinity changes can have a significant effect on density. Figure 7–4 shows the vertical distribution of density in various regions of the ocean. In the equatorial region, there is a shallow zone of low-density water near the surface, and below this lies a zone where the increase in density is very rapid. This zone of rapidly

FIGURE 7–4 Vertical Density Profiles for Various Latitudinal Regions.

After G. L. Pickard, *Descriptive physical oceanography* © 1963. By permission of Pergamon Press Ltd., Oxford, England.

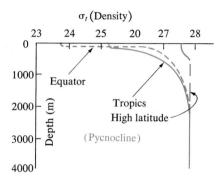

A. Because of natural tendencies toward stability, density can normally be expected to increase with depth. In low-latitude equatorial and tropical regions a thin layer of low-density surface water is separated from high-density deep water by a zone of rapid density change, the pycnocline. The pycnocline is poorly developed or missing in high-latitude waters. It is the absence of the stable pycnocline that facilitates vertical mixing in some high-latitude areas.

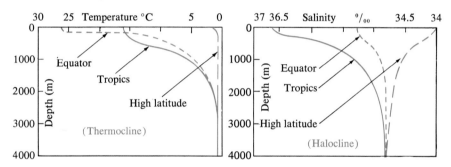

B. The gravitationally stable pycnocline described above is primarily the result of and coincides with a zone of rapid vertical decrease in temperature, the thermocline. The thermocline is also more highly developed in low latitudes. A zone of rapid vertical change in salinity, the halocline, may also affect density. The halocline usually represents a decrease in salinity with increasing depth in low latitudes but may represent the opposite condition in high-latitude areas (note that in B the temperature increases to the left. Although this is unconventional, it was done to make the curves easier to compare).

increasing density is called the *pycnocline* (density slope) and represents a very stable barrier to the mixing of the low-density water above and the high-density water below. The pycnocline is considered to have high *gravitational stability*. It would require a greater amount of energy to move a given mass of water from some point in the pycnocline either up or down than would be required to move an equal mass an equivalent distance at depths where density change occurred very slowly with increasing depth.

The pycnocline is developed as a result of the combined effects of zones of rapid vertical changes in temperature, the *thermocline* (temperature slope), and salinity, the *halocline* (salinity slope). The interrelatedness of these three zones, which determines the degree of separation between the *upper-water* and *deep-water* masses, is shown in figure 7–4.

The pycnocline is lacking in the high latitudes. Because there is not a great difference in the density of the surface waters as compared to the bottom waters in high-latitude areas, the high-latitude water columns are less stable than low-latitude water columns possessing a pycnocline.

Sound Transmission

Sound can be transmitted much more efficiently through water than through air. This characteristic has made it possible to develop systems for determining the positions of objects in the ocean as well as determining the distance to the bottom of the ocean by *sonar* (sound navigation and ranging). The average velocity of sound in the ocean is 1450 m/s (4756 ft/s), compared to 334 m/s (1096 ft/s) in a dry

FIGURE 7–5 Transmission of Sound in the Ocean.
Velocity of sound in the ocean increases with increases in temperature, salinity, and pressure. Generally salinity changes are not as important in affecting the velocity of sound transmission as are the changes in temperature and pressure. Pressure increases steadily with depth, which increases the velocity of sound. However, the rapid decrease in temperature represented by the thermocline offsets the pressure increase and produces a low-velocity sound channel. Wave phenomena, such as sound, bend (refract) into low-velocity areas. This refraction produces shadow zones where velocity maxima exist as well as a channel that traps sound energy in the low-velocity zone, the *sofar* channel.

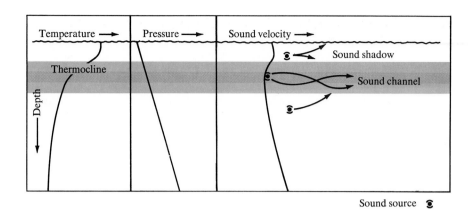

atmosphere at 20°C (68°F). Velocity of sound increases with increases in temperature, salinity, and pressure. Therefore, the velocity with which sound travels through water in a particular area will vary by season. To determine the exact velocity of sound for any point in the ocean, these three variables—temperature, salinity, and pressure—must first be determined, or a *velocimeter* can measure it directly.

At a depth of around 1000 m (3280 ft) there exists a layer of ocean water where these three variables have values that result in relatively low velocity for the transmission of sound. Sound originating above and below this layer will be refracted, or bent, into the low-velocity layer and be trapped there. Thus by following this channel, sound can be transmitted for unusually great distances (fig. 7–5). This channel is called the *sofar* (sound fixing and ranging) channel because of its practical application in distance determination.

Such determinations are made by emitting a sound signal that will travel through the water, bounce off an object, and return to be picked up by a receiver. If the velocity at which the sound will travel in water has been determined, the distance to the object is equal to the velocity times one-half the time required for the sound signal to make the round trip.

SOLAR ENERGY

Distribution of Solar Energy

Essentially all of the energy available to the earth comes from the sun. Solar energy strikes the earth at

an average rate of 2 cal/cm^2/min. Although energy radiated by the sun covers the full electromagnetic spectrum, most of the energy that reaches the earth's surface is contained in wavelengths close to and including the visible portion of the spectrum.

In the upper atmosphere, molecular oxygen (O_2) is broken into individual atoms of oxygen (O) that recombine with molecular oxygen to produce ozone (O_3). This process absorbs most of the solar radiation with wavelengths shorter than 0.29 μm (0.0001 in). Water vapor absorbs a high percentage of solar radiation with wavelengths greater than 0.8 μm (0.0003 in) leaving most of the solar radiation within the visible spectrum 0.38–0.76 μm (0.00015–0.0003 in) to penetrate the atmosphere and reach the earth's surface. After scattering by atmospheric molecules and reflection by clouds in the atmosphere, about 47% of the solar radiation that is directed toward the earth is absorbed by the oceans and continents. Nineteen percent is absorbed by the atmosphere and 34% is lost to space.

If it is true that the earth has maintained a constant average temperature over the years, the earth must be radiating energy back to space at the same rate it is absorbing solar energy. The energy radiating from the earth falls within the infrared range (0.76 μm–0.1 cm, or 0.0003–0.4 in). Figure 7–6 shows the intensity of energy radiated by the sun peaks at 0.48 μm (0.0002 in) in the visible spectrum. It also shows that the radiation from the earth peaks at 10 μm (0.004 in) in the infrared range. While the atmosphere absorbs only 19% of the short-wavelength solar radiation, the water vapor and carbon dioxide in the atmosphere absorb a large amount of the longer-wavelength infrared radiation emitted by the earth. It is this change of wavelength that is the essence of the *greenhouse effect*. Some of the infrared energy absorbed in the atmosphere will be reabsorbed by the earth (the rest being lost to space) to continue the process. Therefore, the solar radiation received is retained for a time within our atmosphere to moderate

the temperature fluctuations between night and day and between seasons. The amount of atmospheric heating varies with latitude because of the unequal distribution of solar radiation on the earth's surface and water vapor in the atmosphere.

Considering the earth as a sphere, figure 7–7 shows that radiation reaching its surface will strike at an angle of 90° near the center of the sphere that is lit by this radiation, while it will be striking at 0° at the edge of this circle (solar rays will be tangent to the earth's surface). If we consider a 1 km² (0.4 mi²) cross-sectional area of solar radiation that falls near the center of the illuminated portion of the sphere, this ray of light will cover exactly 1 km² of the earth's surface. By contrast, if we consider the same cross-sectional area falling on a high-latitude portion of the earth's surface, where it does not strike the earth at a right angle, this square-kilometer beam will be spread out over considerably more than 1 km² of the earth's surface. The intensity of radiation available per unit of surface area will be greatly decreased over that available in the lower latitudes.

As the earth rotates on its axis, which is inclined 23.5° from being perpendicular to the ecliptic (plane of earth's orbit), there will be significant portions of its surface above 66.5°N latitude, the *Arctic Circle*, and 66.5°S latitude, the *Antarctic Circle*, that will spend up to six months in darkness. The direct rays of the sun will migrate back and forth across the equator between the *Tropic of Cancer* and the *Tropic of Capricorn* 23.5° north and south of the equator.

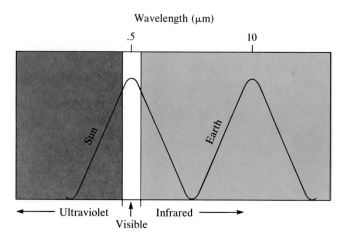

Wavelength (μm)

FIGURE 7–6 Comparison of Spectrums of Energy Radiated by the Sun and the Earth.

Most of the energy coming to the earth from the sun is within the visible spectrum and peaks at 0.48 μm (0.0002 in). The atmosphere is transparent to most of this radiation, and it is absorbed at the earth's surface. When the earth reradiates this energy back toward space, it does so as infrared radiation with a peak at a wavelength of 10 μm (0.004 in). The CO_2 and H_2O in the atmosphere absorb this infrared radiation to produce the greenhouse effect.

FIGURE 7–7 Solar Radiation.

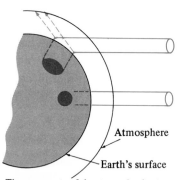

The amount of heat received at higher latitudes is less than that at lower latitudes for three reasons: (1) a ray of solar radiation that strikes the earth at a high latitude is spread over more area than an equal ray that is perpendicular to the earth's surface at a lower latitude, (2) the high latitude ray also passes through a greater thickness of atmosphere, and (3) more of its energy is reflected due to the low angle at which it strikes the earth's surface.

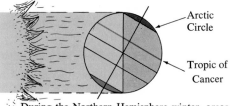

During the Northern Hemisphere winter, areas north of the Arctic Circle receive no direct solar radiation, while areas south of the Antarctic Circle receive continuous radiation for months.

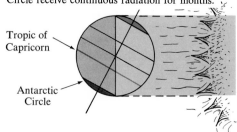

The reverse of the above situation is true during the Northern Hemisphere summer. Throughout the year, the sun is directly over some latitude between the tropics.

The belt between the two tropics will receive a much greater amount of radiation per unit of surface area during a year than will the portions of the earth's surface north of the Arctic Circle and south of the Antarctic Circle.

Heat Balance of the Earth

Because of the different angles at which solar radiation strikes the earth's surface and the very highly reflective characteristics of the ice cover found in high-latitude areas of the earth, there is more energy absorbed than is radiated back into space in latitudes lower than 35°N and 40°S, while there is less energy absorbed than is lost to space in latitudes higher than 35°N and 40°S. Figure 7–8 shows how this phenomenon is manifested in the average daily heat flow of the oceans of the Northern Hemisphere.

As a direct result of this condition, one might expect that the equatorial zone will continually get warmer as the years pass and the polar regions will become progressively cooler. Such is not the case. Although the polar regions are considerably colder than the equatorial zone, the temperature difference does not appear to be increasing with time. In order to explain this condition, we must conclude that the excess heat that is absorbed in the low latitudes is

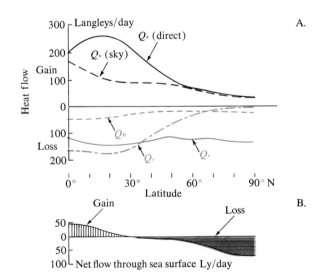

FIGURE 7–8 Heat Flow: Ocean-Atmosphere.
Heat flow through the ocean surface is represented as a function of latitude. At all latitudes there is a gain from solar radiation (Q_s) and a loss from back radiation (Q_r), conduction (Q_h), and evaporation (Q_e). *A*, the amount of heat lost from the ocean by back radiation, conduction, and evaporation is less than that gained from solar radiation below 30° latitude. *B*, the heat loss exceeds heat gain above 30° latitude. The langley (ly) is a calorie of heat flow through 1 cm^2 of the ocean surface. After G. L. Pickard, *Descriptive physical oceanography* © 1963. By permission of Permagon Press Ltd., Oxford, England.

transferred by some mechanism into the higher latitudes. Thus a more or less stable climatic difference is maintained between the two regions. The mechanism by which such transfer is achieved involves both the oceans and atmosphere.

Figure 7–9 shows graphically that the greater heating of the atmosphere over the equator causes air to decrease in density and rise. As it rises, it cools by expansion, and the water vapor contained in the rising air mass condenses and falls as precipitation in the equatorial zone. After losing its moisture, this dry air mass descends. The descents occur in the tropical to subtropical regions of the Northern and Southern Hemispheres. As the descending air approaches the earth's surface, it is warmed by compression. Upon reaching the surface it moves away from the tropics, toward either the equator or higher latitudes. A high-pressure belt develops in the subtropics (30° latitude) owing to the dense air that descends in these regions. Between the *subtropical high-pressure belts*, the horse latitudes, lies an *equatorial low-pressure belt* resulting from the low density of the rising air column above the equator.

The masses of air that move across the earth's surface from the subtropical high-pressure belts toward the equatorial low-pressure belt constitute the *trade winds*. Some of the air that descends in the subtropical regions moves along the earth's surface to higher latitudes as the *westerly wind belts*. These masses rise over the dense, cold air moving away from the *polar high-pressure caps* at the *subpolar low-pressure belts* located near 60°N and 60°S latitude. The air moving away from the poles produces the *polar easterly wind belts*. The air that rises at the 60° latitudes cools, releases precipitation in these regions, and ultimately descends in the polar regions or in the subtropics. These idealized latitudinal patterns are significantly altered by the uneven distribution of land and ocean over the earth's surface. The general effects of this idealized system are, however, clearly visible on a very broad scale. In the following paragraphs, we will examine how these air masses are affected by the earth's rotation as they move across the earth's surface.

CORIOLIS EFFECT

Any freely moving object on the earth's surface moving horizontally through a long distance for a relatively long period of time will veer to the right in the Northern Hemisphere and to the left in the Southern Hemisphere. The magnitude of this effect will increase with the velocity and the latitude of the object.

FIGURE 7-9 Air Masses.

Low-pressure belts of precipitation develop along the equator and the 60° latitude regions owing to rising columns of relatively warm air. Descending cool dry air produces high-pressure belts in the 30° latitude regions. Air movements are primarily vertical in these belts. Strong lateral movements of air between the belts produce the westerlies and trade winds. Because of cold air masses overlying the polar regions, high-pressure conditions exist there also.

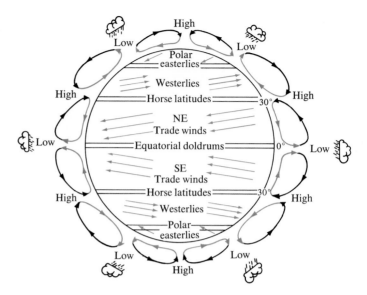

This phenomenon is called the *Coriolis effect*. It is zero at the equator and maximum at the poles.

We will, for the sake of our following discussion, consider only the Northern Hemisphere. The air masses moving away from the subtropics toward the equator and high latitudes are moving basically south in the first case and north in the second. In the case of the trade winds, they do not appear to blow directly out of the north but out of the northeast, while the westerlies, as their name implies, blow out of the southwest. Why do we observe this east-west component of motion in air masses that are moving north and south?

Figure 7-10A shows that as the earth rotates on its axis, points at different latitudes rotate at different velocities. The rotational velocity is proportional to the distance from the earth's surface to the axis of rotation and ranges from 0 km/h at the poles to over 1600 km/h (994 mi/h) at the equator.

The trade winds blow out of the northeast because as the air mass moves south from the subtropical region near 30°N latitude, it is also moving with the rotating earth in an easterly direction at about 1400 km/h (870 mi/h). The air mass starts moving toward a point on the equator at the same longitude that is moving east at 1600 km/h (994 mi/h). The distance that the air mass must cover to reach the equator is about 3200 km (2000 mi). If it moves to the south at 32 km/h (20 mi/h), it will require 100 h to arrive at the equator (fig. 7-10B).

The air mass is moving south at 32 km/h and east at 1400 km/h. For every hour that the air mass moves in the southerly direction, the point on the equator toward which it started moves east 200 km (124 mi) farther than does the air mass (1600 km/h − 1400 km/h). In the period of 100 h it takes the air mass to

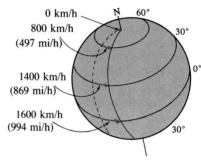

A.

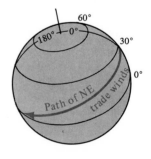

B.

FIGURE 7-10 Coriolis Effect.

A, the speed of rotation for points at different latitudes increases from 0 km/h at the pole to a maximum of 1600 km/h (994 mi/h) at the equator. This results because points at lower latitudes must travel through larger circular paths during each 24-hour rotation. *B,* an air mass representing the northeast trade winds travels south from 30°N latitude to the equator (0° latitude). Observed at the earth's surface, the air mass veers to the right of its southerly course and is observed to come out of the northeast.

make the trip, the point on the equator toward which it started will be 20,000 km (100 h × 200 km/h) (12,420 mi) east of the air mass when it reaches the equator. While the air mass was moving the 3200 km in a southerly direction, it appears, as a result of the earth's rotation, to have moved 20,000 km in a westerly direction. Stated another way, as the air mass covered 30° of latitude from north to south, it appeared to move across 180° of longitude in a westerly direction. Certainly, if we were to encounter this air mass aboard ship at some point between the equator and 30°N latitude, it would be coming out of the east-northeast.

A westerly wind air mass moves from 30° latitude in a northerly direction to 60° latitude. This mass also is moving in an easterly direction at 1400 km/h. The point at 60° latitude toward which the air mass is headed is moving in an easterly direction at only 800 km/h (497 mi/h). Unlike the situation discussed in regard to the trade winds, the westerly air mass is moving in an easterly direction at a velocity greater than that of the point at 60°N latitude toward which it started. For every hour that this mass moves in a northerly direction, it moves 600 km (373 mi) farther east than the point at 60° latitude toward which it started. If we were to encounter this air mass at some point between 30° and 60°N latitude, it would appear to come out of the west-southwest.

If the air mass moves north at the same velocity as a trade wind mass moves south, the amount of deviation to the right in the case of the westerlies will be greater than for the trade winds. This results from the fact that the condition responsible for the intensity of the Coriolis effect is the different rates at which points at different degrees of latitude rotate about the earth's axis. The rotational velocity of these points ranges from 0 km/h at the poles to more than 1600 km/h at the equator, but the rate of change per degree latitude is not constant. As we approach the pole from the equator, the rate of change of rotational velocity per degree of latitude change increases. As a comparison, we can note that there is a difference of 200 km/h in velocity across 30° of latitude from the equator to 30°N latitude, while there is a difference of 600 km/h in the velocity of rotation from 30°N to 60°N latitude. If we carry this consideration over another 30° of latitude from 60°N latitude to the pole, where the velocity is zero, the difference is over 800 km/h. This explains the fact that the Coriolis effect increases with increased latitude.

The primary factor affecting the amount of deflection resulting from the Coriolis effect is not, however, the magnitude of the effect since it is very small, even at high latitudes. It is the length of time a par-

ticle is in motion that will have the greatest influence on how much it is deflected. Thus, even at low latitudes, a large Coriolis deflection is possible if an object is in motion for a long time.

Although we will not attempt to discuss it here, a complete mathematical treatment of the rotating earth shows that the effects described for these north–south moving masses are valid for movements in any direction. Freely moving objects veer to the right of their intended path in the Northern Hemisphere and to the left in the Southern Hemisphere.

HEAT BUDGET OF THE WORLD OCEAN

As a result of variations in the absorption of solar energy, ocean temperatures will vary from place to place and from time to time. We previously discussed the fact that the temperature difference between polar and equatorial regions remains constant as a result of a transfer of heat energy from the equator to the higher latitudes by moving air and ocean masses. Driven by the moving air masses, large surface current systems are set up in the world's oceans and play an important role in transfer of heat energy. The temperatures at various places in the ocean will depend upon the rate at which heat flows in and out of these areas. Quantitative considerations of heat transfer in ocean water masses constitute the *heat budget* of these masses (fig. 7–11).

The heat budget for a given *locality* in the ocean is expressed by the following equation:

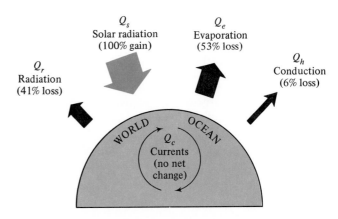

FIGURE 7–11 Avenues of Heat Flow between the Ocean and the Atmosphere.

Solar radiation (Q_s) provides heat to the oceans. It is circulated within the ocean by currents (Q_c). Heat is lost from the ocean through evaporation (Q_e), radiation (Q_r), and conduction (Q_h). Heat loss equals heat gain.

Rate of heat gain − Rate of heat loss
= Net rate of heat loss or gain

$$(Q_s + Q_c) - (Q_r + Q_e + Q_h) = Q_t$$

in which

Q_s = the rate of heat gain from solar radiation

Q_c = the rate of heat gain (or loss) through ocean current

Q_r = rate of heat loss through radiation

Q_e = rate of heat loss through evaporation

Q_h = rate of heat loss through conduction into the atmosphere

Q_t = net rate of heat loss or gain in the locality

If Q_t is a positive value, it means the temperature of the ocean in that locality is rising. If it is a negative value, the temperature of the mass of water is decreasing. If it is zero, the temperature is not changing.

Theoretically, Q_c and Q_t should be zero when we consider the *world ocean* over a long period of time. Current flow is internal and seasonal gains and losses of heat should average out so that Q_t becomes zero. Therefore, the rate at which heat is gained in the world ocean through solar radiation should equal the rate at which heat is lost through radiation, evaporation, and conduction into the atmosphere (fig. 7–11). Symbolically the heat budget of the world ocean is expressed using average values as:

$$Q_s = Q_r + Q_e + Q_h$$

$$100\% = 41\% + 53\% + 6\%$$

Q_s—Heat Gain from Solar Radiation

Comparing daily rates of heat inflow due to solar radiation, some interesting Q_s values develop. At the equator, where the greatest rate of heat inflow occurs, the maximum daily Q_s variation during a year is less than 12%. In the temperate latitudes between 40° and 50° the rate of heat inflow due to solar radiation may be 450% higher during midsummer than during midwinter. At the poles Q_s is essentially zero throughout much of the winter; during midsummer it reaches values approaching the maximum daily heat inflow through solar radiation measured on the equator. This condition results, of course, from the fact that the poles are exposed to 24 hours of solar radiation every day during midsummer.

Heat Loss through Radiation (Q_r), Evaporation (Q_e), and Conduction (Q_h)

The amount of heat lost primarily through infrared radiation is greatest at low latitudes, where the greatest amount of shorter-wavelength solar radiation is absorbed by the earth. It is the heating of the atmosphere by this heat loss at the equator that causes the rising air column and low-pressure belt associated with this region. Forty-one percent of the world ocean heat loss is by radiation *(Q_r)*.

The heat required to evaporate the average 1 m (39 in) of water taken from the ocean surface by this process each year is tremendous. Since much of this heat comes from adjacent water molecules, a significant amount of heat is lost from the oceans through evaporation each year. The net amount of water evaporated from the ocean surface varies considerably—from a latitudinal average of about 5 cm/yr (2 in/yr) in high latitudes to a maximum of 140 cm/yr (55 in/yr) in the subtropics. At the equator, where precipitation is heavy, the net loss is reduced to 110 cm/yr (43 in/yr) (fig. 7–12). At a temperature of 20°C (68°F) the latent heat of evaporation is 585 cal/g, so each gram of water removed from the ocean by evaporation represents a heat loss of 585 cal. Fifty-three percent of the world ocean heat loss is by evaporation *(Q_e)* (plate 16).

Under any stable condition, heat is conducted through molecular movement. When we consider the flow of heat between the ocean and the atmosphere by conduction, *turbulence* becomes a very important factor. With the water mass and air mass in motion, the rate of transfer of heat by conduction is much greater than would occur solely through random molecular movement.

The basic factor determining the direction of heat flow by conduction, ocean to atmosphere or atmosphere to ocean, is the *temperature gradient*. If the air above the ocean is cooler than the ocean, heat will be lost to the atmosphere. As a result, the greatest rates of heat loss through conduction occur along the west-

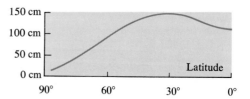

A column of water ranging from less than 5 cm in high latitudes to a maximum of 140 cm in the subtropics is evaporated from the ocean each year.

FIGURE 7–12 Net Water Loss through Evaporation *(Q_e)*.

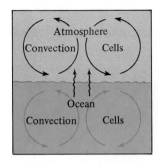 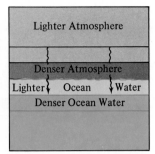

As molecules of atmospheric gases gain heat by conduction from the ocean surface, they rise, setting up convection cells that carry in low energy molecules to absorb more heat. Cooling surface water may also sink producing convection in the ocean. This facilitates the process of heat *loss* from the ocean surface through conduction.

As heat is lost from the atmosphere through conduction to the ocean, the lower layer of the atmosphere becomes denser and settles into a stable horizontal layer immediately above the ocean. The warming upper layer of the ocean becomes lighter and stays at the surface. This results in a low rate of heat *gain* to the ocean by conduction.

FIGURE 7–13 Heat Loss through Conduction (Q_h).
Heat is more readily lost than gained by the ocean surface water through conduction.

ern margins of ocean basins in temperate latitudes, where warm surface currents move into regions of lower air temperature. Heat loss by conduction will be aided by convection as air is heated and rises, pulling denser cooler air in to replace it and absorb additional heat energy from the ocean surface. If the air is warmer than the underlying water and heat is lost from the atmosphere to the ocean, convection currents do not develop because the lower layers of air are losing heat and becoming denser. They tend to settle in a very stable manner to the bottom of the atmosphere column (fig. 7–13). Therefore the rate of heat gain to the ocean will be considerably lower than the rate of heat loss for the same number of degrees of temperature difference between the ocean and atmosphere in the opposite direction. Only 6% of the world ocean heat loss is by conduction (Q_h).

OCEANS, WEATHER, AND CLIMATE

The idealized pressure belts and consequent wind systems previously discussed are significantly modified as a result of two factors. First, the tilt of the earth's axis of rotation produces seasons, and second, the air over continents gets colder in winter and warmer in summer than the air over adjacent oceans. As a result, continents usually develop atmospheric high-pressure cells that reflect the weight of the cold air centered over them during winter and low-pres-

sure cells during the summer (fig. 7–14). As air moves away from the high-pressure cells and toward the low-pressure cells, the Coriolis effect produces a cyclonic (counterclockwise) flow of air around low-pressure cells and an anticyclonic (clockwise) flow of air around high-pressure cells in the Northern Hemisphere (fig. 7–15). Thus, wind patterns associated with continents may reverse themselves on a seasonal basis as winter high-pressure cells are replaced by summer low-pressure cells.

Day-to-day weather may change little at high and low latitudes. Polar regions are usually cold and dry regardless of the season. Near the equator, the air is warm, damp, and still because the dominant direction of air movement in the doldrums belt is up. Midday rains are common. It is at the midlatitudes that weather gets interesting. Owing to the seasonal change of pressure systems over continents, air masses from the high and low latitudes may move into the mid-latitudes, meet, and produce severe storms. In the United States, we may be invaded by three major polar air masses and two tropical air masses (fig. 7–16).

As polar and tropical air masses move into the midlatitudes, they are also gradually moving in an easterly direction. A *warm front* is the contact between a warm air mass moving east into an area occupied by cold air. A *cold front* is the contact between a cold air mass moving east into an area occupied by warm air (fig. 7–17). These eastward movements are brought about by the movement of the *jet stream*, an easterly moving air mass centered at about 10 km (6.2 mi) elevation above the midlatitudes. It usually follows a wavy path and may cause unusual weather by pulling a polar air mass far south or a tropical air mass far to the north. Regardless of whether the colliding air masses create a cold front or a warm front, the warm air rises above the denser cold air, cools, and the moisture in it condenses as precipitation. A cold front will usually be steeper, and the temperature differences across it will be greater. Therefore, rainfall associated with a cold front will usually be heavier and of shorter duration than that resulting from a warm front.

Climate Patterns in the Oceans

The open ocean can be divided into climatic regions with relatively stable boundaries that run generally east–west (fig. 7–18). Temperature and salinity of surface waters are determined by the amount of solar radiation and the amounts of evaporation and precipitation.

In *equatorial* regions the major air movement is vertical as air rises, so winds are weak. Surface wa-

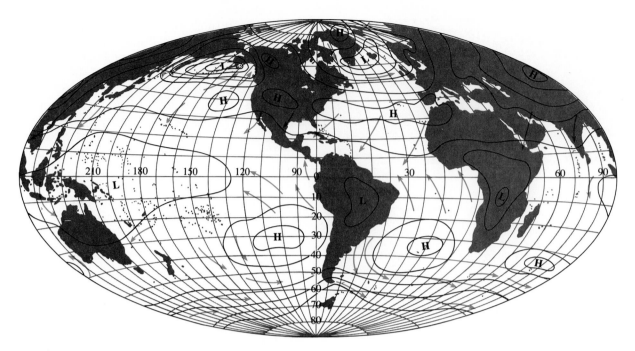

FIGURE 7–14   Sea-Level Atmospheric Pressures and Winds.

The idealized high- and low-pressure belts shown in figure 7–9 are modified by the seasons and distribution of continents. Pressure patterns shown in this figure are the averages for the month of January. Because continental rocks have a lower heat capacity than the water of the oceans, continents get hotter in the summer and colder in the winter than the adjacent oceans. The unusually cold air over the continents of the Northern Hemisphere and unusually warm air over the continents in the Southern Hemisphere produce the high- and low-pressure cells centered over the continents. Courtesy of NOAA.

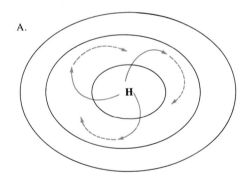
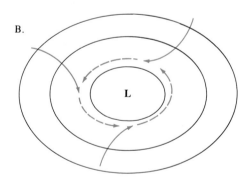

FIGURE 7–15   Coriolis Effect on Low- and High-Pressure Cells.

As air moves away from high-pressure cells (A) and toward low-pressure cells (B), the Coriolis effect causes the air masses (winds) to veer to the right in the Northern Hemisphere. This results in anticyclonic clockwise winds around high-pressure cells and cyclonic counterclockwise winds around low-pressure cells. The concentric lines are called *isobars*, lines of equal pressure. The winds blow fairly parallel to the isobars as they move around the pressure cells. Cold, dry air characteristically descends within high-pressure cells, producing clear skies. Low-pressure cells contain rising air, which cools to produce massive cloud banks as a result of condensation of water vapor.

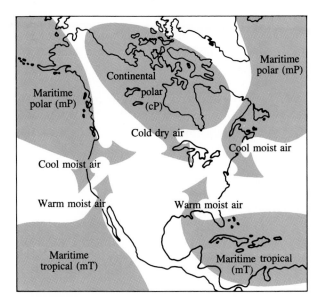

FIGURE 7–16 Air Masses That Affect Weather in the United States.

The polar air masses are more likely to invade the United States during winter, and the tropical air masses tend to move in from the south during summer.

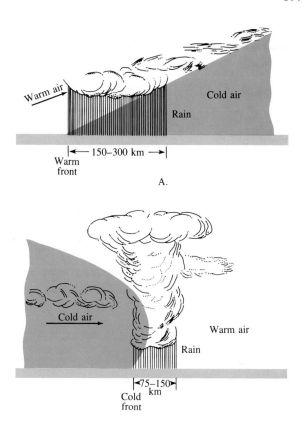

FIGURE 7–17 Warm and Cold Fronts.

Cross sections through a warm front (A) and cold front (B).

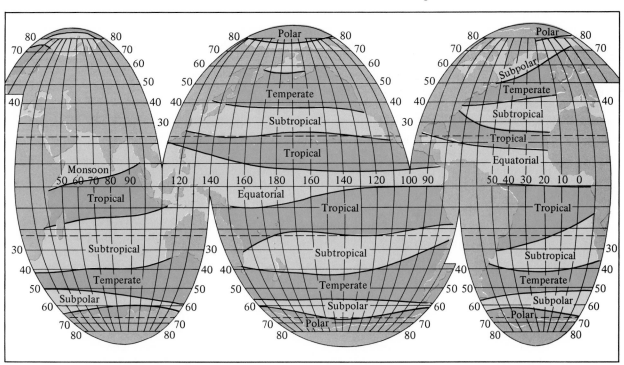

FIGURE 7–18 Climatic Patterns of the Open Ocean.

Base map courtesy of National Ocean Survey.

ters are warm, and heavy precipitation keeps salinity relatively low. Sailors once referred to this region as the *doldrums* because their sailing ships were becalmed by the lack of winds. Meteorologists call it the *Intertropical Convergence Zone* (ITCZ) because it is the zone between the tropics where the trade winds converge.

Tropical regions are characterized by northeasterly trade winds in the Northern Hemisphere and southeasterly trade winds in the Southern Hemisphere. These winds push the equatorial currents and create moderately rough seas. Relatively little precipitation falls at higher latitudes within tropical regions, but precipitation increases toward the equator. Hurricanes and typhoons that carry large quantities of heat into higher latitudes are initiated here as tropical storms.

The belts of high pressure previously described are centered in the *subtropical* regions. The dry air descending on the subtropics results in little precipitation and a high rate of evaporation, producing the highest surface salinities in the open ocean. Winds are weak in the open ocean as are currents. However, strong boundary currents flow north and south, particularly along the western margins of the subtropics.

The *temperate* regions are characterized by seasonal temperature changes and strong westerly winds blowing from the southwest in the Northern Hemisphere and from the northwest in the Southern Hemisphere. Severe storms are common, especially during winter, and precipitation is heavy.

The *subpolar* ocean is covered in winter by sea ice that melts away, for the most part, in summer. Icebergs are common, and the surface temperature seldom exceeds 5°C (41°F) in the summer months.

Surface temperatures remain at or near freezing in the *polar* areas, which are covered with ice throughout most of the year. In these areas, which include the Arctic Ocean and the ocean adjacent to Antarctica, there is no sunlight during the winter and no night during the summer.

Sea Ice

Resulting directly from the low-temperature conditions that are characteristic of high-latitude areas is the development of a permanent to nearly permanent ice cover at the sea surface. The term *sea ice* is used to distinguish these masses of ice from *icebergs,* which may also be found at sea but originate by breaking away from glacial masses of ice that form on land.

The freezing point of water with a salinity of 35‰ is −1.91°C (28.6°F). As water freezes at the ocean surface, the dissolved solids do not fit into the crystalline structure of the ice and are left behind so that the salinity of the surrounding water increases. This greater salinity tends to lower the freezing point of the remaining water. However, the low-temperature, high-density water that is excluded from the ice tends to sink and be replaced by warmer, less dense water at the surface. This circulation enhances the formation of sea ice, since freezing is aided by low salinity and calm water conditions. Sea ice is found throughout the year around the margin of Antarctica, within the Arctic Sea, and in the extreme high-latitude region of the North Atlantic Ocean (fig. 7–19A and 7–19B; plate 4).

Sea ice begins to form as small needlelike crystals of hexagonal shape that eventually become so numerous that a *slush* develops. As the slush begins to form into a thin sheet, it is broken up by wind stress and wave action into small disc-shaped *pancakes* (fig. 7–19C). As further freezing occurs, the pancakes coalesce to form *floes.* The rate at which sea ice forms is closely tied to temperature conditions, and large quantities of ice form in relatively short periods of time when the temperature is very low; for example, −30°C (−22°F). Even at low temperatures, the rate of new ice formation will decrease as ice thickness increases, owing to the poor heat conduction of ice. Newly formed sea ice contains a significant quantity of brine that is trapped during the freezing process.

Depending upon the rate of freezing, newly formed ice (new ice) may have a total salinity that ranges from 4‰ to 15‰. The more rapidly it forms, the more brine will be captured and the higher the salinity. After a period of time, the brine will trickle down through the coarse structure of the sea ice, and the salinity of the sea ice will decrease. Normally by the time it is a year old (old ice), sea ice has become relatively pure.

In the Arctic Sea, ice that forms at the sea surface can be classified into one of three categories—pack ice, polar ice, or fast ice. *Pack ice* forms around the margin of the Arctic Sea, extending through the Bering Strait into the Bering Sea and as far south as Newfoundland and Nova Scotia in the North Atlantic. Pack ice reaches its maximum extent during the month of May and breaks up to cover its least area in September (fig. 7–19A). This ice, which can be penetrated by icebreakers, reaches a maximum thickness of about 2 m (6.6 ft) in the winter period. The pack ice is driven primarily by the winds, although it also responds to surface currents, which produce stresses

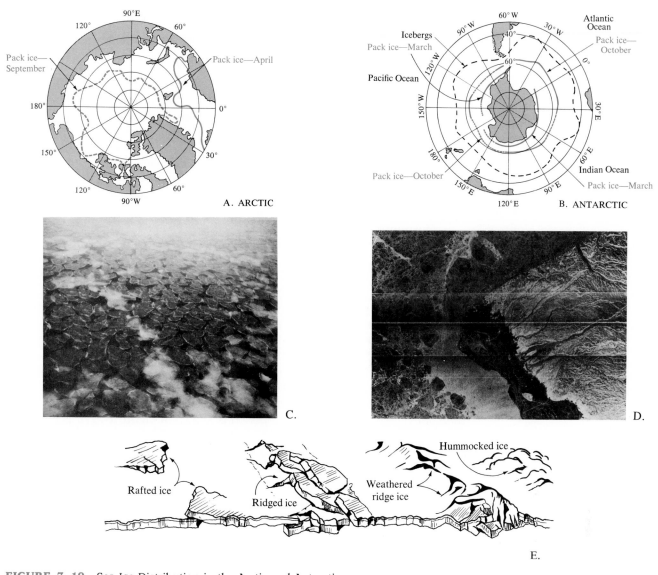

FIGURE 7–19 Sea Ice Distribution in the Arctic and Antarctic.

A, Arctic. *B,* Antarctic. *C,* pancake ice. Photo from Oceanographer of the Navy, Public Affairs Office. *D,* Banks Island, Canada, to the right, is bounded by fast ice (black). A strip of open water (light) separates the fast ice from pack ice along the left margin. Photo courtesy of NASA. *E,* rafted ice.

that continually break and reform its structure. Its formation is achieved by the expansion of floes that begin to raft onto one another as they expand to cover the sea's surface (figs. 7–19D and 7–19E).

The *polar ice* that covers the greatest portion of the Arctic Sea, including the polar region, attains a maximum thickness in excess of 50 m (164 ft). During the summer months, melting may produce enclosed bodies of water called *polynyas,* although the polar ice never totally disappears. Its average thickness during the summer months is over 2 m (6.6 ft). The polar ice is constantly being exchanged as floes from the pack ice are carried into the polar region during

the winter season, and floes break out of the polar ice and reenter the pack ice during the summer. Circling in a clockwise direction around the Arctic Ocean, about one-third of the pack and polar ice is carried into the North Atlantic by the East Greenland Current each year.

Developing in the winter from the shore out to the pack ice is the *fast ice,* which completely melts during the summer. The fast ice, which is firmly attached to the shore, attains a winter thickness in excess of 2 m (6.6 ft).

Figure 7–19D shows a portion of the Beaufort Sea ice pack west of Banks Island, Canada (right), and

SEASAT MEASURES THE OCEAN SURFACE

It has long been the complaint of meteorologists who try to forecast the world's weather that there are never enough data available from the 71% of the earth's surface covered by ocean. They, of course, are joined in their desire to know what is going on "out there" by the seafarers who make their living on the ocean. Many a vessel, from the smallest fishing boat to massive oil tankers, has disappeared with all hands aboard beneath unexpected massive waves that may reach heights of 30 m (98 ft).

Most observations of conditions at sea are provided by ships moving through major shipping lanes—mostly in the Northern Hemisphere. This state of affairs changed for a 100-day period beginning June 28, 1978. On this date, the U.S. satellite *SEASAT* began measuring atmospheric liquid water, water vapor, sea-surface temperature, sea-surface topography, ocean wave height, and wind speed—all on a global basis. Unfortunately, the satellite power supply short-circuited and *SEASAT* stopped gathering data on October 10, 1978. However, the data obtained during the life of *SEASAT* have proven the instrumentation effective in providing useful data on all the atmospheric and oceanographic conditions it was designed to investigate.

The coverage of the world ocean achieved by *SEASAT* during August 9–11, 1978, is shown in figure 7A*A*. For comparison, the distribution of ship observations for the same three-day period is shown in Figure 7A*B*. It is clear that the satellite coverage is much more uniform. Averaging the data collected in strips of the earth's surface 2.5° of latitude by 7.5° of longitude throughout the entire 100-day period resulted in the maps of water vapor, wind speed, and wave height shown on plates 18, 19, and 20.

Average Water Vapor   High concentrations of atmospheric water vapor associated with the southwest monsoon of the northern Indian Ocean exceed 5 g/cm^2 of ocean surface (plate 18). High values are also associated with the Intertropical Convergence Zone, the doldrums, along the meteorological equator a few degrees north of the geographical equator. It is here that the northeast trade winds of the Northern Hemisphere and the southeast trade winds of the Southern Hemisphere converge and rise, producing a low-pressure belt of warm, moist air. Water-vapor concentrations decrease toward higher latitudes. The anomalous apparent high concentrations off Antarctica are thought to be caused by interference effects of ice in the Weddell Sea and Ross Sea.

The lower water-vapor levels in the Southern Hemisphere, compared to those in the same latitudes in the Northern Hemisphere, reflect the colder winter air with its lower capacity for holding water vapor.

Had water-vapor values been available over a longer period of time, it would probably have been possible to greatly increase our understanding of the role of water's latent heat of evaporation in transferring heat from the warm low-latitude ocean to the high-latitude atmosphere.

Average Wind Speed   The prevailing wind systems are evident in plate 19. The northeast and southeast trade winds, with wind speeds within the 6.8 to

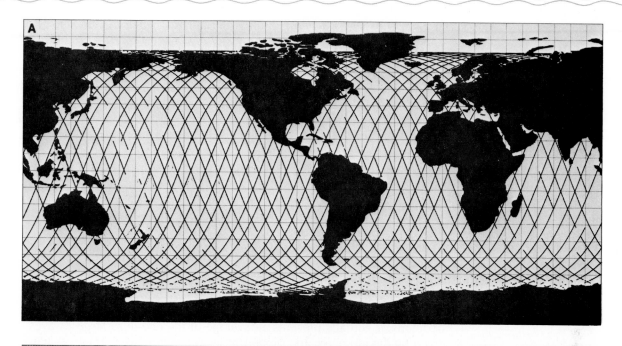

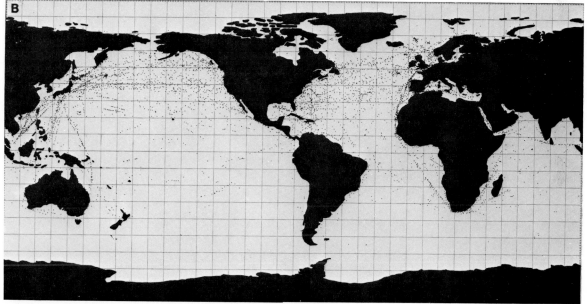

FIGURE 7A Three-Day Global Data Coverage Comparison.

A, from *SEASAT,* compares to *B,* conventional ship weather reports, for August 9–11, 1978.

8.6 m/s (15 to 19 mi/hr) range in the Atlantic and Pacific oceans, are separated by the doldrums along the Intertropical Convergence Zone where horizontal movement (wind speed) drops below 5 m/s (11.2 mi/h). The highest wind speeds at trade wind latitudes occur in the Indian Ocean, where the southeast trade winds of the Southern Hemisphere and the southwest monsoon of the Northern Hemisphere summer season exceed 8.6 m/s (19 mi/h).

The horse latitudes (30° N and S) separating the trade winds from the westerlies also show up as low-speed wind belts. Here dense, dry air is descending in the lower atmosphere to produce high-pressure belts. The westerlies show up as high-speed winds in both hemispheres and are best developed between 50° and 60° latitudes. Within this southern latitudinal belt, the highest wind speeds are recorded. Winds averaging in excess of 10.4 m/s (23 mi/h) occur in the Southern Hemisphere winter storms. The high wind speeds observed within this belt were surprisingly steady, with over 68% of the observations falling between 8 and 12 m/s (17.9 and 26.8 mi/h).

The high wind speeds along Antarctica may represent the polar easterlies moving away from the high-pressure cell of the polar region.

Average Wave Height   Plate 20 gives the first accurate picture of global wave conditions. It actually presents an average of the one-third largest waves observed by the altimeter, the *significant wave height* (SWH). These values are relatively small (about 2 m) in the Northern Hemisphere, where summer wind speed is low. The largest waves are created by the high winter wind speed of the Southern Hemisphere westerlies. Average heights in excess of 4.5 m (14.8 ft) are found in the southern reaches of all oceans, with values in excess of 5.3 m (17.4 ft) observed in the southern Indian Ocean.

The narrowest range of wave conditions occurred at the Intertropical Convergence Zone, where weak wind conditions in the doldrums resulted in a wave field of which over 70% of the waves were less than 2 m (6.7 ft) high, and none exceeded a height of 4 m (13 ft). In contrast, the high speed of the Southern Hemisphere westerlies produced waves of which only 2% were less than 2 m (6.7 ft) high, and over 40% were more than 5 m (16.4 ft) high. In one southern Indian Ocean strip, the average significant wave height was 6.5 m (21.3 ft).

The data indicate that, although the probability of doing so is less than 1%, one might encounter waves as high as 8 m (26.2 ft) at a latitude of 52.5°N and waves of over 11 m (36.1 ft) height at 52.5°S.

Whether or not the data recorded during this three-and-a-half-month period are typical cannot be determined. The quality of the data obtained does, however, demonstrate the unique ability of satellites to measure atmospheric and sea-state conditions on a global basis. Having such data continuously available greatly increases the ability to forecast world weather and sea states.

With such a system of prediction, it is possible that many marine disasters could be prevented. The justification for taking the necessary steps to regain the ability to make such satellite wind and wave observations seems obvious.

Reference: Chelton, Dudley B.; Hussey, Kevin J.; and Parke, Michael E. 1981. Global satellite measurements of water vapour, wind speed and wave height. *Nature*, 294: 5841, 529–532.

covers an area about 100 by 125 km (62.1 by 77.6 mi). The region is northeast of Alaska and some 800 km (497 mi) inside the Arctic Circle. Stream channels, alluvial fans, and beaches are seen on Banks Island. The dark zone adjacent to the island is an area of fast ice composed primarily of first-year sea ice, 1 to 2 m (3.3 to 6.6 ft) thick. Linear *pressure ridges* are seen within the fast zone, and west of this zone is an area of open water, called a *shore lead*. At the western edge of the lead and to the north is a marginal ice zone composed of a mixture of open water and large and small rounded multiyear floes, typically 3 to 5 m (9.8 to 16.4 ft) thick, and some first-year ice. Farther west is the main polar pack, made up of large floes up to 20 km (12.4 mi) in diameter, surrounded by new leads. A random pattern of pressure ridges is visible within the floes. The very bright areas within the floes indicate intensive surface roughness, called *rubble fields.*

In the Southern Hemisphere, where the polar region is covered by a continental mass, we might characterize all of the sea ice that forms around the margin of the Antarctic continent as pack ice and fast ice that have a rather temporary existence. The pack ice rarely extends north of 55°S latitude and breaks up rather completely from October to January except in the very quiet bays (fig. 7–19B). Winds that are frequently very strong help prevent the formation of a greater pack ice accumulation around the Antarctic continent.

Icebergs

Icebergs are derived from land ice in the form of glaciers that cover the continent of Antarctica and most of the island of Greenland and part of Ellesmere Island in the Arctic (fig. 7–20). These vast sheets of ice that grow from the snow accumulation on the landmasses spread outward toward the margins until they push their edges out into the marginal sea. There the ice sheets are buoyed up by the water and break up under the stress of current, wind, and wave action. This breakup that produces icebergs is called *calving.*

In the Arctic, the icebergs that are produced originate primarily from the ice that follows narrow valleys into the sea along the western coast of Greenland. Icebergs are also produced along the east coast of Greenland and the east coast of Ellesmere Island.

A.

B.

C. NORTH ATLANTIC CURRENTS AND ICEBERGS

FIGURE 7–20 **Icebergs.**
A, North Atlantic icebergs off Cape York, Greenland. They are generally smaller and more irregular in shape than Antarctic icebergs. *B,* the U.S.S. *Wyandotte* moves past a tabular iceberg in the Weddell Sea, Antarctica. Many of the icebergs are kilometers in length and have been mistaken for land. Both photographs from Oceanographer of the Navy, Public Affairs Office. *C,* a map showing North Atlantic currents and icebergs (△).

The East Greenland Current and the West Greenland Current carry the icebergs at rates of up to 20 km (12.4 mi) per day into the North Atlantic, where the Labrador Current may move them into North Atlantic shipping channels. They are seldom carried south of 45°N latitude, but during some seasons icebergs will move as far south as 40°N latitude and become shipping hazards.

Such an accumulation of icebergs developed when the *Titanic* was sunk in April 1912. After receiving repeated warning of the existence of such a hazard, the 46,000-ton ship containing 2224 passengers proceeded at an excessive speed of 41 km/h (22.4 kt) until it came to a grinding halt after hitting the iceberg with its starboard bow. The sinking of the *Titanic* at 41°46'N latitude, 50°14'W longitude near the Grand Banks cost the lives of 1517 people and brought about the formation of the ice patrol that has prevented further loss of life from such accidents. The U.S. Navy began this patrol immediately following the *Titanic* disaster; it became international in 1914. The patrol, now maintained by the U.S. Coast Guard, concentrates its efforts between 40°30' and 48°N latitude and 43° and 54°W longitude.

The *shelf ice* that represents the edges of glaciers pushing into the marginal seas of Antarctica produces icebergs that have received less attention than those of the North Atlantic, owing to the fact that they interfere less with shipping. Vast tabular bergs that break from the edges of the shelf ice have lengths over 100 km (62 mi). They may stand as much as 200 m (656 ft) above the ocean surface, though most are probably less than 100 m (328 ft) above sea level. Most of the calving occurs, as it does in the Arctic region, during the summer months. When the sea ice breaks up and allows the swells driven by the strong westerly winds to reach the edge of the shelf ice, large icebergs are calved and move to the north into the warmer and rougher water, in which they disintegrate. Carried by the strong West Wind Drift, these icebergs move in an easterly direction around Antarctica, rarely moving farther north than 40°S latitude.

Renewable Sources of Energy

The potential of extracting energy from the motions and heat distribution patterns of the atmosphere and ocean, which are maintained by the sun, is attractive for the following reasons:

1. Work can be achieved without significant pollution.
2. The amount of energy available at any time is far greater than that in fossil and nuclear fuels.

3. Such forms of energy are renewable and will not be depleted.

Considered in order of decreasing energy potential, the sources of renewable energy are (1) heat stored in the oceans, (2) kinetic energy of the winds, (3) potential and kinetic energy of waves, and (4) potential and kinetic energy of tides and currents. In later chapters we will consider the use or potential use of winds, waves, tides, and currents. Here we will consider only the renewable source with the greatest store of potential energy: the surface layers of the tropical oceans.

Ninety percent of the earth's surface between the Tropic of Cancer and Tropic of Capricorn is ocean. What makes this warm tropical surface water such an important source of energy is the presence of much colder water beneath the thermocline. With a temperature difference as small as 17°C (58.6°F), useful work can be done by Ocean Thermal Energy Conversion (OTEC) systems (fig. 7–21).

The system works in the opposite way from a typical refrigeration system and on a much larger scale. The warm surface water heats a fluid such as propane or ammonia that is under pressure in evaporating tubes. The fluid is vaporized and passes through a turbine that drives an electrical generator. After passing through the turbine, the fluid is condensed by cold water that has been pumped up from the deep ocean. It is again ready for heating by warm surface water that will cause it to vaporize and pass through the turbine.

The only region with such potential along the conterminous United States coast is a strip about 30 km (18.6 mi) wide and 1000 km (621 mi) long. It extends north from southern Florida along the western margin of the Gulf Stream. The National Science Foundation is supporting an experimental OTEC system to be established 25 km (16 mi) east of Miami, Florida.

Mini-OTEC, a unit mounted on a U.S. Navy barge, was tested off Ke-Ahole Point, Hawaii, in 1979. Designed primarily to test equipment under ocean conditions, it generated only 10 net kilowatts of electricity. A much larger unit, OTEC 1, designed to generate 1 megawatt, tested heat exchanger designs and biofouling systems under ocean conditions for 3 months in 1981. This system is installed on the tanker *Chepachet,* operated in Hawaiian waters (fig. 7–22).

Testing is presently under way on the Seacoast Test Facility, a shore-based OTEC installation, on the island of Hawaii. Although it will require a large initial investment, Hawaii has high hopes that OTEC power generation will be a commercially successful means of power generation by the mid-1990s.

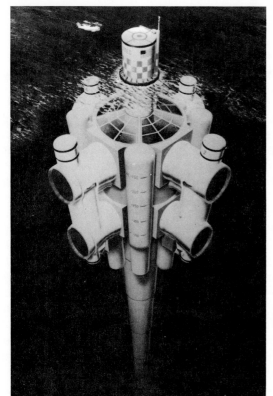

A.   B.

FIGURE 7–21 Proposed Uses of Ocean Thermal Power.

A, Ocean Thermal Energy Conversion (OTEC) system with crew quarters and maintenance facilities. Attached around the outside are turbine-generators and pumps. It is over 75 m (246 ft) in diameter, 485 m (1595 ft) long, and weighs about 300,000 tn. This unit is designed to generate 160 million watts of power, which is enough to meet the needs of a city with a population of 100,000. Photo courtesy of Lockheed Missiles and Space Co., Inc. *B,* OTEC plant for production of ammonia. Moving slowly through tropical waters, this plant could produce 1.4% of the ammonia requirements for the United States each year and save 22.6 billion cubic feet of natural gas. Courtesy of U.S. Department of Energy.

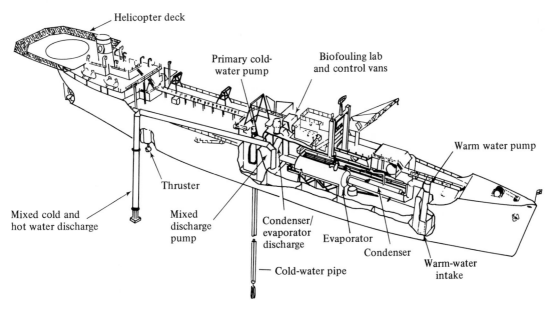

FIGURE 7–22 The *Chepachet.*

The tanker *Chepachet,* converted to become the support vessel for OTEC 1, a 1-megawatt engineering test facility for evaluating heat exchanger designs and biofouling systems. Courtesy of U.S. Department of Energy.

185

Another proposed use of the OTEC concept is producing ammonia. A floating factory that could produce 586,000 tn of ammonia per year has been described in a feasibility report from the Johns Hopkins Applied Physics Laboratory to the U.S. Maritime Administration. The proposed demonstration ship would weigh 68,000 tn, have a width of almost 60 m (197 ft) and have a length of over 144 m (472 ft). It would move at less than 2 km/hr (1 kt) through tropical waters and use energy derived from the temperature difference between the warm surface waters and cold deep waters to produce ammonia. Each such plant could produce about 1.4% of our nation's ammonia requirements—75% of which is used for fertilizer—and save 22.6 billion cubic feet of natural gas each year (fig. 7–21).

SUMMARY

In ocean water of low biological productivity, the molecular-sized particles scatter the short wavelength of visible light, producing a blue ocean. Greater amounts of dissolved organic matter in more productive ocean water scatter more green light which, along with the chlorophyll pigmentation of the plants, produces a green ocean. With increasing depth in the ocean, the colors of the visible spectrum are absorbed by ocean water in a way that removes red, yellow, and violet at relatively shallow depths. The blue and green wavelengths of light are the last to be removed. High turbidity, the measure of the amount of suspended material in water, greatly reduces the depth to which light will penetrate.

Density of ocean water is increased by increased salinity and decreased temperature. Density change per unit of temperature change is greater in warmer water. Density, temperature, and salinity are conservative properties of water, properties that are affected only by mixing and diffusion once a surface water mass sinks. Zones of rapid change in the density, temperature, or salinity in the water column are called a pycnocline, thermocline, and halocline, respectively.

Velocity of sound transmission in the ocean increases with increases in temperature, salinity, and pressure. A low-velocity sound channel is caused by a thermocline, in which temperature decreases with increased depth.

Radiant energy reaching earth from the sun is mostly in the ultraviolet and visible light range, while that radiated back to space from earth is primarily in the infrared spectrum. The atmosphere is unevenly heated and set in motion because more energy is received and radiated back into space at low latitudes and water vapor, which absorbs infrared radiation well, is unevenly distributed in the atmosphere. Belts of low pressure, where air rises, are generally found at the equator and at about 60° latitude. High-pressure regions, where dense air descends, are located at the poles and at about 30° latitude. The air at the earth's surface that is moving away from the subtropical highs produces the trade winds moving toward the equator and westerlies moving toward higher latitudes.

Because the earth's surface rotates at different velocities at different latitudes, increasing from 0 km/h at the poles to over 1600 km/h (994 mi/h) at the equator, objects in motion tend to veer to the right in the Northern Hemisphere and to the left in the Southern Hemisphere. This is called the Coriolis effect.

In the heat budget of the world ocean, the only source of heat gain is the sun, while heat may be lost by radiation, evaporation, and conduction into the atmosphere. The amount of heat lost by radiation, primarily of infrared wavelengths, is greatest near the equator, where the greatest amount of solar energy is absorbed. The greatest heat loss by evaporation is in dry subtropics, while the heat lost through conduction reaches a maximum along the western margins of oceans in temperate latitudes where warm water currents carry water into regions where the atmosphere is much colder than the water.

The tilt of the earth's axis of rotation and the distribution of continents modify the idealized pressure belts discussed above. High-pressure cells form over continents in winter and are replaced by low pressure cells in summer. There is a cyclonic movement of air around the low-pressure cells and an anticyclonic movement around high-pressure cells. Cold air masses of high latitudes meet warm air masses of lower latitudes and create cold and warm fronts that move from west to east across the earth's surface at midlatitudes.

Nonetheless, ocean climate patterns are closely related to the idealized pressure belts previously discussed.

Sea ice forms as sea water is frozen in high latitudes. The process usually involves the formation of a slush, which breaks into pancakes that ultimately grow into floes. Sea ice develops as pack ice that forms each winter and melts almost entirely each summer, polar ice that is a permanent accumulation in polar regions of the Arctic Ocean, and fast ice that forms frozen to shore during the winter. Icebergs form as large chunks of ice break away from the large continental glaciers that form on Ellesmere Island and Greenland in the Northern Hemisphere and Antarctica in the Southern.

From the motions and patterns of heat distribution in the atmosphere and oceans, **renewable, nonpolluting sources of energy** can be exploited. These include *heat stored in the ocean* and potential and kinetic energy stored in the *winds, waves, tides,* and *currents. Ocean Thermal Energy Conversion* (OTEC) is a process developed to use the difference in temperature between warm tropical surface water and the cold water below the thermocline to produce electricity and ammonia.

QUESTIONS AND EXERCISES

1. How is the color of the ocean surface water related to biological productivity? Why does everything in the ocean at depths below the shallowest surface water take on a blue-green appearance?

2. Describe the relative effect of temperature change on water density at high- versus low-temperature ranges.

3. The position of the pycnocline in the water column is determined by the combined effects of the thermocline and halocline. Describe these relationships in tropical waters (*see* figs. 7–3 and 7–4).

4. How does the development of a thermocline produce a *sofar,* or sound channel, below the ocean's surface?

5. Explain why the atmosphere is heated primarily by back radiation from the earth rather than by direct radiation from the sun.

6. Describe the effect on the earth as a result of the earth's axis of rotation being angled 23.5° to the ecliptic.

7. Since there is a net annual heat loss at high latitudes and a net annual heat gain at low latitudes, why does the temperature difference between these regions not increase?

8. Why are there high-pressure caps at each pole and high-pressure belts at 30° latitudes compared to low-pressure belts in the equatorial region and at 60° latitudes?

9. Describe the Coriolis effect in the Northern and Southern hemispheres and include a discussion of why the effect increases with increased latitude.

10. In considering the heat budget of the world ocean over a long period, why should Q_c (rate of heat gain or loss through ocean currents) and Q_t (net rate of heat gain or loss) be considered to be zero?

11. Describe the ocean regions where the maximum rates of heat loss by radiation, evaporation, and conduction occur, and explain why.

12. Discuss why the idealized belts of high- and low-atmospheric pressure shown in figure 7–9 are modified (*see* fig. 7–14).

13. Name the polar and tropical air masses that affect U.S. weather. Describe the pattern of movement across the continent and patterns of precipitation associated with warm and cold fronts.

14. How are the ocean's climatic belts (fig. 7–18) related to the broad patterns of air circulation described in figure 7–9?

15. Describe the formation of sea ice from the initial freezing of the ocean surface water through the development of polar ice in the Arctic Ocean.

16. What is the difference in the average size and shape of icebergs in the Arctic and Antarctic? Why do these differences exist?

17. Construct your own diagram of how an Ocean Thermal Energy Conversion unit might generate electricity, or make a flow diagram presenting the steps of the process.

REFERENCES

Changing climate and the oceans. 1987. *Oceanus* 29:4, 1–93.

Miller, A. 1971. *Meteorology.* Columbus, Ohio: Charles E. Merrill.

Ocean energy. 1979. *Oceanus* 22:4, 1–68.

Oceans and climate. 1978. *Oceanus.* 21:4, 1–70.

Pickard, G. L. 1975. *Descriptive physical oceanography.* 2nd ed. New York: Pergamon Press.

SUGGESTED READING

Sea Frontiers

Boling, G. R. 1971. Ice and the breakers. 17:6, 363–71.
An interesting history of people in icy waters and the development and improvement of ice breakers.

Charlier, R. 1981. Ocean-fired power plants. 27:1, 36–43.
The potential of ocean thermal-energy conversion is discussed.

Land, T. 1976. Europe to harness the power of the sea. 22:6, 346–49.
An article emphasizing the clean, safe, permanent nature of ocean tides and waves as a source of power.

Rush, B., and Lebelson, H. 1984. Hurricane! The enigma of a meteorological monster. 30:4, 233–239.
An overview of the nature of hurricanes and the problem of predicting where they will go.

Smith, F. G. W. 1974. Planet's powerhouse. 20:4, 195–203.
A description of how the earth's "heat engine" works, with an emphasis on the nature of tropical cyclonic storms.

_____. 1974. Power from the oceans. 20:2, 87–99.
A survey of the many tried and untried proposals for extracting energy from the oceans.

Sobey, E. 1979. Ocean ice. 25:2, 66–73.
The formation of sea ice and icebergs as well as their climatic and economic effects are included.

_____. 1980. The ocean-climate connection. 26:1, 25–30.
The increasing amount of knowledge of the effect of the oceans on the earth's climate may be used to predict climatic trends of the future.

Scientific American

Gregg, M. 1973. Microstructure of the ocean. 228:2, 64–77.
A discussion of the methods of studying the detailed movements of ocean water by observing temperature and salinity changes over distances of one centimeter and the motions they reveal.

MacIntyre, F. 1974. The top millimeter of the ocean. 230:5, 62–77.
Processes that are confined to a thin film at the surface of ocean water and their role in the overall nature of the oceans are discussed.

Revelle, R. 1982. Carbon dioxide and world climate. 247:2, 35–43.
Some of the possible effects of increasing atmospheric temperature due to CO_2 accumulation are considered.

Stanley, S. M. 1984. Mass extinctions in the oceans. 250:6, 64–83.
Geological evidence suggests most major periods of species extinction over the last 700 million years occurred during brief intervals of ocean cooling.

CHAPTER EIGHT
OCEAN CIRCULATION

Currents, or water masses in motion, are driven ultimately by energy from the sun. The circulation of these masses can be categorized as being either *wind-driven* or *thermohaline.* The wind-driven currents are set in motion by moving air masses, and this motion is confined primarily to horizontal movement in the upper waters of the world ocean. The thermohaline circulation has a significant vertical component and accounts for the thorough mixing of the deep masses of ocean water. Thermohaline circulation is initiated at the ocean surface by temperature and salinity conditions that produce a high-density mass, which sinks and spreads slowly beneath the surface waters.

The *upper-water mass* of the world ocean includes the well-mixed surface layers of the ocean and the thermocline. The base of the thermocline will be encountered above a depth of 1000 m (3280 ft). The *deep-water mass* includes the rest of the water column below the thermocline to the bottom of the ocean, where the temperatures are relatively low throughout (fig. 8–1).

HORIZONTAL CIRCULATION

Wind-driven horizontal circulation in the surface waters develops from stress at the interface between the ocean and the wind. The trade winds blowing out of the southeast in the Southern Hemisphere and out of the northeast in the Northern Hemisphere provide the backbone of the system of ocean surface currents. Setting the water masses between the tropics into motion, the trade winds develop the equatorial currents that can be found in all of the world's oceans. Owing to the Coriolis effect, these currents move west parallel to the equator, and they are deflected by continents into clockwise *gyres* in the Northern Hemisphere and counterclockwise gyres in the Southern Hemisphere. A gyre is a horizontal circulation cell that refers to the circular motion of water in each of the major ocean basins centered in subtropical high-pressure regions. As these deflected masses, *warm currents,* move along the western margins of their respective ocean basins toward higher latitudes, the westerly wind systems blowing out of the northwest

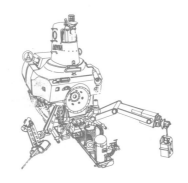

Chart of the Gulf Stream compiled by Benjamin Franklin, published in 1777. Courtesy of U.S. Navy.

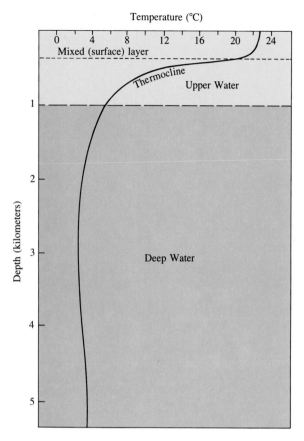

FIGURE 8–1 Major Water Masses of the Open Ocean. Extending from the surface down through the mixed, or surface, layer and the thermocline is the relatively light *upper water.* It is well developed throughout the low latitudes and midlatitudes. It is underlain by the denser, cold *deep-water mass* that extends to the deep-ocean floor. The mixed, or surface, layer represents water of uniform temperature resulting from mixing by waves.

in the Southern Hemisphere and out of the southwest in the Northern Hemisphere add more energy and push the circulating masses on in an easterly direction. Deflection of *cold currents* toward the equator along the eastern margins of their respective basins completes the path of circulation (fig. 8–2).

Ekman Spiral

To explain why water masses move as they do relative to wind direction, we will recall the observations made by Fridtjof Nansen during the voyage of the *Fram.* Nansen determined that the sea ice moved 20° to 40° to the right of the wind blowing across its surface. He passed this information on to V. Walfrid Ekman, a physicist who developed the mathematical relationships that explain the observations.

Ekman developed a circulation model that has been called the *Ekman spiral* (fig. 8–3). The model assumes that a homogeneous water column is being

set in motion by wind blowing across its surface. Owing to the Coriolis effect, the surface current moves in a direction 45° to the right of the wind in the Northern Hemisphere. This surface mass of water, moving as a thin *lamina,* or sheet, sets another layer beneath it in motion. The surface layer moves with a velocity no more than 3% of the wind speed. The energy of the wind is passed through the water column from the surface down, with each successive layer of water being set in motion with a lower velocity than and in a direction to the right of the one that set it in motion. At some depth the momentum imparted by the wind to the moving water laminae will be lost and there will be no motion as a result of wind stress at the surface. The depth at which motion ceases is called the *depth of frictional influence.* Though it depends on wind speed and latitude, this stillness normally occurs at a depth of about 100 m (328 ft). Figure 8–3 shows the spiral nature of movement with increasing depth from the ocean's surface. The length of each arrow in the figure is proportional to the velocity of the individual lamina, and the direction of each arrow indicates its direction of movement.

Under these theoretical conditions, the surface current should flow at an angle of 45° to the direction of the wind. From the surface to the depth of frictional influence, the net water movement, the *Ekman transport,* will be at right angles to the direction of the wind.

There are, however, no ideal conditions in existence in the ocean, and movements actually occurring as a result of wind stress on the ocean surface will deviate from this idealized picture. Generally, we can say that the surface current will move at an angle of less than 45° to the direction of the wind and that the Ekman transport will be at an angle less than 90° to the direction of the wind. This is particularly true in shallow coastal waters, where all of the movement may be in a direction very nearly that of the wind and the turning with increased depth is minor.

Geostrophic Currents

If we consider a gyre in the Northern Hemisphere in the North Atlantic Ocean, for example, and remember the Ekman transport, it can be seen that a clockwise rotation in the Northern Hemisphere will tend to produce a piling up of water in the center of that gyre. We find within all such ocean gyres hills of water that rise as much as 2 m (6.6 ft) above the water level at the margins of the gyres. As Ekman transport pushes water into the "hill" structure, the gravitational force acts to move particles down the surface slope. The Coriolis force deflects the water flowing down the

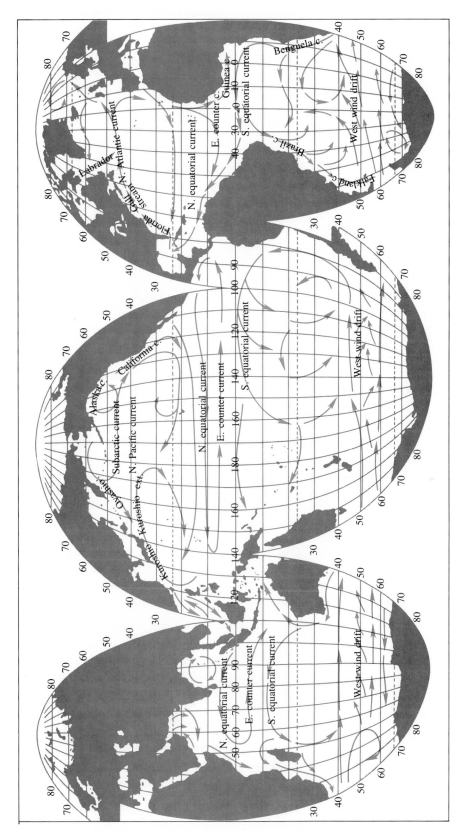

FIGURE 8–2 Wind-Driven Surface Currents in February and March.
After Sverdrup et al., 1942.

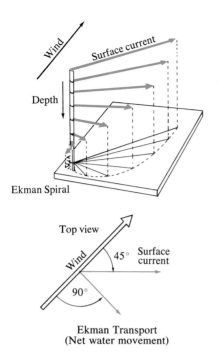

FIGURE 8–3 Ekman Spiral.

Wind drives surface water in a direction 45° to the right of the wind in the Northern Hemisphere. Deeper water continues to deflect to the right and move at a slower speed with increased depth. Ekman transport, the net water movement, is at right angles to the wind direction.

slope to the right in a curved path (fig. 8–4). The water piles up on these hills until the down-slope component of gravitational force acting on individual particles of water balances the Coriolis force. When these two forces finally balance, the net effect is a *geostrophic current* moving around the hill.

Westward Intensification

The apex of the hills formed within the rotating gyres is not in the center of the gyre. The highest part of

each hill is located closer to the western boundary of the gyre. This location is a result of a balance reached among the forces growing out of the changing magnitude of the Coriolis effect with latitude, wind stress that puts energy into the system, and bottom friction that removes energy from it. Assuming we have a steady state with the same volume of water rotating around the apex of the hill, it can be seen in figure 8–4 that the velocity with which the water moves along the western margin will be much greater than that with which it will move around the eastern side of the hill. These relationships can be seen particularly well in the North Atlantic and North Pacific gyres, where the western boundary currents move with speeds in excess of 5 km/h (3.1 mi/h) in a northerly direction, while a flow along the eastern margin of each basin is better characterized as a drift moving at velocities well below 0.9 km/h (0.6 mi/h). Directly related to this difference in speed is the steepness of the hill's slope. The slope of the hill on the east side is quite gentle; the western margin has a steep slope corresponding to the high-velocity current. If the earth did not rotate, there would be no intensification on the western boundary because there would be no Coriolis force, and the effects of wind stress and bottom friction would be balanced on the eastern and western boundaries.

An indirect method of determining current velocity in the upper-water mass is measuring the internal distribution of density and the pressure field it creates. This pattern of distribution is determined by collecting water samples along a number of vertical columns distributed throughout the area to be studied. At each station, temperature and salinity are measured. Average density of the water column is computed from the temperature-salinity characteristics measured above an arbitrary depth at which the surface current is assumed to have died out (possibly 100 m, or 328 ft). No current or horizontal pressure

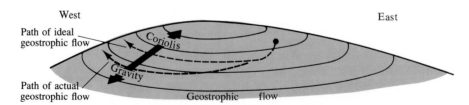

FIGURE 8–4 Geostrophic Current.

As the earth's wind systems set ocean water in motion, circular gyres are produced. Water piles up inside the gyre with the apex of the "hill" closer to its west margin. The geostrophic current flows nearly parallel to the contour of the hill and represents an equilibrium between the Coriolis effect pushing water toward the apex through Ekman transport and the down-slope component of gravity acting to move the water down the slope. Owing to the offset of the apex to the west, the current moves more rapidly along the steeply sloping western margin than along the more gently sloping eastern margin. This phenomenon is referred to as westward intensification of boundary currents.

gradient is thought to exist at this depth. The pressures of the overlying water columns are assumed to be equal.

If our assumption of no horizontal pressure gradient existing at our reference level is correct, all measured water columns above this datum must contain equal mass. However, a water column with a low average density will, because of its greater volume, stand higher above the datum than a water column of high density. By computing the heights of columns necessary at all stations to produce the zero horizontal pressure gradient at the datum depth, a topography for the ocean surface can be determined (fig. 8–5).

Maps showing this type of water motion can be prepared for any depth between the datum and the ocean surface; they represent the *dynamic topography* that is used to compute current velocity. The steeper the slope, the higher the velocity of the geostrophic current flowing in a direction generally parallel to the topography *contours* (lines of equal height above the equal pressure datum). Rapid progress is being made in the use of radar altimeters mounted on satellites to map the dynamic topography of the oceans. Another technique that may replace the laborious procedures involved in the classic method of determining dynamic topography is described in *Ocean Acoustic Tomography* on pages 196–197.

Separating gyres in the northern and southern portions of the Atlantic and Pacific oceans are equatorial countercurrents that flow in an easterly direction. Although such a countercurrent exists throughout the breadth of the Pacific Ocean, it is restricted to the eastern Atlantic. In both of these oceans the current is centered about 5° north of the equator as a result of the northward shift of the ITCZ (Intertropical Convergence Zone) caused by the uneven distribution of land and water in the Northern and Southern hemispheres. In the Indian Ocean, such a current develops during the winter monsoon only. There it is shifted about 7° south of the equator, owing to the fact that the Indian Ocean extends only to about 20°N latitude (*see* fig. 8–2).

VERTICAL CIRCULATION

Lateral movements of water masses may bring about vertical circulation within the upper-water mass. We refer to this shallow vertical circulation system as *wind-induced.* Of greater oceanwide importance in producing the thorough mixing within the ocean is the vertical circulation resulting from density changes at the ocean surface, which cause the sinking of water masses. Since the changes that increase the density are normally changes in temperature and salinity, this circulation is referred to as *thermohaline circulation.*

Wind-Induced Circulation

There are regions throughout the world ocean where water moves vertically to the surface or away from the surface as a result of wind-driven surface currents carrying water away from or toward these regions. *Upwelling* occurs in areas where the surface flow of water is away from the area. If volume is to be conserved and horizontal surface flows bring insufficient water into the area, water must come from beneath the surface to replace that which has been displaced. This condition may occur in the open ocean or along the margins of continents.

As the North Equatorial Current of the Indian Ocean and South Equatorial currents in the Atlantic and Pacific oceans are driven by easterly winds in a westerly direction on either side of the equator, the water on the north side of the equator will tend to veer to the right, moving to a higher latitude, while

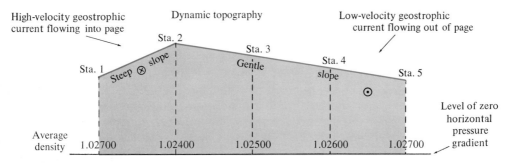

FIGURE 8–5 Dynamic Topography.
Comparison of average densities for the water columns above the arbitrary no-current datum tells us the apex of the hill is near station 2 because the lower the density of the water, the greater the volume a given mass will occupy. Northern Hemisphere geostrophic flow is indicated by circles beneath slopes.

A.

B.

FIGURE 8A

Walter H. Munk of Scripps Institution of Oceanography (A) and Carl Wunsch of Massachusetts Institute of Technology (B) are leaders in the development of ocean acoustic tomography techniques for the syntopic study of mid- to large-scale oceanic circulation patterns. Walter Munk's photo courtesy of University of California at San Diego, Office of Learning Resources S.I.O. Photo Lab.

When we observe the motions of ocean water, we find they range across broad spectrums of size and time. The identifiable patterns of ocean circulation range in size from eddies a few tens of kilometers across, created by water flowing over and around seamounts, to the slowly rotating subtropical gyres the size of ocean basins. Between these extremes are features such as the Gulf Stream rings, which are a few hundred kilometers in diameter, and coastal upwelling such as that responsible for the highly productive waters along the coast of Peru.

Fully understanding these motions and their effects on climate and biological productivity has long been the goal of oceanographers. However, it has not been possible to continuously gather data over sufficient area for the necessary length of time to achieve this goal because of the high cost of sending the required number of ships on long sea voyages.

Recently, Walter Munk of Scripps Institution of Oceanography (fig. 8A*A*) and Carl Wunsch (fig. 8A*B*) of Massachusetts Institute of Technology have led the way in developing a system that may make such data gathering possible—ocean acoustic tomography. "Tomo" is derived from the Greek word *tomos*, meaning a section or slice. A *tomogram* is a picture of a slice. Tomography has already found use in medicine through the use of x-rays. The CAT (computerized axial tomography) scan uses the x-rays to provide tomographic images of parts of the body, usually the brain.

The use of tomography to map the features of ocean circulation involves the transmission of low-frequency sound. Increases in water temperature, salinity, and pressure increase the speed of sound travel in ocean water. As sound travels through features such as cold or warm rings, the sound speed will be increased or decreased. Theoretically, if sound is transmitted across a slice of ocean from a series of transmitters on one side to a series of receivers on the other, features that cause a change in the speed of sound can be mapped.

A 1981 demonstration covering a 300-km (186-mi) square of ocean southwest of Bermuda involved the use of four transmitters and five receivers over a period of six months (fig. 8B). Two sensors of temperature and current flow were included and survey ships periodically entered the area to record data against which the tomography could be checked.

During the demonstration, a large cold eddy moved across the area and a frontal system separating cold and warm water moved into the area (fig. 8C). The data gathered conventionally confirmed the existence of the features and proved the feasibility of such a system to provide instantaneous pictures of large slices of the world's oceans. Subsequent tests (1983) proved the technique could be used to measure current flow and deep-sea circulation at a depth of 760 m (2500 ft). The latter experiment, conducted in the North Pacific Ocean, involved three transceivers arranged in a triangle with sides 1000 km (620 mi) in length. A seven-transceiver system was set up off New England in 1987 to study the Gulf Stream, and a large, five-year study is under way to

During the 1981 demonstration of ocean acoustic tomography, transmitters (T1 to T4) emitted low-frequency sound signals that were received by the receivers (R1 to R5). Temperature and currents were measured by conventional means at moorings E1 and E2.

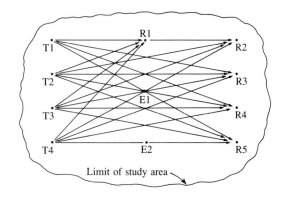

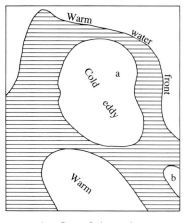

A. Start of observations

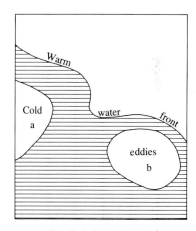

B. Mid-way through observations

C. End of observations

FIGURE 8C

Results of the 1981 ocean acoustic tomography demonstration. Source of data: *A* and *C*, ship surveys; *B*, tomography. During the demonstration, data from a depth of 700 m (2296 ft) showed a large cold eddy in the center of the survey area at the start of the survey. It is separated from warm water to the northeast by a well-defined temperature front. The cold-water eddies observed move off to the northwest, while the front progresses to the southwest during the demonstration. Cold eddies were indicated by the slowing of sound signals that pass through them. Courtesy of Woods Hole Oceanographic Institution.

investigate the contribution of the Greenland Sea to the formation of North Atlantic Deep Water. In the near future it may well be possible to map such features within large ocean basins—something that was only a dream a few years ago.

Reference: Behringer, D.; Birdsall, T.; Brown, M.; Cornuelle, B.; Heinmiller, R.; Knox, R.; Metzger, K.; Munk, W.; Spiesberger, J.; Spindel, R.; Webb, D.; Worcester, P.; and Wunsch, C. 1982. A demonstration of ocean acoustic tomography. *Nature* 299, 121–125.

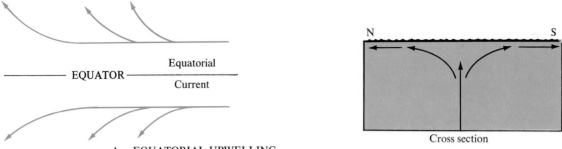

A. EQUATORIAL UPWELLING

The Coriolis effect acting on westward flowing wind-driven equatorial currents pulls surface water away from the equatorial region. This water is replaced by subsurface water.

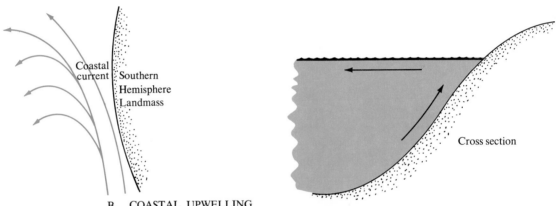

B. COASTAL UPWELLING

In areas where wind-driven coastal currents flow along the western margin of continents and toward the equator, the Ekman transport carries surface water away from the continent. An upwelling of deeper water replaces the surface water that has moved away from the coast.

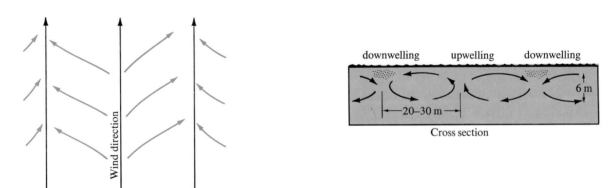

C. LANGMUIR CIRCULATION

Steady winds blowing across the ocean surface create convection cells with alternate right- and left-hand circulation. The axes of these cells run parallel to wind direction. Organic debris accumulates in downwelling zones of convergence and produces windrows that run parallel to the wind direction for great distances.

FIGURE 8–6 Wind-Driven Vertical Circulation.

that in the Southern Hemisphere will veer to the left to a higher latitude (fig. 8–2). The net effect is a water deficiency at the surface between the two currents. Water from deeper within the upper-water mass comes to the surface to fill the void resulting from the deflection of equatorial water to the north and south (fig. 8–6A). This phenomenon is called *equatorial upwelling*.

Coastal upwelling is common along the margins of continents where the wind conditions are such that the surface waters adjacent to the continents are carried out to the open ocean via Ekman transport. The replacement of this water comes from the lower portions of the upper water, as is the case with the Peru Current. Such areas are characterized by low surface temperatures and high concentrations of nutrients that make them areas of high biological productivity (fig. 8–6B). See *Upwelling and Bumps on the Ocean Floor* for information obtained from laboratory research regarding the effect of bottom topography on coastal upwelling (pages 332–333).

Near the center of the rotating gyres of the major ocean current systems the winds are relatively weak, and the water rotates at a very low rate. The winds that do blow in these regions may be steady in direction and are known to set up convection cells in the upper-water mass. This phenomenon was first recognized by Irving Langmuir while crossing the Sargasso Sea in 1938. He observed straight rows of seaweed parallel to the direction of the wind and concluded that the plants were trapped in zones of convergence between cells. Not only is macroscopic plant material concentrated in these regions, but also microscopic plants and dissolved organic material. By contrast, in the regions of divergence, where water is surfacing as a result of convection, the concentrations of organic material are relatively low (fig. 8–6C). This phenomenon is called *Langmuir circulation*.

Thermohaline Circulation

The large-scale vertical circulation in the ocean results primarily from surface-density changes in oceanic water. Vertical mixing of ocean water is achieved primarily through the sinking and rising of water masses in high latitudes. It is only in the high-latitude areas that the water column has a gravitational stability sufficiently neutral to allow vertical movements of large masses of water. Surface masses may sink to the ocean bottom and deep-water masses may rise to the surface. The relatively strong density stratification, or pycnocline, that separates the upper- and deep-water masses throughout the lower-latitude regions does not exist in the high-latitude oceans.

Because the intensity of solar radiation is greater in the equatorial region, we might expect that heating of surface water there may cause the water to expand and move away from the equator toward the poles, spreading over the colder, more dense high-latitude waters (fig. 8–7 and plate 15). We may further expect that at some depth a return flow from the high latitudes toward the equator would be set in motion to replace the surface water moving away from the equatorial region. This exchange does not seem to occur in the oceans, although it is theoretically proper to assume that such a pattern might develop. The energy imparted to the surface waters by winds greatly exceeds the energy that develops as a result of density changes due to temperature differences. Thus the effect of the wind overcomes any tendency that may exist for such a pattern of circulation to develop.

Another variable that affects the density of surface water is salinity (fig. 8–8). Salinity appears to have a very minimal effect on the movement of water masses in the lower latitudes, and density changes resulting from salinity changes are of importance only in the very high latitudes, where low water temperature remains relatively constant. For example, the highest-salinity water in the open ocean is found in the subtropical regions, but there is no sinking water in these areas because water temperatures are high enough to maintain a low density for the surface-water mass and prevent it from sinking. In such areas, a strong halocline, or salinity gradient, may develop with a relatively thin surface layer of water having salinities in excess of 37‰. Salinity decreases rapidly with increasing depth to typical ocean-water salinities below 35‰.

An important sinking of cold surface waters that become deep-water masses occurs in the subpolar regions of the Atlantic Ocean. In the North Atlantic, the major sinking of surface water is thought to occur in the Norwegian Sea. From there it flows as a subsurface current into the North Atlantic. This flow, which becomes part of the *North Atlantic Deep Water*, enters the North Atlantic over the Wyville Thompson Ridge between the Faeroe Islands and the British Isles and through a deep channel between Iceland and Greenland. Additional surface water may sink at the margins of the Irminger Sea off southeastern Greenland and the Labrador Sea. In the southern subpolar latitudes, the most significant area of deep-water-mass formation is the Weddell Sea, where rapid winter freezing produces high-density water that sinks down the continental slope of Antarctica and becomes *Antarctic Bottom Water*, the densest water in the open ocean (fig. 8–9).

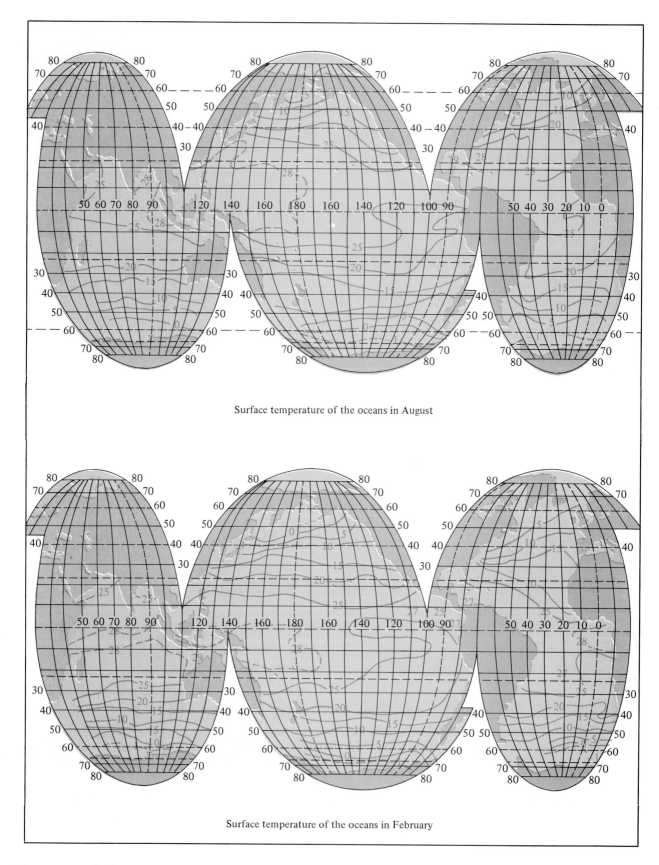

Surface temperature of the oceans in August

Surface temperature of the oceans in February

FIGURE 8-7   Surface Temperature of the World Ocean (°C).
After Sverdrup et al., 1942. Base map courtesy of National Ocean Survey.

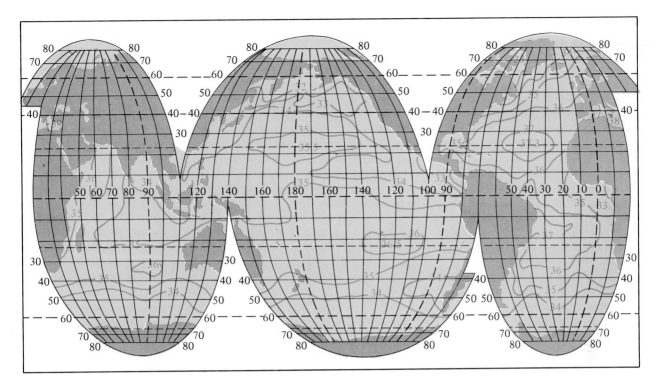

FIGURE 8–8 Surface Salinity of the Oceans in August (‰).

After Sverdrup et al., 1942. Base map courtesy of National Ocean Survey.

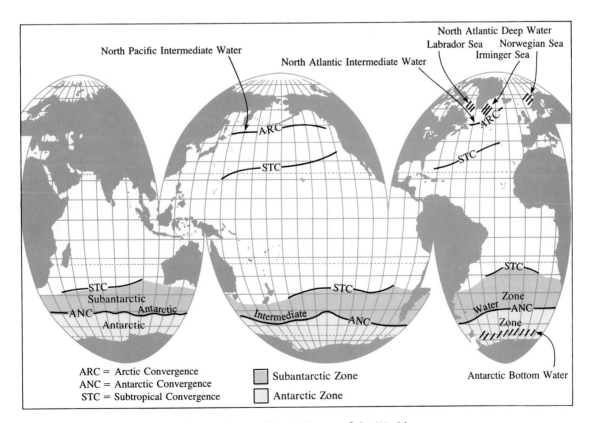

FIGURE 8–9 Regions Where Intermediate and Deep-Water Masses of the World Ocean Sink.

ARC Arctic Convergence Zone. ANC Antarctic Convergence Zone. STC Subtropical Convergence Zone. Base map courtesy of National Ocean Survey.

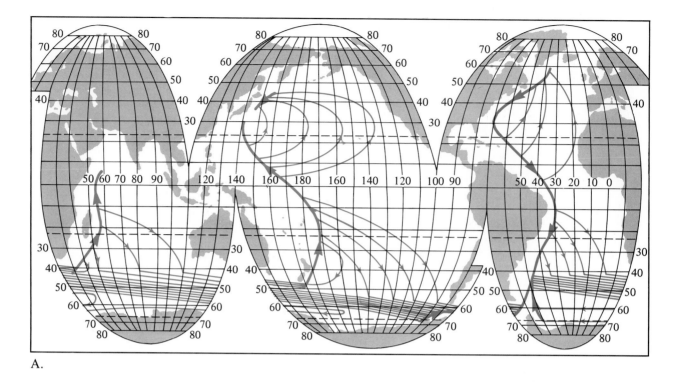

A.

FIGURE 8–10 Stommel's Model of the Deep-Water Circulation of the World Ocean.

A, based on information obtained to date, this highly schematic model of deep-water circulation developed by Henry Stommel in 1958 appears to be reasonably correct. Heavy lines mark the major western boundary currents. They result from the same forces that produce the more intense western boundary currents in the surface circulation. After H. Stommel, 1958. Base map courtesy of National Ocean Survey. *B,* Dr. Henry M. Stommel entered the field of physical oceanography as a research assistant at Woods Hole Oceanographic Institution in 1944. Although his formal education consisted of no more than a B.S. in physics, he soon became prominent in the field. Many of his colleagues consider him to be the world's preeminent physical oceanographer. Photo courtesy of Woods Hole Oceanographic Institution.

B.

On a broad scale, there are latitudinal regions where surface-water masses converge and may cause sinking (fig. 8–9). These regions of convergence are found within the subtropical gyres and in the Arctic and Antarctic. In the subtropics the convergence does not produce sinking because surface waters have relatively low densities as a result of high temperature. Major sinking does occur, however, along the high-latitude convergences. The major water masses formed from sinking at the Arctic and Antarctic convergences are the North Pacific, North Atlantic, and Antarctic Intermediate water masses (fig. 8–9).

Figure 8–10 shows what is thought to be the general pattern of deep-water circulation in the world ocean. The most intense flow occurs as a western boundary current because of the rotation of the earth.

ANTARCTIC CIRCULATION

Although the oceanic mass surrounding the continent of Antarctica is not officially recognized as an ocean, we will first consider this circulation. We begin here because the mightiest of all ocean currents, the Antarctic Circumpolar Current, dominates the movement of water masses in the southern Atlantic, Indian, and Pacific oceans above 50°S latitude. This latitude can be considered the northern boundary of the "Antarctic Ocean."

Surface Circulation

A surface flow called the *East Wind Drift* moves in a westerly direction around the margin of the Antarctic continent (fig. 8–11). This flow is driven by the easterly winds that are manifestations of the polar air masses moving out to sea from Antarctica. The East Wind Drift is most extensively developed to the east of the Antarctic Peninsula in the Weddell Sea region and in the area of the Ross Sea. If this terminology seems confusing, it is because these currents are named after the winds that drive them, and winds are named on the basis of the direction they are coming from. For example, east winds blow out of the east.

The main circulation system in Antarctic waters is the *Antarctic Circumpolar Current,* extending northward to a position of approximately 40°S latitude. This mass is driven by the westerly winds that maintain very great strength throughout much of the year. The surface portion of this flow is named the *West Wind Drift.* As a result of the Coriolis effect deflecting moving masses to the left in the Southern Hemisphere, there is a zone of divergence between the East Wind Drift and the West Wind Drift.

Two major zones of convergence in the high southern latitudes exist: the *Antarctic Convergence* (the northern boundary of the Circumpolar Current) and the *Subtropical Convergence.* The *Antarctic Zone* includes the water from the Antarctic continent to the Antarctic Convergence, and the *Subantarctic Zone* includes the water between the two zones of convergence. The Antarctic Convergence lies between 50°S and 60°S latitude, and the surface water in the Antarctic Zone has a temperature range between −1.9°C (28.6°F) in the winter and 4°C (39.2°F), during the summer. The Subtropical Convergence lies within a few degrees of 40°S latitude, and the Subantarctic Zone is characterized by surface water that ranges between 4° and 14°C (39.2° and 57.2°F) (figs. 8–11 and 8–12A).

The Antarctic Circumpolar Current meets its greatest restriction as it passes through the 1000-km (621-mi) Drake Passage between the Antarctic Peninsula and the southern islands of South America. Although the current does not move at high velocity, reaching a maximum velocity of about 0.5 km/h (0.3 mi/h) just north of the Antarctic Convergence, it does transport more water than any other ocean current because of its breadth and depth.

Water volume transport is usually measured in million cubic meters per second. One of the first oceanographers to study the Antarctic circulation, Harald U. Sverdrup, found that up to 150 million m^3/s (5295 million ft^3/s) were moved by this current. Russian studies from 1956 through 1959 determined that up to 190 million m^3/s (6707 million ft^3/s) of water were transported by the portion of the current running between Africa and Antarctica. This mass of cold water, which flows beneath 22% of the world ocean

FIGURE 8–11 Antarctic Surface Circulation.

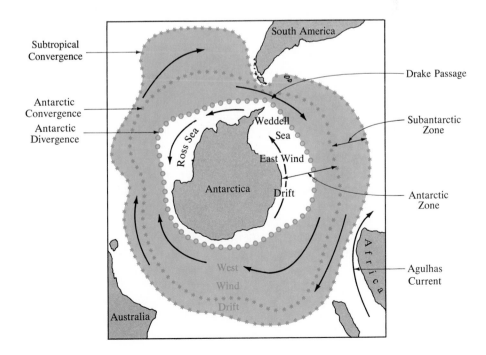

FIGURE 8–12 Antarctic Sub-surface Water Masses.

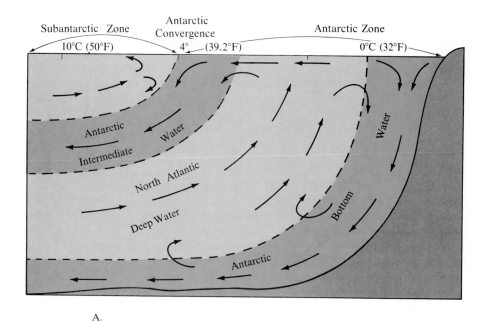

A.

B. Major subsurface water masses may be identified at 40° S latitude by observing vertical temperature and salinity profiles as follows:

1. Antarctic Intermediate Water—Salinity minimum identifies core of mass at about 900 m (2952 ft).

2. North Atlantic Deep Water—Temperature and salinity maxima identify NADW. Core at 2200 m (7216 ft) approaches 34.9 ‰.

surface and moves a volume about twice that of the maximum recorded for the Gulf Stream, has a very significant effect on the climatic patterns on the earth.

As we continue to discuss ocean circulation, we will refer to the amount of water being transported by currents in the various oceans and will use the unit *sverdrup* in place of one million m³/s (35.3 million ft³/s), as has been suggested by Dunbar in honor of Harald Sverdrup.

The fact that ice is freezing in the winter and thawing in the summer has a significant effect on the tem-perature and salinity of surface waters in high-latitude areas. The freezing point of water at the salinity normal for the Antarctic Zone is about −1.9°C (28.6°F). This marks the lowest limit to which water temperatures may fall. Since large volumes of heat energy are required to melt the ice during the summer, very little is left to warm the surface water. Thus the surface temperatures are always relatively low.

All around the Antarctic continent the Antarctic Circumpolar Water has similar characteristics. Especially during winter, a mixture of circumpolar water and shelf water from the Weddell Sea produces the densest water in the world ocean, the *Antarctic Bottom Water*. This mixture of shelf water with circumpolar water produces a water mass with σ_t value of 27.9 (density of 1.02790 g/cm^3). Owing to its high density, the Antarctic Bottom Water flows down the Antarctic continental shelf toward the Atlantic Ocean. As it moves northward along the bottom, it is carried to the east by the Circumpolar Current into the Indian and Pacific portions of the Antarctic Ocean.

The surface water in the Antarctic Zone has a northward component of flow that produces a convergence between this cold-water mass and the warmer water of the Subantarctic Zone. At this convergence, Antarctic surface water sinks and becomes the subsurface mass called *Antarctic Intermediate Water,* which has a temperature of about 2°C (32.4°F) and salinity of 33.8‰. This combination results in a σ_t of 27.0. The water mass can be identified with its core at about 900 m (2952 ft) depth at 40°S latitude by its low salinity and relatively high dissolved oxygen content (fig. 8–12).

Sandwiched between the Antarctic Intermediate and the Antarctic Bottom water masses at 40°S latitude is the *North Atlantic Deep Water*, which moves southward with its core at about 2200 m (7216 ft) depth. It is characterized by a temperature between 2° and 3°C (35.6 and 37.4°F) and salinity that averages 34.7‰. The North Atlantic Deep Water can be identified on the basis of its temperature, which is slightly above that of the Intermediate Water and the Antarctic Bottom Water. This deep water of North Atlantic origin has been away from the surface for at least 300 yr and is sandwiched between two water masses that have more recently left the surface, so it is also possible to identify it on the basis of a dissolved oxygen minimum. Although it is low in dissolved oxygen content, it still has a very positive effect on the biological productivity of the Antarctic Zone. The major contribution made by this deep water to biological productivity as it surfaces within the Antarctic Zone is its high concentration of nutrients that have been accumulating during the hundreds of years it was beneath the photosynthetic zone.

ATLANTIC OCEAN CIRCULATION

Surface Circulation

In the Atlantic Ocean, the basic surface circulation pattern is that of two large gyres. The North Atlantic gyre rotates in a clockwise direction, while the South Atlantic gyre rotates in a counterclockwise pattern. The driving force behind these rotations, which are

separated by the *Atlantic Equatorial Countercurrent,* is that of the northeast and southeast trade winds.

The South Atlantic gyre is composed of the *South Equatorial Current,* which reaches its greatest strength just below the equator and is split in two by the topographic interference of the eastern extension of Brazil. Part of the South Equatorial Current moves off along the northeastern coast of South America toward the Caribbean Sea and the North Atlantic. The rest is turned southward as the *Brazil Current,* which ultimately merges with the *West Wind Drift* and moves across the South Atlantic. The gyre is completed by a slow-drifting movement of cold water, the *Benguela Current,* that flows up the western coast of Africa. There is also a significant flow of cold water moving in a northerly direction along the western margin of the South Atlantic. The *Falkland Current,* an important cold current, moves up the coast of Argentina as far north as 25° to 30°S latitude, wedging its way between the continent and the Brazil Current. The Brazil Current has a much smaller volume than its Northern Hemisphere counterpart, the Gulf Stream. This smaller size results partly from the splitting of the South Equatorial Current by the configuration of South America (fig. 8–13).

The *North Equatorial Current* moves parallel to the equator in the Northern Hemisphere, where it is joined by that portion of the South Equatorial Current that is shunted toward the north along the South American coast. This flow splits into two masses, the *Antilles Current*, which flows along the Atlantic side of the West Indies, and the *Caribbean Current*, which passes through the Yucatan Channel into the Gulf of Mexico. These masses reconverge as the water that entered the Gulf of Mexico exits between Florida and Cuba as the *Florida Current*. The Florida Current flows close to shore over the continental shelf, carrying a volume that at times exceeds 35 sverdrups (hereafter abbreviated as sv). As it moves off Cape Hatteras and flows across the open ocean in a northeasterly direction it becomes the *Gulf Stream*, flowing at velocities up to 9 km/h (5.6 mi/h). The western margin of the Gulf Stream can frequently be defined as a rather abrupt boundary that moves periodically closer to and farther away from the shore. The eastern boundary becomes very difficult to identify, as it is usually masked by filamentous meandering masses that continuously change their positions. Its character gradually merges with that of the water of the Sargasso Sea to the south and east of the Gulf Stream. A volume transport of over 90 sv off Chesapeake Bay indicates a large volume of Sargasso Sea water has been added to the flow provided by the Florida Current. However, south of Newfoundland,

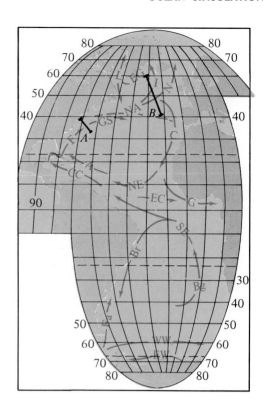

CURRENTS

A–Antillean	G–Guinea
Bg–Benguela	GS–Gulf Stream
Br–Brazil	I–Irminger
C–Canary	L–Labrador
CC–Caribbean	NA–North Atlantic
EG–East Greenland	NE–North Equatorial
EW–East Wind Drift	N–Norwegian
EC–Equatorial Counter	SE–South Equatorial
Fa–Falkland	WW–West Wind Drift
F–Florida	

FIGURE 8–13 Atlantic Ocean Surface Currents.
Base map courtesy of National Ocean Survey.

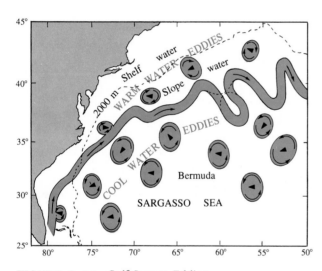

FIGURE 8–14 Gulf Stream Eddies.
Large eddies form to the north and south of the Gulf Stream as meanders pinch off. Those to the north contain warm water and rotate clockwise while those to the south contain cool slope water and rotate counterclockwise as they move southwest to rejoin the mighty current. Also see plate 5. From Philip Richardson, Gulf Stream rings, *Oceanus*, Spring 1976. Reprinted by permission of the author.

the volume carried by the Gulf Stream has been reduced to 40 sv. This lower volume would indicate that most of the Sargasso Sea water that joined the Florida Current and made up the Gulf Stream has returned to the diffuse flow of the Sargasso Sea.

Just how this loss of water occurs is yet to be determined, but much of it may be achieved by meanders (sinuous curves in the course of a current) pinching off to form large eddies. As is shown in figure 8–14, meanders north of the Gulf Stream pinch off and trap warm Sargasso Sea water in eddies rotating in a clockwise direction. These eddies move at speeds of 3 to 7 km/day (1.9 to 4.3 mi/day) southwest toward Cape Hatteras, where they rejoin the Gulf Stream. South of the Gulf Stream and west of 50°W latitude, eddies with a core of cold inshore water rotating counterclockwise move south and west. (See *Gulf Stream Rings and Marine Life*, pp. 339–341.) These large eddies, ranging in diameter from 100 to 500 km (62 to 310 mi) and extending to depths of 1 km (0.62 mi) in the warm rings and as much as 3.5 km (2.2 mi) in the cold rings, could remove large volumes of water from the Gulf Stream. The use of a marvelous new technique for mapping dynamic features such as the Gulf Stream rings is discussed in *Ocean Acoustic Tomography*, on pages 196–197.

At the Tail of the Banks, about 40°N latitude and 45°W longitude, the Gulf Stream becomes the *North Atlantic Current* and continues in a more easterly direction across the North Atlantic. The cross sections in figure 8–15 show the changing character from the Gulf Stream off Chesapeake Bay to the North Atlantic Current northwest of the Azores. The intense temperature slope that shows the 18° northwestern boundary for the Gulf Stream off Chesapeake Bay is not to be found in the less intense structure of the North Atlantic Current, which is composed of numerous branches into which the Gulf Stream flow is ultimately broken (fig. 8–15).

Two branches that are composed of water produced through the mixing of the Labrador Current and the Gulf Stream are the *Irminger Current*, flowing up along the west coast of Iceland, and the *Norwegian Current*, which moves north along the coast of Norway. The other major branch of the North Atlantic Current follows 45°N latitude across the North Atlantic and turns south as the *Canary Current,* passing between the Azores and Spain. This southward flow is very diffuse and spreads over a broad area as it moves southward and eventually joins the North Equatorial Current.

The effects of the westward intensification of current flow in the North Atlantic Ocean can be seen from the surface temperatures shown in figure 8–7. From 20° to 40°N latitude off the coast of North America we can see a 20°C (36°F) temperature change from waters near the Dominican Republic to those near New York in February. By contrast, on the eastern side of the North Atlantic only a 5° to 6°C (9° to 10.8°F) range in temperature can be observed between 20° and 40°N latitude from a point off the coast of Mauritania to waters off the coast of Spain.

Deep-Water Masses

Surface and near-surface water masses are underlain by the Antarctic Intermediate Water throughout the South Atlantic and to some extent in the North Atlantic. This intermediate water mass begins to lose its characteristics in the North Atlantic, and we find that *Mediterranean Water* that has passed into the Atlantic Ocean over the Gibralter Sill underlies much of the Central Water Mass in the North Atlantic. Figure 8–16 shows the temperature, salinity, and dissolved oxygen characteristics of the Atlantic water masses.

The North Atlantic Deep Water, which we discussed in the previous section on Antarctic circulation, forms in the Norwegian Sea and in the North Atlantic off the southern tip of Greenland, where the water temperature is about 3°C (37.4°F) to a depth of

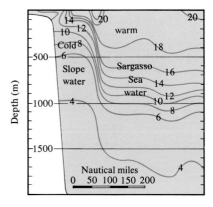

Cross section A

This vertical section cuts the Gulf Stream between Chesapeake Bay and Bermuda (*see* fig. 8–13). The Gulf Stream appears as a very steep slope in the isotherms (lines connecting points of equal temperature). Here, the Gulf Stream is a narrow intense flow bounded on the west by low velocity cold slope water and on the east by the warm Sargasso Sea. The current velocity is proportional to the slope of the isotherms. Intense Gulf Stream flow can readily be identified to a depth of 1000 m.

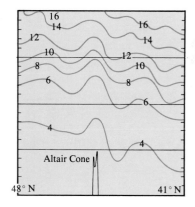

Cross section B

Running from the Azores north-northwest to 48° latitude, this vertical section shows a gently irregular south slope to the isotherms that indicates a moderate rate of flow to the east (see fig. 8–13). The North Atlantic Current splits up in the region of this section and becomes the Irminger, Norwegian, and Canary currents between which significant eddying occurs. The Altair Cone, a volcanic peak, appears to have a significant effect on the distribution of physical characteristics of the overlying water.

FIGURE 8–15 Comparison of Temperature (°C) and Flow Characteristics of the Gulf Stream and the North Atlantic Current.

300 m (984 ft). Owing to low surface temperatures and relatively high salinity during the winter, the water at the surface develops a density great enough to allow it to sink and move off to the south as North Atlantic Deep Water. The North Atlantic Deep Water, transporting 14 sv, can still be identified as far south as 40°S latitude on the basis of temperature and salinity maxima observed at depths around 3000 m (9840 ft). Antarctic Bottom Water, transporting 7 sv, can be identified as far north as 40°N latitude by its low temperature and salinity and high oxygen content. The Mediterranean Water is most easily identified by its salinity maximum at a depth between 1000 and 2000 m (3280 and 6560 ft). The Antarctic Bottom

Water moves northward along the western margin of the Atlantic Ocean in the West Atlantic Basin, which is separated from the East Atlantic Basin by the Mid-Atlantic Ridge. As a result of this movement, the bottom temperatures in the West Atlantic Basin are maintained well below 2°C (35.6°F, while they are around 2.5°C (36.5°F) in the bottom of the East Atlantic Basin.

Although the exact volumes involved are difficult to determine, it is evident that there is a large transfer of water at depth across the equator between the North and South Atlantic. The residence time of deep water in the Atlantic Ocean is approximately 275 years. As we will see in our discussions of the Pacific

FIGURE 8–16 Atlantic Ocean Subsurface Circulation.

From Sverdrup, Johnson & Fleming, *The oceans: their physics, chemistry, and general biology*, © 1942, Renewal © 1970. Reprinted by permission of Prentice-Hall, Inc., Englewood Cliffs, New Jersey.

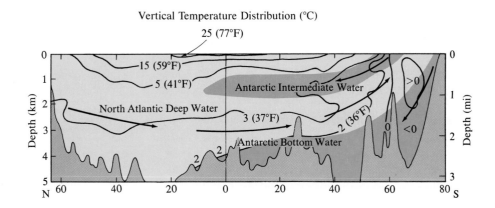

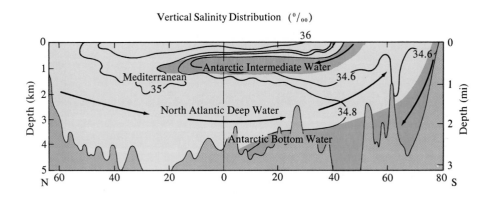

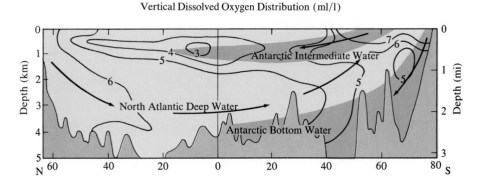

IS THE DEEP OCEAN CHANGING MORE RAPIDLY THAN WE EVER THOUGHT IT COULD?

During a study of deep-ocean circulation in 1981, observations in the North Atlantic showed that below a depth of a few hundred meters water between 50° and 60°N latitude was 0.02 ‰ less salty and 0.15°C (0.27°F) colder than it was in 1972. This finding flew in the face of a long-held conception of a deep ocean that changed significantly only on a scale of thousands of years. Observations conducted as part of the Transit Tracers in the Ocean Program (TTO) were published in 1982 and showed changes at 24° and 36°N latitude in the Atlantic Ocean compared to 1958. The water from 500 to 3000 m (1640 to 9840 ft) is 0.2°C (0.36°F) warmer. Above and below this warmer water, the water is now colder than in 1958.

The cooling and freshening of the 50°–60°N latitude waters may result from an increased rate of sinking of North Atlantic Deep Water during the 1970s. The temperature changes observed at 24° and 36°N latitude may be more closely related to changing rates of formation of intermediate and bottom water masses in the Antarctic.

These changes could result from atmospheric changes in the 1970s, a north-south sloshing of water back and forth in the Atlantic basin, or a number of other possible causes. Although the changes seem small, they are significant and need to be understood. Only a full understanding of ocean circulation may provide the explanation of these changes. We appear to be approaching such an understanding, but it may still be many years away.

Ocean and the Indian Ocean, there is little evidence of significant sinking and transequatorial water transport in either of these oceans. From this we may conclude that the bottom-water masses throughout the world ocean originate in the Atlantic Ocean, particularly from sinking off the coast of Antarctica in the Weddell Sea.

PACIFIC OCEAN CIRCULATION

Surface Circulation

The surface circulation in the Pacific Ocean is generally similar to that which we described for the Atlantic Ocean, except that the Equatorial Countercurrent is much better developed in the Pacific Ocean than in the Atlantic (fig. 8–17). We will first discuss the counterclockwise rotation in the South Pacific and the flow of an undercurrent present within the system.

Southeast trade winds drive the *South Equatorial Current*, which has a lateral dimension from about 10°S to 3°N latitude. The velocity of the South Equatorial Current is approximately 1.8 km/h (1.1 mi/h). Embedded within the South Equatorial Current and flowing in an easterly direction along the equator is a thin ribbonlike *Equatorial Undercurrent*. This undercurrent extends for over 6000 km (3726 mi) from the western Pacific to the Galapagos Islands. Only 0.2 km (0.12 mi) thick and approximately 300 km (186 mi) wide, the Equatorial Undercurrent at depths of 70 to

200 m (230 to 656 ft) flows with velocities approaching 5 km/h (3.1 mi/h) in an easterly direction. Its volume transport is approximately 40 sv.

The South Equatorial Current becomes the western boundary current of the South Pacific gyre—the *East Australian Current*. This current, which is relatively weak for a western boundary current, joins the West Wind Drift that carries the water across the southern Pacific to be pushed up along the western coast of South America as the *Peru Current*, which completes the gyre by turning west near the equator to join the South Equatorial Current.

In the North Pacific, the *North Equatorial Current* flows from east to west between latitudes of 8° and 20°N at maximum velocities slightly in excess of 1 km/h (0.62 mi/h). As the North Equatorial Current approaches the western boundary of the North Pacific, most of the water moves north to form the *Kuroshio Current*, the North Pacific counterpart of the Florida Current in the North Atlantic. Part of the North Equatorial Current water turns south and joins the Equatorial Countercurrent. After the Kuroshio Current, which has a maximum velocity of just under 9 km/h (5.6 mi/h), leaves the coast and begins to flow in an easterly direction, it becomes the *Kuroshio Extension* to a longitude of about 170°E. This segment of the North Pacific gyre corresponds to the Gulf Stream of the North Atlantic and has a volume transport of about 65 sv.

The Kuroshio Extension is met by the cold *Oyashio Current*, which carries water south from the Bering

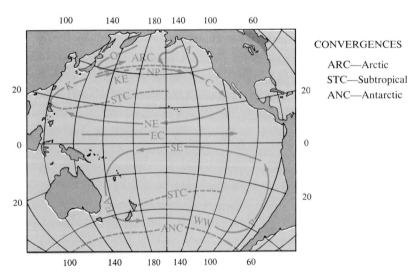

CONVERGENCES

ARC—Arctic
STC—Subtropical
ANC—Antarctic

CURRENTS

A—Alaskan
C—California
EA—East Australian
EC—Equatorial Counter
K—Kuroshio
KE—Kuroshio Extension
NE—North Equatorial
NP—North Pacific
O—Oyashio
P—Peru
SE—South Equatorial
WW—West Wind Drift

FIGURE 8–17 Pacific Ocean Surface Currents.
Base map courtesy of Open University Press.

Sea and the Sea of Okhotsk. Beyond this juncture the flow becomes the *North Pacific Current* that moves toward the North American continent, where it divides. Part of the North Pacific Current turns southward as the *California Current*, which completes the gyre by joining the North Equatorial Current. The remainder flows north to become the *Alaskan gyre*.

Lying between the North and South Equatorial currents is the *Equatorial Countercurrent*, which carries water in an easterly direction at velocities slightly in excess of 2 km/h (1.2 mi/h), except during March and April, when its velocity decreases to approximately 0.5 km/h (0.3 mi/h). The shift of the equatorial system to the north is a result of the trade wind system carrying air masses toward the ITCZ that, because of land distribution, lies approximately 5° north of the geographical equator.

Deep-Water Masses

Surface-temperature distribution with regard to latitude in the Pacific Ocean is similar to that observed in the Atlantic Ocean, but surface salinity in the Pacific is lower than in the Atlantic, with the South Pacific water being slightly more saline than that in the North Pacific.

Underlying the surface and near-surface water masses are the Antarctic Intermediate Water, formed north of the Antarctic Convergence, and the North Pacific Intermediate Water, both of which possess low salinity. These water masses appear to form as a result of subsurface mixing rather than by obtaining their temperature and salinity characteristics at the surface as do most deep-water masses.

The fact that no deep-water masses form in the Pacific Ocean leads to a very uniform temperature and salinity structure in the deep Pacific waters below 2000 m (6560 ft). Possibly the main factor that retards sinking in the Pacific Ocean is the low salinity of the surface water. The Pacific Ocean bottom water is a mixture of the North Atlantic Deep Water and Antarctic Bottom Water. It is called the *Common Water* and is introduced into the Pacific by the Antarctic Circumpolar Current. It moves very slowly toward the north with a volume transport of about 25 sv. The northward movement of this Antarctic mass can be identified by observing the temperature of the bottom water, which increases gradually from south to north. There is also a slight decrease in salinity and dissolved oxygen content from south to north (fig. 8–18).

In contrast to the condition of high volume interchange in the Atlantic Ocean, the rates and volume of water transport between the North and South Pa-

cific are so low that it is difficult to accurately measure them. One of the main criteria for concluding that the deep waters in the Pacific Ocean move at very low rates and are old is the relatively low dissolved oxygen content of these waters. Compared to the Atlantic deep-water masses, where dissolved oxygen content falls within the range of 4.5 to 6.5 ml/l, the Pacific deep-water dissolved oxygen content ranges between 3.5 and 4.5 ml/l (fig. 8–18). The residence time of deep water in the Pacific Ocean is 510 years.

INDIAN OCEAN CIRCULATION

Surface Circulation

Surface circulation in the Indian Ocean, which extends only to about 20°N latitude, varies considerably from that which we have discussed for the Atlantic Ocean and Pacific Ocean. From November to March the equatorial circulation is similar to that of other oceans, with two westward-flowing equatorial currents separated by an *Equatorial Countercurrent*. In contrast to the wind systems in the Atlantic and Pacific oceans, which are shifted to the north of the geographical equator, we find that in the Indian Ocean the ITCZ is shifted to the south of the geographical equator. The Equatorial Countercurrent flows between 2° and 8°S latitude, bounded on the north by the *North Equatorial Current,* which extends as far as 10°N latitude, and on the south by the *South Equatorial Current*, which extends to 20°S latitude. During winter the typical northeast trade winds are developed and are referred to as the *northeast monsoon*. They are reinforced because rapid cooling of the air during the winter months over the Asian mainland has created a high-pressure cell that is forcing atmospheric masses off the continent out over the ocean, where the air pressure is lower (fig. 8–19).

The Asian mainland warms up faster than the oceanic water owing to the relatively lower heat capacity of continental crustal material compared to that of water. As a result, a low-pressure cell develops over the continent during summer and "sucks" in the air masses overlying the ocean. This force gives rise to the *southwest monsoon*, which may be thought of as a continuation of the southeast trade winds across the equator. During this season the North Equatorial Current disappears and is replaced by the *Southwest Monsoon Current*, which flows from west to east across the North Indian Ocean. In September or October, the northeast trade winds are re-established,

FIGURE 8–18 Pacific Ocean Subsurface Circulation.

From Sverdrup, Johnson & Fleming, *The oceans: their physics, chemistry, and general biology,* © 1942, Renewal © 1970. Reprinted by permission of Prentice-Hall, Inc., Englewood Cliffs, New Jersey.

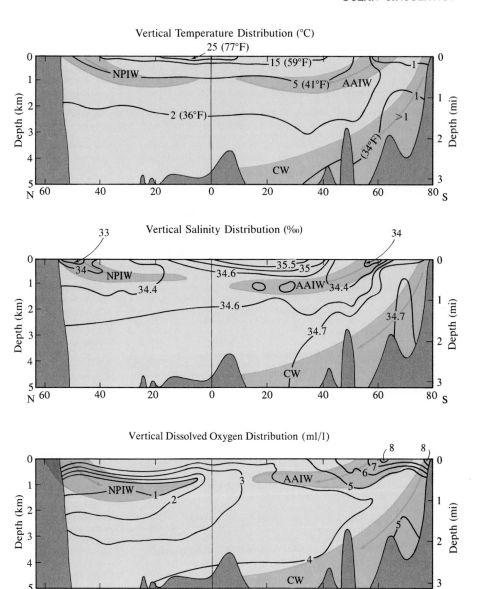

NPIW–North Pacific Intermediate Water   CW—Common Water

AAIW—Antarctic Intermediate Water

and the North Equatorial Current reappears (fig. 8–19).

The surface circulation in the southern Indian Ocean is similar to the counterclockwise circulation observed in other southern oceans. During the time the northeast trade winds blow, the South Equatorial Current provides water for the Equatorial Countercurrent and the *Agulhas Current,* which flows south along the eastern coast of Africa and joins the West Wind Drift. Turning north out of the West Wind Drift is the *West Australian Current,* which completes the gyre by merging with the South Equatorial Current.

During the southwest monsoon, a northward flow from the equator along the coast of Africa, the *Somali Current,* develops with velocities approaching 4 km/h (2.5 mi/h).

DEEP-WATER MASSES

Directly beneath the surface and near-surface water is the Antarctic Intermediate Water, identifiable by its low temperature and high dissolved oxygen content (fig. 8–20). The Red Sea Water is a high-salinity mass

FIGURE 8–19 Indian Ocean Surface Currents.

Base map courtesy of National Ocean Survey.

A. WINTER: November–March, Northeast monsoon wind season

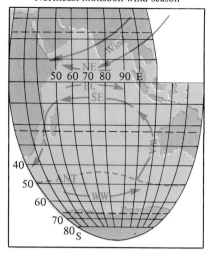

From November to March equatorial circulation in the Indian Ocean is similar to that of other oceans, except that the Equatorial Countercurrent is shifted south.

B. SUMMER: May–September, Southwest monsoon wind season

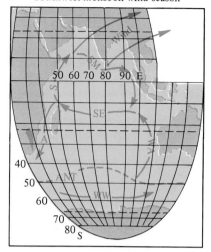

Low pressure over the mainland during summer draws the Southwest Monsoon winds over the North Indian Ocean. This wind produces the Southwest Monsoon Current which flows in an easterly direction and replaces the North Equatorial Current.

CURRENTS   A–Agulhas   EC–Equatorial Counter   NE–North Equatorial   S–Somali
SE–South Equatorial   SM–Southwest Monsoon   WA–West Australian   WW–West Wind Drift

ANT–Antarctic Convergence

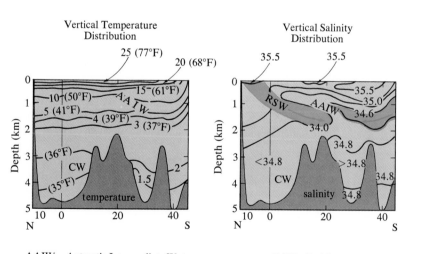

AAIW— Antarctic Intermediate Water

RSW—Red Sea Water

CW—Common Water

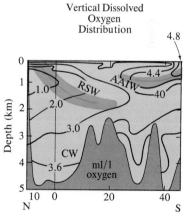

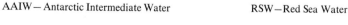

FIGURE 8–20 Indian Ocean Subsurface Circulation.

From A. J. Clowes and G. E. R. Deacon, *The deep-water circulation of the Indian Ocean*, © 1935. Reprinted by permission of Nature, Macmillan Journals, and G. E. R. Deacon.

with low oxygen content that is fed by outflow of the Red Sea and descends to a depth between 1000 and 1500 m (3280 and 4920 ft) in the northern Indian Ocean.

The Common Water of the Indian Ocean has an average temperature of 1.5°C (34.7°F) and an average salinity of 34.7‰. It enters the Indian Ocean at a rate of 20 sv, which results in a residence time for deep water in the Indian Ocean of 250 years.

RESEARCH IN OCEAN CIRCULATION

Investigation of surface circulation was undertaken by the Mid-Ocean Dynamics Experiment (MODE), in which United Kingdom and United States oceanographers attempted to identify short-term, small-scale "weather" phenomena of the North Atlantic in 1973. Instead of analyzing the ocean in terms of long-term average conditions or oceanic "climate," MODE and subsequent POLYMODE investigations have shown that the oceans display "weather" phenomena similar to those of the atmosphere. The velocities of the "winds" may be lower, but the "storms" last much longer. Evidence for this is the discovery of large eddies that may be analogous to hurricanes in the atmosphere.

The Geochemical Ocean Sections (GEOSECS) program has shown that tritium produced in the late 1950s and early 1960s by atmospheric testing of nuclear weapons had been carried to depths approaching 5 km (3.1 mi) in the North Atlantic by early 1972. This datum suggests a surprisingly rapid rate of descent for the North Atlantic Deep Water.

POWER FROM WINDS AND CURRENTS

The winds that transform solar energy into ocean currents have for centuries been used by society to drive its machines. Westerly winds off New England represent a near-shore wind supply that is reasonably sustained and contains a large amount of energy. There is some optimism that Offshore Windpower Systems (OWPS) such as that shown in figure 8–21 could provide electricity or hydrogen generated by electricity to meet the needs of large sections of the North Atlantic coast of the United States.

Many have considered the great amount of energy in the Florida–Gulf Stream Current System and dreamed of harnessing it to do the work of society. A group of scientists and engineers met in 1974 to consider this possibility and concluded that some 2000 megawatts (MW) of electricity could be recovered

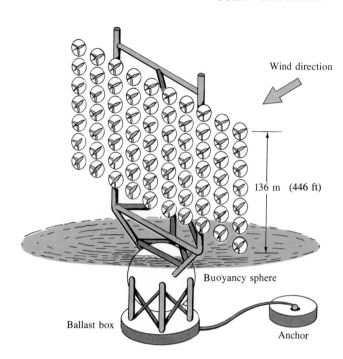

FIGURE 8–21 Offshore Windpower System. Courtesy of Woods Hole Oceanographic Institution.

along the east coast of southern Florida. Devices proposed for extraction range from underwater "windmills" to a Water Low Velocity Energy Converter (WLVEC), which is operated by parachutes attached to a continuous belt (fig. 8–22). Again, calculations indicate such systems can be economically competitive. What remains is to see if the designs for ocean wind and current generation of electricity actually function as expected in the sometimes hostile marine environment. The Coriolis Program was proposed in 1973 by William J. Moulton of Tulane University. Hydroturbines with diameters of 170 m (558 ft) are envisioned that could generate 43 MW of electricity from the movement of ocean currents (fig. 8–23). An array of 242 units covering an area 30 km (18.6 mi) wide and 60 km (37 mi) long could generate 10,000 MW—the equivalent of 130 million barrels of oil. Tests are now being conducted to establish an optimum size, identify suitable sites, and confirm engineering, economic, and environmental estimates.

SUMMARY

Horizontal currents set in motion by wind systems are characteristic of the surface waters of the world ocean. According to the **Ekman spiral** concept, winds will set the surface waters in motion in a direction 45° to the right of the wind in the Northern Hemisphere, and the net water movement will be at right angles

OVERHEAD VIEW

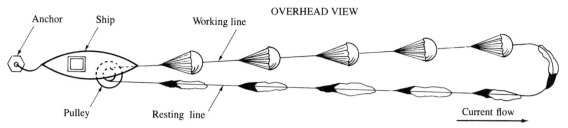

FIGURE 8–22 Water Low-Velocity Energy Converter.
Courtesy of G. E. Steelman.

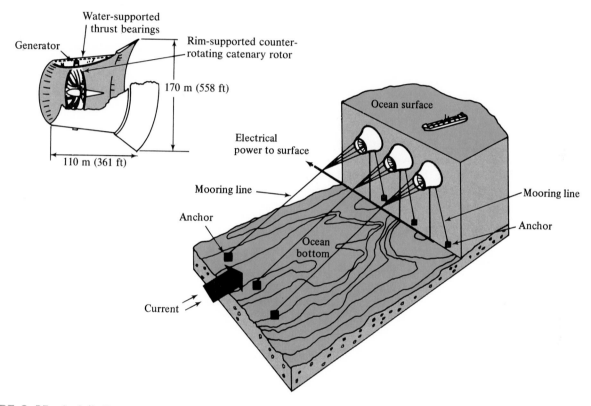

FIGURE 8–23 Coriolis Program.
Buoyant Coriolis hydroturbines anchored in ocean current. It is estimated that an array of 242 such units placed in the Gulf Stream could provide 10% of the present electricity needs of Florida. Courtesy of U.S. Department of Energy.

to the wind direction. As a result of the Ekman spiral phenomenon, water is pushed in toward the center of clockwise gyres in the Northern Hemisphere and counterclockwise gyres in the Southern Hemisphere, forming "hills." As water in the Northern Hemisphere runs downslope on the hills of water that result, the Coriolis effect causes it to turn right and into the clockwise flow pattern. As a result, a **geostrophic current** flowing parallel to the contours of the hill is maintained. The apex of the hill is always located to the west of the geographical center of the gyre. Using **dynamic topography**, oceanographers can map the surface currents in ocean basins by determining the average density of water columns at locations

throughout the basin and mapping this distribution. The speed of the current will increase with increased slope of the hills, which will normally be steepest on the west side.

Vertical circulation produced by wind will be confined to the surface-water mass. An example is equatorial or coastal **upwelling**, which is created by the **Ekman transport** pulling wind-driven water masses away from the equator or coastal areas. Water will surface from a depth of no more than a few hundred meters to replace the surface water. An even shallower vertical movement of water is represented by the **Langmuir circulation** of long, shallow convection cells running parallel to the gentle, steady winds that create them. These cells may extend no deeper than 6 m (19.7 ft). The mixing of the water in the deepest ocean basins is caused by **thermohaline circulation**, which is initiated by the sinking of dense surface water in high latitudes. The **North Atlantic Deep Water** and the **Antarctic Bottom Water** carry most of the oxygen-bearing water to the deep basins of all the oceans to maintain their life-sustaining capacity.

The **Circumpolar Current** flows in a clockwise direction around the continent of Antarctica. It is the largest current in the world ocean, transporting over 190 **sverdrups** of water. The surface portion of this flow is often referred to as the **West Wind Drift** because it is driven by strong westerly winds. Two major deep-water masses form in Antarctic waters. The Antarctic Bottom Water, the densest water mass in the oceans, forms primarily in the **Weddell Sea** and sinks along the continental shelf into the Atlantic Ocean. Farther north, at the **Antarctic Convergence**, the low-salinity **Antarctic Intermediate Water** sinks to a depth of about 900 m (2952 ft). Sandwiched between these two masses is the North Atlantic Deep Water, rich in plant nutrients after hundreds of years

in the deep ocean. The highest-velocity current in the oceans is the **Gulf Stream** as it flows along the southeast Atlantic coast. The best development of an equatorial countercurrent system is found in the Pacific, where large eastern movement of water is achieved by the **Pacific Countercurrent** and the **Equatorial Undercurrent**, a ribbonlike subsurface flow. The strong **Kuroshio Current** is the North Pacific counterpart of the Gulf Stream. The formation of deep-water masses is not pronounced in the Pacific Ocean, as there appears to be no deep sinking of surface water. The Indian Ocean circulation is dominated by the **monsoon** wind systems that blow out of the northeast in the winter and the southwest in the summer.

Research activities such as MODE and GEOSECS rapidly increase our understanding of ocean circulation and mixing processes and have indicated that temperature and salinity conditions in the deep ocean may not be as stable as was once believed. Continued investigation of the energy potential of ocean winds and currents may lead to development of a significant addition of **renewable** energy that can be exploited by society.

QUESTIONS AND EXERCISES

1. Compare the forces directly responsible for creating horizontal and deep vertical circulation in the oceans. What is the ultimate source of energy for driving both circulation systems?

2. On a base map of the world, plot the major surface circulation gyres of the oceans, the meteorological equators of each ocean, and the Subtropical, Arctic, and Antarctic Convergences. Superimpose the major wind belts of the world on the gyres. Label the currents using the symbols used in figures 8–13, 8–17, and 8–19.

3. Diagram and discuss how the Ekman transport produces the "hill" of water within major ocean gyres that causes geostrophic current flow. As a starting place on the diagram, use the prevailing wind belts, the trade winds, and the westerlies.

4. What causes the apex of these geostrophic "hills" to be offset to the west of the center of the ocean gyre systems?

5. Explain how oceanographers determine the dynamic topography of the ocean waters.

6. Describe the relationships among the wind, the surface current it creates, and the development of equatorial and coastal upwellings.

7. Discuss why thermohaline vertical circulation is driven by sinking of surface water that occurs only in high latitudes.

8. Name the two major deep-water masses and give the locations of their formation at the ocean's surface.

9. The largest current in the world ocean in terms of volume transport is the Antarctic Circumpolar Current. Explain why its surface portion is referred to as the West Wind Drift. What is its maximum volume transport as compared to the maximum volume transport measure for the Gulf Stream?

10. Why does the North Atlantic Deep Water enhance biological productivity in the Antarctic Zone as it surfaces between the Antarctic Intermediate Water and the Antarctic Bottom Water?

11. Observing the flow of Atlantic Ocean currents in figure 8–13, compare the Brazil Current to the Gulf Stream and offer some possible explanations for the fact that the Brazil Current has a much lower velocity and volume transport than the Gulf Stream.

12. Explain why the Gulf Stream eddies that develop northeast of the Gulf Stream rotate clockwise and have warm-water cores while those that develop to the southwest rotate counterclockwise and have a core of cold water.

13. Describe the relationship between current velocity and slope of isotherms.

14. The Antarctic Intermediate Water is identifiable throughout much of the South Atlantic on the basis of a temperature minimum, salinity minimum, and dissolved oxygen maximum. Why should it be colder, less salty, and contain more oxygen than the surface-water mass above it and the North Atlantic Deep Water below it?

15. What evidence is there that the bottom water in the Pacific Ocean has been away from the surface much longer than that of the Atlantic Ocean?

16. Discuss the changes in current flow in the North Indian Ocean and their relationship to the monsoon winds.

17. Where is the potential for using ocean windpower systems and ocean current power systems greatest, and why?

REFERENCES

Defant, A. 1961. *Physical oceanography*, 2 vols. New York: Macmillan.

Gill, A. 1982. *Atmosphere-ocean dynamics. Orlando, Florida: Academic Press.*

Hammond, A. L. 1971. Oceanography: Geochemical tracers offer new insight. Science 195:164–66.

MacLeish, W., ed. 1974. Energy and the sea. *Oceanus* 17:52.

Montgomery, R. 1940. The present evidence of the importance of lateral mixing processes in the ocean. *American Meteorological Society Bulletin* 21:87–94.

Pickard, G. L. 1975. *Descriptive physical oceanography*. 2nd ed. New York: Pergamon Press.

Stommel, H. 1955. The anatomy of the Atlantic. *Scientific American* 190:30–35.

———.1958. The abyssal circulation. Letters to the ediitor. *Deep Sea Research*. 5. New York: Pergamon Press.

Stuiver, M.; Quay, P. D.; and Ostlund, H. G. 1983. Abyssal water carbon-14 distribution and the age of the world oceans. *Science* 219:4586, 849–51.

Sverdrup, H. U.; Johnson, M. W.; and Fleming, R. H. 1942. Renewal 1970. *The oceans.* Englewood Cliffs, N.J.: Prentice Hall.

SUGGESTED READING

Sea Frontiers

Frye, J. 1982. The ring story. 28:5, 258–67.
    A discussion of the Gulf Stream, its rings, and how the rings are studied.

Miller, J. 1975. Barbados and the island-mass effect. 21:5, 268–72.
    A discussion of the phenomenon by which waters around tropical islands are much more productive than the surface waters of the open ocean.

Smith, F. G. W. 1972. Measuring ocean movements. 18:3, 166–74.
    Discussed are some practical problems related to current flow and some methods
    used to determine current direction, speed, and volume.

Smith, F. G. W., and Charlier, R. 1981. Turbines in the ocean. 27:5, 300–305.
    A discussion of the potential of extracting energy from ocean currents is pre-
    sented.

Sobey, E. 1982. What is sea level? 28:3, 136–42.
    The factors that cause sea level to change are discussed.

Scientific American

Baker, D. J., Jr. 1970. Models of ocean circulation. 222:1, 114–21.
    This article discusses observations of a model depicting a segment of the surface
    of the earth over which fluids move and helps explain geostrophic flow within
    ocean gyres. Some knowledge of basic physics is necessary for full comprehen-
    sion of material presented.

Hollister, C. D., and Nowell, A. 1984. The dynamic abyss. 250:3, 42–53.
    Submarine "storms" are associated with deep, cold currents flowing away from
    polar regions toward the equator.

Stewart, R. W. 1969. The atmosphere and the ocean. 221:3, 76–105.
    The exchange of energy between the atmosphere and ocean and the resulting
    phenomena, currents, and waves are covered in this readable, comprehensive ar-
    ticle.

Webster, P. J. 1981. Monsoons. 245:5, 108–19.
    The mechanism of the monsoons and their role in bringing water to half the
    earth's population is discussed.

Weibe, P. H. 1982. Rings of the Gulf Stream. 246:3, 60–79.
    The biological implications of the large cold-water rings are considered.

Waves are one of the most obvious phenomena of the ocean. Yet it was not until well into the nineteenth century that some understanding of what caused waves and how they behaved was developed. We will first look at the character of waves in general before proceeding to discuss oceanic waves. Wave phenomena involve the transmission of energy and momentum by means of vibratory impulses through the various states of matter. Theoretically, the medium itself does not move in the direction the energy passes through. The particles that make up the medium simply oscillate in a back-and-forth or orbital pattern, transmitting energy from one particle to another.

Simple *progressive waves,* shown in figure 9–1A, may be described as *longitudinal* and *transverse.* In longitudinal waves, such as sound waves, the particles that are in vibratory motion move back and forth in a direction parallel to the propagation of energy. Energy may be transmitted through all states of matter—gaseous, liquid, or solid—by longitudinal movement of particles.

Transverse wave phenomena involve the propagation of energy at right angles to the direction of particle vibration. An example of such a wave would be that created when one end of a rope is tied to a doorknob while the other end is moved up and down with the hand. A waveform is set up and progresses along the rope, and energy is transmitted from the motion of the hand to the doorknob. If one were to pick out a particular segment of the rope and watch it, the segment would be seen to move up and down at right angles to a line drawn from the doorknob to the hand that is putting energy into the rope. We generally consider that this type of wave can transmit energy only through solids, since it is only in a solid that particles are strongly attached to one another.

Waves on the ocean surface, as do all waves that transmit energy along an interface between any two fluids of different density, have particle movements that are neither longitudinal nor transverse. We may say more correctly that the movement of particles along such an interface involves components of both, since the particles move in circular orbits at the interface between the atmosphere and ocean.

CHAPTER NINE
WAVES

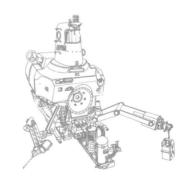

Rough seas in the Gulf of Alaska. Courtesy of Scripps Institution of Oceanography, University of California, San Diego.

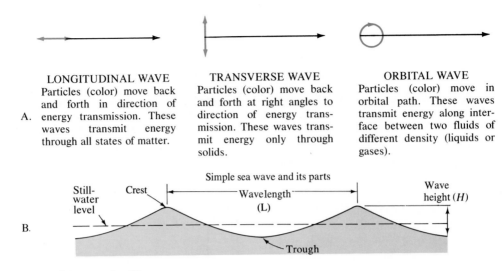

LONGITUDINAL WAVE
Particles (color) move back and forth in direction of
A. energy transmission. These waves transmit energy through all states of matter.

TRANSVERSE WAVE
Particles (color) move back and forth at right angles to direction of energy transmission. These waves transmit energy only through solids.

ORBITAL WAVE
Particles (color) move in orbital path. These waves transmit energy along interface between two fluids of different density (liquids or gases).

Simple sea wave and its parts

Still-water level   Crest   Wavelength (L)   Wave height (H)

B.

Trough

FIGURE 9–1   Types of Progressive Waves.

WAVE CHARACTERISTICS

When we observe the ocean surface, we see that there are waves of various sizes moving in various directions, resulting in a complex wave pattern that is constantly changing. We will, for the sake of introducing some characteristics that must be used to discuss waves, look at a much simpler form of an idealized wave representing the transmission of energy created by a single source and traveling along the ocean-atmosphere interface. Such a series of waves will have very uniform characteristics. We use such an idealized form in figure 9–1B.

As an idealized *progressive* wave (a phenomenon in which the wave form can be observed to move) passes a permanent marker, such as a pier piling, we will notice a succession of high parts of the waves, *crests,* separated by low parts, *troughs.* If we were to mark the water level on the piling when the troughs pass and then do the same for the crests, the vertical distance between the marks would be the *wave height, H.* The horizontal distance between corresponding points on successive waveforms, such as from crest to crest, is the *wavelength, L.* The ratio of H/L is *wave steepness.* The time that elapses during the passing of one wavelength is the *period, T.* Since the period is the time required for the passing of one wavelength, if either the wavelength or period of a deep water wave is known, the other can be calculated, since $L(m) = 1.56T^2$ (fig. 9–2):

$$\text{Speed } (S) = \frac{L}{T}$$

For example:

$$\text{Speed } (S) = \frac{L}{T} = \frac{156 \text{ m}}{10 \text{ s}} = 15.6 \text{ m/s}$$

Another characteristic related to wavelength and speed is *frequency, f.* Frequency is the number of wavelengths that pass a fixed point per unit of time. If 6 wavelengths pass a point in 1 min, and we have

FIGURE 9–2   Speed—Deep-Water Waves.

The theoretical relationships between wave speed, wavelength, and period for deep-water waves. Speed is equal to wavelength divided by period. $L(m) = 1.56T^2$

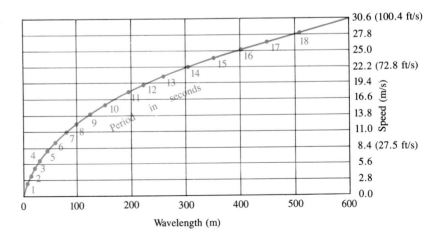

PLATE 15

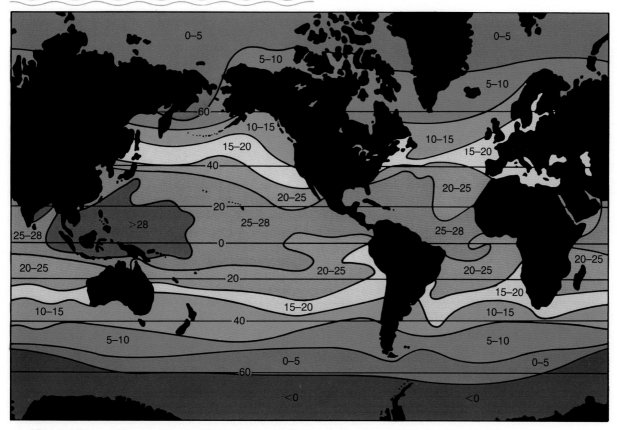

Average surface temperature distribution pattern for July (in °C).

Near the equator, the oceans receive more energy from the sun each year than they release, and the oceans lose more heat than they receive each year in the polar regions. Thus, we would expect equatorial surface water to be warmer than polar surface water. Because this difference between the temperature of equatorial and polar waters is not increasing, the excess heat gained in equatorial regions is apparently transferred toward the heat-deficient polar regions. About half of this transfer is thought to be achieved by ocean currents. The other half must then occur in the atmosphere.

Because the atmosphere is heated primarily by back radiation from the earth's surface, the prevailing wind systems are a direct outgrowth of this uneven distribution of heat in the earth's oceans. The winds, in turn, drive the oceans' surface currents and create ocean waves. Therefore, if oceanographers are to fully understand these dynamic components of the oceans, they must first fully understand the heat budget of the oceans and atmosphere.

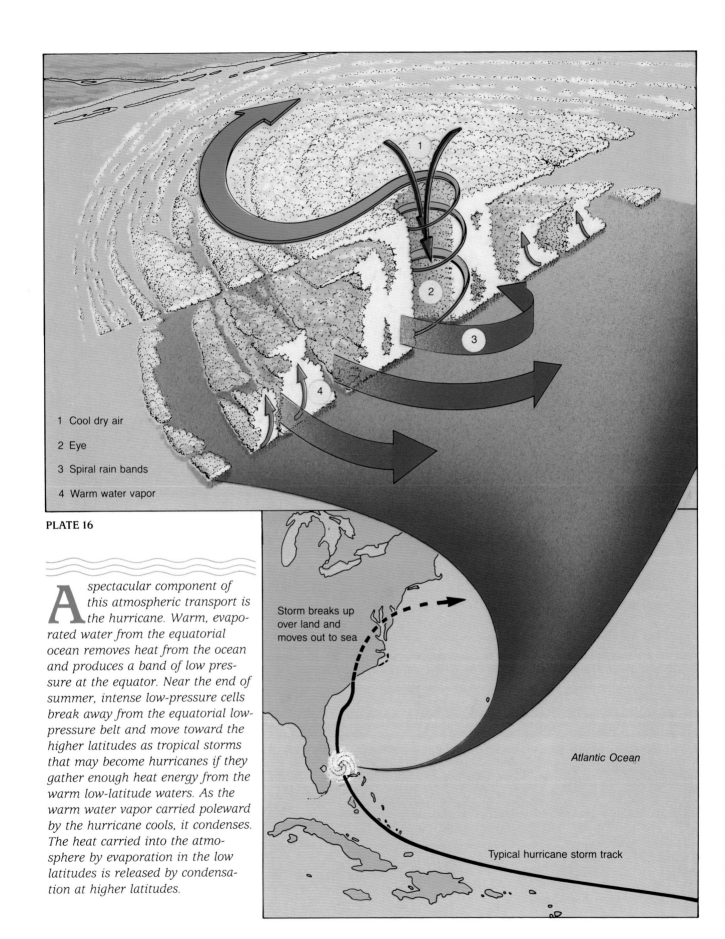

1 Cool dry air

2 Eye

3 Spiral rain bands

4 Warm water vapor

PLATE 16

A spectacular component of this atmospheric transport is the hurricane. Warm, evaporated water from the equatorial ocean removes heat from the ocean and produces a band of low pressure at the equator. Near the end of summer, intense low-pressure cells break away from the equatorial low-pressure belt and move toward the higher latitudes as tropical storms that may become hurricanes if they gather enough heat energy from the warm low-latitude waters. As the warm water vapor carried poleward by the hurricane cools, it condenses. The heat carried into the atmosphere by evaporation in the low latitudes is released by condensation at higher latitudes.

Storm breaks up over land and moves out to sea

Atlantic Ocean

Typical hurricane storm track

HURRICANE DIANA
PERSPECTIVE VIEW FROM TIROS-N
VISIBLE AND IR DATA
11 SEPTEMBER 1984 2000Z

PLATE 17

Hurricane Diana developed winds of up to 217 kph (135 mph) before coming ashore at Cape Fear, South Carolina, in September 1984. The precautions taken by local authorities prevented any deaths occurring as a direct result of the storm, although more than $60 million in property damage occurred.

PLATE 18

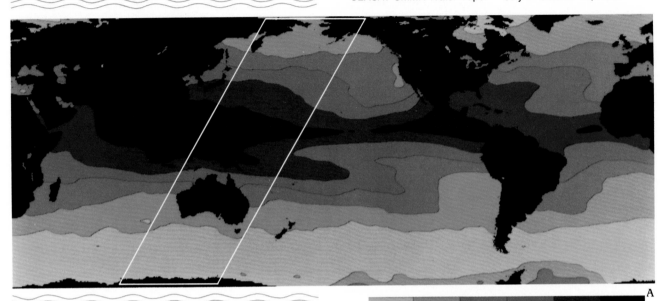

A

1   2   3   4   5

Grams/Centimeter^2

Because the ability of the atmosphere to hold water vapor increases with increased temperature, the greatest amount of atmospheric water vapor is found near the equator. This illustration reflects the great amount of heat being removed from the ocean by evaporation in the low latitudes.

This belt of maximum atmospheric water vapor content is called the doldrums. It is a region where warm air slowly rises and there are minimal horizontal air movements (wind) to propel sailing vessels or cool their crews.

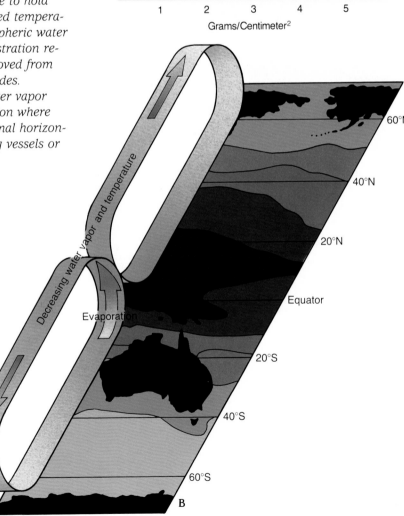

B

PLATE 19

SEASAT Altimeter Wind Speed   July 7–October 10, 1978

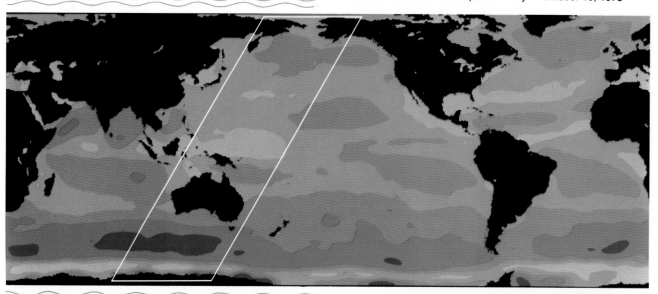

A

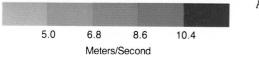

5.0   6.8   8.6   10.4

Meters/Second

Two of the earth's prevailing wind systems are the trade winds that blow from 30° latitudes toward the equator and the westerlies that blow from 30° toward 60° latitudes. Because there are fewer continents in the Southern Hemisphere, the westerly wind belt there contains the strongest year-round winds.

The lowest wind speeds occur at the doldrums, where the air is rising, and the horse latitudes (30°N and S latitudes), where air descends. Wind speeds gradually increase away from these belts of minimum wind speed.

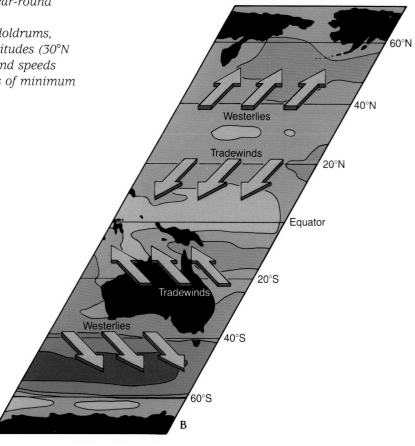

B

PLATE 20

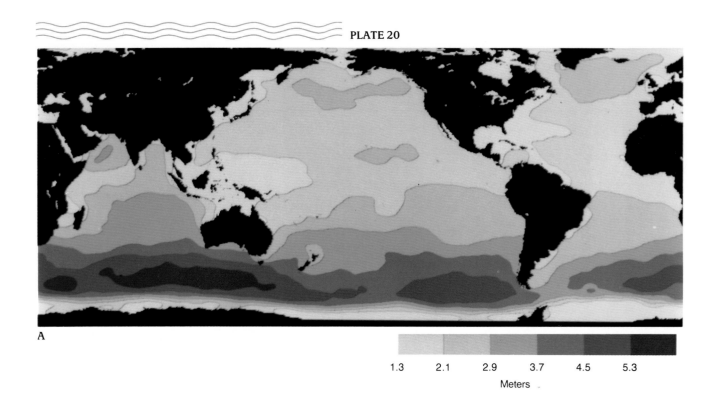

A

1.3   2.1   2.9   3.7   4.5   5.3

Meters

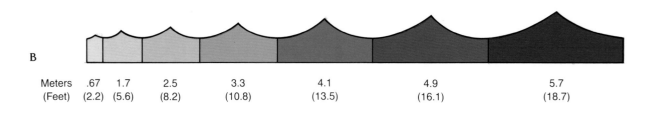

B

| Meters | .67 | 1.7 | 2.5 | 3.3 | 4.1 | 4.9 | 5.7 |
| (Feet) | (2.2) | (5.6) | (8.2) | (10.8) | (13.5) | (16.1) | (18.7) |

B ecause the westerly wind belt in the Southern
Hemisphere reaches the highest average wind
speed, this wind system produces the largest
ocean waves. The largest waves occur in the southern
Indian Ocean.

The wave heights represented here are the averages
for the largest one-third of the waves occurring at any
location.

A true-color image of the eastern Gulf of Mexico as sensed by the Coastal Zone Color Scanner (CZCS) aboard the Nimbus 7 satellite (A). Computer processing of CZCS data produces false-color images to show concentrations of chlorophyll a and phaeopigment a pigments. Increasing levels of biological productivity correlate with increasing pigment concentrations. Images B and C_1 differentiate concentrations up to 1 mg/m³. Greater detail is available in C_2, which can break out pigment concentrations up to 10 mg/m³.

CHLOROPHYLL a + PHAEOPIGMENTS a (MG/M³)

B

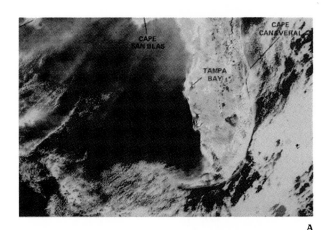

A

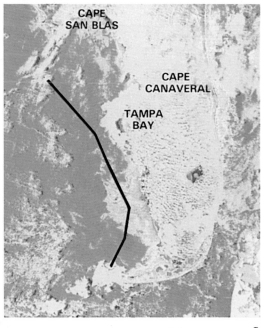

PIGMENT
CONCENTRATION
(mg m⁻³)

C_1	C_2
< 0.050	
0.075	0.1
0.10	
0.15	
0.20	0.2
0.25	
0.30	
0.35	0.3
0.40	0.6
0.45	
0.50	1.0
0.60	2.0
0.70	
0.80	3.0
0.90	6.0
1.00	10.0
>1.00	>10.0

SHIP TRACK_____

$$C_1 = 0.5 \left(\frac{L_{443}}{L_{550}} \right)^{-1.3}$$

$$C_2 = 0.8 \left(\frac{L_{520}}{L_{550}} \right)^{-4.0}$$

C

PLATE 21

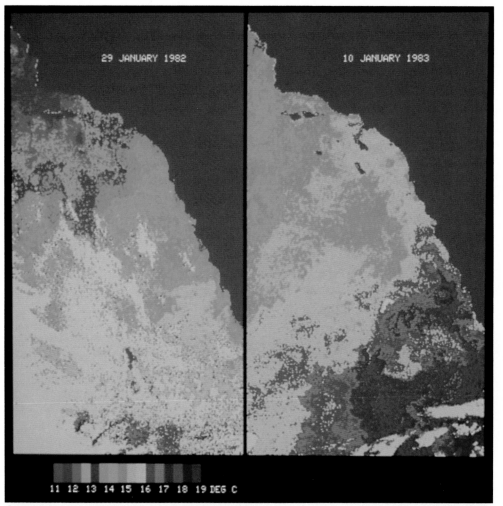

29 JANUARY 1982   10 JANUARY 1983

11 12 13 14 15 16 17 18 19 DEG C

A

19 APRIL 1982   15 MARCH 1983

0   0.50   1.00   1.50   2.00   2.50

PHYTOPLANKTON PIGMENTS, MG/M3

B

*T*he Advanced Very High Resolution Radiometer sea-surface temperature image (top) shows that waters off southern California in January 1983 are an average of 1.72°C warmer than a year before.

The Coastal Zone Color Scanner phytoplankton pigment images (bottom) show that there was a reduction in plant productivity over the same time span.

The strongest El Niño ever recorded occurred in 1982–83 and pushed a layer of warm water north from the equator. Coinciding with this event was a reduction of upwelling of nutrient-rich, cold water along the coast. This reduction in the nutrient level of the surface water reduced the level of biological productivity.

Spectacular physical phenomena such as this and their biological consequences help highlight the close relationship between the physical and biological phenomena of the oceans.

PLATE 22

the same wave system as in our previous example, then:

$$\text{Speed } (S) = Lf = 156 \text{ m} \times \frac{6}{\text{min}} = 936 \text{ m/min}$$

$$\frac{936 \text{ m}}{1 \text{ min}} \times \frac{1 \text{ min}}{60 \text{ s}} = 15.6 \text{ m/s}$$

Since the speed and wavelength of ocean waves are such that less than one wavelength passes a point per second, the preferred unit of time for scientific measurements, *period* (rather than frequency), is the more practical measurement to use when calculating speed.

Returning to the particle motion of ocean waves, the circular orbits followed by the water particles at the surface have a diameter equal to the wave height. While a particle is in the crest of a passing wave, it is moving in the direction of energy propagation. While it is in the trough, it is moving in the opposite direction. The half of the orbit that is accomplished in the trough is at a lower speed than the crest half of the orbit. Therefore, there is a small net transport of water in the direction the waveform is moving. This condition results from the fact that particle speed decreases with increasing depth below the still-water line. Also, the diameters of particle orbits decrease with increased depth until particle motion associated with our idealized wave ceases at a depth of one-half wavelength, $L/2$.

Deep-Water Waves

The ocean waves with the characteristics just discussed belong to a category called *deep-water waves.* Such waves travel across the ocean where the water depth *(d)* is greater than one-half the wavelength (fig. 9–3A). Included in deep-water waves are all wind-generated waves as they move across the open ocean. As can be seen by the speed calculations previously discussed, speed of deep-water waves is related to wavelength *(L)* [S (m/s) = 1.25 $\sqrt{L(\text{m})}$] and period

A. DEEP-WATER WAVE   Wave profile and water-particle motions of a deep-water wave. Note the diminishing size of the orbits with increasing depth below the surface.

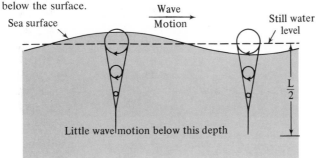

B. SHALLOW-WATER WAVE   Motions of water particles in shallow-water waves. Water motion extends to ocean floor.

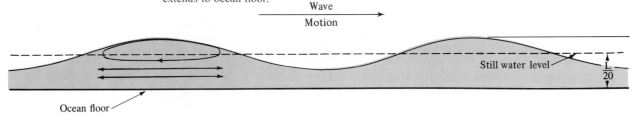

C. RELATIONSHIP OF WAVELENGTH TO WATER DEPTH

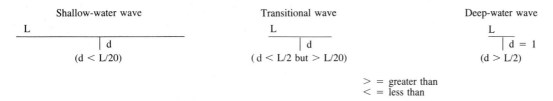

Shallow-water wave	Transitional wave	Deep-water wave
L	L	L
d	d	d = 1
(d < L/20)	(d < L/2 but > L/20)	(d > L/2)

> = greater than
< = less than

FIGURE 9–3   Deep-Water and Shallow-Water Waves.

FIGURE 9–4 Wave Speed.

To determine the wave speed of shallow-water waves, move up from the water depth to the line marked shallow-water waves, then over to the wave-speed scale [*see* example for 50-m (164-ft) water depth]. To determine the wave speed of deep-water waves, continue the wavelength (*L*) line to the left until it intersects the wave-speed scale (*see* example for 50 m wavelength). Combine these procedures for transitional waves (colored lines) [*see* example for 1000-m (3280-ft) wavelength and 100-m (328-ft) water depth]. Note that the transitional waves have a lower wave speed than they would have had as deep-water waves.

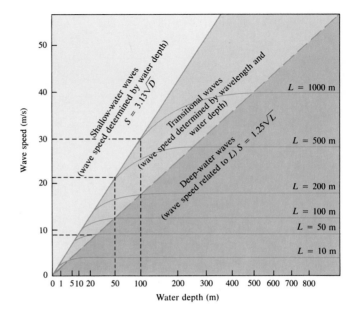

(*T*). The easier of these characteristics to measure is period. Thus the following equation is most commonly used for computing wave speed (*g* is acceleration due to gravity).

$$\text{Speed (m/s)} = \frac{gT}{2\pi} = \frac{9.8 \text{ m/s}^2 \times T \text{ (s)}}{2 \times 3.14} = 1.56T$$

Shallow-Water Waves

Waves in which $d < L/20$ are classified as *shallow-water waves* (long waves). Included in this category are wind-generated waves that have moved into shallow nearshore areas; *tsunami* (seismic sea waves), generated by disturbances in the ocean floor; and *tide waves,* generated by gravitational attraction of the sun and moon (fig. 9–3B). In these waves, the wavelength is very great relative to water depth, and speed is determined by water depth, *d*:

$$\text{Speed (m/s)} = \sqrt{gd} = 3.1 \sqrt{d(\text{m})}$$

Particle motion in shallow-water waves is in a very flat elliptical orbit approaching horizontal oscillation, and the vertical component of particle motion decreases with increasing depth. The presence of shallow-water waves can be detected to the ocean bottom.

Transitional Waves

Transitional waves have wavelengths greater than twice, but less than 20 times the water depth. The speed of transitional waves is determined partially by wavelength and partially by water depth (fig. 9–4). Deep-water waves generated by winds at the ocean

surface usually have periods of 10–12 s. They maintain their periods after encountering shallow coastal water.

Since it is easier to measure the period of a wave than any of its other characteristics, Kinsman has used the system presented in figure 9–5 as a means of classifying waves. The illustration also shows the principal causes of waves of different periods.

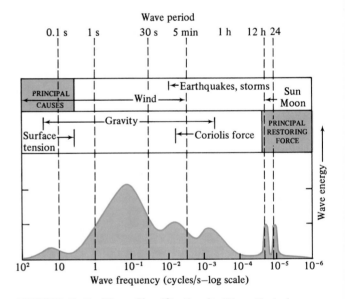

FIGURE 9–5 Wave Classification by Wave Period.

This figure shows that most of the energy possessed by ocean waves is in wind-generated waves with a period of about 10 seconds. The long period peak above the 10^{-3} represents tsunamis, while the two sharp peaks to the right represent the tides with semidaily and daily periods. After Blair Kinsman, *Wind waves: Their generation and propagation on the ocean surface,* © 1965. Reprinted by permission of Prentice-Hall, Inc., Englewood Cliffs, New Jersey.

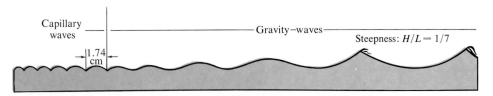

FIGURE 9–6   Capillary and Gravity Waves.

As energy is put into the ocean surface by wind, small rounded waves with V-shaped troughs develop (capillary waves). As the water gains energy, the waves increase in height and length. When they exceed 1.74 cm (0.68 in) in length, they take on the shape of the sine curve and become gravity waves. Increased energy increases the steepness of the waves. The crests become pointed and the troughs rounded. As the steepness reaches 1/7, the waves become unstable, and whitecaps form as they "break."

WIND-GENERATED WAVES

Sea Waves

As the wind blows over the ocean surface, the pressure and stress deform the ocean surface into small rounded waves with wavelengths less than 1.74 cm (0.7 in). These waves are called *capillary waves* (ripples) because the dominant *restoring force* that works to destroy them and smooth the ocean surface is surface tension (capillarity). Capillary waves characteristically have rounded crests and V-shaped troughs (fig. 9–6).

As capillary wave development increases, the sea surface takes on a rougher character, which allows the wind and ocean surface to interact more efficiently. As more energy is transferred to the ocean, *gravity waves* with lengths over 1.74 cm (0.7 in) develop with a shape more like that of a sine curve. Because they reach sufficient height at this stage, the force of gravity becomes the dominant restoring force. The length of "young" waves in a sea is generally 15–35 times their height. As additional energy is gained, the wave height increases more rapidly than wavelength, and the crests become pointed, while the troughs are rounded. When the steepness reaches 1/7, the velocity with which the waves travel is 1.2 times that of typical deep-water waves, or $S = 1.56T \times 1.2 = 1.87T$.

Energy imparted by the wind increases the magnitude of height, length, and speed. When the wave speed reaches that of the wind, neither of these characteristics can change because there is no net energy exchange, and the wave is at its maximum height.

The area in which wind-driven waves are generated is called *sea* and is characterized by a choppy, short wave structure with waves moving in many directions and having many different periods and lengths. The variety of wave periods and wavelengths is caused by frequently changing wind speed and direction. The factors that are important in increasing

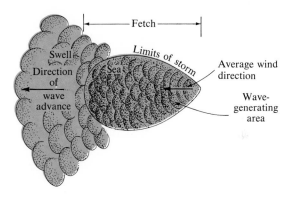

FIGURE 9–7   The Sea and Swell.

the amount of energy that waves obtain are (1) wind speed, (2) the time during which the wind blows in one direction, and (3) the *fetch*—the distance over which the wind blows in one direction. Figure 9–7 shows the relationship among wind direction, fetch, and the sea.

Wave height is a characteristic that is directly related to the amount of energy possessed by the wave. Wave heights in a sea area are usually less than 2 m (6.6 ft), although it is not uncommon to observe waves with heights of 10 m (32.8 ft) and periods of 12 s (plates 19 and 20). The largest wind-generated wave authentically measured was a 34-m (112-ft) wave with a period of 14.8 s seen in the North Pacific in February 1935 by the crew of the U.S. Navy tanker U.S.S. *Ramapo*. As sea waves gain energy, their steepness increases. When it reaches a critical value of 1/7, open ocean breakers called *whitecaps* form. The appearance of a sea surface as it changes from calm to the condition that results from hurricane-force winds is described in table 9–1.

For a given wind speed there is a maximum fetch and duration of wind beyond which the waves will not grow. When both maximum fetch and duration are reached for a given wind velocity, the sea is said to be "fully developed." The reason it can grow no further is the waves are losing as much energy

TABLE 9–1 Beaufort Wind Scale and the State of the Sea.

Beaufort Number	Descriptive Term	Speed m/s	Speed mi/h	Appearance of the Sea
0	Calm	—	—	Like a mirror
1	Light air	0.3–1.5	1–3	Ripples with the appearance of scales; no foam crests
2	Light breeze	1.6–3.3	4–7	Small wavelets; crests of glassy appearance, no breaking
3	Gentle breeze	3.4–5.4	8–12	Large wavelets; crests begin to break; scattered whitecaps
4	Moderate breeze	5.5–7.9	13–18	Small waves, becoming longer; numerous whitecaps
5	Fresh breeze	8.0–10.7	19–24	Moderate waves, taking longer form; many whitecaps; some spray
6	Strong breeze	10.8–13.8	25–31	Large waves begin to form; whitecaps everywhere; more spray
7	Near gale	13.9–17.1	32–38	Sea heaps up and white foam from breaking waves begins to be blown in streaks
8	Gale	17.2–20.7	39–46	Moderately high waves of greater length; edges of crests begin to break into spindrift; foam is blown in well-marked streaks
9	Strong gale	20.8–24.4	47–54	High waves; dense streaks of foam and sea begins to roll; spray may affect visibility
10	Storm	24.5–28.4	55–63	Very high waves with overhanging crests; foam is blown in dense white streaks, causing the sea to appear white; the rolling of the sea becomes heavy; visibility reduced
11	Violent storm	28.5–32.6	64–72	Exceptionally high waves (small and medium-sized ships might be for a time lost to view behind the waves); the sea is covered with white patches of foam; everywhere the edges of the wave crests are blown into froth; visibility further reduced
12	Hurricane	32.7–36.9	73–82	The air is filled with foam and spray; sea completely white with driving spray; visibility greatly reduced

After Bowditch, 1958.

0

1

2

3

4

5

6

7

8

9

10

11

12

TABLE 9–2 Fetch and Duration Required to Produce Fully Developed Sea for a Given Wind Speed.

Wind Speed km/h (mi/h)	Fetch km (mi)	Duration h
20 (12)	24 (15)	2.75
30 (19)	77 (48)	7
40 (25)	176 (109)	11.5
50 (31)	380 (236)	18.5
60 (37)	660 (409)	27.5
70 (43)	1093 (678)	37.5
80 (50)	1682 (1043)	50
90 (56)	2446 (1517)	65.25

TABLE 9–3 Description of Fully Developed Sea for a Given Wind Speed.

Wind Speed km/h (mi/h)	Average Height m (ft)	Average Length m (ft)	Average Period s	Highest 10% Waves m (ft)
20 (12)	.33 (1)	10.6 (34.8)	3.2	.75 (2.5)
30 (19)	.88 (2.9)	22.2 (72.8)	4.6	2.1 (6.9)
40 (25)	1.8 (5.9)	39.7 (130.2)	6.2	3.9 (12.8)
50 (31)	3.2 (10.5)	61.8 (202.7)	7.7	6.8 (22.3)
60 (37)	5.1 (16.7)	89.2 (292.6)	9.1	10.5 (34.4)
70 (43)	7.4 (24.3)	121.4 (398.2)	10.8	15.3 (50.2)
80 (50)	10.3 (33.8)	158.6 (520.2)	12.4	21.4 (70.2)
90 (56)	13.9 (45.6)	201.6 (661.2)	13.9	28.4 (93.2)

through the breaking of whitecaps as they are receiving from the wind.

Table 9–2 shows the fetch and duration of wind required to produce a "fully developed" sea for given wind speeds. Table 9–3 describes the "fully developed" sea at given wind velocities in terms of average height, length, and period of waves. Also, the average height of the highest 10% of the waves is given.

Swell

As waves generated in a sea area move toward the margins of the generating area, where wind speeds are lower, they eventually are moving with a speed greater than that of the wind. When this occurs, the steepness of the waves decreases, and they become long-crested waves called *swell*. The swell moves with little loss of energy over large stretches of the ocean's surface, transmitting energy away from the input area of the sea to where most of it will eventually be released along the margins of continents. As a part of this transfer of energy away from the sea area, the waves with the longer length, traveling at higher speed, leave the sea area first. They are followed by the shorter-length, lower-speed *wave trains* (groups of waves). This progression illustrates the principle of *wave dispersion* (the sorting of waves by wavelength).

In the generating area, waves of many wavelengths are present. In deep water, wave speed is a function of wavelength, so the longer waves "outrun" the shorter ones. To illustrate the distance that can be traveled by swell without having its energy depleted, swell that originated from storms in the Antarctic has been recorded breaking along the Alaskan coast after traveling a distance of more than 10,000 km (6210 mi).

As a set of waves leaves a sea and becomes a *swell wave train,* the group moves across the ocean surface at only half the velocity of an individual wave in the group. Progressively, the leading wave disappears. There will, however, always be the same number of waves in the group. As the front wave disappears, a new wave replaces it at the back of the group.

Interference Patterns Since swells may be moving away from a number of storm areas in a given ocean, it is inevitable that the swell forms will run together and interfere with one another. This gives rise to one of the special features of wave motion—interference patterns. An interference pattern produced by the superposition of two or more wave systems will be the algebraic sum of the disturbance each wave would have produced individually. The result may be a larger or smaller trough or crest, de-

THE FASTNET DISASTER—WAVE INTERFERENCE CREATES A MONSTER SEA

On Tuesday, August 14, 1979, the worst disaster in the history of yachting struck 303 vessels entered in the Fastnet Race, the finale of the five-race Admiral Cup Series. The 1000-km (621-mi) race started on August 11 from the Isle of Wight off the south coast of England, and the course required rounding Fastnet Rock off the southern tip of Ireland with a return to Plymouth, England (fig. 9A).

The day of the start of the race, a low-pressure system that had spawned tornadoes and thunderstorms from Ohio to New England weakened and moved into the Atlantic off the coast of Nova Scotia (fig. 9B). It was of no concern to the participants in Fastnet. By 1200 Greenwich Meridian Time (GMT) on August 12, the system had accelerated into the mid-Atlantic to 40°E longitude and 48°N latitude. Having become no more than a minor low-pressure trough with a minimum pressure of 1004 millibars (as compared to standard sea level pressure of 1013 millibars), it was relatively unnoticed as it reached 19°E longitude and 48°N latitude by 1200 GMT August 13 (figs. 9B and 9CA).

The storm immediately began to develop, and by 1800 GMT, August 13, the pressure had dropped to 996 millibars (fig. 9CB). At 2100 GMT, with the storm center just off the southwest coast of Ireland, the pressure had dropped to 983 millibars (fig. 9CC). Gusts up to 100 km/h (62 mi/h) were being felt south of Ireland. These west-southwest winds represented a force of 10 on the Beaufort Wind Scale. Although the larger, leading yachts had already rounded Fastnet Rock and had the wind on their stern-starboard beam, most of the fleet was struggling into a rising wind and sea.

FIGURE 9A   Course of the Fastnet Race.

The yachts sail from the Isle of Wight, around Fastnet Rock, then back to Plymouth.

FIGURE 9B Course of the Storm.

A weakening storm center moves off the coast of Nova Scotia at 1200 GMT (noon) Saturday, August 11. It races at 80 km/h (50 mi/h) into the Mid-Atlantic by 1200 GMT Sunday. Slowing slightly, the center moved to 19°E longitude and 48°N latitude by 1200 GMT Monday. Here the storm slows appreciably and deepens as it moves northeast toward Ireland. When the storm reaches the Irish Sea at 600 GMT Tuesday, August 14, the trough crosses the race fleet and the west-southwest winds change to northwest winds. The storm has reached its peak. As the storm moves to the north-northeast, the seas in the race course recede. The circled numbers indicating the storm's location are atmospheric pressure in millibars. Velocity values along the course of the storm are the rate at which the storm moved across the Atlantic Ocean. A detailed view of the storm's development from 1200 GMT Monday, August 13, to 1200 GMT Tuesday, August 14, is presented in figure 9C.

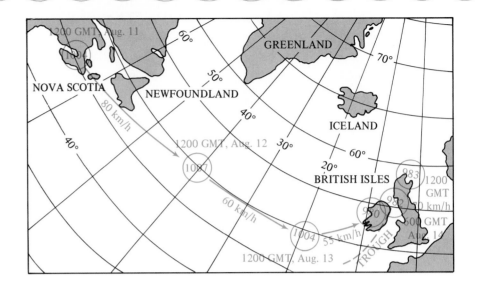

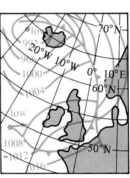 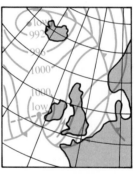 

A. 1200 GMT, August 13. From this location, the weak low-pressure cell (1004 millibars) begins to develop into the Fastnet storm.

B. 1800 GMT, August 13. Storm center pressure drops to 996 millibars, bringing gusts upward to 100 km/h (62 mi/h) in southwest winds south of Ireland. Wind direction is indicated by arrow.

C. 2100 GMT, August 13. Pressure falls to 983 millibars as west-southwest winds increase.

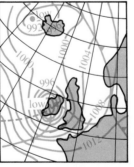 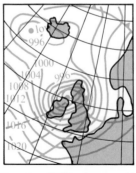 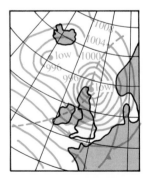

D. 0000 GMT, August 14. Storm reaches maximum intensity as pressure drops to 980 millibars.

E. 0600 GMT, August 14. Trough moves over the race course and winds shift from west-southwest to northwest, producing a cross-sea.

F. 1200 GMT, August 14. Storm moves off to the northeast and winds diminish over the race course.

FIGURE 9C Development of the Storm.

Courtesy of NOAA.

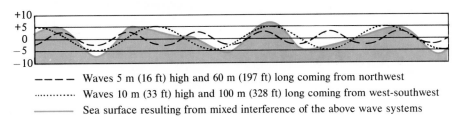

———— Waves 5 m (16 ft) high and 60 m (197 ft) long coming from northwest

············· Waves 10 m (33 ft) high and 100 m (328 ft) long coming from west-southwest

———— Sea surface resulting from mixed interference of the above wave systems

FIGURE 9D Constructive Wave Interference.

Although the exact size of the waves created by each of the wind regimes has not been determined, it seems unlikely that the 15-m (49-ft) waves reported could have developed without constructive wave interference. This condition would result from the combination of the west-southwest wind with the later northwest wind. Here a possible pattern of interference shows the
production of crests of 7.5 m (24.5 ft) above the still-water line and troughs 7.5 m below it. This
gives a maximum height of 15 m for the resulting waves.

By 0000 GMT on August 14, the storm center was in Galway Bay in cen-
tral-western Ireland with a pressure of 980 millibars (fig. 9CD). The trough of
the storm moved into the fleet by 0600 GMT, and the winds shifted from west-
southwest to northwest (fig. 9CE). This right-angle change in the winds caused
the tragedy. As crests of the incoming waves from the northwest merged with
the crests of the waves created by the west-southwest winds, constructive interference produced very short, steep waves as high as 15 m (49.2 ft). Gusts approaching 145 km/h (90 mi/h) were reported. By 1200 GMT Tuesday, the storm
center had moved to Moray Firth in northeastern Scotland, and the seas subsided dramatically along the course of the race (fig. 9CF).

Considering the short duration of the winds—about 12 hours for the west-
southwest winds and less than 6 hours for the northwest winds—the sea was
abnormally developed. The compact, steep, rogue waves had to result from
constructive interference between the two wave systems as they came together
to produce a cross-sea condition (fig. 9D).

When they took a yacht near their crests and on their leading edges,
these short, elevated, massive waves could probably not have been managed
by even the best designed yacht. As the yachts slid down the leading edge of
the breaking wave, they rolled over (fig. 9E). The water particles rising into the
wave crest pushed the keel up and rotated the mast into the sea. The following
wave struck the keel, rotating it back into the water and completing the revolution. Many of the yachts reported being rolled a number of times in such a
fashion. To survive, the crew would have to be wearing safety harnesses. Some
of the yachts may have been thrown onto others as the sea tossed them
around. Fifteen men died and twenty-three yachts were sunk or abandoned
during the fierce encounter (fig. 9F).

Although the weather forecasters were criticized for not giving sufficient
warning to allow the yachts to seek harbor, they did not have adequate advance knowledge that the conditions would be so severe. At any rate, with the
outcome of the race still to be determined, few yachts would probably have
abandoned the race.

The Fastnet disaster is a recent example of the unpredictability of the
oceans and the intense fury that they can develop. During a storm, one is
likely to consider only the waves that will theoretically develop directly from a
single wind direction. Although all mariners are aware of the potential effects

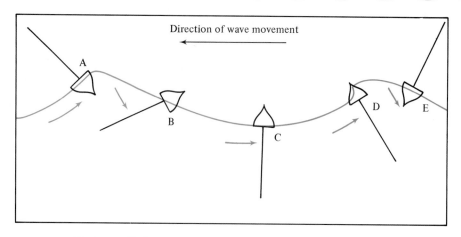

Direction of wave movement

FIGURE 9E Yachts Rolled Over.

As the steep wave takes the yacht near the crest on its leading edge (A), the yacht begins to roll over, aided by the upward motion of water particles moving into the wave crest that push the keel up while the mast descends. As the wave passes, the downward movement of water particles into the trough helps rotate the mast down (B and C). The next wave strikes the upturned keep while water particles moving into the crest help push the mast upward (D). Finally as the wave passes, the rotation is completed (E).

FIGURE 9F Rescue Attempt.

The winchman of a rescue helicopter hangs over an unidentified crewman from the yacht *Ariadne* during a rescue operation off southwest England on August 14, 1979. The *Ariadne* was later reported sunk with two crewman dead. Courtesy of Wide World Photos.

of a cross-sea on wave conditions, it probably did not seem important at the time. Even as they came, most looked upon the rough seas as just another challenge to be met. Such is revealed in the words of Ted Turner, skipper of the race winner *Tenacious*: "But we weren't really concerned with the conditions, we were concerned with winning."

pending on whether the individual disturbances are of the same or opposite sign, in or out of phase (fig. 9–8).

For instance, if swells are moving away from two storm areas and come together as shown in figure 9–8, the interference pattern may be constructive, destructive, or more likely mixed. *Constructive interference* results if the wave trains with the same wavelength came together in phase, crest to crest and trough to trough. Adding the displacements that would result from the waves individually, we will see that the interference pattern produced will be a wave with the same length as the two wave systems that are converging but with a wave height that is equal to the sum of the wave heights of the individual wave systems.

Destructive interference results if the crests produced by the waves from one generating area coincide with the troughs produced by the waves from the second generating area. If the waves had identical characteristics, the algebraic sum of the crest plus the trough would be zero, and the waves would cancel each other out.

It is more likely that the two systems possess waves of different heights and lengths and would come together with both destructive and constructive interference. Thus, a more complex *mixed interference* pattern would develop. It is such interference that explains the occurrence of a sequence of high waves followed by a sequence of lower waves and other irregular wave distribution patterns observed as the swell approaches the margins of the continents.

Free and Forced Waves In the swell we see what may be referred to as a *free wave,* which is moving with the momentum and energy imparted to it in the sea area but is not experiencing a maintaining force that keeps it in motion. In the wave-generating area, free and forced waves are present. A *forced wave* is one that is maintained by force that has a periodicity coinciding with the period of the wave. This force would be the wind; owing to the variability of the wind, many wave systems in the sea area alternate between being forced waves and free waves.

Surf

Most waves that are generated in the sea area by the force of storm-velocity winds move across the ocean as swell. They release their energy at the margins of the continents in the surf zone, where the swell forms breakers. As the deep-water waves making up the swell move toward the margin of the continent over gradually shoaling water, they eventually encounter water depths which are less than one-half wavelength.

These shoaling depths interfere with the particle movement at the base of the wave, and the wave slows down. As one wave is slowed down, the following waveform, which is still moving at an unaffected speed, tends to "catch up" with the wave that is being slowed, thus reducing the wavelength. Wave height increases, and the crests become narrow and pointed while the troughs become wide curves, a form that was previously described for high-energy waves in the sea. The increase in wave height accompanied by a decrease in wavelength increases the steepness (*H/L*) of the waves. As the wave steepness reaches 1/7, the waves break as *surf.*

If the surf is composed of swell that has traveled from distant storms, breakers will develop relatively near shore in shallow water, the shoaling of which is primarily responsible for their breaking. By the time waves break, they have become shallow-water waves.

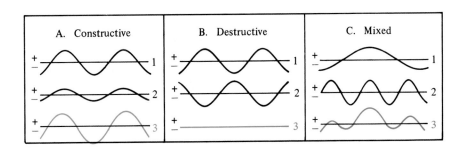

FIGURE 9–8 Wave Interference.

As wave trains come together from different sea areas, three possible interference patterns may result. Very rarely, if the waves have the same length and come together in phase (crest to crest and trough to trough), totally constructive interference may occur. The algebraic sum of amplitudes (*A*1 & 2) produces a wave of the same length but of greater height (*A*3). If two sets of waves have identical characteristics (*B*1 & 2) and come together 180° out of phase, the result would be destructive interference—no wave at all (*B*3). More commonly, waves of different lengths and heights (*C*1 & 2) encounter one another and produce a mixed interference pattern (*C*3).

The horizontal water motion associated with such waves translates water toward and away from the shore. The surf will be characterized by parallel lines of relatively uniform breakers. However, if the surf is composed of waves that have been generated by local wind, the waves may not have been sorted out into swell. The surf may be more nearly characterized by unstable, deep-water, high-energy waves with steepness already near 1/7. They will break shortly after feeling bottom some distance from shore and the surf will be rough and choppy with an irregular nature.

Ideally, when the water depth is about 1.3 times

the breaker height, the crest of the wave breaks, producing surf. When the water depth becomes less than 1/20 the wavelength, the waves in the surf zone begin to behave as shallow-water waves. Particle motion is greatly interfered with by the bottom, and a very significant transport of water toward the shoreline occurs (fig. 9–9).

The breaking experienced in the surf results from the fact that the particle motion near the bottom of the wave has been interfered with greatly by contact with the ocean bottom, and this tends to slow down the waveform. However, the individual particles that

FIGURE 9–9 Surf Zone.

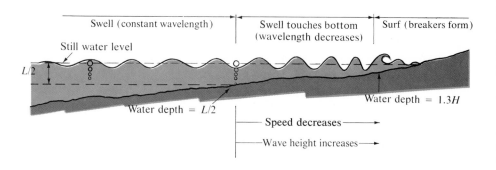

FIGURE 9–10 Breakers.
A, spilling breakers. *B*, plunging breakers. Photos courtesy of Charles H. V. Ebert.

A.

B.

are orbiting near the ocean surface have not been slowed down as much as the entire waveform because they have experienced no contact with the bottom. This causes the top of the waveform to lean toward the shore as the bottom of the waveform is held back. The slowing down of the entire waveform compared to the increasing speed of motion in the circular orbits near the ocean surface because of increasing wave height results in the water particles at the interface between the ocean and the atmosphere moving faster toward shore than the waveform itself.

This motion can be seen particularly well in *plunging breakers* (fig. 9–10), which have a curling crest that moves over an air pocket resulting from the fact that the curling particles have outrun the wave and there is nothing beneath them to support their motion. Plunging breakers form on moderately steep beach slopes. The more commonly observed breaker is the *spilling breaker* that results from a relatively gentle slope of the ocean bottom, which more gradually extracts the energy from the wave, producing a turbulent mass of air and water that runs down the front slope of the wave instead of a spectacular cresting curl. Because of the gradual extraction of energy of spilling breakers, they have a longer life span and give surfers a longer, if less exciting, ride than do the plunging breakers.

Recalling the particle motion of ocean waves in figure 9–3, we can see that in front of the crest the water particles are moving up into the crest. It is this force along with the buoyancy of the surfboard that helps maintain a surfer in front of a cresting wave. As the individual water particles move past the surfer into the crest of the wave, more water is carried from the shoreward side of the crest to maintain the hill along the front of the wave. When this upward motion of water particles is interrupted by the wave passing over water that is too shallow to allow this movement to continue, the ride is over. A skillful surfer, by positioning his board properly on the wave front, can regulate the degree to which the gravitational forces that propel him exceed the buoyancy forces, and high speeds can be obtained while moving along the face of the breaking wave.

Wave Refraction

We previously discussed that waves begin to bunch up and wavelengths become shorter as swell begins to "feel bottom" upon approaching the shore. It is seldom that the swell will approach the shore at right angles. Some segment of the wave can be expected to feel bottom first and therefore be slowed down be-

fore the rest of the wave. This produces *refraction,* or bending, of the waves as they approach the shore. The slowing down of a portion of a wave may result from its angular approach to a straight shoreline. As shown in figure 9–11, an irregular shoreline might result from an irregular bottom topography that could slow down portions of a wave that was approaching the shore at right angles.

The effect of wave refraction is to unevenly distribute wave energy along the shoreline (fig. 9–11). *Orthogonal lines* are constructed perpendicular to the wave fronts and spaced in such a way that the amount of energy between lines is equal at all times. They are of great assistance in illustrating how energy is distributed along the shoreline by breaking waves. Orthogonals indicate the direction the wave is traveling and can be seen to converge on headlands jutting out into the ocean and diverge in bays. A concentration of energy is released against the headlands, while energy released in the bays is spread more thinly. This condition produces erosion on the headlands, while deposition may occur in the bays. The increased energy of the waves breaking on headlands is reflected in an increased wave height.

Wave Diffraction

Wave *diffraction* can be considered as the bending of waves around objects. It is this kind of movement that allows waves to move past barriers into harbors as energy moves laterally along the crest of the wave, as shown in figure 9–12. This bending is on a much smaller scale and is less easily explained than the bending discussed in refraction, which is a simple response to changes in velocity. Diffraction results because any point on a wave front can be a source from which energy can propagate in all directions.

Wave Reflection

Swell can be reflected back into the ocean with little loss of energy from a vertical barrier such as a sea wall. For this ideal reflection to occur without energy loss, the wave would have to strike the barrier at a right angle. Such a condition would be rare in nature. Less ideal reflections will nonetheless produce *standing waves,* which are the product of two waves of the same length moving in opposite directions.

The standing wave is a special interference pattern where no net momentum is carried because the waves are moving in opposite directions. The particles continue to move vertically and horizontally, but there is no more of the circular motion that we saw

FIGURE 9–11 Refraction.
A, as the waves (color) "feel bottom" first in the shallow areas off the headlands, they are slowed. The segments of the waves that move through the deeper water leading into the bay are not slowed until they are well into the bay. As a result, the waves are *refracted* (bent) so the release of wave energy is concentrated on the stacks and headlands. Erosion is active on the headlands, while deposition occurs in the bay, where the energy level is low. Orthogonal lines (black), spaced so that equal amounts of energy are between each line, help to show the distribution of energy along the shore. *B,* wave refraction along the California coast. Note the changes in wavelength that occur in the bay on the right. Photo from U.S. Department of Agriculture.

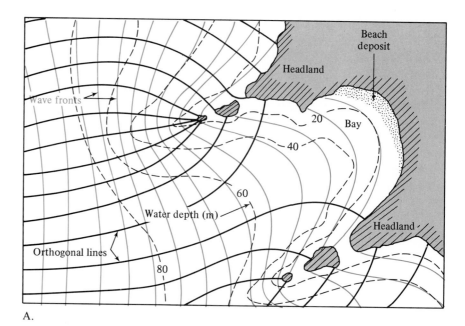

A.

B.

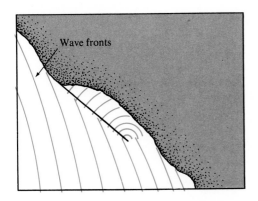

FIGURE 9–12 Diffraction.

Diffraction is bending caused by waves passing an obstacle; it is not related to refraction. By diffraction, wave energy may spread to the most protected areas within a harbor.

in the progressive wave. Standing waves are characterized by lines along which there is no vertical movement. There may be one or more such *nodal* lines. *Antinodes,* crests that alternately become troughs, represent the points of the greatest vertical movement within a standing wave (fig. 9–13). There is no particle motion when an antinode is at its greatest vertical displacement, and the maximum particle movement occurs when the water surface is level. At this time, the maximum movement of the water is in a horizontal direction directly beneath the nodal lines. The movement of water particles beneath the antinodes is entirely vertical. We will consider standing waves further when we discuss the tides in the following chapter, as there are certain conditions under which the development of standing waves has a significant effect on the tidal character of coastal regions.

For the most part, reflection of wind-generated waves from coastal barriers will be at an angle equal to the angle at which the wave approached the barrier, as shown in figure 9–14. This type of reflection may produce a small-scale interference pattern similar to those we previously discussed.

An outstanding example of reflection phenomena is the Wedge that develops west of the 400-m (1312-ft) jetty at Newport Harbor, California. As the incoming waves strike the jetty and are reflected, a constructive interference pattern develops. When the crests of the incoming waves merge with the crests of the reflected waves, plunging wedge-shaped breakers that may reach heights in excess of 8 m (26 ft) form. These waves present a fierce challenge to the most experienced body surfers. Attracting some who were not up to the challenge, the Wedge has killed and crippled many who have come to try it.

Storm Surge

Large cyclonic storms that develop over the ocean will, owing to their low pressure, produce a hill of water beneath them that moves with the storm across the open ocean. Additionally, as the storm approaches shallow water near shore, the portion of the hill over which the wind is blowing shoreward frequently will produce a *storm surge,* a mass of wind-driven water that produces an abnormal increase in sea level that may be extremely destructive to low-lying coastal areas. Storm surges can be particularly destructive when accompanied by high tide. The coincidence of a storm surge with high tide in areas that have particularly high tides, such as the North Sea coastal region, frequently produces major catastrophes, with great loss of life and property damage (fig. 9–15).

One of the most outstanding examples of storm surge is that which relates to the windows of a lighthouse atop a 90-m (295-ft) cliff at Dunnet Head, Scotland, being broken by stones tossed by waves break-

FIGURE 9–13 Reflection— Standing Waves.

An example of water motion at quarter-period intervals. Water is motionless when antinodes reach maximum displacement. Water movement is maximum when the water is level. Movement is totally vertical beneath the antinodes, and maximum horizontal movement occurs beneath the node. The circular motion of particles in progressive waves does not exist in standing waves.

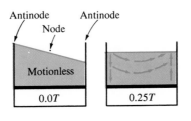

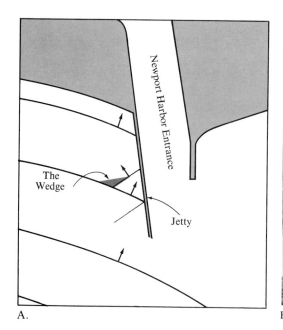

A.   B.

FIGURE 9–14 Reflection—The Wedge.

A, the Wedge, a wedge-shaped crest that may reach heights in excess of 8 m (26 ft), develops as a result of interference between incoming waves and reflected waves near the jetty protecting the entrance to Newport Harbor, California. *B,* a view of a wedge crest taken from the landward end of the jetty. The three dots in the water in front of the wave are the heads of body surfers waiting to catch the wave.

FIGURE 9–15 Storm Surge.

Low pressure and high onshore winds of Hurricane Camille produce storm surge along Mississippi coast in 1969. Photo courtesy of NOAA.

ing on the cliff. The lighthouse marks the western entrance to Pentland Firth, which connects the Atlantic Ocean with the North Sea between the Orkney Islands to the north and Scotland to the south.

TSUNAMI

The large waves referred to by the Japanese word *tsunami* originate as a result of disturbances within the earth's crust. We commonly hear them called *tidal waves,* which implies that they are related to the tides. They have no relationship to the tides, however. Tsunami are usually caused by fault movement, a displacement in the earth's crust along a fracture, that causes a sudden change in water level at the ocean surface. Secondary events, such as underwater avalanches, produced by the faulting may also produce tsunami (fig. 9–16). The ocean that is most plagued by the tsunami is the Pacific. It is ringed by a series of trenches that are unstable margins of crustal plates along which large-magnitude earthquakes occur.

One of the most destructive tsunami ever generated came from the greatest release of energy from the earth's interior observed during historical times. On August 27, 1883, the island of Krakatoa in the Sundra Strait, between Sumatra and Java, exploded

FIGURE 9–16 Tsunami.
Movement along a fault in the earth's crust sends energy to the ocean surface. This energy then is distributed laterally along the atmosphere-ocean interface in the form of a tsunami. The energy is transmitted across the open ocean by undetectable waves over 200 km (124 mi) long and about 0.5 m (1.6 ft) high. They release their energy after reaching the shore and developing heights that may be in excess of 30 m (98 ft). Base map courtesy of Open University Press.

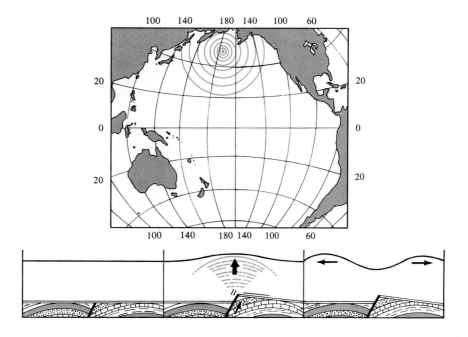

and essentially disappeared. The sound of the explosion was heard 4800 km (2981 mi) away at Rodriguez Island in the western Indian Ocean, and the dust that ascended into the atmosphere circled the earth and produced unusual and beautiful sunsets for nearly one year. A tsunami that rose to heights of more than 30 m (98 ft) devastated the coastal region of the Sundra Strait, taking more than 36,000 lives. The energy carried by this wave was still detectable after crossing the Indian Ocean, passing the southern tip of Africa, and moving north in the Atlantic Ocean into the English Channel.

Since the wavelength of a typical tsunami will be in excess of 200 km (124 mi), it is obviously a shallow-water wave, the speed of which is determined by water depth. Moving at speeds well in excess of 700 km/h (435 mi/h) and with wave heights of approximately 0.5 m (1.6 ft) in the open ocean, tsunami are not readily observable until they reach shore. In shallow water, they are slowed down, and the water begins to pile up to form crests that may exceed heights of 30 m (98 ft).

Unlike the hurricane that represents a great hazard to ships at sea and may send them seeking the protection of a coastal harbor, a tsunami sends them from their coastal moorings into the open ocean. Until 1948, there was no possible warning of the coming of a tsunami that would be adequate to allow shippers to take the appropriate precautions to avoid the destructive waves. Since that time and as a result of a destructive wave that struck Hawaii in 1946, a tsunami warning system has been established throughout the Pacific Ocean. With every seismic disturbance that occurs beneath the ocean surface with the potential of producing a tsunami, observations are made at the closest tide-measuring stations to see if there is any indication that such a wave has been created. Should one be detected, warnings are sent to all the coastal regions that might conceivably encounter a destructive wave along with the time at which the wave is to be expected. This warning allows for the removal of ships from harbors before the waves arrive if the disturbance has occurred at a great enough distance.

Prior to such a warning system, the first notice that most observers had of an impending tsunami would be the rapid seaward recession of the shoreline. The recession would then be followed in a few minutes by the destructive wave (fig. 9–17). Such a recession was observed in the port of Hilo, Hawaii, on April 1, 1946, as a result of an earthquake in the Aleutian Trench off the island of Unimak over 3000 km (1863 mi) away. The recession created by the formation of a trough preceding the first tsunami wave crest was followed by a wave that carried the waters to heights nearly 8 m (26 ft) above the normal high tide level. The tsunami also struck Scotch Cap, Alaska, on Unimak Island. The lighthouse on the island, the base of which stood 14 m (46 ft) above sea level, was totally destroyed by a wave that must have reached a height of 36 m (119 ft).

It was the wave produced by this disturbance in the Aleutian Trench that was recorded throughout the coastal regions of the Pacific Ocean at tide-recording stations and led to the development of what is now the International Tsunami Warning System (ITWS).

A.

B.

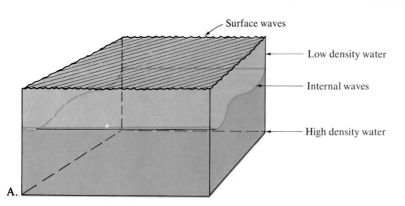

A.

B.

FIGURE 9–18 Internal Wave.

A, a simple internal wave moving along the density interface below the ocean surface. *B,* SEASAT Synthetic Aperture Radar image of internal waves in the Gulf of California. They are thought to be associated with tidal currents in the shallow water around the island of Angel de La Guarda located in the lower right. The image is about 100 km (62 mi) on each side.

INTERNAL WAVES

We have to this point discussed waves that occur at an obvious density discontinuity existing between the atmosphere and the ocean. From our previous discussions we should be aware that there are also such discontinuities within the ocean water column itself. For instance, the pycnocline represents a density interface that is typically associated with the thermocline. Just as energy can be transmitted along the interface between the atmosphere and the ocean, it can also be transmitted along density interfaces beneath the ocean surface in what are called *internal waves*. Internal waves may have heights above 100 m (328 ft) (fig. 9–18). The greater the difference in density between the two fluids, the faster the waves will move.

There is still much to be learned about internal waves, but their existence is well documented, and it is thought that many causes for them may be identified. Internal waves are known to have periods related to tidal forces, indicating that these forces may be a significant cause. Underwater avalanches in the form of turbidity currents, wind stress, and energy put into the water by moving vessels may also be causes of internal waves. It is thought that the parallel slicks seen on the surface waters may overlie the troughs of internal waves. The slicks are caused by a film of surface debris accumulating and dampening surface waves.

Internal waves reach greater heights from a smaller energy input than do the waves resulting from very large energy input observed at the ocean's surface. This is because they move along interfaces, such as the pycnocline, across which the density difference is considerably less than that which exists between the ocean surface and the atmosphere. They are thought to move as shallow-water waves at speeds considerably less than those of surface waves, with periods of 5–8 min and wavelengths of 0.6–0.9 km (0.37–0.56 mi).

The loss of the U.S.S. *Thresher,* a nuclear-powered submarine, which occurred 350 km (217 mi) off the coast of Massachusetts on April 10, 1963, may have been related to internal waves (fig. 9–19). Nobody knows what happened to cause the loss of the submarine and the crew of 129 men. The last message indicated "minor difficulties" were being experienced and they were trying to blow fluids out of ballast tanks with high-pressure air. This communication could mean they were experiencing a descent they could not stop.

If the *Thresher* was near its maximum operating depth at the time it encountered an internal wave, it

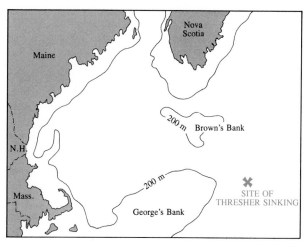

A.

B.

C.

FIGURE 9–19 U.S.S. *Thresher.*

A, site of *Thresher* sinking. *B,* U.S.S *Thresher,* which sank during diving trials April 10, 1963. *C,* mosaic of the site. The conning station is at arrow (1). It is upside down with the leading edge to the left. The sail planes (2) are completely reversed. Below the sail is a torpedo shutter door (3), an air bottle (4), and actuating gear for torpedo shutter door (5). Photos courtesy of U.S. Navy.

could conceivably have been quickly pulled to depths that would have crushed it before the descent could be stopped. This would have been especially probable if mechanical difficulties were being experienced, and there is evidence that a ballast pump was installed backwards.

POWER FROM THE WAVES

Extracting power from waves as they approach the shore is a possibility, although there are significant problems to be overcome. Owing to the refraction of waves, a large half-cone could be constructed to focus the energy of waves at the apex of the structure. Such a device might extract up to 10 MW of power per kilometer of shore, but could produce significant power only when large storm waves broke against it. Such a system would operate only as a power supplement, and a series of such structures along the shore would be required. Structures of this type could have a serious effect on longshore transport of sediment, which could lead to serious coastal erosion problems in areas deprived of sediment.

Along shores endowed with proper topography of the ocean floor, internal waves with their great heights may be effectively focused by refraction and

induced to "break" against an energy-conversion device.

A small pilot project to harness wave energy was developed in Mauritius, an island state of the Commonwealth of Nations. Located off Africa in the Indian Ocean, Mauritius is experimenting with a design it is hoped will provide electricity and a wave-free area for commercial mariculture.

Two engineers at Lockheed Corporation have developed an interesting ocean wave energy device called Dam-Atoll, which is shown in figure 9–20. Waves enter the top of the unit, just at the ocean surface. Guide vanes at the opening cause the entering water to spiral into a whirlpool within an 18-m (60-ft) central core. The swirling column of water turns a turbine, the unit's only moving part. It can provide a continuous electrical power output of 1 to 2 MW according to the inventors, Leslie S. Wirt and Duane L. Morrow. In particularly good wave areas, such as the Pacific Northwest, they believe it would be possible to anchor 500 to 1000 units that would provide power in quantities comparable to those provided by Hoover Dam. These units, 76 m (250 ft) in diameter and made of concrete, have many potential uses other than the generation of electricity. They could be used in cleaning up oil spills, protecting beaches from wave erosion, creating calm harbors in the open sea, and desalinating seawater through reverse osmosis.

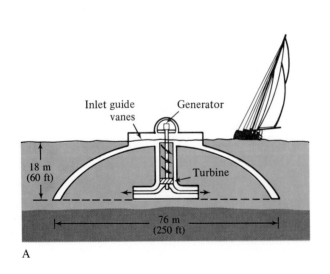

FIGURE 9–20 Dam-Atoll.
Lockheed Corporation's Dam-Atoll designed to generate electricity from wave action. *A,* water entering at the surface spirals down an 18-m (60-ft) central cylinder to turn a turbine located at the bottom of the cylinder. Each unit is designed to produce from 1 to 2 MW. *B,* view from above. Courtesy of Lockheed Corporation.

SUMMARY

Wave phenomena transmit energy through various states of matter by setting up patterns of oscillatory motion in the particles that make up the matter. **Progressive waves** may be described as **longitudinal, transverse, or orbital,** depending on the pattern of particle oscillation. Particles related to ocean waves move primarily in orbital paths.

Characteristics used to describe waves are **wavelength (L), wave height (H), period (T),** and **wave speed (S).** If water depth is greater than 1/2 wavelength, a progressive wave travels as a **deep-water wave** with a speed that is directly proportional to wavelength. If water depth is less than 1/20 wavelength, it will move as a **shallow-water wave,** the speed of which increases with increased water depth.

As wind-generated waves form in the **sea area, capillary waves** with rounded crests and wavelengths less than 1.74 cm (0.7 in) form first. As the energy of the waves increases, **gravity waves** with increased wave speed, wavelength, and wave height form. Energy is transmitted from the sea area across the ocean by low, rounded waves called **swell.** The swell releases its energy in the **surf** that forms as the waves break in the shoaling water near shore. If the waves break on a relatively flat surface, the result is usually a **spilling breaker,** while breakers forming on steep slopes have spectacular curling crests and are called **plunging breakers.**

When swell approaches the shore, the segments of the waves that encounter shallow water first will be slowed, and those parts of the wave that have not yet been interfered with by shallow water will move ahead, causing the wave to refract, or bend. **Refraction** causes a concentration of wave energy on the headlands, while low-energy breakers are characteristically found in bays. **Reflection** of waves off sea walls or other barriers can cause an interference pattern called a **standing wave.** In standing waves, crests do not move laterally as in progressive waves but form alternately with troughs at locations called **antinodes.** Separating the antinodes are locations where there is no vertical movement of the water—**nodes.**

During storms, the combination of low air pressure and onshore winds may produce **storm surge** that raises the water level at the shore many meters above normal sea level. Such surges are particularly destructive if they coincide with high tide.

Tsunami, or seismic sea waves, are generated by seismic disturbances beneath the ocean floor. Such waves have lengths in excess of 200 km (124 mi) and travel across the open ocean with undetectable heights of about 0.5 m (1.6 ft) at speeds in excess of 700 km/h (435 mi/h). On approaching shore, they may increase in height to over 30 m (98 ft). Single-wave tsunami have been known to cause millions of dollars worth of damage and take tens of thousands of lives. **Internal waves** are not well understood but are thought to form at density interfaces beneath the ocean surface, especially in connection with the pycnocline. They may be expected to have heights of up to 100 m (328 ft), with periods from 5 to 8 min.

Ocean waves can potentially be focused to produce hydroelectric power, but there are significant environmental considerations that will have to be incorporated into any such plan.

QUESTIONS AND EXERCISES

1. Discuss longitudinal, transverse, and orbital wave phenomena including the states of matter in which they transmit energy.

2. Calculate the speed (S) for deep-water waves with the following characteristics:
   A. $L = 351$ m, $T = 15$ s
   B. $L = 351$ m, $f = 4$ waves/min   Express speed (S) in meters per second (m/s).
   C. $T = 11$ s

3. Calculate the speed with which a shallow-water wave will travel across an ocean basin 4 km (2.5 mi) deep.

4. Describe the change in the shape of waves that occurs as they progress from capillary waves to increasingly larger gravity waves until they reach a steepness ratio of 1/7.
   A change in which variable will make gravity the dominant restoring force, $H, L, S, T,$ or f?

5. What is the minimum fetch and duration of wind required to produce a fully developed sea with a wind speed of 45 km/h (27.9 mi/h)? What would be the average height, length, and period of waves in this fully developed sea?

6. Waves from separate sea areas move away as swell and produce an interference pattern when they come together. If the waves from Sea A and Sea B have wave heights of 1.5 m (5 ft) and 3.5 m (11.5 ft), respectively, what would be the height of waves resulting from constructive interference and destructive interference? Illustrate your answer with a diagram (refer to fig. 9–8).

7. Describe changes, if any, in wave speed *(S),* length *(L),* height *(H),* and period *(T)* that occur as waves move across shoaling water to break on the shore.

8. Using orthogonal lines, illustrate how wave energy can be distributed along the shore. Identify areas of high and low energy release.

9. On the basis of the fundamental characteristics of standing waves shown in figure 9–13, construct a similar diagram of a standing wave in which two nodes and three antinodes exist.

10. List three factors that may affect the height of storm surge. Make a diagram of a hurricane coming ashore from the south along an east-west shore and indicate the segment of shore along which you think the storm surge will reach maximum height. Explain why.

11. Why is it more likely that a tsunami will be generated by faults beneath the ocean along which vertical rather than horizontal movement has occurred?

12. How long will it take for a tsunami to travel across 2000 km (1242 mi) of ocean if the average depth of ocean is 4500 m (14,760 ft)?

13. What ocean depth would be required for a tsunami with a wavelength of 220 km (136 mi) to travel as a deep-water wave? Is it possible that such a wave could become a deep-water wave any place in the world ocean?

14. Why is the development of internal waves likely within the thermocline?

15. Discuss some environmental problems that might result from the development of facilities for the conversion of wave energy to electrical energy.

REFERENCES

Bascom, W. 1959. Ocean waves. *Scientific American* 201:89–97.

Bowditch, N. 1958. *American Practical Navigator.* Rev. ed. H. O. Pub. 9. Washington, D.C.: U.S. Naval Oceanographic Office.

Iselin, C. 1963. The loss of the *Thresher. Oceanus* 6:4–6.

Kinsman, B. 1965. *Wind waves: Their generation and propagation on the ocean surface.* Englewood Cliffs, N.J.: Prentice-Hall.

Melville, W., and Rapp, R. 1985. Momentum flux in breaking waves. *Nature* 317:6037, 514–516.

Pickard, G. L. 1975. *Descriptive physical oceanography: An introduction.* 2nd ed. New York: Pergamon Press.

Sverdrup, H. U.; Johnson, M. W.; and Fleming, R. H. 1942. Renewal 1970. *The oceans: Their physics, chemistry, and general biology.* Englewood Cliffs, N. J.: Prentice-Hall.

van Arx, W. S. 1962. *An introduction to physical oceanography.* Reading, Mass.: Addison-Wesley.

SUGGESTED READING

Sea Frontiers

Land, T. 1975. Freak killer waves. 21:3, 139–41.
    The British design a buoy that will gather data in areas where 30-m (98-ft) waves, which may be responsible for the loss of many ships, occur.

Mooney, M. J. 1975. Tragedy at Scotch Cap. 21:2, 84–90.
    A recounting of the events resulting from an earthquake off the Aleutians April 1, 1946. The resulting tsunami destroyed the lighthouse at Scotch Cap, Alaska.

Pararas-Carayannis, G. 1977. The International Tsunami Warning System. 23:1, 20–27.
   A discussion of the history and operations of the International Tsunami Warning
   System.

Robinson, J., P., Jr. 1976. Newfoundland's disaster of '29. 22:1, 44–51.
   A description of the destruction caused by a tsunami that struck Newfoundland
   on November 18, 1929.

————. 1976. Superwaves of southeast Africa. 22:2, 106–16.
   A discussion of the formation and destruction caused by large waves that strike
   ships off the southeast coast of South Africa.

Smail, J. 1982. Internal waves: The wake of sea monsters. 28:1, 16–22.
   An informative discussion of the causes of internal waves and their effect on surface ships and submarines.

Smail, J. R. 1986. The topsy-turvy world of capillary waves. 32:5, 331–37.
   Capillary waves are clearly described, and their role in transmitting wind energy to
   the motion of waves and currents is discussed.

Smith, F. G. W. 1970. The simple wave. 16:4, 234–45.
   This is a very readable explanation of the nature of ocean waves. It deals primarily with the characteristics of deep-water waves.

————. 1971. The real sea. 17:5, 298–311.
   A comprehensive and readable discussion of wind-generated waves.

————. 1985. Bermuda mystery waves. 31:3, 160–63.
   A discussion of the possible source of large waves that struck Bermuda on November 12, 1984.

Truby, J. D. 1971. Krakatoa—the killer wave. 17:3, 130–39.
   The events leading up to the 1883 eruption of Krakatoa and the tsunami that followed are described.

Scientific American

Bascom, W. 1959. Ocean waves. 201:2, 89–97.
   An informative discussion of the nature of wind-generated waves, tsunami, and
   tides.

Koehl, M. A. R. 1982. The interaction of moving water and sessile organisms. 274:6,
   124–35.
   A discussion of the adaptations of benthic shore-dwelling animals to the stresses
   of strong currents and breaking waves.

CHAPTER TEN
TIDES

People have undoubtedly observed the rise and fall of the tides since they first inhabited the coastal regions of the continents. But until Herodotus (450 B.C.) observed tides on the Mediterranean, there was no known record of them. The first Greek observers concluded that the tides were related to the motion of the moon since both followed a similar cyclic pattern, but it was not until Sir Isaac Newton (1642–1727) developed his universal law of gravitation that the tides could be explained adequately.

The tides are the ultimate manifestation of shallow-water wave phenomena, possessing lengths measured in thousands of kilometers and heights ranging from zero to more than 15 m (49 ft). The ocean tides generated by the gravitational attraction of the sun and moon on the mass of the ocean affect every particle of water from the surface to the deepest part of the ocean basin. Thus, the tide undoubtedly has a much more far-reaching effect on ocean phenomena than we can observe at the ocean surface.

TIDE-GENERATING FORCES

Law of Gravitation

Isaac Newton published his *Philosophiae naturalis principia mathematica* in 1686 and stated in his preface, ". . . I derive from the celestial phenomena the forces of gravity with which bodies tend to the sun and several planets. Then from these forces, by other propositions which are also mathematical, I deduce the motions of the planets, the comets, the moon, and the sea." What followed is our first understanding of why tides behave as they do.

Newton's law of gravitation states that every particle of mass in the universe attracts every other particle of mass with a force that is proportional to the product of their masses and inversely proportional to the square of the distance between the masses. The greater the mass of the objects and the closer they are together, the greater will be the gravitational attraction. Mathematically this can be expressed as:

$$\text{Gravitational force} = G\,\frac{m_1 m_2}{r^2}$$

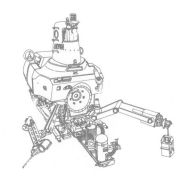

La Rance tidal power station, the only large tidal power-generating plant in the world. Courtesy of U.S. Department of Energy.

Here, G is the universal gravitational constant, m_1 and m_2 are the masses, and r is the distance between the masses. For spherical bodies, all of the masses can be considered to exist at the center of the sphere. Thus, r will always be the distance between the centers of bodies being considered.

Actually, tide-generating forces vary inversely as the cube of the distance from the center of the earth to the center of the tide-generating object, instead of varying inversely to the square of the distance as does the gravitational attraction. Therefore, the tide-generating force, although it is derived from the force of gravitational attraction, is not proportional to it. Distance becomes a more highly weighted variable in the tide-generating forces than it is in gravitational attraction force:

$$\text{Tide-generating force} \propto \frac{m_1 m_2}{r^3}$$

Although the gravitational attraction between the earth and sun is over 177 times that between the earth and moon, the moon dominates the tides, as will be seen from the following comparison of the tide-generating force of the sun with that of the moon. Since the sun is 27 million times more massive than the moon, it should, solely on the basis of comparative masses, have a tide-generating force 27

million times greater than that of the moon. However, we must also consider the distance between the earth and the moon as compared to the distance between the earth and the sun. Since the sun is 390 times farther from the earth than the moon, its tide-generating force is reduced by 390^3, or about 59 million times compared to that of the moon. These conditions result in the sun's tide-generating force being $^{27}\!/_{59}$, or about 46%, that of the moon (fig. 10–1).

Since the moon is the dominant force producing tides on earth, we will first consider the tide-generating forces resulting from the earth-moon system only. We will later discuss the modifications in this dominant pattern that result from the tide-generating force of the sun.

As the daily rotation of the earth about its polar axis has no tide-generating effect, we will ignore this motion in our initial considerations. The tidal pattern that we see on the earth is primarily the result of the rotation of the earth and moon about their common center of mass, which is shown in figure 10–2A to be about 4700 km (2918 mi) from the earth's center. As the two bodies rotate as a system around this point, all particles that make up the earth follow circles of equal radius. The condition described in the last statement may be hard to visualize, but a little thought and perhaps some experimentation may convince you it is true.

FIGURE 10–1 Comparison of Sun and Moon as Tide-Generating Bodies.

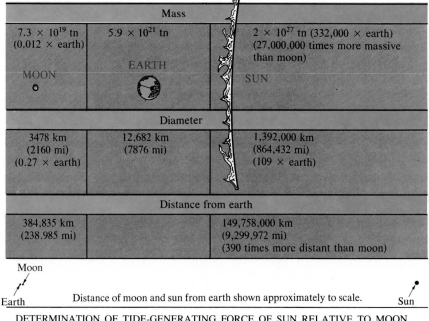

Mass		
7.3×10^{19} tn (0.012 × earth)	5.9×10^{21} tn	2×10^{27} tn (332,000 × earth) (27,000,000 times more massive than moon)
MOON	EARTH	SUN
Diameter		
3478 km (2160 mi) (0.27 × earth)	12,682 km (7876 mi)	1,392,000 km (864,432 mi) (109 × earth)
Distance from earth		
384,835 km (238,985 mi)		149,758,000 km (9,299,972 mi) (390 times more distant than moon)

Moon

Earth   Distance of moon and sun from earth shown approximately to scale.   Sun

DETERMINATION OF TIDE-GENERATING FORCE OF SUN RELATIVE TO MOON

$$\text{Tide-generating force} \propto \frac{\text{Mass}}{(\text{Distance})^3} \propto \frac{\text{Sun—27 million times more mass}}{(\text{Sun—390 times farther away})^3}$$

$$(390)^3 = 59,000,000 \quad \text{Thus,} \quad \frac{27 \text{ million}}{59 \text{ million}} = 0.46 \text{ or } 46\%$$

The sun has 46% the tide-generating force of the moon.

FIGURE 10–2 Earth-Moon Rotation.

A.

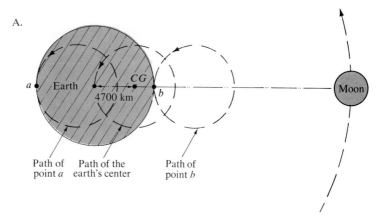

The dashed line through the center of the earth is the path of the earth's center as it moves around the common center of gravity (*CG*) in the earth-moon system. The circular paths followed by points *a* and *b* have the same radius as that followed by the earth's center.

B.

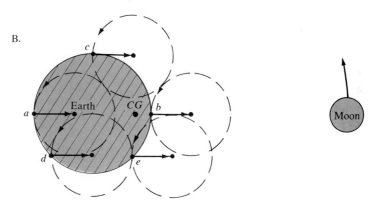

Arrows from points *a*, *b*, *c*, *d*, and *e* to the center of the circular orbit they follow as a result of earth-moon rotation around *CG* represent the magnitude and direction of centripetal force required to hold them in their orbital paths. Note that all of the arrows have the same length, indicating they represent the same amount of force. The direction of the required force is also the same for all the points—parallel to a line connecting the centers of the earth and moon.

If we divide the earth into a great number of particles of equal mass, the *centripetal* (center-seeking) acceleration required to keep each particle of the earth following an identical orbit is the same (fig. 10–2B). The centripetal force that provides the acceleration is supplied by the gravitational attraction between the particle and the moon.

Although the average gravitational attraction per unit of mass must equal the average centripetal acceleration for different particles of earth mass to keep the earth in its proper path, the two are not equal for all points of the earth. The centripetal acceleration *required* for all particles is the same and is directed toward the center of each particle's orbit. The gravitational attraction of the moon that supplies this acceleration is greater for particles close to the moon and is directed, for all particles, toward the center of the moon. Figure 10–3A shows the required centrip-

etal force and gravitational attraction of the moon acting on points along the earth's surface as vector lines. The arrow points in the direction of the force, and the length of the vector is proportional to the magnitude of the force.

The difference in the magnitude and direction of the gravitational attraction and the required centripetal force may be determined by vector subtraction. The tides result from this *net force*. The moon is at *zenith* (directly overhead) at point *Z* and at *nadir* (on the opposite side of the earth) at point *N* (fig. 10–3B). The net forces at both of these points are vertical and away from the earth's surface. This force is extremely small, however, and if you were to step on scales at the location of the moon's zenith, your weight would be reduced by about 0.11 mg (0.000035 oz).

A plane through the center of the earth and perpendicular to a line through the centers of the earth

A.

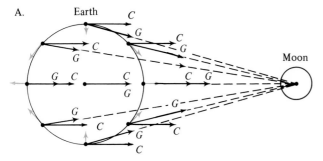

Arrows marked *C* represent identical centripetal forces required to keep all particles of equal mass in their orbits. This force is supplied by the gravitational attraction between the particles and the moon, arrows marked *G*. For the center of the earth the gravitational force is identical to the force required. For points closer to the moon the gravitational force is greater than required, and for points farther away it is less than needed. For points not on a line running through the centers of the earth and moon, the gravitational force is directed at a slight angle to the required centripetal force. These conditions produce the resulting net forces indicated by the short arrows. They are determined by drawing a line from the end of the *C* arrow to the end of the *G* arrow.

B.

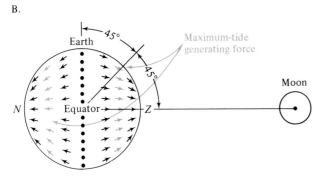

Along the intersection of the earth's surface and a plane running through the earth's center perpendicular to a line connecting the centers of the earth and moon, all resulting net forces are directed vertically toward the earth's center. Horizontal components of the net forces are zero along this line and at the zenith (*Z*) and nadir (*N*) where they are directed vertically upward. Elsewhere on the earth's surface there is a horizontal component to the net forces. The magnitude of the horizontal component, which is the tide-generating force, varies. It reaches maximum values along two circles which lie 45° on either side of the previously described circle of zero magnitude horizontal components.

FIGURE 10–3   Vector Forces Generating Tides.

and moon intersects the earth's surface along a circle where all of the net forces are directed toward the center of the earth, or vertically downward (fig. 10–3). These downward-directed forces are also of small magnitude. Because they are so small compared to the earth's gravitational attraction, which also acts vertically, none of the vertical forces along this line or at the moon's zenith or nadir are of consequence in creating the tides.

At all other points on the earth's surface there is a horizontal component to the net force, and it is this horizontal component that is responsible for creating the ocean tide. The horizontal forces are also very small, but since they are more nearly of the magnitude of other horizontal forces in existence at the earth's surface, they can induce movement. Thus all of the particles of water that make up the oceans are moved away from the line where the resultant forces are directed inward and toward points *N* and *Z*, where the resultant forces are directed away from the earth's center. This phenomenon produces the two bulges that give the tide we will refer to as the *equilibrium tide* its characteristics (fig. 10–4). These lateral, or tractive, forces increase from zero at the circumference line, where the force is directed down, to a maximum value at an angle of 45° away from this line of circumference and decrease again to zero at the zenith and nadir.

EQUILIBRIUM THEORY OF TIDES

We have established that the dominant force causing the tides is the horizontal component of the resulting net force applied at points throughout the earth's surface as a result of the earth-moon system rotating

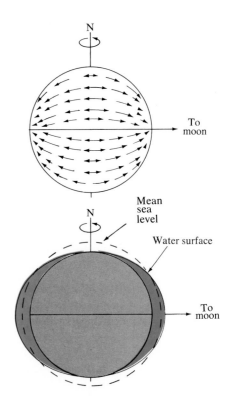

FIGURE 10–4   Equilibrium Tide.
Assuming an ocean of uniform depth covering the earth, the tide-generating forces will produce two bulges on the ocean surface. One will extend in the direction of the moon and the other away from the moon. Arrows indicate direction of tide-generating forces.

about its center of mass and the mutual attraction of the earth and moon. We will consider the kinds of tide observations that would be predictable on an earth with two tidal bulges, one toward the moon and one away from the moon as the earth rotates on its axis. In the following discussion, we assume an ideal ocean of uniform depth covers the earth and that there is no friction between the water in the ocean and the ocean floor. This theoretical tide is the *equilibrium tide*. Although this oversimplification will not give us an accurate means of predicting the types of tides that will occur throughout the earth's surface, it will give us the gross characteristics of tides in the world ocean. Only after we have established an understanding of the equilibrium theory will we attempt to consider the *dynamical theory* of tides that deals with changes in ocean depth, the existence of continents, and friction between the ocean water and ocean floor.

The Rotating Earth

We will now consider the basic effect of the earth's rotation on the prediction of tides in a world ocean that covers the entire surface of the earth to a uniform depth. This ideal ocean is modified only by the tide-producing forces that cause bulges on opposite sides of the earth as shown in figure 10–4. Let us assume that a stationary moon is directly above the equator so that maximum bulge will occur on the equator on opposite sides of the earth. Since the earth requires 24 h for one complete rotation, an observer on the equator would experience 2 high tides per day. The time that would elapse between high tides, the *tidal period*, would be 12 h. An observer at any latitude north or south of the equator would experience a similar period, but the high tides would not be so high at higher latitudes since the observer would be at the edge of the bulge rather than the apex.

But high tides do not occur every 12 h on the earth's surface. This is due to the fact that the earth-moon system is rotating about its center of mass while the earth is rotating on its axis.

The *lunar day*, the time that elapses between successive passages of the moon across the meridian (longitude line) of an observer, must be somewhat longer than the solar day of 24 h. It is actually 24 h 50 min. If one observes the time at which the moon rises on successive nights, it can be seen to rise 50 min later each night. Figure 10–5 shows that this results from the fact that as the earth is making its rotation on its axis in 24 h, the moon has moved 12.2° to the east. The earth must rotate another 50 min to have the moon again on the meridian of the observer. With the knowledge that the moon completes a 360° revolution in 29.53 days, this 12.2° of eastward revolution for the moon can be computed by

$$\frac{360°}{29.53 \text{ days}} = 12.2° \text{ per day}$$

We have so far considered the effects of the earth's rotation and the revolution of the earth-moon system about its center of mass on the prediction of tides. We have ignored the effect of the sun. In the following discussion we will consider the combined effect of the moon and sun on the earth's tides.

Combined Effects of Sun and Moon

Figure 10–6 shows the path of the earth and moon as the earth-moon system revolves around the sun. It can be seen that approximately every 29½ days the moon is in the same phase. When the moon is between the earth and the sun it is said to be in *conjunction*, producing a new moon. The moon is in *opposition* when it is on the opposite side of the earth from the sun, causing a full moon. A quarter moon results when the moon is in *quadrature*, at right angles to the sun relative to the earth.

FIGURE 10–5 The Lunar Day. A lunar day is the time that elapses between successive appearances of the moon on the meridian of a stationary observer. As the earth rotates on its axis, the earth-moon system rotates in the same direction (to the east). During one complete rotation of the earth on its axis (the 24-h solar day), the moon moves east 12.2°, and the earth must rotate another 50 min to put the observer in line with the moon.

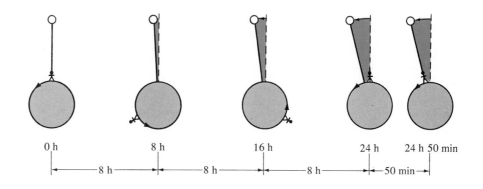

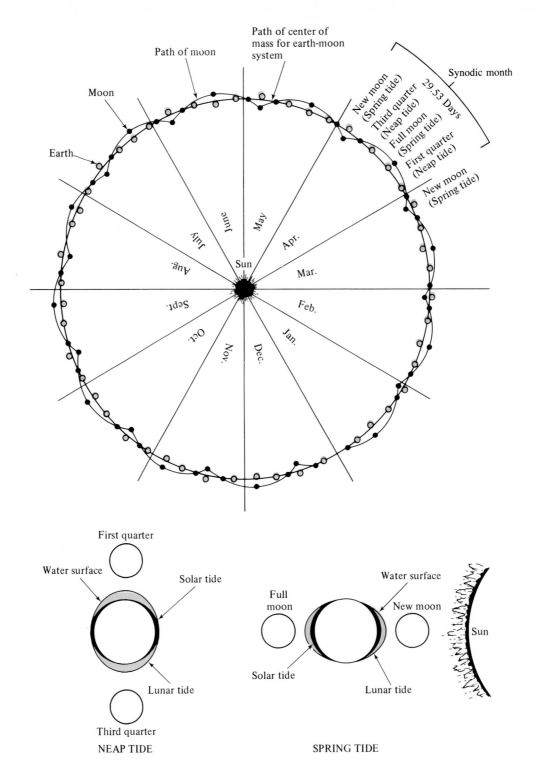

FIGURE 10–6 Sun-Moon-Earth Motion and Tides.

As the earth-moon system moves around the sun, the center of each body follows a wavy path because each is also rotating about the system's center of mass. The wobble of the moon is more pronounced and is shown in exaggerated form in the figure. During a synodic month, the moon moves from a position between the sun and the earth (new moon) to a position that puts the earth between the moon and the sun (full moon) and back to its original position. When the moon is in the new or full position, the tidal bulges created by the sun and moon are aligned, producing a large bulge. When the moon is in the positions halfway between the new and full phases, the first and third quarters, the tidal bulge produced by the moon is at right angles to the bulge created by the sun. The bulges tend to "cancel" each other, and the resulting bulge is smaller. New moon and full moon phases produce spring tides with maximum tidal ranges, while the first and third quarter phases of the moon produce neap tides with minimal tidal ranges.

When the sun and moon are in opposition or conjunction, the tide-generating forces of the sun and moon are additive, and we experience maximum *tidal ranges*, the vertical difference between high and low tide. During quadrature, the tide-generating force of the sun is working at right angles to the tide-generating force of the moon, and we have minimum tidal range. The maximum tidal range condition that exists during the new and full moon phases is referred to as the *spring tide*, and during the quadrature phases we have *neap tide* (fig. 10–6). The time that elapses between successive spring tides (full

moon and new moon) or neap tides (first and third quarters) is about two weeks.

In figure 10–7A we see the spring-tide condition associated with the new moon (the same conditions would exist with a full moon). As is the case in all the diagrams in the figure, the tide generated by the sun is indicated by a dash-and-dot line and the tide generated by the moon is indicated by a dashed line. The net tide of both tide-generating forces is depicted by the colored line. Since the moon and sun are on the same side of the earth and are more or less aligned along a line connecting the center of the earth with

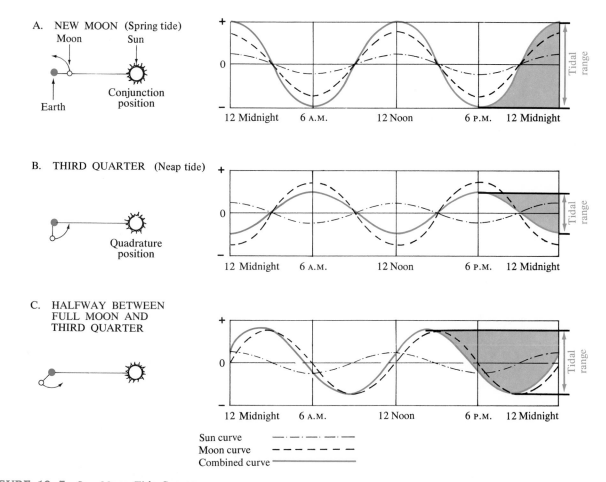

FIGURE 10–7 Sun-Moon Tide Curves.

A, **spring tide.** During a new moon, the sun and moon tidal bulges are centered at the same longitude, so their effects are added to produce maximum high tides and minimum low tides. This produces maximum tidal range. *B,* **neap tide.** During a third quarter moon, the sun and moon tidal bulges are at right angles to each other. Thus, the sun tide reduces the effect of the moon tide. For example, at 6 A.M. the observer is passing through the moon bulge and the sun trough. This produces a high tide in which the effect of the moon bulge is reduced by an amount equal to the sun trough. At 12 noon the moon trough is reduced by the sun bulge, so a reduced low tide occurs. This produces minimum tidal range. *C,* in a situation where the sun and moon bulges are separated by an angle of 45° the sun bulge is alternately added to and subtracted from the moon bulge. This produces a tidal range halfway between that of the spring and neap tides.

the center of the sun, maximum tide heights occur at midnight and noon. We have extremely high tides and extremely low tides under this arrangement of the earth, moon, and sun—a spring-tide condition.

Part B of the figure illustrates the tide that results from earth-moon-sun positions during quadrature. During third quarter, the tide-generating effect of the moon is maximum at 6:00 A.M. and 6:00 P.M., when it is at zenith and nadir, respectively, in relation to the observer. The sun is at nadir at 12:00 midnight and zenith at 12:00 noon. These are the times of maximum tide-generating effect of the sun. In this instance, we can see that the maximum lunar tides correspond with the minimum solar tides, producing a tide pattern with relatively small tidal range—a neap tide. The same conditions would also exist when the moon is in its first quarter.

Figure 10–7C shows a situation that would develop halfway between neap-tide and spring-tide alignments of the earth, moon, and sun. For example, the moon is in a position halfway between the full moon and the quadrature that occurs at the end of the third quarter. When the moon is at zenith at 3:00 A.M., the observer still has to rotate 135° before the sun will be at zenith at 12:00 noon. The sun and moon are at nadir at 12:00 midnight and 3:00 P.M., respectively. The pattern is a little more complicated than that which occurs during spring and neap tides. Nevertheless, it can still be seen that the net tide curve, which shows a tidal range that is less than that of spring tides and greater than that of neap tides, is still determined by the combined effect of the solar and lunar tide-generating forces. *Grunion and the Tides*, a special feature on pages 256–257, discusses the close association between the spawning of these small fish and the spring neap-tide sequence.

Effects of Declination

Up to this point we have considered that the moon and the sun remained at all times above the equator, or at least we did not describe specifically any deviation from such a position. We now must consider the fact that this is not the case. As it revolves around the sun, the earth's axis of rotation is tilted 23.5° from vertical relative to the *ecliptic*, the plane of the earth's orbital path about the sun. It is owing to this tilt that we experience the seasons—spring, summer, fall, and winter. To further complicate our consideration, the plane of the moon's orbit is at an angle of 5° to the ecliptic. This angular distance of the sun or moon

above or below the equatorial plane of the earth is called *declination*.

The tilt of the earth's axis relative to its plane of revolution around the sun is shown in figure 10–8A. It can be seen that the tilted axis assumes a constant direction in space throughout the yearly cycle that includes the equinoxes and solstices. At the *vernal equinox*, which occurs about March 21, the sun is directly above the equator and is moving from the Southern Hemisphere to the Northern Hemisphere. On about June 21 the *summer solstice* occurs. At this time the sun reaches its most northerly point in the sky, directly above the Tropic of Cancer, which is at 23.5°N latitude. Following this occurrence, the sun moves farther south in the sky each day, and on about September 23 it is directly above the equator again and produces the *autumnal equinox*. During the next three months the sun appears to be more southerly in the sky until the *winter solstice* on about December 22, when it is directly over the Tropic of Capricorn at 23.5°S latitude. Thus the sun may be found at declinations between 23.5° north and 23.5° south of the equator on a yearly cycle.

Since the plane of the moon's orbit intercepts the plane of the ecliptic at an angle of 5°, and since the plane of the moon's orbit *precesses*, or rotates, while maintaining its 5° angle with the ecliptic with a precessional cycle of 18.6 yr, we have a relatively complex consideration regarding the declination of the moon relative to the plane of the earth's equator. In part 1 of figure 10–8B, the declination of the moon's orbit relative to the earth's equator is 28.5°. The declination will change from 28.5° south to 28.5° north and back to 28.5° south of the equator in a period of one month. Part 2 of the illustration shows the relationship of the ecliptic, the plane of the moon's orbit, and the plane of the earth's equator after one-fourth of the precession, or 4.65 yr later. The maximum declination of the moon's orbit relative to the earth's equator still approaches 28.5°. However, in part 3, when one-half precession is completed, which is 9.3 yr after the condition observed in part 1, it can be observed that the maximum declination of the moon relative to the earth's equator is 18.5°.

On the basis of these considerations we must alter our previous concept of the predicted equilibrium tide. We must now expect that tidal bulges will rarely be aligned with the equator and will occur for the most part either north or south of the equator. Since the moon is the dominant force that creates tides in the earth's oceans, we would suspect that the bulges

FIGURE 10–8 Orbital Planes of Earth and Moon.

A, as the earth orbits the sun during one year, the axis of rotation is tilted 23.5° from perpendicular relative to the ecliptic. The sun shines down directly over the Tropic of Cancer (23.5°N) on the day of the summer solstice, June 21. Three months later the sun is directly above the equator (0°) during the autumnal equinox, September 23. The sun is directly over the Tropic of Capricorn (23.5°S) during the winter solstice, December 22, and returns to a position above the equator on the vernal equinox, March 21. Three months later, the yearly orbit is completed, and the sun is again directly over the Tropic of Cancer.

B, the plane of the moon's orbit (grey plane) is tilted at an angle of 5° relative to the plane of the ecliptic (colored plane) and rotates with a clockwise precession that has a period of 18.6 years. In B1, the declination of the moon's orbit is the sum of the angle of intersection of the plane of the earth's equatorial plane (black plane) with the plane of the ecliptic (23.5°) plus the angle of intersection between the plane of the moon's orbit and the ecliptic (5°). This produces the maximum declination of the moon relative to the earth's equator of 28.5°. B2 shows the positions of the planes 4.65 years later when the moon has achieved one-fourth of its precessional rotation. B3 shows the relative positions of the planes after 9.3 years or one-half of the precession. The maximum declination of the moon relative to the earth's equator is now 18.5° or 23.5° less 5°. From C. Hauge, Tides, currents, and waves, *California Geology*, July 1972. Reprinted by permission of the author.

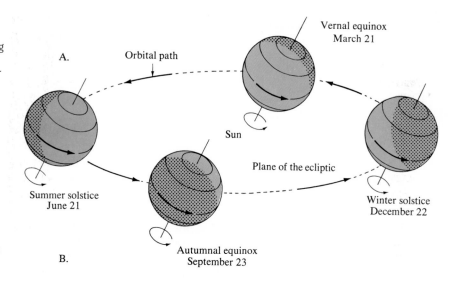

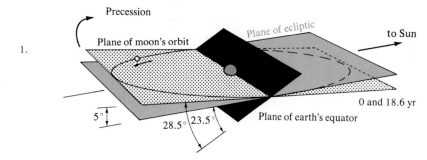

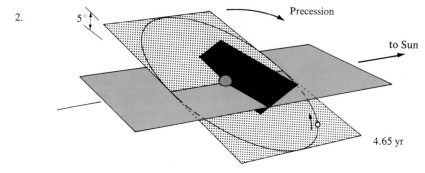

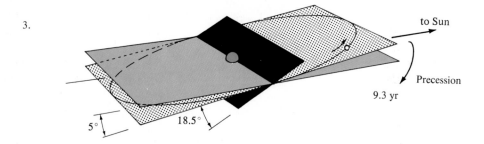

GRUNION AND THE TIDES

Along the beaches of southern California and Baja California from March through September a very unusual fish-spawning behavior can be observed. Shortly after the maximum spring tide has occurred, small silvery fish come ashore to bury their fertilized eggs in the sand. They are the grunion (*Leuresthes tenuis*), slender little fish 12 to 15 cm (4.7 to 6 in) in length. The name grunion comes from the Spanish *grunon,* which means grunter. The early Spanish settlers gave the fish this name because of the faint noise they make during spawning.

The type of tide that occurs along southern California and Baja California beaches is a mixed tide. On most tidal days (24 h 50 min) there are two high and two low tides. There is usually a significant difference in the heights of the two high tides that occur each day. During the summer months, this higher high tide occurs at night. As the higher high tides become higher each night while the maximum spring-tide range is approached, sand is eroded from the beach (fig. 10A). After the maximum height of the spring tide has occurred, the higher high tide that occurs each night will be a little lower than the one of the previous night. During this sequence of decreasing heights of the tides as neap-tide conditions approach, sand is deposited on the beach. The grunion depend greatly on this pattern of beach sand deposition and erosion in their spawning process.

Grunion spawn only after each night's higher high tide has peaked on the three or four nights following the night of the occurrence of the highest spring high tide. This behavior assures that the eggs will be covered deeply by sand deposited by the receding high tides. The fertilized eggs buried in the sand are ready to hatch nine days after spawning. By this time, another spring tide is approaching, and the higher high tide that occurs each night will be higher than that of the previous night. This condition causes the beach sand to erode, exposing the eggs to agitation by waves that break ever higher on the beach. The eggs hatch about three minutes after being freed in the water. Tests done in laboratories have shown that the eggs will not hatch until agitated in a manner that simulates the agitation of the eroding waves.

The spawning begins as the grunion come ashore immediately following an appropriate high tide, and it may last from 1 to 3 hours. Spawning activity usually peaks about an hour after its start and may last from 30 minutes to an hour. Thousands of fish may be on the beach at this time. During a run, females move high on the beach. If no males are near, a female may return to the water without depositing her eggs. In the presence of males, she will drill her tail into the semifluid sand until only her head is visible. The female continues to twist, depositing her eggs 5 to 7 cm (2 to 3 in) below the surface. The male curls around the female's body and deposits his milt against it (fig. 10B). The milt runs down the body of the female to fertilize the eggs. The spawning completed, both fish return to the water with the next wave.

As soon as the eggs are deposited, another group of eggs begins to form within the female. They will be deposited during the next spring-tide run.

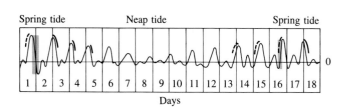

Spring tide   Neap tide   Spring tide

1 2 3 4 5 6 7 8 9 10 11 12 13 14 15 16 17 18
                    Days

⋀ Flood tides erode sand and free grunion
   eggs during higher high tide as
   maximum spring tidal range is
   approached.

⋀ Grunion deposit eggs in beach sand
   during early stages of the ebb of higher
   high tides on the three or four days
   following maximum spring tidal range.

▮ Maximum spring tidal range

FIGURE 10A Grunion and the Tidal Cycle.

Eggs deposited in the beach during ebb tides on days 3, 4, and 5
will be ready to hatch when eroded by the flood tides of days 13,
14, 15, and 16. The lines separating days on the tide curve chart
represent midnight.

FIGURE 10B Grunion Spawning.

A male grunion wraps around the female that has burrowed into
the sand to deposit her eggs beneath the surface. The male releases sperm-laden milt, which runs down the female's body to
fertilize the eggs. Photo by Bill Beebe.

Larger females are capable of producing up to 3000 eggs for each series of
spawning runs, which are separated by the 2-week period between spring-tide
occurrences. Early in the spawning season, only older fish spawn, but by May
even the 1-year-old females are in spawning condition.

Young grunion grow rapidly and are about 12 cm (5 in) long when they
are a year old and ready for their first spawning. They usually live 2 or 3 years,
but 4-year-olds have been recovered. The age of a grunion can be determined
by the scales. After growing rapidly during the first year, they grow very slowly.
There is no growth at all during the 6-month spawning season, which causes
marks to form on each scale.

How the grunion are able to time their spawning behavior so precisely is
not known. Some investigators believe the grunion are able to sense very small
changes in the hydrostatic pressure caused by the changing level of the water
associated with the tidal ebb and flow. Certainly some very dependable detection mechanism keeps the grunion accurately informed of the tidal conditions,
because their survival depends on a spawning behavior precisely tuned to tidal
motions.

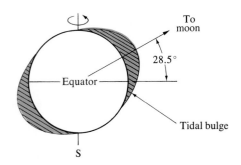

FIGURE 10–9 Maximum Declination of Tidal Bulges from Equator.

The center of the tidal bulges may lie at any latitude from the equator to a maximum of 28.5° on either side of the equator.

would follow the moon as it moves on its monthly journey across the equator, being found at a maximum of 28.5° north and south of the equator (fig. 10–9).

Effects of Distance

Additional considerations that will affect the tide-generating force of the sun and moon on the earth are their changing distances from the earth. The sun ranges from *perihelion* with the earth, when the distance between the bodies is 148.5 million km (92.2 million mi) during the winter months in the Northern Hemisphere, to *aphelion,* when the distance between the bodies is 152.2 million km (94.5 million mi) during the summer months. The moon moves from *perigee,* its closest approach to the earth, of 375,200 km (233,000 mi) to *apogee* of 405,800 km (252,000 mi) and back to perigee during a period of 27½ days— the *anomalistic month.* (See fig. 10–10.)

Because of these movements, spring tides have greater ranges during the Northern Hemisphere winter than in the summer. Also, as the result of changing distance between the earth and moon we will find that tidal ranges will become greater at perigree each anomalistic month. Keeping in mind that tide-generating forces vary inversely with the cube of the distance from the center of the earth to the center of the tide-generating body, it can be appreciated that these changes resulting from the elliptical nature of the earth's orbit around the sun and the moon's orbit around the earth can be readily observed in the tides.

Equilibrium Tide Prediction

To consider the tidal patterns we would predict for an idealized water-covered earth, let us return our attention to the effect of declination. Let us assume that the declination of the moon, which will determine the

alignment of the tidal bulges, is 28° north of the equator. If we position an observer at this latitude on a permanent point, the observation of the tides that occurs here will be different from observations at the equator.

If the observations begin when the moon is directly over the observer's head, he will record high tide. Six lunar hours later a low tide will be recorded, followed by another high tide that will be much lower than the initial high tide (fig. 10–11A–D). At the end of a 24-lunar-hour period, the observer will have passed through a complete lunar day cycle of two low tides and two high tides. Thus, the predicted period of the equilibrium tide is *semidiurnal,* half a lunar day. A representative curve of the type of tide observed can be seen in figure 10–11E. Tide curves showing the heights of the same tides during one lunar day at the equator and at 28°S latitude are also provided. The difference in heights of successive high tides or successive low tides that occurs as a result of the declination of the moon and the sun relative to the earth's equator is called the *diurnal inequality.* When the moon is at its maximum northern declination and again about two weeks later when it is at its maximum southern declination, the diurnal inequality reaches a maximum. Since the moon is generally above the tropics at these times, we call these tides *tropical tides.* Separating the tropical tides are tides that have a minimum diurnal inequality when the moon is over the equator, *equatorial tides.*

To summarize the tides that we might predict as a result of the tide-generating forces of the moon and sun on an earth covered with a uniform depth of wa-

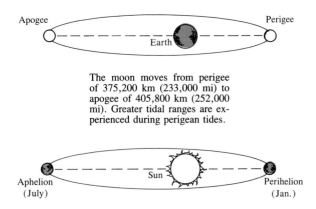

The moon moves from perigee of 375,200 km (233,000 mi) to apogee of 405,800 km (252,000 mi). Greater tidal ranges are experienced during perigean tides.

Perihelion brings the earth within 148,500,000 km (92,200,000 mi) of the sun. Aphelion distance is 152,200,000 km (94,500,000 mi). Greater tidal ranges are experienced during perihelion tides.

FIGURE 10–10 The Effects of Elliptical Orbits.

FIGURE 10–11 Predicted Equilibrium Tides.

F, from Anikouchine & Sternberg, *The world ocean: An introduction to oceanography,* © 1973. Reprinted by permission of Prentice-Hall, Inc., Englewood Cliffs, New Jersey.

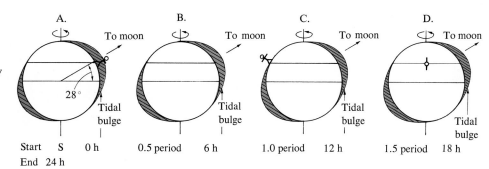

A. *At start of lunar day the observer is at zenith in the center of the tidal bulge. He experiences a high tide.* B. *Six lunar hours later (0.5 period) the observer experiences low tide while located on the back side of diagram.* C. *After twelve lunar hours (1.0 period or 0.5 lunar day) the observer again experiences high tide. It is much lower than the first high tide, however, because the observer is now passing through the edge of the tidal bulge.* D. *Observer experiences low tide again eighteen lunar hours (1.5 period) after start. At the end of one lunar day (24 hours) the observer returns to A and experiences a high high tide.*

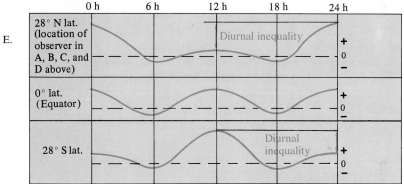

Tide curves for 28°N, 0°, and 28°S latitudes when the declination of the moon is 28° N (all curves for same longitude). Note that tide curves for 28°N and 28°S have identical highs and lows, but are out of phase by 12 hours. This results from the fact that the bulges in the two hemispheres occur on opposite sides of the earth.

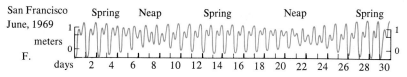

Along with the unequal heights for the two high and low tidal extremes occurring each lunar day, we expect to observe spring and neap tides. Thus the tide curve for the month of June 1969 demonstrates the general character of the predicted equilibrium tide.

ter for any point not on the equator, we would predict that the equilibrium tide would generally be observed at any location to have two high tides and two low tides per lunar day. Except for the rare occasions when the sun and moon are simultaneously above the equator, we would expect that neither the two high tides nor the two low tides would be of the same height because of the changing declination of the moon and the sun. We would expect yearly and monthly cycles of tidal range related to the changing distances of the earth from the sun and moon. Lastly, we would expect that each fortnight, half a lunar month, we would experience spring tides that would be separated by neap tides (fig.10–11F).

Considering the great number of variables that are involved in predicting tides, it is interesting to consider when the conditions might be right to produce the maximum possible tide-generating force. The maximum tidal range will be produced when the sun is at perihelion and in conjunction or opposition with the moon at perigee and when both the sun and moon have a zero declination. This condition occurs only once each 1600 years, and the next occurrence is predicted for A.D. 3300.

An indication of the significance of even a near coincidence of the earth's perihelion with the perigee of the moon during a spring tide was made woefully clear during the January 1983 storms in the North Pacific. The introduction of slow-moving low-pressure cells that developed over the Aleutians caused strong northwest winds to blow across the ocean from Kamchatka to the U.S. coast. Averaging about 50 km/h (31 mi/h), the winds produced a near fully developed 3-m (10-ft) swell along the coast from Oregon to Baja California (*see* tables 9–2 and 9–3). The storm surge from this condition would have been trouble enough under average conditions, but the situation was made worse by the 2.25-m (7.4-ft) high spring tides and a slight increase in sea level along the coast that resulted from the influx of warm water driven north by the 1982–83 El Niño (see *El Niño-Southern Oscillation Events,* pp. 312–316).

The unusually high tides occurred because the earth was still near perihelion (January 2) when the moon reached perigee on January 28, 1983. Some of the largest waves came ashore January 26–28, when over $100 million in damage was done (fig. 10–12). Twenty-five homes were destroyed, over 3500 homes were seriously damaged, and many commercial and municipal piers collapsed. At least a dozen lives were lost. With each such occurrence, we learn more of the cost of development on and near the shore.

DYNAMICAL THEORY OF TIDES

In our previous discussion of the equilibrium tide, we considered that the tidal bulges were directed at the moon and away from the moon on opposite sides of the earth. Maintaining this relationship with the moon as the earth rotates beneath the moon, the bulges (or wave crests), which would be separated by a distance of half the earth's circumference (about 20,000 km, or 12,420 mi), would be moving across the earth at a

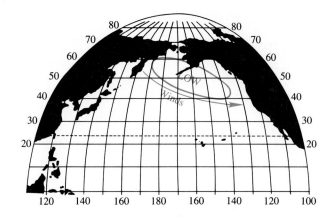

A.

B.

FIGURE 10–12 High Tides of January 1983.
A, January 1983 storm winds blow uninterrupted across the North Pacific Ocean from Kamchatka to the U.S. coast. *B,* homes threatened by storm waves and unusually high tides on January 27, 1983, at Stinson Beach north of San Francisco. Wide World Photos.

speed of more than 1600 km/h (994 mi/h). We previously stated that the tides were an extreme example of shallow-water waves, whose velocity is proportional to the square root of the water depth. In order for the tidal wave to travel at 1600 km/h, the depth of the idealized ocean would have to be 22 km (13.7 mi). Since the average depth of the ocean is less than 3.9 km (2.4 mi), the tidal bulges move at speeds slightly under 700 km/h (435 mi/h). The actual position of the bulges relative to the tide-generating forces is determined by a balance that is reached between these forces and the shallow-water wave movement of the bulges provided by the depth of the ocean.

We must also consider the effect of the continents, which interrupt the free movement of the tidal bulges across the ideal unobstructed ocean surface we considered to exist in our discussion of the equilibrium tide. The ocean basins between continents have set up within them free oscillatory waves whose character modifies the forced astronomical tidal waves that develop within the basin. It is impossible for us to explain here the various tidal phenomena that occur throughout the world. For instance, high tide rarely occurs at the time the moon is at zenith, and the amount of time that elapses between the passing of the moon and the occurrence of high tide varies from place to place as a result of the many factors that determine the characteristic of the tide at any given location.

In order to develop a maximum understanding of tides, a combination of mathematical analysis and observation is required. When we consider the *harmonic analysis,* a mathematical approach to the study of tides which embraces all the tide-generating variables that possess a *periodicity* (cyclic pattern), there are almost 400 tide-generating variables, called *partial tides,* identified. The actual tide observed at any given location is the combined effect of all the partial tides at that point. To simplify things somewhat, a relatively accurate model of the actual tide can be computed considering only the seven major partial tides. They are listed with their characteristics in figure 10–13. Combining the periods of each of the partial tides with the amplitudes and phases that can be obtained from observation, relatively accurate predictions of the tide at any location can be made. The observations must have been made throughout a period of at least 18.6 yr to make the predictions as accurate as possible, as this is the period of the precession of the plane of the moon's orbit through the ecliptic (fig. 10–8).

Types of Tides

It was previously mentioned that everywhere on the earth we should expect two high tides and two low tides of either equal or unequal heights during a lunar day. Owing to modifications resulting from varying depths, sizes, and shapes of ocean basins, the tide predicted by the equilibrium theory is replaced in many parts of the world. In reality one finds either a *diurnal* (daily), *semidiurnal* (twice daily), or *mixed* tide (fig. 10–14).

The diurnal tide is characterized by a single high and low water each lunar day. These tides are common in the Gulf of Mexico and along the coast of Southeast Asia. Such tides have a tidal period of 24 h 50 min.

The semidiurnal tide has two high and two low waters each lunar day, and the heights of successive

A. The seven most important partial tides.

	Symbol	Period in solar hours	Amplitude $M_2 = 100$	Description
Semidiurnal tides	M_2	12.42	100.00	Main lunar (semi-diurnal) constituent
	S_2	12.00	46.6	Main solar (semi-diurnal) constituent
	N	12.66	19.1	Lunar constituent due to monthly variation in moon's distance
	K_2	11.97	12.7	Soli-lunar constituent due to changes in declination of sun and moon throughout their orbital cycle
Diurnal tides	K_1	23.93	58.4	Soli-lunar constituent
	O	25.82	41.5	Main lunar (diurnal) constituent
	P	24.07	19.3	Main solar (diurnal) constituent

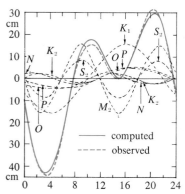

B. Partial tides, computed tide, and observed tide at Pula, Yugoslavia (January 6, 1909). Note close fit of computed and observed tides.

FIGURE 10–13 Partial Tides.

The semidiurnal and diurnal partial tides described in A are shown in tide curve form in B, along with the computed and observed tides that resulted from their combined effects at Pula, Yugoslavia. A and B, from A. Defant, *Ebb and flow,* © 1958. Reprinted by permission of The University of Michigan Press, Ann Arbor.

high waters and successive low waters are approximately the same. Since tides are always getting higher or lower at any location, owing to the spring-neap–tide sequence, successive high tides and suc-

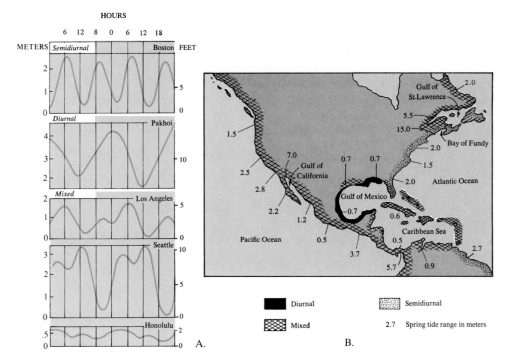

FIGURE 10–14 Types of Tides.

A, types of tides. In a **semidiurnal** (twice daily) type of tide, there are two highs and lows during each tidal day and the heights of each successive high and low are about the same. In the **diurnal** (daily) type of tide, there is only one high and one low each tidal day. In the **mixed** type of tide, both diurnal and semidiurnal effects are detectable and the tide is characterized by a large difference in the high water heights, the low water heights, or both, during one tidal day. Even though a tide at a place can be identified as one of these types, it still may pass through stages of one or both of the other types. For example, the tide along the North Carolina coast changes from semidiurnal to mixed, but is predominantly semidiurnal. *B,* map showing the types of tides that have been observed along portions of the coasts of North and South America. The numbers give the spring tide range in meters and are therefore near the maximum tidal range that can be expected. Storm waves, lower barometric pressure, ocean currents, and the alignment of additional celestial bodies could increase the range. After C. Hauge, Tides, currents, and waves, *California Geology,* July 1972. Reprinted by permission of the author.

cessive low tides can never be exactly the same at any location. Semidiurnal tides are common along the Atlantic coast of the United States. The tidal period is 12 h 25 min.

The mixed tide may have characteristics of both diurnal and semidiurnal tides. The diurnal inequality discussed earlier is a characteristic of this tide, as successive high tides and/or low tides will have significantly different heights. Mixed tides commonly have a tidal period of 12 h 25 min, which is a semidiurnal characteristic, but may also possess diurnal periods. This is the tide that is most common throughout the world and the type that is found along the Pacific coast of the United States. Note that the tide curve for Los Angeles, shown in figure 10–15, is mixed because it is semidiurnal in period during all days except September 15 and 16, when the period is diurnal.

Tides in Lakes

Before discussing the open ocean, we will consider the effects of the tide-generating forces on a body of water in a small rectangular closed basin. In general, lake basins are too small to have any appreciable tide setup by the tide-producing forces, but they may be significant where the long axis of the basin runs parallel to lines of latitude. In any such basin, very small standing waves will be generated with a period equal to that of the tide-generating force. This wave is referred to as a *forced standing wave* (fig. 10–16).

Of much greater importance is the *free standing wave* initiated by strong winds at the surface or by some seismic disturbance that may be set up in the basin (fig. 10–16). The period of a free standing wave is determined by the length and depth of the basin, and this period is termed the *characteristic period* for

FIGURE 10–15 Tidal Curves for Four Locations of the Earth.

The declinations of the moon and the sun, phase of the moon, and position of the moon in its orbit as noted across the top of the chart all contribute to the tidal variations that are depicted. Additional contributing factors include the position of the earth in its orbit around the sun and the configuration of the sea bottom and basin boundaries. Some names of water levels are listed along the right margin. Those that are used as the chart datum for a place are marked with an asterisk. MHWS, *mean high water springs*—the average height of the high water of the spring tides; MHW, *mean high water*—the average height of all the high tides at a place; MLW, *mean low water*—the average height of all the low tides at a place; MLWS, *mean low water springs*—the average height of all low waters of the spring tides; MHHW, *mean higher high water*—the average height of the higher high tides at a place where the tide is the mixed type and displays an inequality during a tidal day; MLLW, *mean lower low water*—the average height of the lower low tides at a place where the tide is of the mixed type. From U.S. Naval Oceanographic Office, *American Practical Navigator,* rev. ed., Bowditch, H.O. Pub. No. 9, Washington, D.C. 1958.

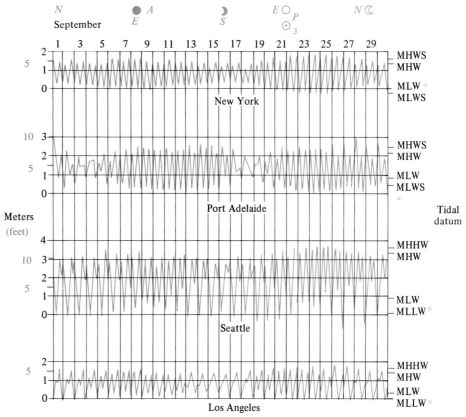

●, new moon; ☽, first quarter; ○, full moon; ☾, last quarter; *E*, moon on the equator; *N, S,* moon farthest north or south of the equator; *A, P,* moon in apogee or perigee; ₁⊙₃, sun at autumnal equinox; *, chart datum.

FIGURE 10–16 Tides in Lakes.

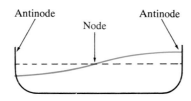

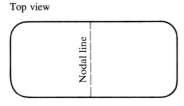

A forced standing wave generated by tide-generating forces. It will have a period of 12 h 25 min.

A free standing wave (seiche) generated by atmospheric disturbance. Its period is determined by the size and shape of the basin of the lake. The period (*T*) is related to length and depth of the lake as is shown by the equation:

$$T = \frac{2L}{\sqrt{gh}}$$

L = length of lake (m)
g = acceleration due to gravity (9.8 m/s²)
h = depth of lake (m)

If the lake has dimensions that produce a period for the seiche that is approximately equal to the forced tide generated wave, they will be resonant and produce greater displacements at the antinodes.

the basin. If the characteristic period of the free wave is very nearly that of the period of a forced wave resulting from the tide-generating force, the oscillations may reinforce each other, and *resonance tides* with greater range may develop as they do in Lake Ontario.

Free oscillations of the type described above are common on large lakes and were given the name *seiche* by the inhabitants of the region near Lake Geneva, Switzerland. For seiches that have a single nodal line, the formula for the period in figure 10–16 gives a close approximation to the actual period that develops. For rectangular basins with two and three nodal lines the periods would be approximately one-half and one-third those that would be computed for a mononodal seiche.

Tides in Narrow Open Basins

As discussed above, even under conditions of resonance between free and forced oscillations in the small closed bodies of water, the maximum tides are not great. Seldom do they reach ranges in excess of a few centimeters. By contrast, if we consider similarly sized basins that are open at one end to the ocean, we find that the tides can become large in comparison. Why the difference?

To help answer this question, let us consider a rectangular bay aligned east-west with one end open to the ocean (fig. 10–17). Even though one end of the basin is open, we will still have to consider the effect of free standing waves developing much as we might conceive of the standing waves that develop in the pipe of an organ that has one end open. Without attempting to explain the mathematics, it is a fact that free standing waves can be reflected from the open end of the basin.

At the open end of the bay, the water level must always be the same as the height of the ocean at the location. Because of this, tidal range will be of greater magnitude than it is in closed basins, which have relatively small tidal ranges. If we think of the free standing wave that forms as a result of reflection of wave energy from the closed end of the basin as superimposed on the forced wave produced by tidal forces, we can see that combining the energy of both waves could produce a standing wave with increased amplitude within the embayment (fig. 10–17).

Tides in Broad Open Basins

In large, broad basins, a manifestation of the earth's rotation causes a rotary movement of the wave crest around the margins of the embayment. Since there is

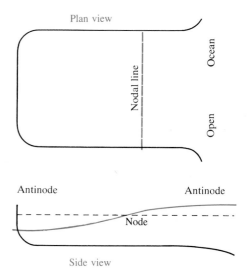

FIGURE 10–17 **Tides in Seas, Bays, and Gulfs.**
In seas, bays, and gulfs the forced standing waves will have a greater height than those created in lakes. This is because the height of the tide at the open end of the basin must be the same as the open ocean. Therefore, the development of a resonant condition between the free and forced standing waves in such basins will produce much greater displacements at the antinodes.

a significant horizontal transport of water associated with standing waves, the Coriolis effect becomes important in embayments that are wide and will thus allow lateral movement of the water.

Let us assume an embayment in the Northern Hemisphere that has developed a mononodal standing wave is experiencing the development of a crest at its closed end, as illustrated in figure 10–18A. As the water begins to move toward the open end of the embayment (10–18B), the Coriolis effect diverts the water to the right, forcing it to pile up along the left-hand side of the embayment as we look at it from the open end. One-half period after the crest at the closed end of the basin begins to fall, a crest will form at the open end (10–18C). As the water moves toward the closed end of the basin, again the Coriolis effect will pile it up along the right-hand margin of the embayment (10–18D). If a basin is wide, this produces an *amphidromic point,* a point of no tidal range, around which the crest of the wave rotates in a counterclockwise direction instead of in a mononodal line as might have developed in a narrow basin.

Radiating from the amphidromic point and connecting points at which high water will occur simultaneously are *cotidal lines.* Figure 10–18E shows the cotidal lines at one-hour intervals for a tide in a rectangular basin with a period of 12 lunar hours. The cotidal lines are designated by Roman numerals that show that the crest of the tide wave moves in a counterclockwise direction. The crest makes one complete

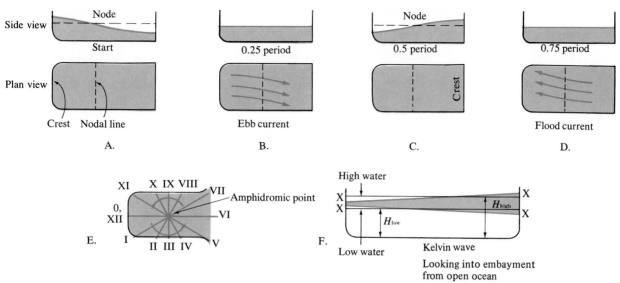

A–D. As the crest forms in the head of the bay there is no horizontal water movement. In *B* the surface is in an equilibrium (flat) condition and water moves horizontally toward the open ocean. The Coriolis force causes the water to veer to the right in the Northern Hemisphere. After the crest forms at the mouth of the bay (*C*), the same effect causes water to veer toward the opposite side of the bay when water moves toward the head of the bay in *D*.

E. In bays with sufficient width, the nodal line of the standing wave disappears and is replaced by an amphidromic point, a point of no tide. The standing wave has taken on a progressive character, and the crest moves in a counterclockwise path around the bay. The lines labeled with Roman numerals are cotidal lines that connect all points experiencing high tide at the indicated lunar hour. The tide crest rotates around the amphidromic point, and tidal range increases with increasing distance along the cotidal lines from the amphidromic point.

F. Kelvin wave—Along with the effects shown in *A–E*, the Coriolis force acting on a progressive wave moving along any channel or embayment produces a tilted water surface instead of a flat high and low water at tidal extremes. This results in a greater tidal range on one side of a basin than the other. On the right side, the tidal range is $H_{high} - H_{low} + 2x$. On the left side the range is $H_{high} - H_{low} - 2x$.

FIGURE 10–18   Tides in Wider Embayments.

rotation around the amphidromic point during the tidal period. The greater width of the basin combined with the Coriolis effect prevents the perfect back-and-forth reflection needed to produce pure standing waves. In wide basins a progressive wave moves around the basin in a counter-clockwise direction. Figure 10–19 shows the cotidal rotations that have replaced the equilibrium-theory bulges in the world ocean.

Figure 10–18F shows that as progressive waves move through broad basins there will be a much greater tidal range on the right-hand side of the basin, looking into the basin, than on the left. Assuming a theoretical high- and low-tide level for the system and a tilt in the surface of the wave caused by the Coriolis effect as it moves into and out of the basin, it can be seen that along the right-hand side of

the basin the tidal range is equal to the theoretical tidal range plus $2x$. The x value results from the Coriolis force tilting the wave, a *Kelvin wave*, as it moves through the basin. On the left-hand side of the basin the tidal range, as a result of this tilting, will be equal to a theoretical difference between high and low tide of minus $2x$.

Tide Currents   The current that accompanies the slowly turning tide crest in a Northern Hemisphere basin will turn in a counterclockwise direction, producing a *rotary current* in the open portion of the basin. Near shore, except where the shore is straight and steeply sloping, the rotary current is changed to an alternating or *reversing current* that moves in and out rather than along the coast as would a rotary current. These reversing currents are of the greatest con-

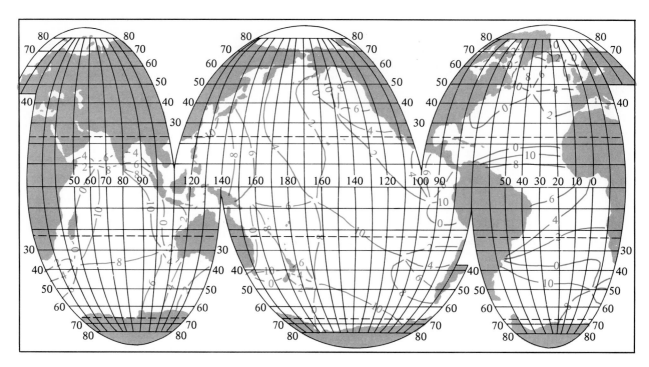

FIGURE 10–19 Cotidal Map of the World.

Numbers indicate times of M_2 high tide in lunar hours after the moon has crossed the Greenwich Meridian. Tidal ranges generally increase with increasing distance along cotidal lines away from the amphidromic points. Where cotidal lines terminate at both ends in amphidromic points, maximum tidal range will be near the midpoints of the lines. Base map courtesy of National Ocean Survey. After von Arx, 1962; original by H. Poincaré, 1910, Leçons de Mécanique Céleste, Gauther-Crofts, Vol. 3.

cern to navigators, since they may reach velocities near 20 km/h (12.4 mi/h) in restricted channels, while velocities of the rotating currents in the open ocean are usually well below 1 km/h (0.62 mi/h).

The reversing tidal current that develops throughout a lunar day for a mixed tide is depicted in figure 10–20. Beginning at high tide, the current velocity is zero because the water has just reached its highest stage and is momentarily to begin its outward flow. Following this *high slack water,* the lowering of the tide begins, and the *ebb current* velocity increases and reaches a maximum about 3 lunar hours after high slack water. The velocity decreases and eventually reaches zero again at the first *low slack water.* Following the change in current velocity associated with the tidal phase throughout the day, it can be seen that the maximum current velocity is reached midway through the ebb current that occurs between the higher high water and lower low water.

The illustration shows that the lower low water (LLW) is below the datum of the chart, the zero mark. How can the tide be less than zero? The datum that is commonly used for mixed tides is the mean lower low water, the average height of the lower low tides at the locality. Since the lower low tide that we are

recording is below the average lower low tide for this locality, this tide will have a negative value. In areas where mixed tides are not observed, the tide datum is very commonly the mean low tide recorded at that place—the average low tide. Thus most tide extremes recorded, even low tides, will have positive values. Only during spring tides are negative low tides observed (*see* fig. 10–15).

TIDES OBSERVED THROUGHOUT THE WORLD

Tides in a Narrow Basin

Bay of Fundy With a length of 258 km (160 mi), of Fundy has an extremely wide opening into the Atlantic Ocean. The Bay of Fundy splits into two narrow basins at its northern end, Chignecto Bay and Minas Basin (fig. 10–21). The period of free oscillation in the Bay of Fundy is very nearly that of the tidal period. This condition brings about resonance, which along with the narrowing of the bay toward the north end and the shoaling in that direction, produces maxi-

FIGURE 10–20 Reversing Current.

Note that tidal current velocity is zero at high and low tidal extremes. Maximum tidal current velocity occurs midway between tidal extremes.

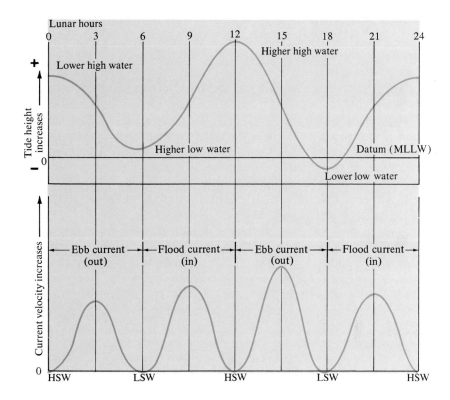

HSW—High slack water (velocity zero)

LSW—Low slack water (velocity zero)

mum tidal ranges in the extreme northern end of Minas Basin. The maximum perigean tidal range of about 17 m (56 ft) occurs at the north end of Minas Basin, and the minimum tidal range of about 2 m (6.6 ft) is found at the opening into the bay. There are no nodes within the bay, and the tidal range progressively increases from the mouth of the bay northward. Along with the standing wave phenomena, a progressive component exists that causes the tidal-range on the south shore to be greater than that experienced on the north shore (see fig. 10–18F).

Tides in a Broad Basin

North Sea Tide The North Sea is a very broad basin that is relatively shallow and open to the Atlantic Ocean for a great distance along its northern margin. Bounded on the west by the British coast, and to the south and east by the coast of northern Europe, the North Sea experiences maximum tidal ranges approaching 5 m (16 ft). As the semidiurnal tide in the form of a progressive wave enters the North Sea from the Atlantic Ocean, the maximum tidal range within the sea is produced along the Scottish and English

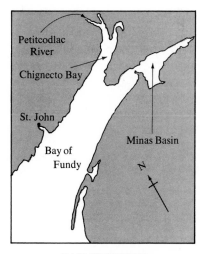

BAY OF FUNDY

Because of the great width at its mouth a Kelvin wave creates a larger tide range on the south shore than is observed on the north shore.

FIGURE 10–21 Tides in a Narrow Basin.

coasts. In contrast to the 3 to 4 m (9.8 to 13.1 ft) tidal range seen along the western coast of the North Sea, the tides recorded at the southern tip of Norway are frequently of a range less than 25 cm (9.9 in).

Owing to the earth's rotation, the mass of water that moves with the progressive wave is deflected to the right as it enters the North Sea. The wave is reflected off the coast of Belgium and the Netherlands and travels north with the right-hand deflection, this time pushing the mass of moving water against the coasts of Denmark and Norway as the wave moves toward the Atlantic Ocean. The reason tidal ranges are so much greater along the coast of Scotland and England is twofold. First, the wave possesses more energy as it moves in from the Atlantic Ocean. As the wave moves southward toward the European coast it gradually loses energy to the gently shoaling bottom, and the reflected wave contains considerably less energy than that possessed by the southbound wave. Secondly, the Kelvin wave that develops in any situation where a progressive wave travels through a broad channel has contributed to the effect. The effect of the loss of energy as the wave moves into and out of the North Sea and the tilting of the Kelvin wave can be seen in the corange lines of figure 10–22A.

Theoretically, for a basin with the dimensions of the North Sea, the tidal pattern should be approximately that shown in figure 10–22B. Two rotary tides should develop around amphidromic points, one in the north end and one in the south end of the basin. The tidal wave should travel from the north along the western margin, around the southern end of the basin, which is closed, then back along the eastern side to the north. The actual pattern that develops in the North Sea is shown in figure 10–22A and can be seen to be similar to the predicted tidal picture, with the exception that the amphidromic points are offset considerably to the east.

Tides in the Open Ocean

Tidal observations in the open oceans of the world are insufficient to give a detailed picture of the total tide that exists in each, since observations are centered on the most important harbors and on islands scattered throughout the ocean. Figure 10–19 shows the general pattern of amphidromic rotations throughout the world ocean.

Atlantic Ocean The most detailed study of any ocean tidal pattern has been done for the Atlantic Ocean. We may think of the Atlantic as a large North Sea with the opening at the south end instead of the north end. A progressive wave enters the South Atlantic from the Antarctic region and moves in a northerly direction until it encounters the shallow barriers that

FIGURE 10–22 North Sea Tide.

In *A* the cotidal lines show a much higher tidal range along the coast of the British Isles than along the coasts of Denmark and Norway. Also, the amphidromic points are offset to the east, because the tide wave enters the North Sea from the North Atlantic, and the incoming Kelvin wave is much more energetic than the outgoing Kelvin wave that moves north along the coasts of Denmark and Norway. Much of the incoming wave energy is lost as it moves into the shallow water at the south end of the North Sea. *B* shows the theoretical pattern that would develop without the energy loss described in a bay the size and shape of the North Sea. *A* and *B* from A. Defant, *Ebb and flow*, © 1958. Reprinted by permission of the University of Michigan Press, Ann Arbor.

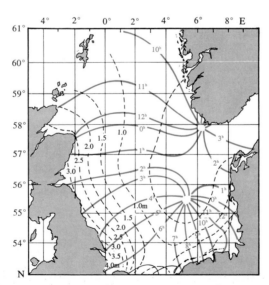

A. Cotidal and corange lines of the North Sea M_2 tide, ——: time of high water after moon's transit through Greenwich meridian; – – –: mean corange lines.

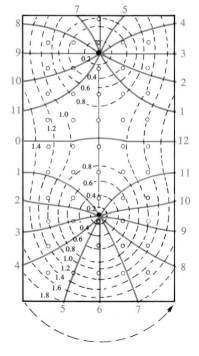

B. Theoretical cotidal lines and corange lines of the semi-diurnal tide in a bay whose length is twice its breadth. The arrow shows the direction of the earth's rotation.

FIGURE 10–23 Atlantic Ocean Tide.

From A. Defant, *Ebb and flow,* © 1958. Reprinted by permission of The University of Michigan Press, Ann Arbor.

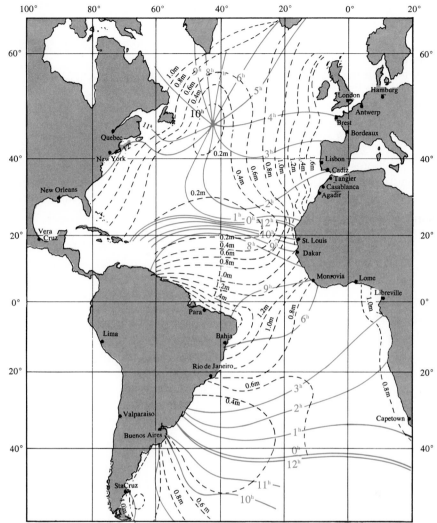

Cotidal (color) and corange lines of the M_2 Atlantic tide. The times of high water refer to the moon's transit through the Greenwich meridian; the corange lines are in meters.

separate the North Atlantic from the Arctic region. A great deal of energy is lost in these shallow waters and that which is reflected in a southerly direction produces a much lower-energy wave. As a result of energy loss and the Kelvin wave formation, the tides are slightly greater along the African and European coasts than along the North American and South American Coasts.

There are indications of three nodal regions developing. One is at about 40°S latitude, another at 20°N latitude, and the third at about 50°N latitude. The two southerly indications are minimal tidal ranges developing at these latitudes, while the third is identified by the location of an amphidromic point seen in figure 10–23. The M_2 semidiurnal lunar partial tide cotidal and corange lines are shown for the Atlantic Ocean in this illustration. The total tide picture in the Atlantic Ocean is certainly not as simple as we have

presented it here and must include various combinations of progressive and standing waves.

Tides in Rivers

The Amazon River probably possesses the longest estuarine stretch that is affected by oceanic tides. Tides can be measured as far as 800 km (546 mi) from the mouth of the Amazon, although the effects are quite small at this distance. Tidal waves that move up river mouths lose their energy owing to the decreasing depth of water and the flow of the river water against the tide during the flood interval. As the wave moves up the river it develops a steep front (fig. 10–24). This front produces a rapidly rising tide. An extreme development of this type produces a *tidal bore* in which a very steep wave front surges up the river. In the Amazon it is called *pororoca* and appears as a water-

FIGURE 10-24   River Bores.
As the tidal crest moves upriver it
develops a steep forward slope
through resistance to its advance by
the river flowing to the ocean. Such
crests (bores) may reach heights of
5 m (16.4 ft) and move at speeds
up to 22 km/h (13.6 mi/h).

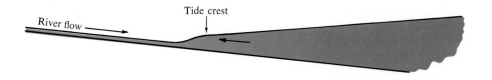

fall up to 5 m (16.4 ft) in height moving upstream at
speeds up to 22 km/h (13.6 mi/h). Other rivers that
experience bores are the Chientang in China, where
they may reach 8 m (26 ft); the Petitcodiac in New
Brunswick, Canada; the Seine in France; and the
Trent in England.

TIDES AS A SOURCE OF POWER

Another Look at the Earth-Moon System

We do not doubt that the tides possess energy, but
before we discuss this energy let us examine the
earth-moon system again so that we better understand the source of the energy of the tides. In figure
10-25 we have shown a distorted view of the earth-
moon system in which the tidal bulges of the idealized ocean are greatly exaggerated. Points a and b
mark the centers of mass for these bulges and lines
R_a and R_b the distance from each of these centers of
mass to the center of mass for the moon. As the earth
rotates in a counterclockwise direction indicated by
the arrow, it can be seen that point a is closer to the
center of the moon than is point b. Therefore, the
gravitational attraction of the moon on bulge a is
greater than that for bulge b, and the force F_a repre-

senting this attraction is greater than F_b. A *torque*,
turning force, exists that tends to slow the earth's rotation. The slower rotation gradually increases the
length of the day and results in the earth losing angular momentum.

If in any ideal system we assume conservation of
angular momentum, the momentum lost by the earth
must be gained by the moon. To compensate for this
loss of angular momentum experienced by the earth,
the moon increases its velocity, which causes it to
move farther from the center of the earth-moon center of mass, increasing its angular momentum. However, not all of the energy lost by the slowing of the
earth's rotation goes into increasing the velocity of the
moon; some goes into the kinetic energy of the tides.
We don't know how much of the total tidal energy
this is, but it is considerable, and it is all eventually
converted to heat energy through the force of friction
on the tides. Could we harness some of this energy
before it is dissipated through tidal friction?

Some Basic Considerations   The most obvious
benefit of using the tides to generate electrical power
would be in reduced operating costs as compared to
the conventional thermal power plants that require
radioactive isotopes or fossil fuels. Even though the
initial cost of the tidal power-generating plant may be
higher, there would be no ongoing fuel bill.

**FIGURE 10-25   Source of En-
ergy That May Be Extracted
from the Tides.**
Since the oceans are not deep
enough to allow the tidal bulges to
travel across the earth's surface at
the velocity in excess of 1600 km/h
(992 mi/h) necessary to keep the
bulges directly in line with the
moon, the bulges assume an equilibrium position. Frictional drag of
the ocean bottom tends to carry the
bulges in the direction of the earth's
rotation. As a result, the bulges assume positions similar to those
shown relative to the moon. Since
the center of bulge a is closer to the
moon than the center of bulge b,
the moon exerts a greater force on
bulge a than on bulge b. This creates a torque that tends to slow the
rotation of the earth.

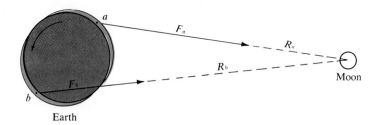

A negative consideration involves the periodicity of the tides. As generation of electricity would be our primary concern, power would be generated only throughout a portion of a 24-h day unless special design features were included in the construction of the facility. Since the tides operate on a lunar period and our energy demand operates on a solar period, the energy available through generating power from the tides would only accidentally coincide with need.

Generators must run at constant speed, and we must utilize the flow of the tidal current in two directions. It is therefore obvious that we would benefit by using turbine blades whose angle could be varied and reversed. Adjusting the pitch of the turbine blades would allow them to be set flatter when the

head of water was greatest and the angle of the blades increased as the head of water decreased, so that the generators could be maintained at a constant speed. Since the maximum head would usually not be great, the flow channels constructed through dams would have to be of small diameter. This would require a series of small channels and small turbines, which would be less efficient than one large machine.

Let us examine the tidal power plant constructed in the estuary of La Rance River off the English Channel in France (see chapter opening photograph). The estuary shown in figure 10–26A has a surface area of approximately 23 km² (8.9 mi²) and the tidal range at La Rance reaches a maximum of 13.4 m (44 ft). Usable tidal energy is proportional to the area of the

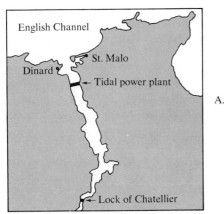

La Rance Estuary

A.

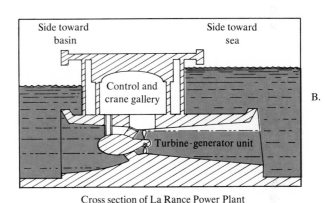

Cross section of La Rance Power Plant

B.

C. Turbine generating unit with variable-angle turbine blades.

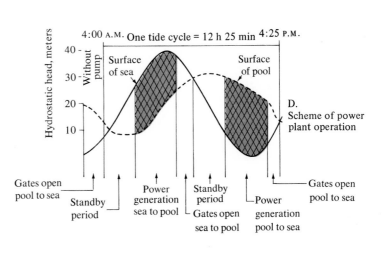

D. Scheme of power plant operation

FIGURE 10–26 La Rance Tidal Power Plant.

A, B, and *D* from *The tides* by Edward P. Clancy, © 1968 by Doubleday & Company, Inc. Reprinted by permission of Doubleday & Co., Inc. Photo courtesy Michel Brigaund, French Embassy.

basin and to the square of the amplitude of the tide. A barrier was built across the estuary a little over 3 km (1.8 mi) upstream where it is 760 m (2493 ft) wide to protect it from storm waves. The deepest water ranges from just over 12 m (39.4 ft) at low tide to more than 25 m (82 ft) at high tide. To allow water to flow through the barrier when the generating units are shut down, sluices (artificial channels) with adjustable vanes 10 m (32.8 ft) high and 15 m (49.2 ft) wide have been built into the barrier.

The bottoms of the conduits containing the generating units are 10 m (32.8 ft) below the surface of the water at the lowest tide, as can be seen in figure 10–26B. The conduits containing the units are 53 m (174 ft) long, and the cross-sectional area at each end is about 93 m^2 (995 ft^2). The turbine-generator units resemble large fat torpedoes that are surrounded by water and held in place by radial struts. Each unit has a generating capacity of 10,000 kW at 3.5 kV (fig. 10–26C).

Figure 10–26D shows the operational scheme of the installation during a complete tidal period. The plant generates electricity only during about one-half of the tidal period when sufficient head exists between the pool and the ocean. Annual power production of about 540 million kWh without pumping can be increased to 670 million kWh by using the turbine-generators as pumps at the proper times to increase the head of water.

Although some engineers think a tide-generating plant proposed for Passamaquoddy Bay near the United States-Canadian border at the south end of the Bay of Fundy could be made to generate electricity constantly, others who have studied the project are less optimistic. Potentially, the usable tidal energy seems great compared with La Rance because the volume of flow is about 117 times greater than that of La Rance. Regardless of whether tide-generating units are ever constructed on a large scale, this potential source of energy will receive increased attention as the cost of generating electricity by conventional means increases. Figure 10–27 shows the locations of some sites around the world that have potential for tidal electrical power generation.

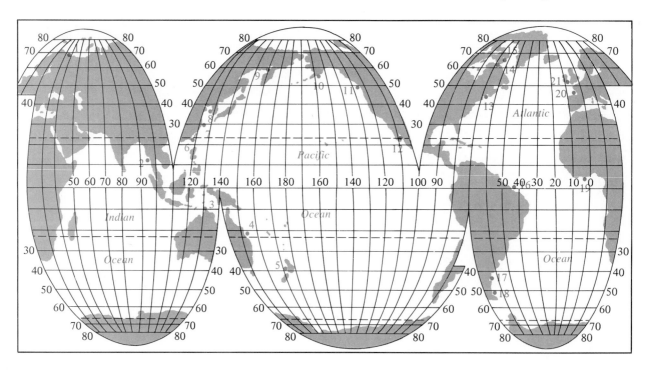

1. Mezan/Kislaya	8. Asan Bay	15. Frobisher Bay
2. Rangoon	9. Sea of Okhotsk	16. Sao Luis
3. Darwin	10. Cook Inlet	17. Golfo San Jorge
4. Broad Sound	11. Strait of Georgia	18. Straits of Magellan
5. Aukland	12. Gulf of California	19. Abidjan
6. Amoy	13. Bay of Fundy/Passamaquoddy Bay	20. Rance River/Chausey I.
7. Shanghai	14. Ungava Bay	21. Severn River

FIGURE 10–27 Sites with Major Potential for Tidal Power Generation.
Base map courtesy of National Ocean Survey.

SUMMARY

The tides of the earth are derived from the gravitational attractions of the sun and moon. The moon has about twice the tide-generating effect of the sun. Small horizontal forces tend to push water into two bulges, one at the earth's **zenith** and one at the **nadir** relative to the **tide-generating body**—the sun or moon. Since the moon bulges are dominant, the tides we observe on earth have periods dominated by lunar motions and modified by the changing position of the sun bulges.

If the earth were a uniform sphere covered with an ocean of uniform depth, and if we could ignore some of the considerations of physical motion, the tides on earth would be those predicted by the **equilibrium theory** of tides. Such a tide would have a **tidal period** of 12 h 25 min, or half a lunar day. A tide with maximum **tidal range** would occur each **new moon** and **full moon**, and tides with minimum range would occur with the **first and third quarter** phases of the moon. These would be the **spring** and **neap tides**, respectively. Since the moon may be as much as 28.5° north or south of the equator on a monthly cycle and the sun is directly over the equator only two times per year, the tidal bulges would usually be located so as to create two high tides of unequal height per lunar day. The same may be said for the low tides. Tidal ranges are greater when the earth is at **perihelion** in its orbit around the sun and when the moon is at **perigee** in its orbit around the earth because the tide-generating bodies are then closest to the earth.

Since the earth has an irregular surface with continents dividing the world ocean into irregularly shaped basins, the tides we actually observe on earth are explained by the **dynamical theory** of tides. The basic types of tides observed on the earth are a **diurnal tide** with a period of one lunar day, a **semidiurnal tide** with a period of half a lunar day (like that predicted for the equilibrium tide), and a **mixed tide** with characteristics of both. Mixed tides are usually dominated by semidiurnal periods and display a significant **diurnal inequality**. The diurnal inequalities are greatest when the moon is over the tropics and least when it is over the equator.

All basins have a characteristic free standing wave, or **seiche**, depending on their length and depth. It usually is not very high for small basins, but if it is in phase with the forced standing wave created by the tide-generating bodies, the height can be significant. If the basin is wide enough and in the Northern Hemisphere, the standing wave may be converted to a progressive wave rotating in a counterclockwise direction around an **amphidromic point**, a point of zero tidal range. Tidal currents follow this rotary pattern in open-ocean basins but are converted to reversing currents at the margins of continents. The maximum velocity of reversing currents occurs during ebb and flood currents when the water is halfway between high and low standing waters. The effects of shoaling and narrowing on the tide can be observed in the **Bay of Fundy**. The development of multiple amphidromic points can be observed in the **North Sea. Tidal bores,** tide waves that force their way up rivers, are common in such rivers as the Amazon, Chientang, Seine, and Trent.

Since tides can be used to generate power without the requirement of fossil or nuclear fuel, the possibility of constructing such generating plants has always been attractive to engineers. One such plant is located in the estuary of **La Rance River** in France and is operating satisfactorily.

QUESTIONS AND EXERCISES

1. Explain why the sun's influence on the earth's tides is only 46% that of the moon, even though the sun exerts a gravitational force on the earth 177 times greater than that of the moon.

2. Describe how the centripetal force *required* to keep particles of the earth rotating in their identical orbits within the earth-moon system varies from the centripetal force *provided* by the gravitational attraction between the particles and the moon. Note: Consider strength and direction of the force.

3. Construct a diagram of vector arrows to show how the horizontal tide-generating force results from the difference in the *required* centripetal force, C, for a particle on the earth's surface at 45° from the zenith, and the centripetal force *provided* by the gravitational attraction between the moon and that particle, G.

4. Discuss why the length of the lunar day is 24 h 50 min of solar time.

5. Explain why the maximum tidal range (spring tide) occurs during new and full moon phases and the minimum tidal range (neap tide) with quadratures.

6. Discuss the length of cycle and degree of declination of the moon and sun relative to the earth's equator. Include a discussion of the effects of precession of the plane of the moon's orbit through the ecliptic.

7. Describe the effects of the declination of the moon and sun on the world ocean tides.

8. Diagram the moon's orbit around the earth and the earth's orbit around the sun. Label the positions on the orbits at which the moon and sun are closest to and farthest from the earth, stating the terms used to identify them. Discuss the effects that the moon and earth being in these positions have on the earth's tides.

9. Define tropical and equatorial tides. Include the concept of diurnal inequality (*see* the days of September 1, 2, 22, and 23 in fig. 10–15).

10. Describe the period and diurnal inequality of the following: diurnal tide, semidiurnal tide, and mixed tide.

11. . What forces produce forced and free standing waves in lakes and narrow ocean embayments?

12. In narrow embayments, standing waves may develop as a result of ocean tides entering from the ocean. In wider embayments, a rotating progressive wave may develop instead. Discuss how the increased width of the embayment allows this rotary wave to develop.

13. Describe the velocity of tidal currents in relationship to high and low tidal extremes.

14. Using a diagram, explain how development of a Kelvin wave could cause the tidal range on the south side of the Bay of Fundy to be greater than that on the north side.

REFERENCES

Clancy, E. P. 1969. *The tides: Pulse of the earth.* Garden City, N.Y.: Doubleday.

Defant, A. 1958. *Ebb and flow: The tides of earth, air and water.* Ann Arbor: University of Michigan Press.

Pond, S., and Pickard, G. L. 1978. *Introductory dynamic oceanography.* Oxford: Pergamon Press.

Sverdrup, H. U.; Johnson, M. W.; and Fleming, R. H. 1942. Renewal 1970. *The oceans: Their physics, chemistry, and biology.* Englewood Cliffs, N.J.: Prentice-Hall.

von Arx, W. S. 1962. *An introduction to physical oceanography.* Reading, Mass.: Addison-Wesley.

SUGGESTED READING

Sea Frontiers

Sobey, J. C. 1982. What is sea level? 28:3, 136–142.
The role of tides and other factors in changing the level of the ocean surface is discussed.

Zerbe, W. B. 1973. Alexander and the bore. 19:4, 203–8.
An account of Alexander the Great's encounter with a tidal bore on the Indus River.

Scientific American

Goldreich, P. 1972. Tides and the earth-moon system. 226:4, 42–57.
The tide-generating force of the sun and moon on the earth and the effect of transfer of angular momentum from the earth to the moon as a result of tidal friction are discussed. Also considered are theories of lunar origin.

Lynch, D. K. 1982. Tidal bores. 246:4, 146–57.
The tidal bore phenomenon is explained.

Moving from the oceans onto the continent, one encounters the *shore,* which is the zone that lies between the low-tide *shoreline* and the highest elevation on the continent that is affected by storm waves, the *coastline* (fig. 11–1). The *coast* extends from the landward limit of the shore inland as far as features that seem to be related to marine processes can be found. The width of the coast may vary from less than one kilometer to many tens of kilometers. As the waves beat against the shore they cause erosion, which produces sediment that will be transported along the shore and deposited in the low-energy areas.

CHAPTER ELEVEN GEOLOGY OF THE COASTAL REGIONS

GENERAL DESCRIPTION OF THE COASTAL REGION

The physiographic features of a coastal region can be classified as being primarily the products of either erosion or deposition, although most coasts possess features produced by both processes. Coasts dominated by erosional features are usually exposed to high-energy wave action, while those displaying well-developed depositional features are low-lying protected areas.

Erosional Shore Features

The landward limit of a shore that is dominated by erosion will normally be marked by a cliff. The coastline, which marks the boundary between the shore and the coast, will be a line along the cliff, connecting points at which the highest effective wave action takes place.

The shore is divided into the *foreshore,* that portion exposed at low tide and submerged at high tide, and the *backshore,* which extends from the normal high-tide shoreline to the coastline. The shoreline migrates back and forth with the tide and is the water's edge. The *nearshore* zone is that region between the low-tide shoreline and breakers. Beyond the low-tide breakers is the *offshore* zone (fig. 11–1).

The *beach* consists of the wave-worked sediment that moves along the shore. It may extend from the coastline to the low-tide line of breakers (fig. 11–1).

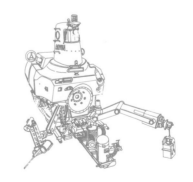

The California coast at Big Sur. Photo by Michael DiSpezio.

FIGURE 11–1 Landforms and Terminology of Coastal Regions.

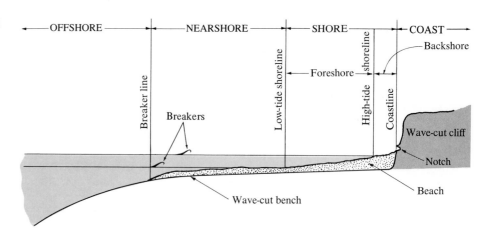

Wave Erosion Owing to refraction (bending of waves discussed in chapter 9), wave energy is concentrated on *headlands* that jut out from the continent, while the amount of energy reaching the shore in *bays* is reduced. As the waves concentrate their energy on the headlands, erosion occurs and the shoreline retreats. The greatest concentration of wave energy, on a day-to-day basis, is in the foreshore region. However, during the rare periods when storm waves batter the shore, more erosion may occur across the entire shore in one day than may be achieved by average wave conditions in a year.

The cliff shown in figure 11–1 is referred to as a *wave-cut cliff,* which has been produced by wave action cutting away its base. The cliff develops as the upper portions collapse after being undermined by wave action. The undermining may be evident in the form of a notch at the base of the cliff that may be characterized by *sea caves.* Such caves are most commonly cut into hard sedimentary rock. Further wave action eroding the softer portions of the rock outcrops may develop the caves into openings running through the headlands, *sea arches.* Continued erosion and crumbling of arches will produce *stacks.* Such remnants rise from the relatively smooth *wave-cut bench* cut into the bedrock by wave erosion (fig. 11–2). The rate of erosion by waves is determined by a number of variables:

1. One of the most important variables is the *degree of exposure* of the coastal region to the open ocean. Coasts that are fully exposed receive higher-energy wave action and are likely to have rugged cliffs in areas of high topographic relief.

2. The *tidal range* is also an important variable affecting the amount of wave erosion. Given the same amount of wave energy, a region with a small tidal range will erode much more rapidly than one with a large tidal range that allows the wave energy to be spread over a much broader shore. Although high-velocity tidal currents may develop in areas where a large tidal range exists, these currents are of limited importance in eroding the coastline.

3. The *composition of coastal bedrock* is very significant. Crystalline igneous rocks, such as granite, as well as metamorphic and hard sedimentary rocks, are relatively resistant and generally produce rugged shoreline topography. Weak sedimentary rocks, such as sandstone and shale, are more easily eroded, and a gentler topography associated with more extensive beach deposits is produced by their erosion.

Regardless of the rate of erosion, all coastal regions follow the same developmental path. As long as

FIGURE 11–2 Coastal Erosional Features.

there is no change in the elevation of the landmass relative to the ocean surface, the cliffs will continue to retreat, the benches will widen, and the eroded material will be carried from the high-energy areas and deposited in the low-energy areas.

Depositional Shore Features

The coastal erosion we have just discussed, as well as erosion being carried on by running water inland, produces large amounts of sediment that must be distributed along the continental margin. As waves strike the shore at an angle, they set up a longshore movement of water—the *longshore current.* The velocity of the longshore current increases with increasing beach slope, angle of breakers with the beach, wave height, and decreasing wave period.

This current moves parallel to the shore between the shoreline and the breaker line, carrying with it the materials that make up the beach. At the landward margin of the surf zone the *swash,* a thin sheet of water, moves sediment onto the exposed beach at an angle, but the force of gravity causes the backwash

to carry the sediment straight down the beach face. As a result, the pebbles and sand grains that are transported by the swash move in a zigzag pattern along the shore in the same direction as the longshore current within the surf zone (fig. 11–3).

Longshore drift is a term that has been applied to the movement of sediment by the processes just described. The amount of longshore drift in any coastal region is determined by an equilibrium between erosional and depositional forces. Any interference with the movement of sediment along the shore will destroy this equilibrium and result in a new erosional and depositional pattern determined by the nature of the interference.

Rip Currents As longshore-current water moves onto the shore it must eventually run back into the ocean. This backwash of water finds its way into the open ocean as a thin sheet flow across the ocean bottom or in local *rip currents* that occur perpendicular to or at an angle to the coast where topographic lows or other conditions allow their formation. Rip currents may be less than 25 m (82 ft) wide and can attain

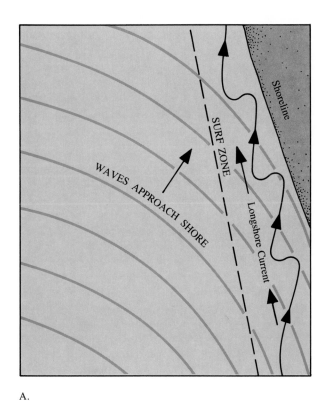

A.

B.

FIGURE 11–3 Longshore Drift and Rip Currents.

A, longshore current. As waves approach shore from a southerly direction, they produce a longshore current that flows north. Water in the current follows a zigzag path as waves push it up the beach slope from the direction of approach. The water runs back down the slope under the influence of gravity. *B,* rip currents moving seaward along the California coast. Photo courtesy of Scripps Institute of Oceanography, University of California/San Diego.

velocities of 7 to 8 km/h (4 to 5 mi/h) but do not travel far from shore before they break up. If a light moderate swell is breaking, numerous rip currents moderate in size and velocity may develop. A heavy swell will usually produce fewer, more concentrated rips (fig. 11–3).

Beach Composition The material found in a beach deposit will depend on the source of the sediment that is locally available or transported by the longshore drift. In areas where the sediment is provided by coastal mountains, the beaches will be composed of minerals contained in the rocks of those mountains and may be of relatively coarse texture. If the sediment is provided primarily by rivers that drain lowland areas, sediment that reaches the coastal regions will normally be finer in texture; in many cases mud flats develop along the shore because only clay- and silt-sized particles are emptied into the ocean. In low-relief low-latitude areas such as southern Florida, where there are no mountains or other sources of rock-forming minerals nearby, most of the material of the beaches is derived from the remains of the organisms that live in the coastal waters. Beaches in these areas will be composed predominantly of calcareous algae and shell fragments as well as the remains of microscopic animals, particularly foraminifers. Many volcanic island beaches in the open ocean will be composed of dark-colored fragments of the basaltic lava that makes up the islands or of coarse fragments of coral debris from the reefs that develop around the margins of islands in low latitudes.

Beach Slope The slope of beaches is closely related to the size of the particles of which they are composed. Waves washing onto the beach carry sediment, thereby increasing the slope of the beach. If the backwash returns as much sediment as the waves carried in, the beach has reached equilibrium and will not steepen. A beach composed of fine-grained sand that is relatively angular will have a gently-sloping, firm surface. Because these small grains interlock closely, little of the swash sinks down among the grains. Most of it runs back down the slope to the ocean, possessing enough energy to maintain equilibrium on a gentle slope. The backshores of such beaches are usually nearly horizontal.

Beaches composed of coarse sands or pebbles will usually contain more-rounded particles that are more loosely packed. The swash can quickly percolate into such deposits when it moves up the beach slope. Deposition of particles by the swash will continue until the beach slope becomes steep enough that the

TABLE 11–1 The Relationship of Particle Size to Beach Slope.

Wentworth Particle Size	(mm)	Mean Slope of Beach
Cobble	256	24°
Pebble	64	17°
Granule	4	11°
Very coarse sand	2	9°
Coarse sand	1	7°
Medium sand	0.5	5°
Fine sand	0.25	3°
Very fine sand	0.125	1°
	0.063	

Source: After Table 9, "Average beach face slopes compared to sediment diameters" from *Submarine geology*, 2nd ed. by Francis P. Shepard (Harper & Row, 1963), p. 171.

backwash running to the ocean has sufficient energy to maintain equilibrium. Such beaches are usually much less firm than beaches composed of finer material, and the backshores will slope significantly toward the coastline (table 11–1).

Special Sedimentary Features Numerous depositional features that are partially or wholly separated from the shore are deposited by the longshore drift and other processes that are not well understood. A *spit* is a linear ridge of sediment attached at one end to land. The other end of the deposit points in the direction of longshore drift and ends in the open water. Spits are simply extensions of beaches into the deeper water near the mouth of a bay. The open-water end of the spit will normally curve into the bay as a result of current action.

If tidal currents or the currents initiated by river runoff are too weak to keep the mouth of the bay open, the spit may eventually extend across the bay and tie to the mainland to completely separate the bay from the open ocean. The spit then has become a *bay barrier*. A *tombolo* is a sand ridge that connects an island with another island or the mainland. Such a feature probably results from the growth of a spit (fig. 11–4).

Barrier islands are long offshore deposits of sand lying parallel to the coast. The origin of a barrier island is complex, and it may be that there are several explanations for its existence. It appears, however, that many such structures developed during the rise in sea level that began with the last melting of the glaciers some 18,000 years ago.

Barrier islands are nearly continuous along the Atlantic Coast of the United States. They extend around Florida and along the Gulf of Mexico coast, where they may be found well past the Mexican border. Bar-

FIGURE 11–4 Coastal Depositional Features.

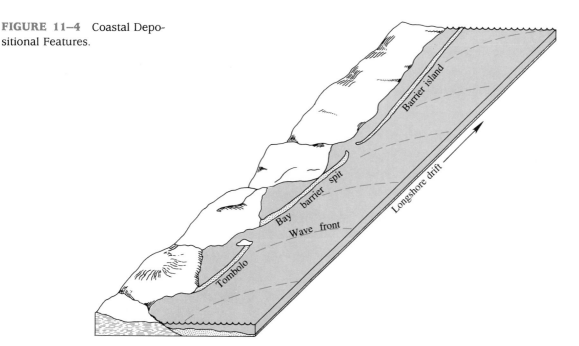

rier islands may attain lengths in excess of 100 km (62 mi) and have widths of several kilometers. Examples of such features are Fire Island off the New York coast and Padre Island off the coast of Texas. The seaward side of a well-developed barrier island will have a beach structure typical of the continental shore, and a dune development usually exists between the seaward shore and the barrier flat that leads into the lagoon behind the island. Barrier islands are an integral part of some estuarine environments. They will be described in more detail as their relationship to this environment is discussed in chapter 12.

Deltas Some rivers carry more sediment than can be distributed by the longshore current. Such rivers develop a *delta deposit* as sediment settles out at the river mouth. One of the largest such features is that produced by the Mississippi River. Deltas are fertile, flat areas that are subject to periodic flooding.

Delta formation begins when a river has filled its estuary with sediment. Once the delta forms, it grows through the distribution of sediment by *distributaries,* branching channels that radiate out over the delta. Distributaries lengthen as they deposit sediment and produce fingerlike extensions to the delta. When the fingers get too long, they become choked with sediment. At this point, a flood may easily cause a shift in the distributary course and provide sediment to the low-lying areas between the fingers.

In areas where depositional processes are dominant over coastal erosion and transportation processes, the "bird foot" Mississippi-type delta results.

Where erosion and transportation processes exert a significant influence, the shoreline of the delta will be smoothed to a gentle curve like that of the Nile Delta (fig. 11–5). The Nile Delta is presently eroding owing to the entrapment of sediment behind the Aswan High Dam, which was completed in 1964. Erosion is only one of the negative effects of the construction of this dam. This erosion, however, may have made possible the discovery of the ruins of ancient Alexandria beneath the Mediterranean waters at the edge of the delta.

CHANGING LEVELS OF THE SHORELINE

Along some coasts there are flat platforms, called *marine terraces,* backed by cliffs. Stranded beach deposits and other evidence of marine processes may exist many meters above the present shoreline. These features characterize *shorelines of emergence* that have reached their present positions relative to the existing shoreline by an uplift of the continent, by the lowering of sea level, or by a combination of the two.

In other areas one may find beneath the water overlying the continental shelf *drowned beaches* and *submerged dune topography*. These features along with the *drowned river mouths* along the present shoreline indicate *shorelines of submergence*. This submergence must have been caused by a subsidence of the continent, a rise in sea level, or a combination of the two.

Attempts to determine the causes of the change in relative level of the ocean and the continent have not

FIGURE 11–5 Deltas.

A, photographed from *Gemini IV.* The relatively smooth, curved surface of the Nile Delta can be seen. The Mediterranean Sea is to the left, and the northern Red Sea is in the upper right. Photo courtesy of NASA. *B,* digitate structure of the Mississippi River Delta results from low-energy environment. Location of present main channel and older main channels (Atchafalaya River and Bayou Lafourche) show how river shifts position with time.

A.

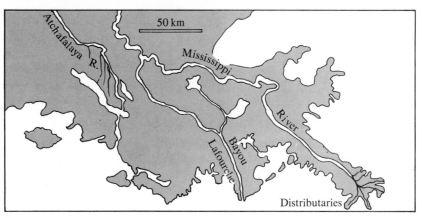

B.

met with great success. Whether the shoreline has become submerged because of a rising sea level or a subsiding continent in a particular region cannot be determined by examining the coastal features in that area, since both processes produce the same end results (fig. 11–6).

Tectonic Movements

Changes in sea level relative to the continent may occur because of movement of the land—tectonic movement. Such movement includes large-scale uplift or subsidence of large portions of continents or ocean basins or more localized deformation of the continental crust involving folding, faulting, and tilting. The earth's crust also responds isostatically to the accumulation or removal of heavy loads of ice, sediment, or lava.

There is evidence that during the last 2.5 to 3 million years at least four major accumulations of glacial ice developed in high latitudes. Although Antarctica is still covered by a very large glacial accumulation, much of the ice cover that once existed in northern Asia, Europe, and North America has disappeared. The most recent period of melting began about 18,000 years ago. Accumulations of ice that were up

FIGURE 11–6 Evidence of Changing Levels of the Shoreline.

Marine terraces resulting from ancient sea cliffs and wave-cut benches being exposed above present sea level mark ancient shorelines, as do drowned beaches that lie below sea level.

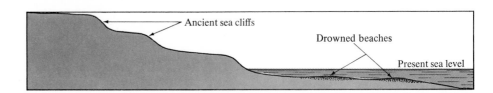

to 3 km (2 mi) in thickness have disappeared from northern Canada and Scandinavia. While these areas were beneath the thick ice sheet, they were pushed down and are still in the process of recovery after the melting of the ice. Some geologists believe that by the time the *isostatic rebound* is finished, the floor of Hudson Bay, which is now about 150 m (492 ft) deep, will be above sea level. There is evidence in the Gulf of Bothnia, between Sweden and Finland, of 275 m (902 ft) of isostatic rebound during the last 18,000 years. Generally, tectonic changes in the level of the shoreline are confined to a segment of the shoreline of a given continent.

Eustatic Movements

Changes in the level of the shoreline that can be measured on a worldwide basis because they are caused by the increase or decrease of water volume in the ocean or the capacity of the ocean basin are termed *eustatic*. This term refers to the highly idealized situation in which all of the continents remain static while the sea rises or falls. Small changes in sea level could be created by the formation or the destruction of large inland lakes. Probably more important is the locking into or the release from continental glaciers of the earth's water during glacial and interglacial stages.

Changes in sea-floor spreading rates can change sea level. Fast spreading produces larger rises like the East Pacific Rise that will displace more water than slow-spreading ridges like the Mid-Atlantic Ridge. Thus, fast spreading produces a rise in sea level.

During the Pleistocene Epoch, when the previously mentioned glacial advances were occurring, the amount of water in the ocean basin fluctuated considerably. Since the climate was colder during ice advances, we might account for some of the lowering of the shoreline by the contraction of the ocean volume as its temperature decreased. It has been calculated that for every 1°C (1.8°F) decrease in the mean temperature of the ocean water, sea level would drop 2 m (6.6 ft). Temperature indications derived from study of fossils from Pleistocene ocean sediments indicate the ocean surface temperature may have been as much as 5°C (9°F) lower than at present. Therefore, contraction of the ocean water may have lowered sea level by about 10 m (33 ft).

Although it is difficult to say with certainty what the range of shoreline fluctuation was during the Pleistocene, there is cause to believe that the shoreline was at least 120 m (394 ft) below the present shoreline. It is also estimated that if all the remaining glacial ice on earth were to melt, sea level would rise by another 60 m (197 ft). This would give a minimum possible range of sea level during the Pleistocene of 180 m (590 ft), most of which must be explained through the capture and release of the earth's water by glaciers. Such changes in sea level are termed *glacioeustatic oscillations.*

During the last 18,000 years the ocean volume has been increasing owing to the expansion of the water resulting from warmer temperature and the melting of polar sea ice and glacial ice. Since the combination of tectonic and eustatic changes in sea level may be very complex and difficult to identify, it is hard to classify coastal regions as purely emergent or submergent. Most coastal areas show evidence of having experienced both submergence and emergence in the recent past. It is believed, however, that sea level has not risen significantly as a result of melting glacial ice during the last 3000 years.

THE UNITED STATES COASTS

If there has been no significant change in sea level due to glacioeustatic oscillations during the last 3000 years, we must look to tectonic processes to explain the changes in the shoreline relative to the continents during the recent past. To understand the underlying tectonic cause of emergence or submergence, we must return to the concept of global plate tectonics.

Atlantic-type (Passive) Margins

Considering only the coasts of North America, we find the Atlantic Coast is subsiding and the Pacific Coast is emerging. Beginning with the breakup of Pangaea and the presently forming Atlantic Ocean, this emergence-submergence pattern can be readily understood. It was previously discussed that the lithosphere thickened, the ocean deepened, and heat flow decreased as the lithospheric plates cooled with age or with increasing distance from the spreading centers. As a result of this process, the ocean floor of the North American Atlantic Coast is thought to have subsided about 3 km (2 mi) over the last 150 million years.

Erosion of the continent produced sediment that has accumulated to a maximum thickness of 15 km (9.3 mi) along the Atlantic Coast. This thick deposit of sediment has been made possible by the plastic asthenosphere allowing the lithosphere to flex downward as the sediment burden is increased. It is this thick sedimentary wedge underlying the continental shelf, slope, and rise that has been exploited for its large petroleum reserves in the Gulf of Mexico.

As Pangaea initially split, the granitic, continental-crustal rocks near the split were heated enough to stretch. This heating produced a gradually thinning continental or transitional crust that gave way to oceanic crust 100 to 150 km (62 to 93 mi) from the present shore beneath the continental rise (fig. 11–7). Thus, the thick wedge of sediment underlying the continental shelf, slope, and rise is deposited on marginal continental crust, transitional crust, and oceanic crust.

This describes the general conditions producing the subsidence of the Atlantic and Gulf coasts. Although the rate of thermal subsidence decreases with time and the rate of subsidence due to sediment loading depends on sediment supply, it is never reversed.

Pacific-type (Active) Margins

In contrast to the *passive* and subsiding Atlantic-type margin, the Pacific-type margin, or *active* margin, is the scene of intense tectonic activity. Along such margins, the eruption of volcanoes and shock of earthquakes are common. The thick, broad sediment wedge characteristic of the Atlantic Coast is not well developed. Instead of being the product of the creation of a new ocean basin, the Pacific-type margin is the scene of lithospheric plate destruction.

The characteristic alignment of mountain ranges parallel to the continental margin can be seen. In all cases of continental margins associated with plate convergence, the compressive forces produce uplift. Along the California coast, the process has been further complicated with the development of the San Andreas Fault, a transform fault. This fault has made possible the local subsidence of the depositional basins in the Los Angeles area. But on the broader scale, emergence along Pacific-type margins is the characteristic condition.

A discussion of the Gulf of Mexico, Atlantic, and Pacific coasts will combine the effects of submergence and emergence that result from the plate tectonic interactions with the conditions of erosion and deposition along these coasts. The sum of these processes along these coastal regions will provide a partial explanation of how the features found along specific segments of the U.S. coast have developed.

The Gulf Coast

The Louisiana-Texas coastline is dominated by the Mississippi River Delta, which is being deposited in a microtidal and generally low-energy environment. The tidal range is normally less than 1 m (3.3 ft), and with the exception of the hurricane season, wave action is generally low. The sediments that have been

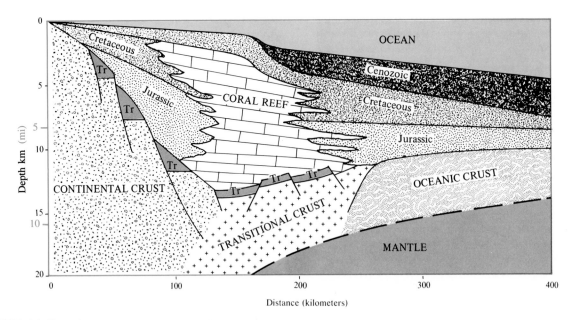

Distance (kilometers)

FIGURE 11–7   Atlantic-type Margin Crustal and Sediment Units.
As Pangaea was rifted apart by upwelling magma, the continental crust was heated enough to flow plastically and thinned at the edge of the continent, producing a transitional crust that gave way to oceanic crust. The Triassic sediments filling the initial rift valleys are overlain by Jurassic, Cretaceous, and Cenozoic sediments that reach thicknesses of 10 to 15 km (6.2 to 9.3 mi) as the lithosphere flexes down in the asthenosphere under the increasing burden of the overlying sediment.

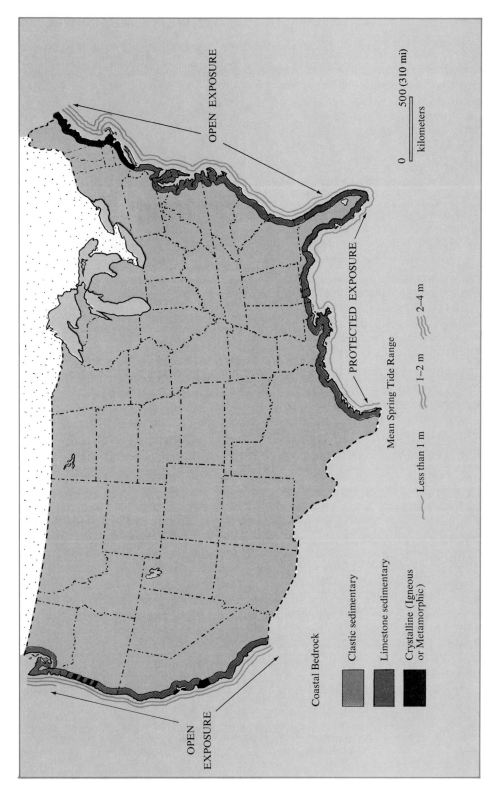

FIGURE 11—8 Distribution of Conditions Related to Coastal Erosion.

deposited in the Gulf Coast region indicate that this area has been generally subsiding for approximately 150 million years. The sedimentary section that can be found beneath the coast is essentially continuous from the present back to the Cretaceous Period. For these sediments to be deposited, preserved, and overlain by younger sediment, at least a low rate of subsidence was required. Even though the Gulf Coast is subsiding, the shoreline is building seaward in some regions. The tremendous volume of sediment that is being carried by the Mississippi River and deposited in its delta has extended the delta completely across the continental shelf (fig. 11–8).

Along the Louisiana Gulf Coast a nip-and-tuck battle is going on between the forces of deposition, attempting to advance the coastal region, and submergence, threatening to make the coastline retreat inland. Locally, one process may be winning for a short period of time. Areas some distance from the delta cease to receive new sediment and begin to subside, while others are being actively advanced by deposition. In the city of New Orleans, it obviously would not be desirable to allow the Mississippi River to continue to add sediment. Without this deposition, the city must subside. Therefore, a constant battle must be waged to keep the area sufficiently diked to keep out the ocean and the river.

A chain of barrier islands extends along the coast east of the Mississippi Delta to the Florida panhandle and west of the delta to the Mexican border. These deposits, some of which are over 100 km (62 mi) long and more than 2 km (1.2 mi) wide, are separated from the mainland by shallow lagoons and indicate the subsidence (fig. 11–9).

The Atlantic Coast

Moving east to the Florida coastline, we encounter a gradually subsiding area with very low relief. The bedrock of most of the state of Florida is a resistant impervious limestone that was alternately inundated and totally exposed numerous times during the Pleistocene. The area of the greatest ecological significance lies at the southern tip of the state—the Everglades. This coastal region is also a relatively low-energy area. It is protected by the Florida Keys from large waves and receives high-energy assault only during local thunderstorms and hurricanes, as is generally the case for the Mississippi Delta region.

Drainage of surface water across the Everglades from northern Florida once occurred as a thin sheet less than 1 m (3.3 ft) thick that flowed south along the gentle slopes of the Everglades to the ocean. This

FIGURE 11–9 Galveston Bay.
Apollo 9 view from an altitude of 191 km (119 mi) of the Texas coast from Freeport to Sabine Lake. The Bolivar Peninsula and Galveston Island extend, respectively, to the north and south of Galveston Bay. Courtesy of NASA.

flow was not restricted to river channels but covered essentially the breadth of the state. Within this unique environment developed an assemblage of plant and animal life that is currently threatened.

To control flooding, the U.S. Army Corps of Engineers restricted the sheet runoff to flood control canals that have left portions of the Everglades short of water. This shortage of water has endangered much of the plant and animal life. Also resulting from this controlled flow of water is a threat that was previously seldom known in the Everglades. Fire has become a serious problem. Since the total relief in the Everglades area is less than 10 m (33 ft), it is probable that natural changes in sea level will, in time, exert stress on the organisms found in the Everglades.

Moving north along the Atlantic Coast from Florida to Maine, we encounter coastal regions where the sea level is rising an average of 0.3 m (1 ft) per century relative to the subsiding land. Although some rise in the continent may well have occurred along the coast from New York through Maine as a result of the melting of the continental glacier that once covered this region, evidence indicates a gradually retreating shore to the south.

The coastal rocks from Florida through New Jersey are predominantly poorly consolidated sedimentary rocks formed in the recent geologic past. These rocks do not provide great resistance to erosion and are

CAN WE LEARN TO LIVE WITH OUR CHANGING SHORES?

Recent studies have shown that there has been a 10-cm (4-in) rise in sea level during the last century (fig. 11A). Investigators believe most of this rise comes from thermal expansion of ocean water as a result of increasing surface temperatures. This temperature increase may well be the result of increasing concentrations of CO_2 in the atmosphere. A continued rise of sea level by as much as another 30 cm (12 in) *could* occur in the next 70 years. Should temperatures rise significantly, the Antarctic ice sheet could begin to melt and increase the rate of rise in sea level.

If, as some investigators predict, sea level does continue to rise, it could create major problems for low-lying coastal areas. Fresh groundwater supplies could become salty, and the threat of damage from storm surge could increase dramatically. Many coastal communities along the Atlantic and Gulf Coast shores would be seriously threatened, because in addition to the possible danger from a rise in sea level, these coastal areas have been subsiding at an average rate of 3 cm/1000 yr (1.2 in/1000 yr) for the last 150 million years.

The city of New Orleans, Louisiana, provides us with an example of the problems that may face many low-lying coastal cities. Although the city is now protected from direct ocean waves by 48 km (30 mi) of salt marsh and has a levee and sea wall to keep out water from the Mississippi River and Lake Pontchartrain, respectively, a rising sea could be devastating to the city (fig. 11B). The subsiding marshes are getting little sediment to build them up because they are separated from the flow of the Mississippi River by levees. Should the marshes become submerged, the ocean would be at New Orleans' door. This is a serious problem because parts of the city are already 2 m (6.6 ft) below sea level.

FIGURE 11A Sea-Level Change from 1880 to 1980. Tide-gauge data averaged over 5-year periods show that global mean sea level has increased about 10 cm (4 in) over the last 100 years. After V. Gornits, S. Lebedeff, and J. Hausen, *Science* 215, pp. 1611–14, 1982.

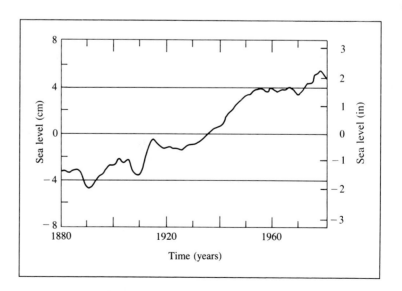

To keep the city dry, 22 pumps 4.3 m (14 ft) in diameter and 12 pumps 3.7 m (12 ft) in diameter must pump a minimum of 113,599 m^3 (4,010,045 ft^3) of water per day. In time, with continued subsidence of the land and rise of sea level, maintaining the pumping rate necessary to keep the city from flooding could become increasingly difficult. Many other coastal cities from Brownsville, Texas, to Long Island, New York, would also suffer damage from encroachment of the ocean (fig. 11C).

Although the Pacific Coast is rising tectonically, open exposure to the storm waves of the Pacific Ocean results in heavy erosion and property damage there, too.

A major part of the problem along the Atlantic and Gulf coasts has resulted from the development of barrier islands, which undergo natural shoreward migration as the sea rises relative to the continent along the Atlantic and Gulf coasts. Sea walls have been built to stabilize the islands, but they are only a temporary solution to protecting developed areas. Unless sand is continually replenished on the seaward side of the sea wall, the beaches will erode and the sea walls will tumble into the ocean.

The general problem of maintenance of coastal construction is becoming a very sensitive issue because a large amount of public money is being spent to protect private property. In a sense, all areas lived in and built upon by humans are on loan from nature, but in the case of barrier islands that loan is very short-term. As the cost of protecting developed areas increases, a point

FIGURE 11B New Orleans Area.

New Orleans, parts of which lie 2 m (6.6 ft) below sea level, is protected from direct inundation by the waters of the Gulf of Mexico by salt marsh to the south and east. The waters of the Mississippi River and Lake Pontchartrain are kept out of the city by a levee and seawall, respectively.

may be reached where the public decides it is time for the private interests to assume the expense of protecting their investments or take their losses.

Whether or not sea level continues to rise, there is no doubt that most of the Atlantic and Gulf coasts will continue to subside. At the very minimum, it seems prudent for the development policy of our coastal regions to include a prohibition against further development of barrier islands. A very cautious approach to the development of mainland coasts should also be followed. Such a policy may be very distasteful to many of us who love to be near the ocean. However, if the reported rise in sea level continues, there may be no alternative.

FIGURE 11C Coastal Cities Subject to Damage by an Encroaching Sea.

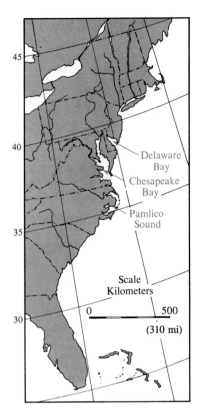

FIGURE 11–10   Drowned River Valleys along the Atlantic Coast.

sources of sand that is deposited as barrier islands and other depositional features common along this coast.

The recent increase in the volume of water in the ocean with the melting of the glaciers at the close of the Pleistocene has caused a significant "drowning" of river valleys that once flowed into an Atlantic Ocean that was a considerable distance seaward of the present shoreline. Delaware Bay, Chesapeake Bay, and the irregular landward margin of Pamlico Sound are of this origin. These estuaries are in the process of being filled by sediment carried to the coastal region by the rivers that drain the inland areas (fig. 11–10).

Because these sediments remain for the most part in the estuaries, there is little sediment made available to the longshore currents by river runoff. This condition has produced locally severe coastal erosion problems. The wave energy released in these coastal areas from the open Atlantic Ocean is greater than in the Gulf of Mexico. High-energy conditions develop during storms and when hurricanes strike the coastal area.

From New York north there is considerable evidence of continental glaciers having affected the coastal region directly. Many coastal features, including Long Island and Cape Cod, are the products of glacial deposition. They are moraines that were deposited when the glacier melted.

This area is subject to even higher-energy conditions than the coastal region farther south, particularly during the fall and winter. The "nor'easters" affect the coast from Cape Hatteras north. The high energy of these storms is manifested in up to 6-m (20-ft) waves with a 1-m (3.3-ft) rise in sea level that follows the low pressure as it moves northward. Such high-energy conditions will significantly change coastlines that are predominantly depositional and may be expected to cause considerable erosion along the coastal region where this process is dominant. The resistant rocks of the Maine coast have resisted erosion.

The Pacific Coast

Unlike the Gulf Coast and Atlantic Coast regions, the Pacific Coast of the United States is characterized by a continental margin that is rising rapidly relative to sea level. The folding, faulting, and volcanic activity associated with this uplift can be observed from southern California to Washington.

Throughout the Pacific Coast, the shoreline is characterized by cliffs that are being actively eroded. Old cliffs and wave-cut terraces have been elevated above the present sea level during the past million years or so. They are well developed in the Palos Verdes Hills in the Los Angeles area (fig. 11–11). Here 13 uplifted wave-cut terraces can be counted, and the highest one is approximately 400 m (1300 ft) above sea level. Because of the rapid coastal uplift, it is difficult to determine the effect of sea-level fluctuation due to the melting and forming of glaciers.

Rocks outcropping along the coast are, for the most part, relatively weak marine deposits that are geologically young. Locally, more resistant granite, metamorphic, and volcanic rocks may form the bedrock along the coast. The effects of glaciation, characterized by the glacially scoured fjords along the Canadian coasts, extend southward into the Washington coastal region and produce the Strait of Juan de Fuca and Puget Sound (fig. 11–12).

The Pacific Coast of the United States is characterized by high-energy conditions. The waves are normally about 1 m (3.3 ft) in height. Frequently, the wave height will increase to 2 m (6.6 ft), and a few times a year 6-m (20-ft) waves hammer the coast. In addition to the Pacific storms that generate these large waves, *tsunami* periodically strike the coastal region. The strong winter waves that generally ap-

FIGURE 11–11   Marine Terraces of Palos Verdes Hills, California.
Photo by John S. Shelton.

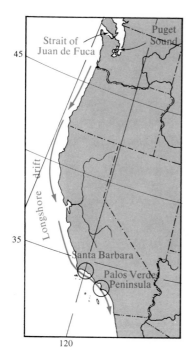

FIGURE 11–12   Features of Special Interest along the
Pacific Coast.

proach the coast from a west-northwesterly direction
produce a strong longshore current that carries millions of tons of sediment in the longshore drift.

During the winter, many beaches have the sand
eroded from the foreshore area by high-energy storm
waves. The exposed beaches, which are composed
primarily of pebbles and boulders during the winter
months, regain their sand as lower-energy summer
waves beat more gently against the shore. Damming
of many of the rivers along the Pacific Coast for flood
control has reduced the amount of sediment for the

longshore transport. Because of this sediment deficiency, erosion along the coast is becoming a significant problem as the sand-starved beaches disappear.

Effect of Constructed Barriers

Jetties that have been constructed to protect harbor
entrances from wave action are the most common
barriers to the longshore transport of sediment. Many
examples of the destruction resulting from the construction of such features along the shore can be
cited. Along the southern California coast, excessive
beach erosion related to the development of local
harbor facilities is especially common. The longshore
drift in California is predominantly from north to
south, and the construction of a jetty results in the
entrapment of sediment on the north side, causing
increased erosion on the south side. A jetty constructed as a breakwater to protect harbor waters at
Santa Barbara had a very significant effect upon the
equilibrium established in this coastal region. The
barrier created by the breakwater on the west side of
the harbor caused an accumulation of sand that was
moving in an easterly direction along this portion of
the California coast, which trends generally east-west.
The beach to the west of the harbor continued to
grow until finally the sand moved around the breakwater and began to fill in the harbor (fig. 11–13).

While deposition was occurring at an abnormal
rate to the west, erosion was proceeding at an alarm-
ing rate east of the harbor. The energy available east
of the harbor was no greater than it had been previously, but the sand that had formerly moved down
the coast to replace sand removed was no longer
available owing to its entrapment behind the break-

FIGURE 11–13 Santa Barbara Harbor.

Construction of a breakwater to the west of Santa Barbara Harbor interfered with the eastward-moving longshore drift, creating a broad beach. Sand being deposited against the breakwater was no longer available to replace sand being removed by wave erosion to the east. As the beach extended around the breakwater into the harbor, dredging operations were initiated to keep the harbor open and to put the sand back into the longshore drift. This helped reduce the coastal erosion east of the harbor. U.S. Coast and Geodetic Survey Chart 5161.

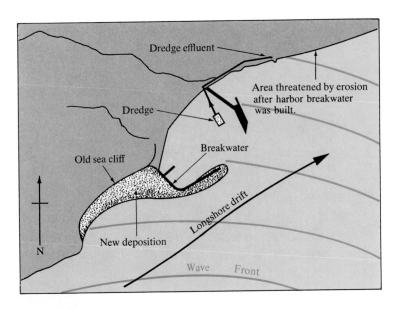

FIGURE 11–14 Groins.

As sand drifts along the shore of Miami Beach, it accumulates on the upcurrent side of the groins to aid in increasing the width of the beach. Arrow indicates direction of longshore transport of sediment. Photo courtesy of U.S. Army Corps of Engineers.

water. To compensate for this deficiency of sand downcurrent from the harbor, dredging had to be initiated within the harbor. The dredging keeps the harbor from filling in, and the sand is pumped down the coast so it can re-enter the longshore drift and replenish the eroded beach. The dredging operation has stabilized the situation, but at a considerable expense. It seems obvious that any time human beings interfere with natural processes in the coastal region, they will have to provide the energy needed to replace that which they have misdirected through modification of the shore environment.

Along the New York, New Jersey, and Florida coasts, small jettylike structures called *groins* have been constructed at intervals along the beach (fig. 11–14). They have helped beach development by causing sand to accumulate on the north side of these structures. Being shorter than jetties built to protect harbors, the groins eventually allow the sand to migrate around their ends. An equilibrium may be reached that allows sufficient sand transport along the coast before excessive erosion occurs downcurrent from the last groin. Although some serious erosional problems in these regions have developed, construction of a series of groins usually results in much less beach erosion than the construction of one large jetty. Groins are, at any rate, highly effective in widening and protecting beaches, although they will ultimately cause erosion down the coast.

Because of the dynamic characteristics of our shores, described in this chapter, serious consideration has been given to changing the policy of the federal government, which now tends to encourage coastal development. *Can We Learn to Live with Our Changing Shores?* on pages 287–289 focuses on this problem.

SUMMARY

The region of contact between the oceans and the continents is marked by the **shore**, lying between the lowest low tides and the highest elevation on the continents affected by storm waves. The **coast** extends inland from the shore as far as land features related to marine processes can be found. The shore is divided into the **foreshore**, extending from low tide to high tide, and the **backshore**, extending beyond the high-tide line to the **coastline** that separates the shore from the coast. Seaward of the low-tide line are the **nearshore zone**, extending to the breaker line, and the **offshore zone** beyond.

Wave erosion of the shore produces a **wave-cut cliff** that constantly retreats, leaving behind such features as **wave-cut benches, sea caves, sea arches,** and **stacks.** The rate of wave erosion increases with increased **exposure** of the shore to the open ocean, decreasing **tidal range,** and decreasing **strength of bedrock.**

As waves break at an angle to the shore, a **longshore current** is set up that produces a **longshore drift** of sediment along the shore. Deposition of sediment transported by the longshore current may produce such features as **beaches; spits,** deposits attached at one end to land; **tombolos,** sand ridges connecting an island to another island or the mainland; and **barrier islands,** long offshore deposits lying parallel to the shore. **Deltas** form at the mouths of rivers that carry more sediment to the ocean than can be distributed by the longshore current.

A drop in sea level may be indicated by ancient wave-cut cliffs and stranded beaches well above the present shoreline, and a rise in sea level may be indicated by submerged **relict beaches, wave-cut cliffs,** and **drowned river valleys.** Such changes in sea level may result from **tectonic** processes causing local movement of the land mass or from **eustatic** pro-cesses changing the amount of water in the oceans or the capacity of ocean basins, thus causing world-wide changes in sea level. Melting of continental ice caps during the last 10,000 years has caused a **glacioeustatic rise** in sea level of about 120 m (394 ft).

The United States **Gulf Coast** is undergoing tectonic subsidence. Therefore, the shoreline tends to move inland. This tendency is reversed only in areas where deposition of deltas, particularly the **Mississippi River Delta,** is occurring. Although the tidal range is small and the bedrock easily eroded, coastal erosion is minimal because the shore is not exposed to the open ocean. Tidal range increases as we travel north along the **Atlantic Coast,** and the bedrock becomes more resistant to erosion. This coast is slowly retreating in the south while it may still be undergoing isostatic uplift in the north. Open exposure to attack by severe storms results in some significant erosion problems. The **Pacific Coast** is definitely rising tectonically, while tidal ranges from 1 to 2 m (3.3 to 6.6 ft) and relatively soft bedrock along much of the openly exposed shore cause erosion to dominate over deposition.

Structures constructed along the shore, such as **jetties** to protect harbors and **groins** used to widen beaches, will trap sediment on the upcurrent side, but erosion may then become a problem downcurrent.

QUESTIONS AND EXERCISES

1. To help you reinforce your knowledge of the shore, construct and label your own diagram similar to that in figure 11–1.

2. Discuss the formation of such erosional features as sea cliffs, sea caves, sea arches, and stacks.

3. List and discuss three factors in the rate of wave erosion.

4. What variables affect the velocity of the longshore current?

5. What is the longshore drift, and how is it related to the longshore current?

6. Discuss the composition of beaches and the relationship of beach slope to particle size.

7. List and define the depositional features spit, tombolo, bay barrier, and barrier island.

8. Discuss why some rivers have deltas and others do not. Also include the factors that determine whether a "bird foot" or a smoothly curved Nile-type delta will form.

9. Compare the causes and effects of tectonic versus eustatic changes in sea level.

10. List the two basic processes by which coasts advance seaward, and list their counterparts that lead to coastal retreat.

11. Describe the tectonic and depositional processes causing subsidence along Atlantic-type margins.

12. How does the Pacific-type margin differ from the Atlantic-type margin?

13. Discuss the Gulf Coast, Atlantic Coast, and Pacific Coast by describing the conditions and features of emergence-submergence and erosion-deposition that are characteristic of each.

14. Describe the effect on erosion and deposition caused by putting a structure such as a breakwater or jetty across the longshore current and drift.

REFERENCES

Bird, E. C. F. 1985. *Coastline changes: A global review.* Chichester, U. K.: John Wiley & Sons.

Burk, K. 1979. The edges of the ocean: An introduction. *Oceanus* 22–3:2–9.

Coates, R., ed. 1973. Coastal geomorphology. *Publications in Geomorphology.* Binghamton, N.Y.: State University of New York.

Ingmanson, D. E., and Wallace, W. J. 1973. *Oceanology: An introduction.* Belmont, Calif.: Wadsworth.

Kuhn, G. G., and Shepard, F. P. 1984. *Sea cliffs, beaches, and coastal valleys of San Diego County: Some amazing histories and some horrifying implications.* Berkeley: University of California Press.

Leatherman, S. P. 1983. Barrier dynamics and landward migration with Holocene sea-level rise. *Nature* 301:5899, 415–17.

Shepard, F. P. 1959. *The earth beneath the sea.* Baltimore: Johns Hopkins Press.

_____. 1977. *Geological oceanography.* New York: Crane, Russak & Company.

SUGGESTED READING

Sea Frontiers

Carr, A. P. 1974. The ever-changing sea level. 20:2, 77–83.
   A discussion of the causes of sea-level change is very well presented.

Emiliani, C. 1976. The great flood. 22:5, 256–70.
   An interesting discussion of the possible relationship of the rise in sea level resulting from the melting of glaciers 11,000–8,000 years ago and biblical and other ancient accounts of a great flood.

Feazel, C. 1987. The rise and fall of Neptune's kingdom. 33:2, 4–11.
   A discussion of factors that change the level of the sea, including atmospheric conditions, currents, climate, ocean topography, and sea-floor spreading.

Fulton, K. 1981. Coastal retreat. 27:2, 82–88.
   The problems of coastal erosion along the southern California coast are considered.

Grasso, A. 1974. Capitola Beach. 20:3, 146–51.
   The destruction of the beach of Capitola, California, shortly after the construction of a harbor by the U.S. Army Corps of Engineers at Santa Cruz to the north.

Mahoney, H. R. 1979. Imperiled sea frontier—barrier beaches of the east coast. 25:6, 329–37.
   The natural alteration of barrier beaches is considered.

Schumberth, C. J. 1971. Long Island's ocean beaches. 17:6, 350–62.
   This is a very informative article on the nature of barrier islands. The specific problems observed on the Long Island barriers serve as examples.

Westgate, J. W. 1983. Beachfront roulette. 29:2, 104–109.
   The problems related to the development of barrier islands are discussed.

Scientific American

Bascom, W. 1960. Beaches. 203:2, 80–97.
A comprehensive consideration of the relationship of beach processes, of both large- and small-scale, to release of energy by waves.

Fairbridge, R. W. 1960. The changing level of the sea. 202:5, 70–79.
A discussion of what is known of the causes of changing level of the sea, which seems to be related mostly to the formation and melting of glaciers and changes in the ocean floor.

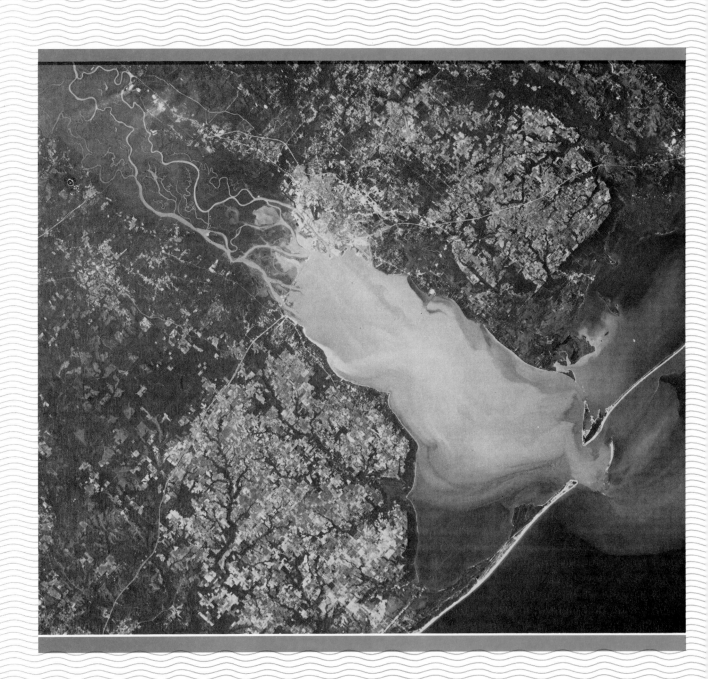

CHAPTER TWELVE
THE COASTAL OCEAN AND MARGINAL SEAS

In previous chapters we discussed the characteristics of the open ocean, the topography of the ocean floor, the generation of tides and currents, and the effects of time and climate. In chapter 11 we discussed the geology of the coastal regions. In this chapter, before turning to a discussion of the biology of the ocean, we examine the features that lie between the coast and the open ocean.

COASTAL OCEAN

The primary difference between the coastal ocean and the open ocean is that of depth. Generally, changes in the nature of water with respect to distance and time are much greater in the shallower coastal ocean. Because of the shallowness of the coastal ocean, river runoff and tidal currents have a very significant effect on the nature of coastal water.

Salinity

River runoff has the direct effect of reducing salinity of the surface layer in areas where mixing is not significant and throughout the water column where mixing does occur. In areas where the precipitation on the landmass is predominantly rain, the runoff of the rivers will be at the maximum during the season of maximum precipitation. However, if the runoff is fed to a great extent by the melting of snow and ice, the season of maximum runoff will always be the summer. In general, salinity will be lower in coastal regions than in the open ocean owing to the runoff of fresh water from the continents (fig. 12–1).

Counteracting the effect of runoff in some coastal regions will be the presence of prevailing offshore winds that will usually have lost most of their moisture over the continent. These winds evaporate considerable quantities of water as they move across the surface of the coastal ocean. The increase in the evaporation rate in these areas tends to increase the surface salinity, as shown in figure 12–1C.

Temperature

In coastal regions of the ocean where the water is relatively shallow, very great ranges in temperature may

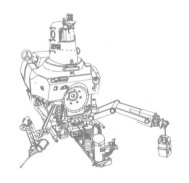

Turbid, sediment-laden water discharged from Mobile Bay, Alabama, is moved along the coast by currents. Courtesy of NASA.

FIGURE 12–1 Temperature and Salinity Changes in the Coastal Ocean.

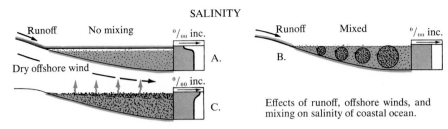

SALINITY

A. Freshwater runoff is not mixed into water column and forms surface layer of low salinity. A well-developed halocline occurs.

B. Runoff is mixed with deeper water, producing an isohaline water column of generally lower salinity than the open ocean.

C. Warm dry offshore winds cause a high rate of evaporation which may offset the effects of runoff, producing a halocline with a gradient which is the reverse of that seen in A.

Effects of runoff, offshore winds, and mixing on salinity of coastal ocean.

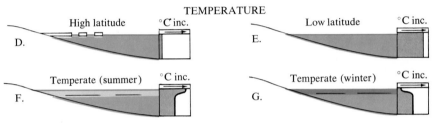

TEMPERATURE

D. In high latitudes where sea ice is forming or thawing throughout the year, the temperature of coastal water will remain uniformly near the freezing point.

E. Water in shallow low-latitude coastal regions protected from free circulation with the open ocean may develop a high-temperature isothermal condition.

F. In the mid-latitudes coastal water will undergo a significant warming during summer. A strong seasonal thermocline may develop.

G. Winter may produce a layer of low-temperature water at the surface. The thermocline shown may develop; however, cooling may cause the surface water to sink because of increased density. This would result in a well-mixed isothermal water column. Mixing due to strong winds may drive the thermoclines shown in F and G deeper and may even cause mixing of the entire water column, producing an isothermal condition.

occur on a yearly basis. Sea ice forms in many of the high-latitude coastal areas in which temperatures are determined by the freezing point of the water, which will generally be above −2°C (28.4°F). Maximum surface temperature in low-latitude coastal water may approach 45°C (113°F) in areas where the coastal water is somewhat restricted in its circulation with the open ocean and protected from strong mixing. The seasonal change in temperature can be most easily detected in the coastal regions of the midlatitudes, where surface temperatures are at a minimum in the winter and reach maximum values in the late summer.

Figure 12–1 shows how strong thermoclines may develop in areas where mixing does not occur. Very high-temperature surface water may form a relatively thin layer. Mixing reduces the surface temperature by distributing the heat through a greater vertical column of water, thus pushing the thermocline deeper and making it less pronounced. One major factor in mixing coastal water is the effect of tidal currents,

which can have a considerable influence on the vertical mixing of shallow water near the coast. Prevailing winds can also have a significant effect on surface temperatures if they blow from the continent. These air masses we previously mentioned in regard to salinity will usually be of relatively high temperature during the summer and cause an increase in the surface temperature and rate of evaporation of ocean water. They will be of much lower temperature than the ocean surface during the winter and will cause a heat loss and cooling of the surface water near shore.

Coastal Geostrophic Currents

The geostrophic effect that was discussed in association with the current gyres in chapter 8 also develops in the coastal ocean. The causes are basically the same. Wind blowing parallel to the coast in a direction that would cause water to pile up along the shore under the influence of the Coriolis effect produces a situation in which that water must eventually, under the influence of gravity, run back down the slope toward the open ocean. As the water runs down the slope away from the shore, the Coriolis effect causes it to veer to the north on the western coast and to the south on the eastern coast of continents in the Northern Hemisphere.

Another condition that produces geostrophic flow along the margins of a continent is the runoff of large quantities of fresh water that gradually mix with the oceanic water. This produces a surface slope of water away from the shore. The seaward slope is associated with salinity and density gradients, as both increase seaward (fig. 12–2). These variable currents, which depend upon the wind and the amount of runoff for their strength, are bounded on the ocean side by the more steady boundary currents of the open-ocean gyres.

These local geostrophic currents frequently flow in the opposite direction of the boundary current, as is the case with the Davidson Current that develops along the coast of Washington and Oregon during the

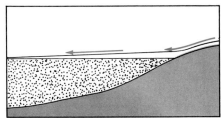

A. During the winter rainy season, fresh runoff water produces a seaward slope away from low-salinity surface water near the shore. A surface flow of low-salinity water from the shore toward the open ocean occurs.

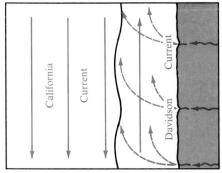

B. As the flow of surface water is acted on by the Coriolis effect, it veers right. This creates a north-flowing current (Davidson Current) along the coast of Washington and Oregon between the shore and the southbound California Current.

FIGURE 12–2 Davidson Current.

winter. A very great amount of precipitation occurs in the Pacific Northwest during the winter months, and the winds that generally blow out of the southwest are the strongest during these same months. Their combined effect produces a relatively strong northward-flowing geostrophic current between the southward-flowing California Current, which is part of the open-ocean circulation, and the continent of North America.

ESTUARIES

Estuaries (fig. 12–3) are semienclosed coastal bodies of water in which the ocean water is significantly diluted by fresh water from land runoff. The mouths of large rivers throughout the world form the most economically significant estuaries, since many serve as ports and centers of ocean commerce. Many estuaries support important commercial fisheries as well, and the environmental changes that are occurring as a result of the commercial importance of the estuaries will be a major concern as the future development of these areas is planned. Many bays, inlets, gulfs, and sounds may be considered estuaries on the basis of their structure.

Origin of Estuaries

Essentially all estuaries in existence today owe their origin to the fact that in the last 18,000 years sea level has been raised approximately 120 m (394 ft) owing to the melting of much of the major continental glaciers that covered portions of North America, Europe, and Asia during the Pleistocene Epoch, more commonly referred to as the Ice Age. Four major classes of estuaries can be identified on the basis of their origin (fig. 12–3):

1. *Coastal plain estuaries* were formed as the rising sea level caused the oceans to invade the existing river valleys. These estuaries are sometimes referred to as *drowned river valleys.* Chesapeake Bay is an example.
2. *Fjords* are glaciated valleys that are U-shaped with steep walls. They usually have a glacial deposit forming a sill near the ocean entrance. Fjords are common along Canadian coasts.
3. *Bar-built estuaries* are shallow estuaries separated from the open ocean by bars composed of sand deposited parallel to the coast by wave action. Lagoons separating barrier islands from the mainland are bar-built estuaries.

FIGURE 12–3 Classification of Estuaries on the Basis of Origin.

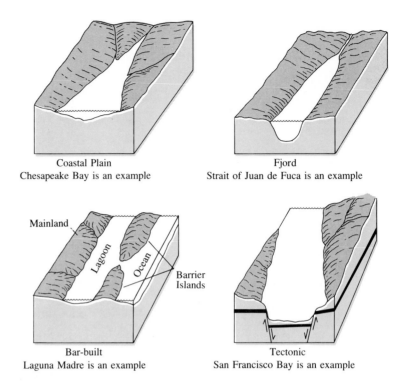

Coastal Plain
Chesapeake Bay is an example

Fjord
Strait of Juan de Fuca is an example

Bar-built
Laguna Madre is an example

Tectonic
San Francisco Bay is an example

4. *Tectonic estuaries* are produced by faulting or folding, which causes a restricted down-dropped area into which rivers flow. San Francisco Bay is in part a tectonic estuary.

Water Mixing in Estuaries

Generally, the freshwater runoff that flows into an estuary moves as an upper layer of low-density water across the estuary toward the open ocean. An inflow from the ocean takes place below the upper layer, and mixing takes place at the contact between these water masses. Stommel has classified estuaries on the basis of the degree of mixing as determined by the distribution of water properties into the following categories, which are shown in figure 12–4:

1. *Vertically mixed*—shallow, low-volume estuaries where the net flow always proceeds from the river at the head of the estuary toward the mouth. Salinity at any point in the estuary will be uniform from the surface to the bottom owing to the even mixing of the river water with ocean water by eddy diffusion at all depths. Salinity increases from the head to the mouth of the estuary.

2. *Slightly stratified*—a relatively shallow estuary in which the salinity increases from the head to the mouth at any depth. Two basic water

layers can be identified: the less saline upper water provided by the river and the deeper marine water separated by a zone of mixing. It is in this type of estuary that we begin to see the typical estuarine circulation pattern develop, as there is a net surface flow of low-salinity water into the ocean and a net subsurface flow of marine water toward the head of the estuary.

3. *Highly stratified*—typical of deep estuaries in which the salinity in the upper layer increases from the head of the estuary to the mouth, where it reaches a value close to that of open-ocean water. The deep-water layer, however, has a rather uniform marine salinity at any depth throughout the length of the estuary. The net flow of the two layers is similar to that described for the slightly stratified estuary, except that the mixing that occurs at the interface of the upper water and the lower water is such that the net movement is from the deep-water mass into the upper water. The less saline surface water does not seem to dilute the deep-water mass and simply moves from the head toward the mouth of the estuary, having its salinity increased as water from the deep mass joins it. Relatively strong haloclines develop in such estuaries at the contact between the upper and lower water masses. It is not

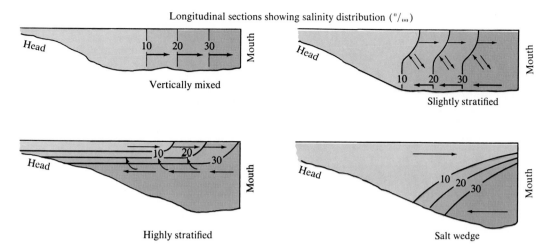

Longitudinal sections showing salinity distribution (°/₀₀)

FIGURE 12–4   Classification of Estuaries on the Basis of Degree of Mixing.

unusual for these haloclines to reach magnitudes approaching 20‰ near the head of the estuary at times of maximum river flow.

4. *Salt wedge*—resulting where a saline wedge of water intrudes from the ocean below that of the river water; typical of the mouths of deep, large-volume-transport rivers. There is no horizontal salinity gradient at the surface in these deep estuaries, the water being essentially fresh throughout the length of, and even beyond, the estuary. There is, however, a horizontal salinity gradient at depth and a very pronounced vertical salinity gradient manifested as a strong halocline at any station throughout the length of the estuary. This halocline will be shallower and more highly developed near the mouth of the estuary.

The mixing patterns described often cannot be applied to an estuary as a whole. Mixing within an estuary may change with longitudinal distance, season, or tidal conditions.

That much is yet to be learned about estuarine circulation is illustrated by the following observations made over an extended period in Chesapeake Bay. Just the opposite of the expected pattern, intervals of upstream surface flow accompanied by downstream deep flow have been recorded. Also, periods when all flow was upstream as well as periods of total downstream flow have been observed. Even more complex patterns known to develop involve a surface and bottom flow in one direction separated by a middepth flow in the opposite direction or landward flows along the shores and seaward flows in the central portions of the estuaries.

Understanding the dynamics of estuarine circulation is essential if desired advances are to be made

in our ability to meaningfully describe (1) parameters of water quality such as dissolved oxygen and coliform bacteria, (2) suspended particulate matter and bottom sediment transport, and (3) biological activity, particularly in relation to microscopic plants and animals, fish eggs, and larvae.

WETLANDS

Bordering estuaries and other shore areas protected from the open ocean are *wetlands,* biologically productive strips of land delicately in tune with natural shore processes. They are of two types: *salt marshes* and *mangrove swamps.* Both are intermittently submerged by ocean water and are characterized by oxygen-poor mud and peat deposits. Marshes, characteristically inhabited by a variety of grasses, are known to occur from the equator to latitudes as high as 65°. Mangrove trees are restricted to latitudes below 30° (fig. 12–5). Once mangroves colonize an area, they can normally outgrow and replace marsh grasses. With our strong desire to live near the oceans, marsh and mangrove wetlands have suffered from systematic filling and development for housing, industry, and agriculture.

Research is showing that wetlands have a very high economic value when left alone. Salt marshes are believed to serve as nursery grounds for over half the species of commercially important fishes in the southeastern United States. Other fishes such as flounder and bluefish use them for feeding and overwintering, and fisheries of oysters, scallops, clams, and fishes such as eels and smelt are located directly in the marshes. The rate at which wetlands are being destroyed is illustrated in the following ex-

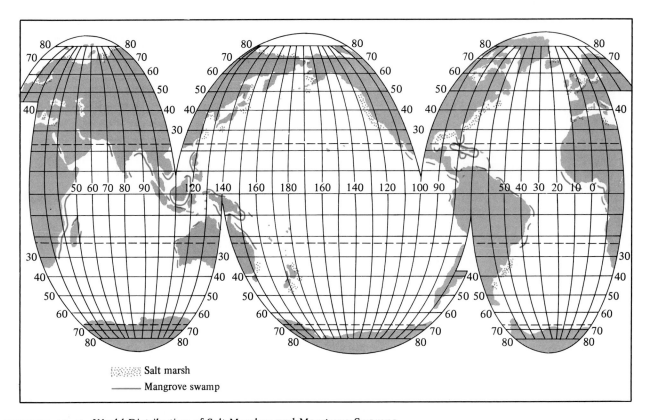

FIGURE 12–5 World Distribution of Salt Marshes and Mangrove Swamps.
Note dominance of mangrove swamps in low latitudes and salt marshes in high latitudes. Base
map courtesy of National Ocean Survey.

amples. Of the nearly 30,000 acres of salt marsh once present in Connecticut less than half remains today. As a result, a $20 million shellfish industry in existence in 1900 has been reduced to less than $2 million in equivalent dollars. Of the 200,000 acres of original marshland around San Francisco Bay, only 40,000 acres remain today.

A very important characteristic of wetlands is their ability to remove inorganic nitrogen compounds and metals from groundwater polluted by land sources. Most of the removal is probably achieved through adsorption on clay-sized particles. Some of the nitrogen compounds trapped in the sediment are decomposed by denitrifying bacteria releasing the nitrogen to the atmosphere as nitrogen gas. Much of the remaining compounds is used for plant production in this environment, which is one of the most productive in the world. With the death of the plants, the organic nitrogen compounds are either incorporated into the sediment and converted to peat or are broken up and become food for bacteria, fungi, or detritus-feeding shell- and fin fish.

Even though marshlands are considered to provide an annual return of over $5500 per acre on the basis of their natural role in supporting fisheries and re-

moving pollutants from groundwater, preserving them will surely be difficult. In many areas, one acre of marshland may sell for over $100,000 to industrial developers.

BARRIER ISLANDS

Long deposits of sediment extend across the seaward ends of many estuarine environments. Such features are called *barrier islands* and are characteristic of the east coast of North America, absent only in very large estuarine bodies such as Chesapeake Bay. Barrier islands, discussed in chapter 11, are products of marine deposition, and protect estuaries as well as the mainland from wave attack during storms. The origin of barrier islands is not fully understood, but they depend on the longshore drift for their maintenance. Some, though not all, may have developed from the extension of spits.

A typical barrier island has the following physiographic features from the ocean to the lagoon behind it: (1) ocean beach, (2) dunes, (3) barrier flat, and (4) salt marsh (fig. 12–6). The *ocean beach* is typical of the beach environment discussed in chapter 11. During the summer, as gentle waves carry sand to the beach, it widens and becomes steeper. Higher-energy winter waves carry sand offshore and produce a narrow, gently sloping beach.

Winds blow sand inland during dry periods to produce *dunes,* which are stabilized by dune grasses that can withstand salt spray and burial by sand. Dunes are the estuary's primary protection against excessive flooding during storm-driven high tides. Numerous passes exist through the dunes, particularly along the southeast Atlantic Coast, where dunes are less well developed than to the north.

Behind the dunes the *barrier flat* forms as the result of deposition of sand driven through the passes during storms. These flats are quickly colonized by grasses. If for some reason the frequency of overwash by storms decreases, the grasses will successively be replaced by thickets, woodlands, and forests.

Salt marshes typically lie inland of the barrier flat. They are divided into the *low marsh,* extending from about mean sea level to the high neap-tide line, and the *high marsh,* extending to the highest spring-tide line. The low marsh is by far the most productive part of the salt marsh. New marsh is formed as overwash carries sediment into the lagoon, filling portions so they are intermittently exposed by the tides. Marshes may be poorly developed on parts of the island that are far from flood-tide inlets. Their development is greatly restricted behind barrier islands where artificial dune enhancement and inlet filling prevent overwashing and flooding.

Because of the gradual rise in sea level relative to the eastern North American coast, barrier islands are migrating landward, a fact clearly visible to those who build structures on these islands. Evidence for such migration can be seen in peat deposits cored beneath the barrier islands. Since the only barrier island environment in which peat forms is the salt marsh, the island must have moved inland over previous marsh development.

A possible cycle of salt marsh formation and destruction by landward migration of barrier islands is presented in figure 12–7.

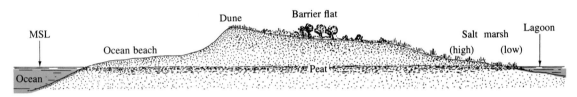

FIGURE 12–6 **Cross Section through a Barrier Island.**
The major physiographic zones of a barrier island are the (1) ocean beach, (2) dunes, (3) barrier flat, and (4) high and low salt marsh. MSL is mean sea level. The peat bed represents ancient marsh environments that have been covered by the island as it migrates toward the mainland as sea level rises.

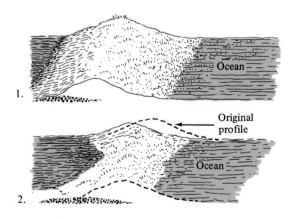

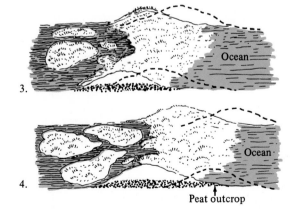

1. Barrier island before overwash with salt marsh development on mainland side.

2. Overwash erodes seaward margin of barrier island, cuts inlets, and carries sand into the lagoon, covering much of the existing marsh.

3. New marsh forms in the lagoon that has retreated toward the mainland.

4. With continued rise in sea level and migration of the barrier island toward the mainland, peat formed from previous marsh deposits may be exposed by erosion of the foreshore of the ocean beach.

FIGURE 12–7 Landward Migration of Barrier Islands.

LAGOONS

Landward of the barrier island salt marshes lie protected, shallow bodies of water called *lagoons*. Because of very restricted circulation between lagoons and the ocean, three distinct zones can usually be identified within a lagoon. There will usually be a freshwater zone near the mouths of rivers that flow into the lagoon, a transitional zone of brackish water (water with a salinity between that of fresh water and ocean water) and a saltwater zone close to the entrance, where the maximum tidal effects within the lagoon can be observed. Moving away from the saltwater zone near the mouth of the lagoon, the tidal effects diminish and are usually undetectable in the freshwater region that is well protected from tidal effects. Figure 12–8 shows the general characteristics of a typical lagoon.

The salinity of lagoons is also determined by factors other than position relative to the mouth of a lagoon. In latitudes with seasonal variations in temperature and precipitation, ocean water will flow through the entrance during a warm, dry summer to compensate for the volume of water that is lost through evaporation. This flow results in an increased salinity within the lagoon. Lagoons may actually become hypersaline in arid regions where the inflow of seawater is not sufficient to keep pace with evaporation that is taking place at the surface of the lagoon. During the

rainy season the lagoon will become much less saline as the amount of freshwater runoff increases.

Laguna Madre

A hypersaline lagoon along the coast of Texas between Corpus Christi and the mouth of the Rio Grande is Laguna Madre, which is a long, narrow body of water protected from the open ocean by Padre Island, a 160-km(100-mi)-long offshore island. Much of the lagoon, which probably formed about 6000 years ago as sea level was approaching its present height, is less than 1 m (3.3 ft) in depth. The tidal range of the Gulf of Mexico in this area is about 0.5 m (1.6 ft), and the inlets, one of which is shown in figure 12–9, at each end of the barrier island are quite small. There is, therefore, very little tidal interchange between the lagoon and the open ocean.

The shallowness of the water in the lagoon makes possible a very great seasonal range of temperature and salinity in this semiarid region. The water temperatures are high in the summer and may fall below 5°C (41°F) in the winter. Salinities range from 2‰ to over 100‰ because of infrequent local storms that provide large volumes of fresh water in a short period of time and high evaporation that keeps the salinity generally well above 50‰. Because even the salt-tolerant marsh grasses cannot withstand such high sa-

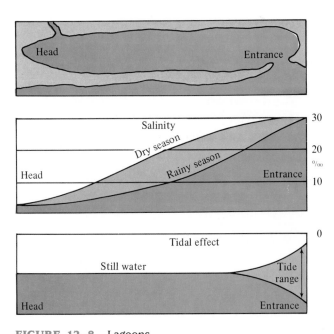

FIGURE 12–8 Lagoons.
Typical configuration, salinity conditions, and tide conditions of a lagoon.

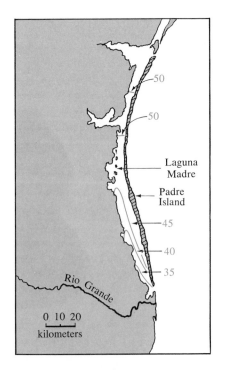

FIGURE 12–9 Laguna Madre.
Typical summer surface salinity distribution (‰).

inities, the marsh is replaced by an open sand beach on Padre Island. At the inlets, the inflow from the ocean is as a surface wedge over the more dense water of the lagoon. In turn, the outflow from the lagoon occurs as a subsurface flow, just the opposite of the circulation described for estuaries.

MARGINAL SEAS

At the margins of the ocean bodies are relatively large bodies of water, *marginal seas,* that have varying degrees of restriction in the circulation that exists between them and the open oceans. We will now discuss the circulation that exists between some of the major marginal seas and the open-ocean bodies they border and the water characteristics that develop within these seas as a result of these circulation patterns.

Marginal Seas of the Atlantic Ocean

Mediterranean Sea The Mediterranean Sea is an east-west-trending body of water that has a very irregular coastline dividing it into subseas with separate circulation patterns. Bounded by Europe and Asia Minor on the north and east and Africa to the south, it is surrounded by land except for a narrow connection with the Atlantic Ocean through the Strait of Gibraltar and the very narrow Bosphorus to the Black Sea. A man-made canal, the Suez Canal, connects the Mediterranean Sea to the Red Sea. The Mediterranean is divided into two major basins separated by a sill with a 400-m (1312-ft) depth, extending from Sicily to the coast of Tunisia. To the north of Sicily, strong currents run between Sicily and the Ital-

ian mainland through the Strait of Messina. The much broader connection between Sicily and Tunisia is the Strait of Sicily (fig. 12–10).

Atlantic water enters the Mediterranean through the Strait of Gibraltar as a surface flow that is carried in to replace water evaporated at a high rate in the very arid eastern end of the Mediterranean Sea. The water level in the eastern Mediterranean is generally 15 cm (6 in) lower than at the Strait of Gibraltar. This surface flow follows the northern coast of Africa throughout the length of the Mediterranean and spreads north across the sea (fig. 12–11).

As the Atlantic water moves to the northern coast of the western basin, the cold, dry polar air that descends over Europe during the winter cools it by evaporation, causing it to sink and form a deep-water mass. This deep water has a temperature of about 12.6°C (54.7°F) and a salinity of 38.4‰. The Atlantic water that finds its way to the northern coast of the eastern basin is cooled by evaporation in the Adriatic Sea and sinks to the bottom with a temperature of 13.2°C (55.8°F) and a salinity of 38.6‰. The water apparently is not trapped in these deep basins because of the formation of new bottom water each winter. This conclusion is reached on the basis of the relatively high dissolved oxygen content that may reach 4.7 ml/l in the western basin and 5.0 ml/l in the eastern basin.

The remaining Atlantic water continues eastward to Cyprus, where during the winter it sinks to form the Mediterranean Intermediate Water, with a temperature of 15°C (59°F) and a salinity of 39.1‰. This water flows west along the North African coast at a depth ranging from 200 to 600 m (656 to 1968 ft) and passes into the North Atlantic as a subsurface flow through the Strait of Gibraltar. By the time it passes

FIGURE 12–10 General Bathymetry and Subseas of the Mediterranean Sea.

From *Encyclopedia of oceanography,* edited by Rhodes Fairbridge, © 1966. Reprinted by permission of Dowden, Hutchinson, & Ross, Inc., Stroudsburg, Pa.

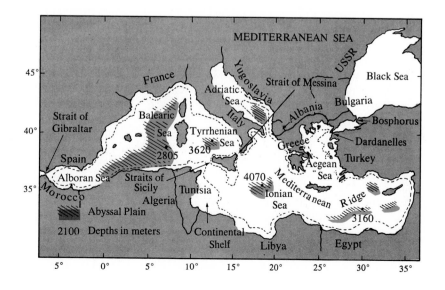

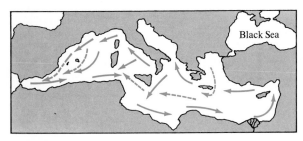

Surface flow ⟶
Intermediate flow   (200–600 m; 650–2,000 ft) ----➤

A. SURFACE AND INTERMEDIATE CIRCULATION

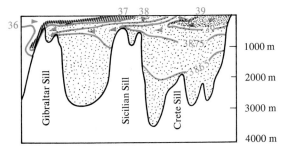

B. VERTICAL DISTRIBUTION OF SALINITY (°/₀₀)
Most of the mass of Mediterranean Water has salinity ranging
between 38 °/₀₀ and 39 °/₀₀. Maximum salinities in excess of
39 °/₀₀ are found in the surface waters at the east end of the sea.
After sinking and moving as Intermediate Water toward the Atlantic Ocean, salinity is reduced to about 37.3 °/₀₀ as it flows over
the sill at Gibraltar.

FIGURE 12–11   Mediterranean Circulation.

through the strait, its temperature has dropped to
13°C (55.4°F) and the salinity is 37.3‰. It is still
much denser than water at this depth in the Atlantic
Ocean, so it moves down the continental slope until
it reaches a depth of approximately 1000 m (3280 ft),
where it encounters Atlantic water of the same density. It is at this level that it spreads out in all directions into the Atlantic water and becomes a detectable water mass over a broad area because of its high
salinity.

The circulation between the Mediterranean Sea
and the Atlantic Ocean is typical of closed restricted
basins in areas where evaporation exceeds precipitation. In such situations the restricted basins will always lose water at a very high rate from the surface
through evaporation, and this water will have to be
replaced by surface inflow from the open ocean. The
evaporation of the water that flows in from the open
ocean increases its salinity to very high values, causing eventual sinking and the return to the open ocean
as a subsurface flow. Since the Mediterranean Sea is
the classic area for this type of circulation it is given
the name *Mediterranean circulation.* You will recognize it as being opposite the pattern we observed to
be characteristic of estuaries, where the surface flow

goes into the open ocean and subsurface flow into
the estuary. In areas where *estuarine circulation* develops between the marginal body of water and the
ocean, the availability of fresh water from runoff and
precipitation is greater than the rate of water loss to
evaporation.

The "American Mediterranean"   The "American
Mediterranean" includes the four basins of the Caribbean Sea and the Gulf of Mexico. The Caribbean Sea
is separated from the Atlantic Ocean by an island arc
called the Antillean Chain. Composed of the islands
of Cuba, Hispaniola, Puerto Rico, and Jamaica, the
Greater Antilles constitute most of the northern
boundary of the Caribbean Sea. The Lesser Antilles
extend from the Virgin Islands in an arc to the continental shelf of South America near the island of Trin-
idad. The deepest connection between the Caribbean
and the Atlantic Ocean is through Anegada Passage
east of the Virgin Islands. This passage has a maximum depth near 2300 m (7544 ft). Other channels
through which communication occurs have depths
ranging between 1500 and 2000 m (4920 and 6560
ft). The Caribbean Sea is divided into four major basins from east to west—Venezuela, Colombia, Cayman, and Yucatan—all of which reach depths in excess of 4000 m (13,120 ft) (fig. 12–12).

The Gulf of Mexico, or the Mexican Basin of the
American Mediterranean, is much less complex than
the Caribbean Sea in that it is a simple structure with
broad continental shelf surrounding much of its margin. It slopes off into a relatively broad basin that
reaches a maximum depth in excess of 3600 m
(11,800 ft). The Gulf of Mexico is connected to the
Caribbean Sea by the Yucatan Strait, which reaches a
maximum depth of 1900 m (6232 ft). The only connection with the Atlantic Ocean is through the Straits
of Florida, which reach depths approaching 1000 m
(3280 ft).

Surface Currents.   Entering through the Lesser Antilles, where sill depths less than 1000 m generally
prevail, the Guiana Current, which represents the portion of the South Equatorial Current that moves
northwest along the Guiana Coast of South America,
brings water with a temperature between 26° and
28°C (78.8° and 82.4°F) and a salinity between 35.0‰
and 36.5‰ into the Caribbean Sea. This relatively
thin mass of water moves with velocities in excess of
4 km/h (2.5 mi/h) at the surface but cannot be detected at a depth of 140 m (460 ft). It passes into the
Caribbean Sea through the shallow channels north
and south of St. Lucia Island and mixes in a 1:3 ratio

FIGURE 12–12 "American Mediterranean" Surface Currents and Bathymetry.

Vertical salinity profiles for numbered stations are shown in figure 12–14.

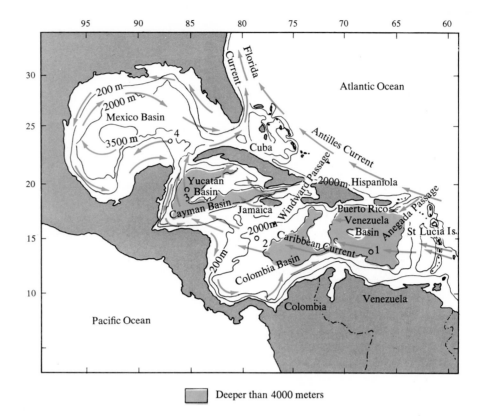

Deeper than 4000 meters

with North Atlantic Water. Becoming the Caribbean Current, whose axis of flow ranges between 200 and 300 km (124 and 186 mi) north of the Venezuelan coast, surface flow continues generally west over the deepest portion of the Caribbean Sea before it turns north and passes through the Yucatan Strait to the Gulf of Mexico. Surface velocities as high as 4.5 km/h (2.8 mi/h) have been observed in the main axis of the Caribbean Current, but the maximum surface velocities usually average less than 2 km/h (1.2 mi/h). The current can be detected as deep as 1500 m (4920 ft) in the eastern basins. Volume transport into the Caribbean Sea by the Caribbean Current is about 30 sv.

The easterly component of the trade winds blowing along the coast of Venezuela and Colombia sets up a surface flow away from the coast that produces a shallow upwelling. Most of the water rising to the surface comes from depths of less than 250 m (820 ft), but nonetheless contains high concentrations of nutrients and lower temperatures that result in relatively high biological productivity in the coastal region compared to that of the deeper ocean.

The Caribbean Current passes through the Yucatan Strait, causing a "dome" of water in the Gulf of Mexico that stands 10 cm (4 in) higher than the Atlantic water southeast of Florida. This hydraulic head forces an intense flow known as the Florida Current to pass through the Straits of Florida. It joins with water car-

ried north by the Antilles Current and flows north along the coast of Florida (fig. 12–12). The Loop Current is a significant feature of the Gulf of Mexico's surface circulation. Figure 12–13A shows the relationship of the current flow direction to the topography of the 20°C (68°F) surface in the southeastern Gulf of Mexico. Much of the surface water enters the Gulf of Mexico through the Yucatan Strait, loops around the temperature "bowl" in a clockwise flow and heads toward the Straits of Florida. Figure 12–13B shows that the Loop Current can be detected by remote sensing on the basis of surface temperature distribution. Like the Gulf Stream, the Loop Current sheds clockwise-rotating eddies that move westward in the Gulf of Mexico.

Water Masses in the Caribbean Sea. There are four readily identifiable water masses in the Caribbean Sea, two of which are relatively warm surface masses found above 200 m (656 ft) depth. The two deeper masses are characterized by lower temperature. The salinity of the Caribbean Surface Water mass is determined by the rate of evaporation versus the amount of precipitation and runoff. It generally ranges above 36‰ during the winter, partially due to the upwelling of high-salinity water. The surface salinity decreases to the north to values generally less than 35.5‰.

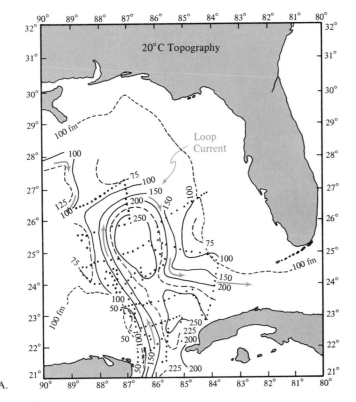

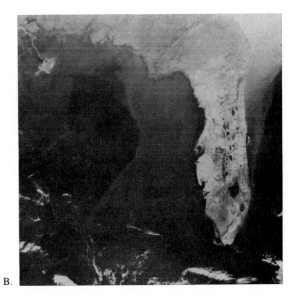

FIGURE 12–13 Loop Current.

A, the position and strength of the Loop Current can be deduced from the depth to the 20°C (68°F) temperature surface. Flowing clockwise, its velocity increases with increased slope of the temperature surface. What may be a clockwise-rotating eddy is indicated northwest of the main loop current. Courtesy of NOAA. *B,* the Loop Current, a mass of warm water driven into the eastern Gulf of Mexico, is identified by this infrared image taken by the *NOAA-3* satellite. Note the white flow of cold Mississippi River water entering the Gulf to the north and the darker shading representing the Florida Current flowing around the tip and up the east coast of Florida. Courtesy of NOAA.

Extending from the southeast to the northwest of the Caribbean Sea near the Yucatan Strait is a thin, sheetlike, high-salinity layer—the Subtropical Underwater. Located at depths as shallow as 50 m (164 ft) in the southeast, it dips to a depth of 200 m near the Yucatan Strait. The maximum salinity within this sheet seems to follow the axis of flow for the Caribbean Current and exceeds 37‰ in the Yucatan Strait. Salinity decreases away from the flow axis.

Also following the main flow axis is a low-salinity water mass, the Intermediate Water, located directly beneath the Subtropical Underwater. A salinity minimum that falls below 34.7‰ in the southeast and becomes less detectable at the Yucatan Strait identifies this mass (fig. 12–14).

Entering the Caribbean primarily through the Anegada Passage between the Virgin Islands and the Leeward Islands of the Lesser Antilles and through the Windward Passage between Cuba and Hispaniola is North Atlantic Deep Water. Characterized by a salinity slightly less than 35‰, this water spreads as Bottom Water and can be identified by an oxygen maximum that reaches values in excess of 5 ml/l in the Venezuela and Colombia basins and goes even higher, to values exceeding 6 ml/l, in the Cayman and Yucatan basins.

Water Masses in the Gulf of Mexico. After passing through the Yucatan Strait, the surface water characteristics can be identified down to a depth of 90 m (295 ft) during the winter and 125 m (410 ft) during the summer months, this being the depth through which seasonal temperature changes extend. Generally, the surface temperature just off the Yucatan coast ranges from 24° to 27°C (75° to 81°F), while that along the northern Gulf coast ranges between 18° and 21°C (64° and 70°F).

The salinity of the surface water generally ranges between 36.0 ‰ and 36.3‰. Runoff from the Mississippi River significantly decreases salinity to depths of 50 m (164 ft) and as far as 150 km (93 mi) from the coast. The influence of the runoff can be identi-

FIGURE 12-14 "American Mediterranean" Water Masses. The cores of the Subtropical Underwater and the Intermediate Water can be identified on vertical salinity profiles as maxima and minima, respectively, at the stations 1, 2, 3, and 4. The locations of the stations are indicated on figure 12–12. The Caribbean Surface Water salinity is generally above 36‰ while the Bottom Water salinity is uniformly between 34.9 and 35.0‰.

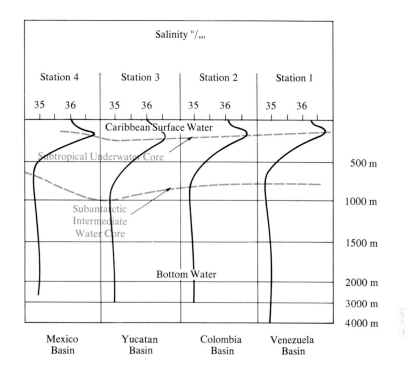

fied by salinity values as low as 25.0‰ near the surface.

The Subtropical Underwater is still identifiable as salinity maximum north of the Yucatan Strait at a depth of 100 to 200 m (328 to 656 ft). The boundary between the upper water, containing the surface water and Subtropical Underwater, and the deep water is marked by the 16°C (61°F) isotherm, which is at a depth of approximately 200 m.

The Intermediate Water, which enters the Gulf through the Yucatan Channel below 850 m (2788 ft), can be identified as a salinity minimum throughout the Gulf of Mexico. The core of the Intermediate Water can be identified throughout the Gulf at depths as shallow as 550 m (1804 ft), but its identity is lost passing through the Florida Straits. The salinity and temperature increase very slightly to the bottom of the basin.

Marginal Seas of the Pacific Ocean

Gulf of California This northwest-southeast-trending sea extends from the Tropic of Cancer at its open south end to the mouth of the Colorado River, which is the main river system draining into it (fig. 12–15). Some half-dozen smaller rivers also empty into the Gulf from the east, carrying water from the Sierra Madre Occidental across a broad coastal plain that forms the western margin of the Gulf. From this coastal plain, which is on the North American Plate, a

continental shelf extends into the Gulf to distances from 5 to 50 km (3 to 30 mi). The shelf terminates at an average depth somewhat in excess of 100 m (328 ft). This depositional feature does not exist on the western side of the Gulf, which is characterized by steep rocky slopes and is on the Pacific Plate. The Colorado River has developed a significant delta at the northern end, which is relatively shallow, with depths rarely exceeding 200 m (656 ft). Two exceptions are 1500-m (4920-ft) and 550-m (1804-ft) basins on the west and east side, respectively, of Angel de la Guarda Island. The depth gradually increases to the south through a series of basins, the deepest of which is 3700 m (12,136 ft). The basins are separated by sills that generally extend from 100 to 400 m (328 to 1312 ft) above the floors of the basin. The character of the Gulf has changed somewhat since the completion of Hoover Dam in 1935. Until that construction was completed, the Colorado River provided an annual average flow of almost 18 billion m³ of water, which carried 161 million tn of sediment. Since 1935 the average annual flow has been reduced to less than 8 billion m³, which carries about 15 million tn of sediment. The drainage systems on the east side of the basin have relatively small flow volumes, as many of them are intermittent owing to the arid nature of their drainage basins.

Surface winds control the surface circulation, which has a seasonal pattern. A low-pressure system located over the northern end of the peninsula develops southerly summer winds that drive the surface

FIGURE 12–15 Gulf of California Bathymetry and Surface Circulation.

Basins increase in depth from the 980 m (3214 ft) depth of the Delfin Basin in the north to more than 3700 m (12,136 ft) in the Pescadero Basin. Winds that reverse on a seasonal basis produce upwelling on the east side of the Gulf in the winter and on the west side during the summer. From *Encyclopedia of oceanography,* edited by Rhodes Fairbridge, © 1966. Reprinted by permission of Dowden, Hutchinson, & Ross, Inc., Stroudsburg, Pa.

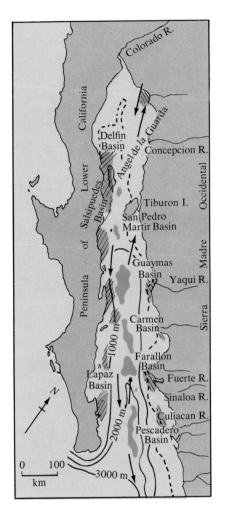

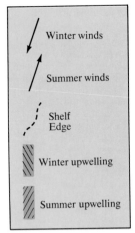

water from the Pacific into the Gulf. This flow produces upwelling along the steep rocky coast of the peninsula. During the winter months, the low-pressure system is found on the mainland east of the Gulf, and the winter winds are northerly. The wind change produces upwelling along the mainland side of the Gulf, so there is a rich plankton bloom of diatoms and dinoflagellates throughout the year. This high plant productivity supports a significant zooplankton population that ultimately supports a large fish population.

The tidal range increases from about 1 m (3.3 ft) in the south to more than 10 m (33 ft) during spring tides at the mouth of the Colorado River. The tidal currents that develop in the north, along with convective mixing, produce an isothermal water column during the winter. Temperatures may drop as low as 16°C (61°F), as opposed to summer surface temperatures that may reach 30°C (86°F). A high rate of evaporation produces a marked stratification, with surface water being warmer and more saline. The water below the thermocline in the central southern portion of the Gulf possesses an oxygen minimum as low as

0.01 ml/l between the depths of 400 m (1312 ft) and 800 m (2624 ft).

A joint Mexican-American expedition identified a warm-spring biological community in the Guaymas Basin during the summer of 1980. In 1982 the biological community was observed and sampled during a dive of the submersible *Alvin*. Although they are covered by sediment, the vents are on a segment of a spreading center. For a description of such communities, see *Alvin Explores Spreading-Center Median Valleys* on pages 81–87.

Bering Sea This sea on the northern margin of the Pacific Ocean extends to latitude 66°N and has the shape of a triangle with a curved base represented by the Aleutian Islands (fig. 12–16). A broad continental shelf with depths less than 200 m (656 ft) extends off the Siberian and Alaskan coasts, but most of the rest of the basin is found to have depths in excess of 1000 m (3280 ft). The deep basin is located in the western half of the sea, and fully 90% of the total area of the sea is either less than 200 m in depth or more than 1000 m in depth. Except where it is cut by the Bering

FIGURE 12–16 Bering Sea Bathymetry and Surface Circulation.

From *Encyclopedia of oceanography,* edited by Rhodes Fairbridge, © 1966. Reprinted by permission of Dowden, Hutchinson, & Ross, Inc., Stroudsburg, Pa.

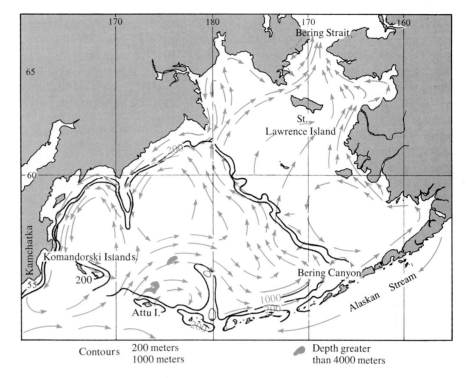

Canyon at the end of the Alaska Peninsula, the continental shelf slopes off very abruptly at 4° to 5° into the deep basin.

Circulation. The major surface flow of water into the Bering Sea occurs between the Komandorski Islands of Russia and Attu Island, the most westerly of the Aleutian Islands belonging to Alaska. Here the Alaskan Stream, flowing in a westerly direction south of the Aleutian chain, converges with northward-moving water of the western Pacific and flows through the passages into the Bering Sea. A counterclockwise gyre is set up north of the Komandorski Islands, while a clockwise rotation develops to the north of the eastern end of the American islands. The main flow passes on to the east until it reaches the broad Alaskan shelf, then follows this shelf edge to the north. A portion of this northward flow passes between St. Lawrence Island and the east Siberian coast before crossing the Bering Strait into the Arctic Ocean.

Tidal currents dominate the shallow Alaskan shelf region, but a persistent northward-flowing current resulting from the runoff of fresh water from the Alaskan coast flows at speeds up to 10 km/h (6 mi/h) along the coast and through the Bering Strait.

Marginal Seas of the Indian Ocean

Red Sea The Red Sea extends over 1900 km (1180 mi) north of the narrow Straits of Bab-el-Mandeb (Gate of Tears) to the northern tip of the Gulf of Suez,

the western branch of the Red Sea that separates the Sinai Peninsula from the African mainland. Forming the eastern boundary of the Sinai Peninsula is the Gulf of Aqaba, which is an eastern branch at the northern end of the Red Sea (fig. 12–17). Extending from 12° to 20°N latitude, the Red Sea lies in a highly

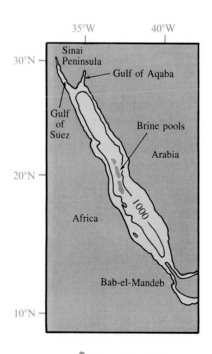

FIGURE 12–17 Bathymetry of the Red Sea.

During the 1920s, G. T. Walker identified what he called the Southern Oscillation (SO), associated with a condition in which the summer high-pressure system in the southeastern Pacific occurs in conjunction with a low-pressure system over the Indo-Australian region (*see* figs. 12A and 7–14). This pressure difference lessens when trade winds diminish, when the temperature of surface waters in the eastern Pacific increases, and when the Equatorial Countercurrent flow increases. The average period of this oscillation is three years, but it ranges from two to ten years. When the oscillations are extreme—producing very widespread warm surface-water conditions and minimal pressure differences across the tropical Pacific—they are called El Niño–Southern Oscillations (ENSO). There were seven ENSO events between 1950 and 1983.

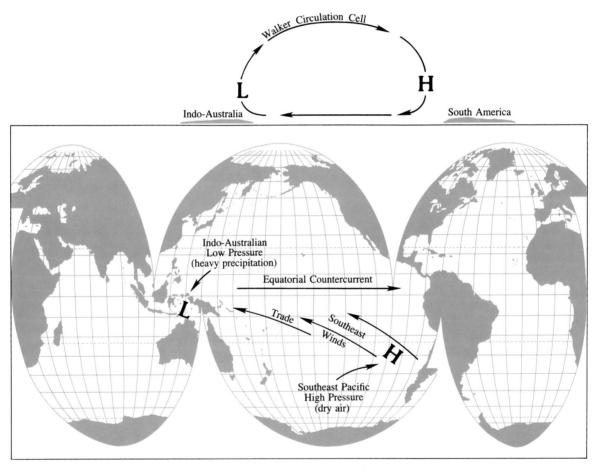

FIGURE 12A   Walker Circulation.

Normally a circulation cell called the Walker Circulation affects the southeast trade winds, which converge on the Indo-Australian low-pressure cell, rise, and produce high rates of precipitation in the low-pressure area. The dry air then descends within the southeastern Pacific high-pressure cell off the west coast of South America. This coast is characterized by high rates of evaporation.

One precursor of an ENSO event is the movement of the Indo-Australian low-pressure cell to the east beginning in October or November. In extreme cases such as the 1982–83 event, severe droughts can occur in Australia because the low-pressure cell moves so far east.

Concurrent with the eastward shift of the Indo-Australian low-pressure cell, the meteorological equator or Intertropical Convergence Zone (ITCZ), where the northeast trade winds and southeast trade winds meet and rise, moves south. Its normal seasonal migration is from 10°N latitude in August to 3°N in February, but during ENSO events it may move south of the equator in the eastern Pacific. Associated with this shift are weak trade winds, a decrease in coastal upwelling, and an unusually thick column of abnormally warm surface water in the eastern Pacific (fig. 12B*A*). These initial events are amplifications of normal seasonal fluctuations.

As the ENSO develops, the weakened trade winds and anomalous warmth of surface waters observed in the eastern Pacific spread toward the west (fig. 12B*B*). The coming of unusually warm surface waters to Kiritimati (Christmas

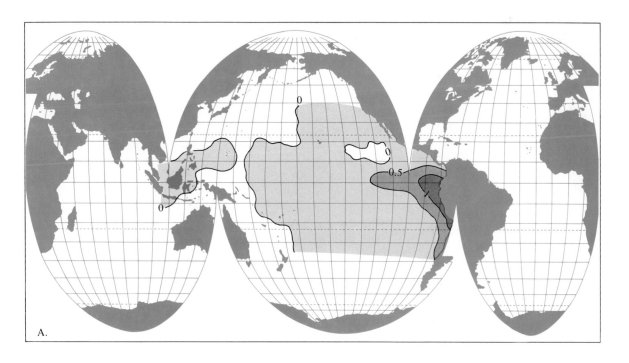

A.

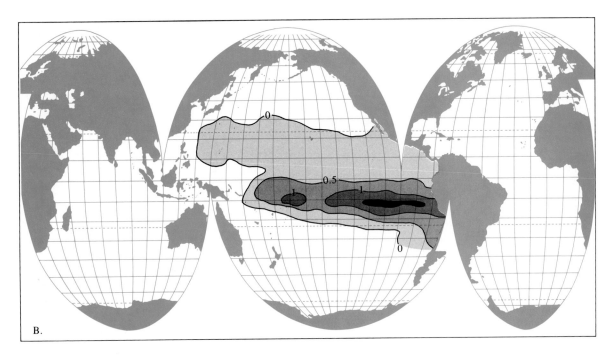

B.

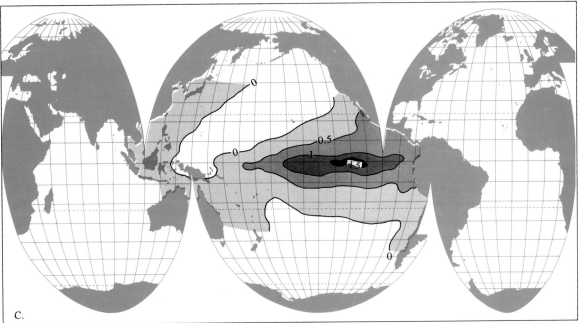

C.

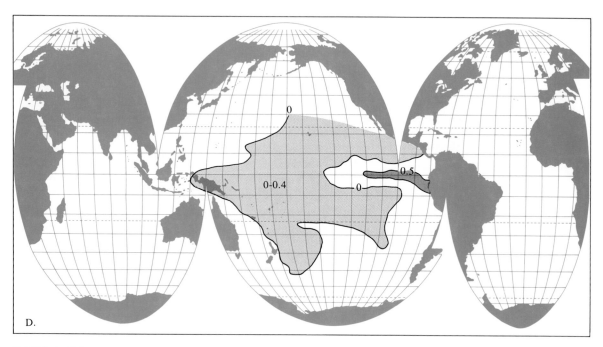

FIGURE 12B Sea Temperature Anomaly (°C) Resulting from the Averaging of ENSO Event Temperature Anomalies from 1950 through 1973.

A, after onset: average of March, April, and May temperature anomalies. *B,* during development: average of August, September, and October anomalies. *C,* maximum development: average of December, January, and February anomalies. *D,* ending of event: average of May, June, and July anomalies. The typical progress of an ENSO event shows that the abnormally high surface-water temperatures first appear in the eastern Pacific off the coast of Ecuador and Peru *(A).* This condition begins to be observable by December or January along the coast of South America and is well developed during the March–May period shown. In *B* and *C,* it can be seen to develop in a westerly direction along the equator and reaches maximum development in the central Pacific by the following February. By July *(D),* conditions return to near normal, with the exception of a significant negative temperature anomaly in the eastern Pacific. Courtesy of NOAA.

0 to 0.5°C (0.9°F) above the average surface-water temperature for non-ENSO years

0.5° to 1.0°C (0.9° to 1.8°F) above average

1.0° to 1.5°C (1.8° to 2.7°F) above average

More than 1.5°C (2.7°F) above average

Island; 2°N, 157°W) can be predicted by earlier observation of increased surface temperatures off the coast of Peru. The event is fully developed by January (fig. 12B*C*). Heavy rainfall from the southward shift of the ITCZ strikes the coast of Ecuador and Peru, which is usually arid, and spreads west across the tropical Pacific. The intense eastward flow of the Equatorial Countercurrent causes a rise in sea level along the western coast of the Americas that progresses poleward in both hemispheres (plate 22). The event ends 12 to 18 months after it starts, with a gradual return to normal conditions that begins in the southeastern tropical Pacific and spreads to the west (fig. 12B*D*).

A very intense 1982–83 ENSO event that caused a severe drought in Australia and Indonesia was anomalous in that it was initially confined to the central and western tropical Pacific and spread to the east late in its development.

Well before the overall pattern of development of the ENSO was recognized, a periodic movement of warm waters over the rich, cold fishing waters off the coast of Peru was recognized. This event, called El Niño, causes severe coastal rains and drives away the anchovy that are the basis of fishing there and serve as a food supply for a large bird population. El Niños have especially disastrous effects on the economy of Peru because of its heavy dependence on this fishery and the guano industry.

In November 1982, it was observed that virtually all of the 17 million adult birds that normally inhabited Kiritimati had abandoned their nestlings. This event indicated the severity of the latest ENSO in the central Pacific, as such an abandonment is not known to have occurred before. It is assumed that the spread of a thick layer of warm water over the surrounding ocean prevented the rise of nutrients into the surface waters, causing the fish to leave in search of better feeding grounds. The birds, in turn, were forced to leave in search of the fish. If birds did not find their necessary supply of fish, a large percentage of the adults may have died in addition to the nestlings.

Plate 22A shows the warming that occurred off southern California from January 1982 to January 1983 as a result of the 1982–83 ENSO. This warming was accompanied by major reduction in biological productivity, as is shown by the Coastal Zone Color Scanner phytoplankton pigment images for April 1982 compared to those of March 1983 (plate 22B).

Also associated with the 1982–83 and other ENSO events are unusual weather conditions throughout the world. Although it is not yet possible to find a close correlation between such events, further study may well uncover some definite relationships between world weather patterns and the ENSOs. So far, the relative roles of the atmosphere and oceans are not clear in the production of these events. They may well be tied to some degree to another outstanding phenomenon of ocean-atmosphere interrelationships, the monsoons of the northern Indian Ocean.

Reference: Philander, S. G. H. 1983. El Niño Southern Oscillation phenomena. *Nature* 302, 5906:295–301.

arid region and is characterized by surface waters of unusually high temperature and salinity. The sea is bordered by broad reef-covered shelves no more than 50 m (164 ft) deep that drop off sharply to a depth of about 500 m (1640 ft) to another gently sloping surface, which eventually leads into a deep central trough where depths are generally in excess of 1500 m (4920 ft) and extend to depths in excess of 2300 m (7545 ft) in the central region.

Geological evidence indicates the Red Sea has formed primarily during the past 20 million years, although the rifting may have begun 180 million years ago as a result of the Arabian Peninsula separating from the African mainland through the processes of sea-floor spreading discussed in chapter 4.

Circulation.   Sill depth at the Straits of Bab-el-Mandeb is only 125 m (410 ft) compared to a maximum depth over part of the Red Sea that is in excess of 1000 m (3280 ft). Across this shallow sill, the basic circulation seems to be dominated by the high rate of evaporation, which exceeds 200 cm/yr (79 in/yr). The surface loss in the Red Sea must be replaced by Indian Ocean water from the Gulf of Aden that enters as a surface flow. As this surface flow moves north, its density is increased as evaporation increases salinity. The dense water sinks and returns as a subsurface flow to the sill and out into the Gulf of Aden. This outflowing warm saline water sinks rapidly until it finds its equilibrium depth and then spreads out into the Indian Ocean. This pattern of circulation is similar to that of the Mediterranean Sea (fig. 12–18).

Because of the arid condition of the region, the surface water in the Red Sea reaches a salinity of 42.5‰ and a temperature of 30°C (86°F) during the summer months. Below a depth of 200 m (656 ft), a very uniform mass of water extends to the bottom throughout most of the Red Sea. This deep-water mass has a temperature of 21.7°C (71°F) and a salinity of 40.6‰.

In 1966, investigators aboard the Woods Hole Oceanographic Institution vessel *Chain* studied a series of deep basins in the central Red Sea region that had been previously noted to contain extremely high-salinity and high-temperature water masses. Between 21°15′ and 21°30′N latitude, two major basins were found: the Discovery Deep to the south and the Atlantis II Deep to the north. These and other similar basins possess brine with temperatures in excess of 36°C (96.8°F) and salinity ranging as high as 257‰. These brines, because of their high salinity, have densities great enough to keep them in their basins and prevent them from mixing with the overlying Red Sea water. The top of the brine accumulations is slightly below 2000 m (5650 ft).

Water may enter the porous oceanic crust beneath the Red Sea, be heated, and then dissolve the salts and metals that are found in the brines. Sediments found in the central deeps occupied by the brine possess concentrations of salts and metals that give them a great potential for economic development. Quite probably the enrichment in the crust beneath the sediment is also great and will enhance the economic potential of this area. These were the first hydrothermal springs to be discovered. This discovery provided the impetus to initiate the search that led to the discovery of the hydrothermal vents in the eastern Pacific.

Arabian Sea   The Arabian Sea is the northward extension of the Indian Ocean between Africa and India (fig. 12–19, p. 320).

Circulation.   Surface currents in the Arabian Sea are dominated by the monsoon winds that blow from the northeast from November until March, when the southwest monsoon begins to develop. The air carried onto the continent during this southwest monsoon contains large quantities of water and produces heavy precipitation in the coastal regions during this season.

During the northeast monsoon the surface current moves south along the west coast of India and turns west at about 10°N latitude, where some of the surface water flows into the Gulf of Aden and the rest turns south along the Somali coast and converges with the North Equatorial Current. When the southwest monsoon begins, the North Equatorial Current disappears, and a portion of the South Equatorial Current flows north along the Somali coast as the Somali Current. This strong seasonal current that flows with velocities in excess of 11 km/h (7 mi/h) continues along the coast of Arabia and India in a clockwise pattern until it reaches 10°N latitude. Here it joins the

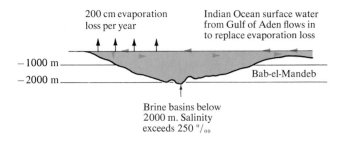

200 cm evaporation loss per year

Indian Ocean surface water from Gulf of Aden flows in to replace evaporation loss

−1000 m

−2000 m

Bab-el-Mandeb

Brine basins below 2000 m. Salinity exceeds 250 ‰

FIGURE 12–18   Circulation between the Red Sea and the Indian Ocean.

Since October 1982, French and American oceanographers have been observing the equatorial Atlantic and have recorded an El Niño-like event with a magnitude somewhat less than half the intensity of the 1982–83 Pacific Ocean El Niño. However, it was a very significant event that has not been matched since a similar occurrence in 1963.

The Atlantic Ocean phenomenon has many features in common with El Niños of the Pacific Ocean. The southeast trade winds weaken, the Intertropical Convergence Zone (ITCZ) is displaced to the south, a significant eastward flow of equatorial water occurs and an anomalous warming of the eastern equatorial water spreads into higher latitudes along the eastern boundary continents, where normal coastal upwelling ceases.

The Atlantic event occurred as follows:

During the height of the Pacific El Niño in early 1983, Atlantic southeasterly trade winds were very strong, forcing the ITCZ (equatorial rain belt) far north of the equator (fig. 12C). This produced a drought in equatorial northeast Brazil. These strong Atlantic trade winds may have resulted from the anomalous low atmospheric pressure over the eastern Pacific at this time. In November 1983, the trade winds weakened and even reversed, allowing the ITCZ to move south of the equator across the entire Atlantic until June 1984. During this time, the eastern equatorial Atlantic warmed in excess of 2°C (3.6°F) above normal (fig. 12D) as the eastward-flowing current south of the equator pumped in warm water that spread south along the coast as far as Namibia (20°S).

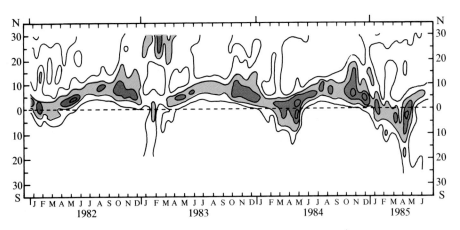

FIGURE 12C   The Intertropical Convergence Zone.

This time vs. latitude plot of the Intertropical Convergence Zone illustrates how the ITCZ usually starts the year near or on the equator and migrates to a maximum northward displacement by the end of the Northern Hemisphere summer. During the Northern Hemisphere winters in early 1984 and 1985, the ITCZ progressed much farther south than usual. Courtesy of Scripps Institution of Oceanography.

FIGURE 12D Intertropical Temperature Comparison

Comparison of sea-surface temperature (°C) in the intertropical region in June 1983 (relatively normal) and June 1984. In 1984 there is a significant warming in the eastern equatorial region and to the south along the coast of Africa. Courtesy NOAA.

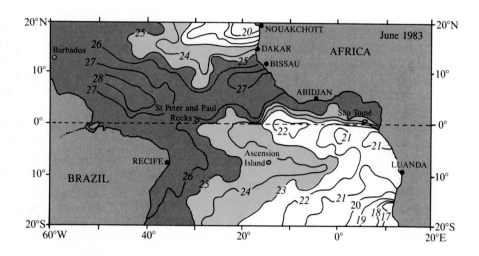

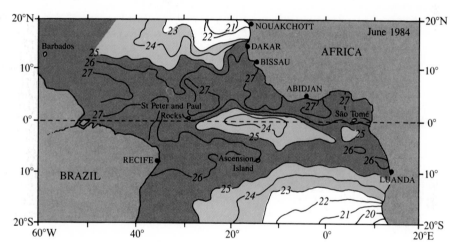

This southward shift of the ITCZ was accompanied by twice the normal rainfall in northeast Brazil and an intensification of the drought in subsaharan West Africa, where rainfall has been decreasing since the 1950s. Early in 1985, the ITCZ again migrated south of the equator, causing another unusually wet rainy season in northeast Brazil, but things returned to normal by May and subsaharan West Africa received much more rain than in 1983 and 1984.

The apparent relationship between marine and continental phenomena is intriguing and convinces oceanographers and meteorologists that they will some day be able to predict such devastating events as droughts. To do so, however, they will have to have worldwide data on meteorological and oceanic phenomena.

FIGURE 12–19 Bathymetry and Surface Circulation of the Arabian Sea and the Bay of Bengal.

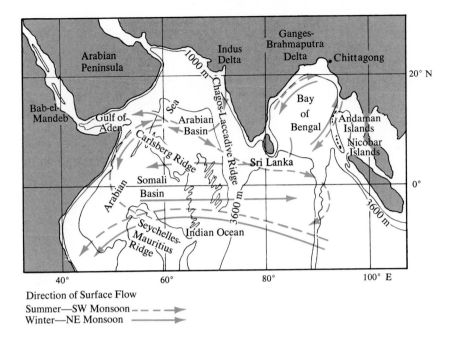

Direction of Surface Flow
Summer—SW Monsoon - - - →
Winter—NE Monsoon ———→

Southwest Monsoon Current, which has replaced the North Equatorial Current. Because of the alignment of winds relative to the African and Arabian coasts, significant upwelling occurs during the southwest monsoon.

Water Masses. Surface salinities north of 5°N latitude in the Arabian Sea are generally above 36‰ during the northeast monsoon. The Somali coastal region surface salinity may fall below 35.5‰ because of dilution by the South Equatorial Current and upwelling during the southwest monsoon. During the rainy season, surface salinities of less than 35‰ can be found as a result of dilution due to precipitation and runoff. Surface temperatures in the central ocean reach a maximum of 28°C (82°F) in June and a minimum temperature of 24°C (75°F) in February.

Below 200 m (656 ft), the salinity decreases to values near 35‰ until an abrupt salinity maximum of 35.4‰ to 35.5‰ develops at depths above 1000 m (3280 ft). This salinity maximum represents the flows of water into the Arabian Sea from the Persian Gulf and the Red Sea. The temperature of the Red Sea Water ranges from 9° to 10°C (48 to 50°F). The Red Sea Water can also be identified on the basis of 0.45 ml/l concentration of dissolved oxygen at about 790 m (2591 ft) off the Somali coast. Hydrogen sulfide has actually been observed on the continental slope at depths where this oxygen minimum occurs in the northern Arabian Sea.

Bay of Bengal The Bay of Bengal is bounded by India and the island of Sri Lanka on the west and by the Burma Peninsula and the Andaman-Nicobar Island Ridge to the east (fig. 12–19). A significant continental influence is exerted on this body of water by the runoff from the Ganges and Brahmaputra rivers at the extreme north end of the bay.

The currents in the Bay of Bengal are also dominated by the monsoon wind system. During the southwest monsoon, a clockwise rotation is established within the bay, and this circulation is accompanied by upwelling along the east Indian coast. With the development of the northeast monsoon in November, the circulation reverses and forms a counterclockwise gyre. At Chittagong in southeast Pakistan, a seasonal change in the sea level of 1.2 m (4 ft) occurs as a result of monsoon wind changes.

The surface salinity in the Bay of Bengal seldom exceeds 34‰. During the southwest monsoon, particularly during late summer when the rainfall is the greatest, the runoff from the Ganges, Brahmaputra, and other rivers along the coast of Burma and India dilute the water and reduce surface salinity. Values as low as 18‰ may be observed at the extreme north end of the bay. The major influence of the dilution is observed along the Indian coast; it is less significant away from shore.

SUMMARY

Because of the shallow depth of the coastal ocean, river runoff, tidal currents, and seasonal changes in solar radiation, physical characteristics of the water such as temperature and salinity vary over a greater range than in the open ocean. **Coastal geostrophic currents** are produced as a result of runoff of fresh water and winds blowing in the proper direction along the coast. **Rip currents** and **longshore currents** are produced by waves breaking on the shore.

Estuaries are semienclosed bodies of water where freshwater runoff from the land and ocean water mix. They are classified on the basis of origin as **coastal plain**, **fjord**, **bar-built**, and **tectonic**. Mixing patterns range from that of a **vertically mixed** shallow estuary through **slightly stratified** and **highly stratified** to the **salt wedge** characteristic of large-volume river mouths. The typical pattern of **estuarine circulation** consists of a surface flow of low-salinity water toward the mouth and a subsurface flow of marine water toward the head.

Wetlands are of two basic types: **salt marsh** and **tropical mangrove swamp**. These biologically productive regions are alternately covered and exposed by the tide. Marshes are also important in their ability to adsorb land-derived pollutants before they reach the ocean. Despite their ecological importance, wetlands are fast disappearing from our coasts as they are encroached upon by human activities.

Long offshore deposits called **barrier islands** protect marshes and lagoons. From the ocean side to the lagoon side, barrier islands commonly exhibit the following divisions: **ocean beach**, **dunes**, **barrier flat**, and **salt marsh**. Some lagoons have very restricted circulation with the ocean and may exhibit a great range of salinity and temperature conditions as a result of seasonal change.

Marginal seas are relatively large bodies of marine water overlying the continental margins. They are shallower than and have varying degrees of restricted circulation with the open ocean. The **Mediterranean Sea** has a classic circulation characteristic of restricted bodies of water in areas where evaporation greatly exceeds precipitation. It is the reverse of estuarine circulation. The "American Mediterranean" includes the **Caribbean Sea** and **Gulf of Mexico**. The major surface currents are the **Caribbean Current** flowing in from the southeast and the **Florida Current** flowing out between Cuba and Florida.

The **Gulf of California** experiences an extreme tidal range at the north end and has high productivity resulting from seasonal upwelling on opposite sides of the Gulf on a seasonal basis. An extensive shelf on the northeast side and a deep basin to the southwest are characteristic of the **Bering Sea** lying north of the Aleutian Islands. Many important commercial fisheries exist in this sea.

The formation of a new sea is seen in the **Red Sea** between Africa and Arabia. Mineral-rich brine basins are found in its median rift. The **Arabian Sea** to the west of India and the **Bay of Bengal** to the east have circulation patterns determined by the **monsoon winds**.

QUESTIONS AND EXERCISES

1. For coastal oceans where deep mixing does not occur, discuss the effect offshore winds and freshwater runoff have on salinity distribution. How will the winter and summer seasons affect the temperature distribution in the water column?

2. How does coastal runoff of low-salinity water produce a longshore geostrophic current?

3. Describe the difference between vertically mixed and salt wedge estuaries in terms of salinity distribution, depth, and volume of river flow. Which displays the more classic estuarine circulation pattern?

4. Name the two types of wetland environments and the latitude ranges where each will likely develop. How do wetlands contribute to the biology of the oceans and the cleansing of polluted river water?

5. How do peat deposits develop beneath barrier islands?

6. What factors lead to a wide seasonal range of salinity in Laguna Madre?

7. Describe the circulation between the Atlantic Ocean and the Mediterranean Sea, and explain how and why it differs from estuarine circulation.

8. How does the flow of the Loop Current relate to the slope of the 20°C (68°F) temperature surface?

9. On what basis can the Subtropical Underwater and the Subtropical Intermediate Water be identified in the "American Mediterranean"?

10. Describe how coastal upwelling in the Gulf of California is related to seasonal winds.

11. Discuss the unusual depth distribution of the Bering Sea and its relationship to biological productivity.

12. Compare the circulation between the Red Sea and the Indian Ocean to that between the Mediterranean Sea and the Atlantic Ocean. Discuss why they are or are not similar.

13. Explain why the Red Sea water that flows into the Indian Ocean at a depth of 125 m (410 ft) sinks to 1000 m (3280 ft) before spreading throughout the Arabian Sea.

14. Describe the relationship between the surface circulation and monsoon winds in the Bay of Bengal.

REFERENCES

Fairbridge, R. W., ed. 1966. *Encyclopedia of oceanography.* New York: Van Nostrand Reinhold.

Godfrey, P. J. 1976. Barrier beaches of the East Coast. *Oceanus* 19:5, 27–40.

Molinari, R. L.; Baig, S.; Behringer, D. W.; Maul, G. A.; and Legeckis, R. 1977. Winter intrusions of the Loop Current. *Science* 198: 505–6.

Officer, C. B. 1976. Physical oceanography of estuaries. *Oceanus* 19:5, 3–9.

Pickard, G. L. 1964. *Descriptive physical oceanography: An introduction.* New York: Macmillan.

Stommel, H. M., ed. 1950. Proceedings of the colloquium on "The flushing of estuaries." Woods Hole: Woods Hole Oceanographic Institution.

Valiela, I., and Vince, S. 1976. Green borders of the sea. *Oceanus* 19:5, 10–17.

SUGGESTED READING

Sea Frontiers

Baird, T. M. 1983. Life in the high marsh. 29:6, 335–41.
The ecology of the salt marsh is discussed.

Baker, R. D. 1972. Dangerous shore currents. 18:3, 138–43.
A discussion of the hazards to swimmers of various types of currents found near the shore.

de Castro, G. 1974. The Baltic—to be or not to be? 20:5, 269–73.
A review of the pollution threat to life in the Baltic Sea and what has been done to solve the problem.

Edwards, L. 1982. Oyster reefs: Valuable to more than oysters. 28:1, 23–25.
The value of oyster reefs to various estuarine life forms is detailed.

Heidorn, K. C. 1975. Land and sea breezes. 21:6, 340–43.
A discussion of the cause of land and sea breezes.

Osing, O. 1974. The meeting of the seas. 20:1, 21–24.
A description of the conditions at the northern tip of Denmark, where the North Sea and the Baltic Sea meet.

Sefton, N. 1981. Middle world of the mangrove. 27:5, 267–73.
The life forms associated with Cayman Island mangrove swamps are described.

Smith, F. 1982. When the Mediterranean went dry. 28:2, 66–73.
The role of global plate tectonics in causing the Mediterranean Sea to dry up some 5 million years ago.

Scientific American

Degens, E. T., and Ross, D. A. 1970. The Red Sea hot brines. 222:4, 32–53.
    Describes the hot brine pools in the Red Sea. The source of brines and the economic potential resulting from metal enrichment associated with the brine deposits are considered.

Hsu, K. H. 1972. When the Mediterranean dried up. 227:6, 26–45.
    Evidence is presented that shows the Mediterranean Sea was a dry basin 6 million years ago.

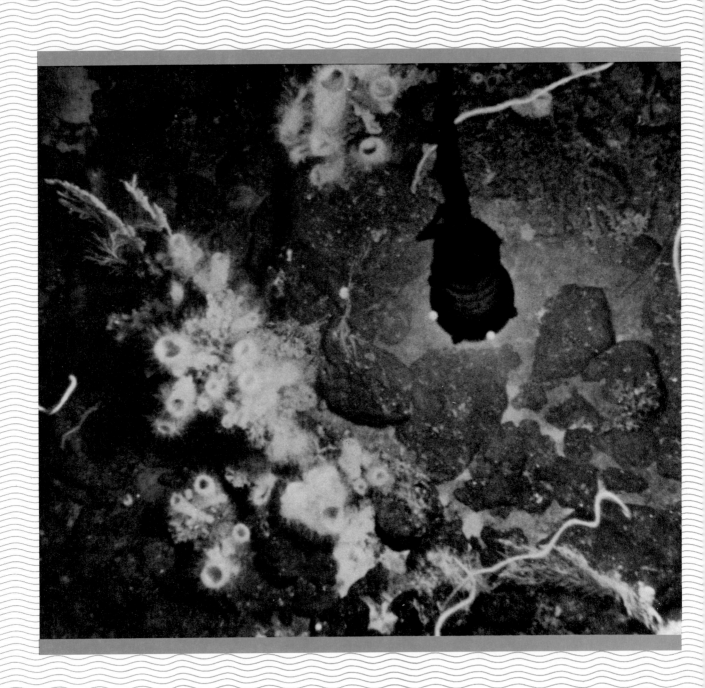

In chapter 2 we discussed some evidence for believing that life originated in the oceans. Primary among these considerations was the fact that the ocean constitutes the major part of the water environment. Considering the organisms' need for water, we can see that life would be totally impossible on the continents were it not for the hydrologic cycle that provides water that runs across and percolates through their surfaces, and is temporarily stored in lakes and rivers before it returns to the oceans. Although marine organisms rarely want for water, the success of individual organisms living in the oceans depends on their ability to adjust to many variables that we will consider in this chapter.

GENERAL CONDITIONS

The fact that representatives of all the animal *phyla* live in the marine environment attests to the ocean's being a far more fit biological realm than land. In fact, all phyla are thought to have originated in the ocean, and five are exclusively marine. Although it may seem inconsistent at this point, the fact that only simple one-celled plants inhabit the open oceans while the more complex and highly developed plants are restricted to its margins and the continents is further evidence pointing to the unique fitness of the oceans as a biological home.

The ocean environment is far more stable than the terrestrial environment; as a result, the organisms that live in the oceans have generally not developed highly specialized regulatory systems to combat sudden changes that might occur within their environment. They are, therefore, affected to various degrees by quite small changes in salinity, temperature, turbidity, and other environmental variables.

Water constitutes over 80% of the mass of *protoplasm,* the substance of living matter. Over 65% of the weight of a human being and 95% of that of the jellyfish is accounted for by the presence of water (fig. 13–1). Water carries dissolved within it the gases and minerals needed by organisms and is itself one of the raw materials required by plants in the production of

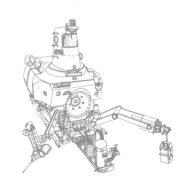

Sponges and spiral black corals at a depth of 600 m (1,968 ft) on Jasper Seamount, 550 km (342 mi) southwest of San Diego, California. Photograph by the Deep Tow Instrument Package of Scripps Institution of Oceanography. Courtesy of Amatzia Genin, Scripps Institution of Oceanography, University of California, San Diego.

FIGURE 13–1   Water Content of Organisms.

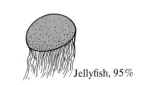

Human, 65%   Herring, 67%   Lobster, 79%   Jellyfish, 95%

food through photosynthesis. Land plants and animals have developed very complex systems and devices to distribute water throughout their bodies and prevent it from being lost. The threat of *desiccation* (drying out) through atmospheric exposure does not exist for the inhabitants of the open ocean, as they are abundantly supplied with water.

Support

Another necessity of plants and animals is support. Land plants have become complex multicellular organisms with vast root systems that fasten them securely to the earth. A number of support systems are used in the animal kingdom, but each requires some combination of appendages that must totally support the land animal's weight.

In the ocean this is not the case. The water of the oceans, as it so lavishly bathes organisms with the gases and nutrients they need, serves as their basic support as well. Organisms that live in the open ocean depend primarily upon buoyancy and frictional resistance to sinking to maintain them in their desired position. This is not to say that they do not have problems maintaining their position within the ocean; some of the special adaptations that are developed toward increasing their efficiency in this respect will be discussed in this and succeeding chapters.

Effects of Salinity

Marine animals are significantly affected by relatively small changes in their environmental conditions. One highly important condition is change of salinity. Some oysters that live in estuarine environments at the mouths of rivers are capable of withstanding a considerable range of salinity. During floods the salinity will be extremely low. On a daily basis, as the tides force ocean water into the river mouth and draw it out again, the salinity changes considerably. The oysters, and most other organisms that inhabit coastal regions, have a tolerance for a wide range of salinity conditions. They are *euryhaline*. By contrast, other marine organisms, particularly those that inhabit the open ocean, can withstand only very small changes in salinity; these organisms are *stenohaline*.

Certain plants and animals cause important changes in the amount of dissolved material in ocean water by extracting minerals to construct hard parts of their bodies that serve as protective covering. The compounds primarily used for this purpose are silica (SiO_2) and calcium carbonate ($CaCO_3$). The silica is used by the most important plant population in the ocean, the *diatoms,* and the microscopic animals called *radiolarians*. Calcium carbonate is used by the *foraminifers,* most members of the phylum Mollusca, corals, and some algae that secrete a calcium carbonate skeletal structure.

Cell membranes are semipermeable and will allow the passage of molecules of some substances while screening out others. By diffusion, cells may take in the nutrient substances they need from the surrounding fluid medium that contains them in high concentration. The nutrient compounds, to which the cell wall is permeable, pass from the surrounding medium into the cell, where they are in lesser concentrations (fig. 13–2).

The waste materials that a cell must dispose of after it has utilized the energy held in these nutrients are passed out of the cell by the same method. As the concentration of waste materials becomes greater within the cell than within the fluid medium surrounding the cell, these materials will pass out of this area of high concentration within the cell into the surrounding fluid. The waste products will then be carried away by the circulating fluid that services the cells in higher animals or by the surrounding water medium that bathes the simple one-celled organisms of the oceans.

Osmotic Pressure   When aqueous solutions of unequal salinity are separated by a water-permeable membrane, there will be diffusion of water molecules through the membrane into the more concentrated solution. This process is called *osmosis* (fig. 13–3). Such solutions can be compared in terms of their osmotic pressures. The *osmotic pressure* of a solution is that which must be applied to keep pure water that is separated from it by a water-permeable membrane from passing through the membrane and diluting the solution. The higher the salinity of the solution, the higher its osmotic pressure. We can compare the os-

A.   DIFFUSION

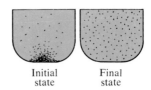

Initial   Final
state   state

If a substance soluble in water is placed in a pile on the bottom of a beaker of water, it will eventually become evenly distributed through random molecular motion. Diffusion means molecules of a substance move from areas of high concentration of the substance to areas of low concentration.

B.

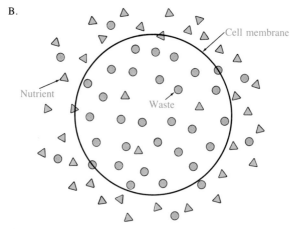

Nutrients are in high concentration outside the cell and diffuse into the cell through the cell membrane. Waste is in higher concentration inside the cell and diffuses out of the cell through the cell membrane.

FIGURE 13–2 Diffusion.

OSMOSIS

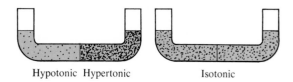

Hypotonic Hypertonic   Isotonic

Separating two water solutions of different salinities by a membrane that will allow the passage of water molecules but not the molecules of dissolved substances sets the stage for osmosis to occur. Water molecules will achieve a net motion by diffusion through the membrane into the more concentrated (hypertonic) solution from the less concentrated (hypotonic) solution. Osmosis will **not** produce a net movement of water through the membrane when the salinity of the two solutions is the same. They are then isotonic.

FIGURE 13–3 Osmosis.

of the external medium, water from the cells will pass through the cell membranes into the external medium. This organism is described as *hypotonic* relative to the external medium.

In an oversimplification, we will consider osmosis as simply a diffusion process. If we consider the relative concentration of water molecules on either side of the semipermeable membrane, we will see that the net transfer of water molecules is from the side where the greatest concentration of water molecules exists through the membrane to the side that contains a lesser concentration of water molecules. The processes of water molecules moving through the semipermeable membrane (osmosis), nutrient molecules moving from the external medium where they are more concentrated into the cell where they will be used to maintain the cell, and the movement of molecules of waste materials from the cell into the surrounding medium that carries them away are all going on at the same time.

We must keep in mind that during this process molecules of all of the substances in the system are passing through the membrane in both directions. Nevertheless, there will be a net transport of molecules of a given substance from the side on which they are most highly concentrated to the side where the concentration is less.

In the case of the marine invertebrates, the body fluids and the external medium in which they are immersed are in a nearly isotonic state. Since no significant difference exists, they do not have to develop special mechanisms to maintain their body fluids at a proper concentration. Thus they have an advantage over their freshwater relatives, whose body fluids are

motic pressure of solutions by comparing their concentrations of dissolved solids—their salinity. This comparison readily is done by determining their freezing points, since the addition of dissolved solids to an aqueous solution lowers its freezing point (fig. 13–4). This effect is due to the interference of the dissolved particles with the formation of ice crystals. The lower the freezing point, the higher the osmotic pressure of a solution.

If the salinity of body fluid and the external medium are equal, they are said to be *isotonic* and have equal osmotic pressure. No net transfer of water will occur through the membrane. If the external fluid medium has a lower osmotic pressure and lower salinity than the body fluid within the cells of an organism, water will pass through the cell walls into the cells. This organism is *hypertonic* relative to the external medium. Should the salinity or osmotic pressure within the cells of an organism be less than that

FIGURE 13–4 Osmotic Pressure and Freezing vs. Salinity. Osmotic pressure increases and the freezing point decreases with increased salinity.

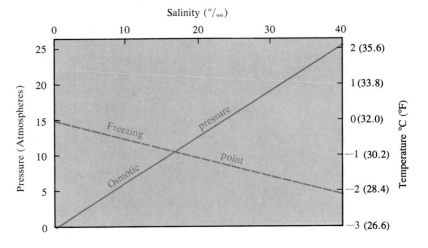

hypertonic in relation to the low-salinity external medium (table 13–1).

Marine fish have body fluids slightly more than one-third as saline as ocean water. This level may be the result of their having evolved in low-salinity coastal waters. They are, therefore, hypotonic in relation to the surrounding medium. This difference produces a problem in that the saltwater fish, without some means of regulation, would be continually losing water from its body fluids to the surrounding ocean environment. This loss is counteracted by the fact that marine fish drink ocean water and secrete the salts through "chloride cells" located in the gills (fig. 13–5).

Freshwater fish are hypertonic in relation to the very dilute external medium in which they live. The

osmotic pressure of the body fluids of such fish may be 20 to 30 times greater than that of the fresh water that surrounds them. Their problem is to avoid taking into their body cells large quantities of water that would eventually rupture the cell walls. In order to meet this problem, the freshwater fish does not drink water. Its cells have the capacity to absorb salt. The water that may be gained as a result of body fluids being greatly hypertonic is disposed of with large volumes of very dilute urine (fig. 13–5).

We should all be aware that we cannot drink salt water when at sea because the osmotic pressure of our body fluids is about one-fourth that of ocean water. To drink ocean water would cause a drastic loss of water through the wall of the digestive tract by osmosis, causing eventual dehydration.

TABLE 13–1 Osmotic Pressure of Body Fluids in Various Organisms.

A comparison of the freezing points of the body fluids of some marine and freshwater organisms and the waters in which they live. The lower the freezing point of the fluid, the higher the salinity and osmotic pressure. Marine invertebrates and sharks are essentially isotonic, while most marine vertebrates are hypotonic. Essentially all freshwater forms are hypertonic.

Marine Animals				Freshwater Animals		
	Freezing Points (°C)		I N V E R T E B R A T E S		Freezing Points (°C)	
Animals	Body Fluid	Water		Animals	Body Fluid	Water
Annelid worm	−1.70	−1.70		Mussel	−0.15	−0.02
Mussel	−2.75	−2.12		Water flea	−0.20	−0.02
Octopus	−2.15	−2.13		Crayfish	−0.80	−0.02
Lobster	−1.80	−1.80				
Crab	−1.87	−1.87				
Cod	−0.74	−1.90	V E R T E B R A T E S	Eel	−0.62	−0.02
Shark	−1.89	−1.90		Carp	−0.50	−0.02
Turtle	−0.50	−1.90		Water snake	−0.50	−0.02
Seal	−0.50	−1.90		Manatee	−0.50	−0.02

FRESHWATER FISH
(hypertonic)

Do not drink
Cells absorb salt
Large volume of dilute urine

MARINE FISH
(hypotonic)

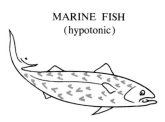

Drink large quantities of water
Secrete salt through special cells

FIGURE 13–5 Freshwater and Marine Fish.

Availability of Nutrients

The distribution of life throughout the ocean's breadth and depth basically depends on the availability of plant nutrients. In areas where the conditions are right for the supply of large quantities of nutrient materials, marine populations will reach their greatest concentration. Where are these areas in which we find the greatest concentration of biomass? To answer this question, we must consider the sources of nutrients.

Previously, we discussed the role played by water in eroding the continents, carrying the eroded material to the oceans, and depositing it as sediment at the margins of the continents. During this process, the water will dissolve inorganic compounds in the form of nitrates, phosphates, and potassium compounds that serve as the basic nutrient supply for plants. Because the sea bottom is relatively shallow near the continents, even nutrients that have settled out there can be returned to the surface waters by the actions of waves and tides. Through the process of photosynthesis, plants combine these with carbon dioxide and water to produce the carbohydrates, proteins, and fats that store the energy on which the rest of the biological community depends.

If the continents are the major sources of these inorganic nutrients, then we might expect that the greatest concentrations of marine life will be found at their margins. This is the case. As we go to the open sea from the margins of the continents, we will travel through waters containing progressively lower concentrations of marine life. This situation is basically due to the vastness of the world's oceans and the vertical descent of nutrients over the great distance between the open ocean and the coastal regions where the nutrients are supplied.

Availability of Solar Radiation

To make use of the available nutrients, plants must have another requirement met. Photosynthesis cannot proceed without the appropriate energy provided by solar radiation. The radiation from the sun penetrates the atmosphere quite readily so plants on the continents would seem to have an advantage over those in the ocean, as there is hardly ever any shortage of solar radiation to drive the photosynthetic reactions for these land-based plants.

By contrast, the ocean water does represent a significant barrier to the penetration of solar radiation, although in the clearest water solar energy may be detected to depths in excess of about 1 km (0.6 mi). However, the amount of solar energy reaching these depths is extremely small and would not be sufficient to support the photosynthetic process. Photosynthesis in the ocean is commonly restricted to a very thin layer, approximately the uppermost 100 m (328 ft). Near the coast, this zone will be much thinner because the water contains more suspended material that will restrict the light penetration. In the clearest ocean water, the base of the photosynthetic zone may be at a depth as great as 200 m (656 ft).

Considering these two factors necessary for photosynthesis, the supply of nutrients and the presence of solar radiation, we find that in the open ocean away from the continental margins the column of water having solar energy available extends deeper, but that this column contains only a small concentration of nutrients. Conversely, in the coastal regions, where the water is more turbid, light penetrates to much shallower depths, but the nutrient supply is quite rich. Since the coastal zone is by far the most productive zone, we can see manifested in this comparison the overriding importance of nutrient availability to life in the oceans.

Margins of the Continents

It has been mentioned that the stable condition of the ocean environment made it ideal for the continuation of life processes, but actually the richest concentration of marine organisms is found in the very margins of the oceans, where the conditions are the least stable. When we consider the physical conditions that would appear at first to be deterrents to the establishment of life, we will see that many are present in the coastal regions.

The water depths are shallow, allowing seasonal variations in temperature and salinity that are much greater than would be found in the open ocean. A varying thickness of water column exists in the near-shore region as a result of the tidal movements that periodically cover and uncover a thin strip along the margins of the continents, where there is a constant beating of the surf. The surf condition represents a sudden release of energy that has been carried for great distances across the open ocean. During this transit, little energy was lost and little effect on the marine environment was observed as a consequence of its transfer.

The development of such a great concentration of biomass in an environment containing so many factors that would seem to inhibit the development of life serves to highlight the importance of the evolutionary process in developing new species by natural selection. Over the eras of geologic time, new life forms have developed to fit into all biological niches. Many of these forms have adapted to live under conditions that would seem deleterious to the conservation of life. This fact attests to the great range of physical conditions within which life can exist so long as those basic requirements for the production of the food supply are met.

We find that along continental margins there are those areas where life is more abundant than others. What are the characteristics of these areas that result in such an uneven distribution of life along the continental margins? Again, we need consider only those basic requirements for the production of food for the answer. If we measure the various properties of the water along the coast of the continents and compare those measurements from place to place, we will find that the areas containing the greatest concentrations of biomass are those where the water temperatures are lower. Because of the low temperature, it is possible for the water to maintain greater amounts of the important gases, oxygen and carbon dioxide, in solution than would be possible in warmer waters. Of particular importance is the increased availability of carbon dioxide, and we see again an example of the greater availability of the basic requirements of plants affecting the distribution of life in the oceans.

Upwelling   In certain areas of the coastal margins we find an additional factor that enhances the conditions for life—upwelling. Upwelling occurs most commonly along the western margins of continents where surface currents are moving toward the equator. It consists of a flow of subsurface water to the surface and grows directly out of the relationship existing between the wind direction and the continental margin.

In the case of northward-flowing currents (driven by southeasterly winds) along the western margin of the Southern Hemisphere continents or southward-flowing currents (driven by northeasterly winds) along the western margin of Northern Hemisphere continents, the Ekman transport tends to move the surface water away from the coast. As the surface water moves away from the continents, water will surface from depths of 200 to 1000 m (656 to 3280 ft) to replace it. This water that surfaces from depths below those where photosynthesis occurs is rich in plant nutrients because there are no plants in these deeper waters to utilize them. This constant replenishing of nutrients at the surface enhances the conditions for life in these areas. Upwelling water is usually of low temperature, giving such water the additional benefit of having a high capacity for dissolved gases (fig. 13–6).

Water Color and Life in the Oceans

It is usually possible to determine visually areas of high and low organic production in the ocean from the color of water. Coastal waters are almost always greenish in color because they contain more dissolved organic matter, which disperses solar radiation in such a way that the wavelengths most scattered are those that represent organic greenish or yellowish light. This condition is also partly the result of the presence of yellow-green microscopic marine plants in these coastal waters. In the open ocean, where particulate matter is relatively scarce and marine life exists in low concentration, the water will appear blue owing to the size of the water molecules and the scattering they effect on the solar radiation. This is similar to the process that produces the apparent blueness of the skies.

Green color in water usually indicates the presence of a lush biological population such as might correspond to the jungle environments on the continents. The deep indigo blue of the open oceans, particularly between the Tropics of Cancer and Capricorn, usually indicates an area that lacks abundant life and could be considered a biological desert.

Size

Earlier in this chapter we made reference to the fact that marine plants were quite simple in comparison to the specialized forms we find throughout the con-

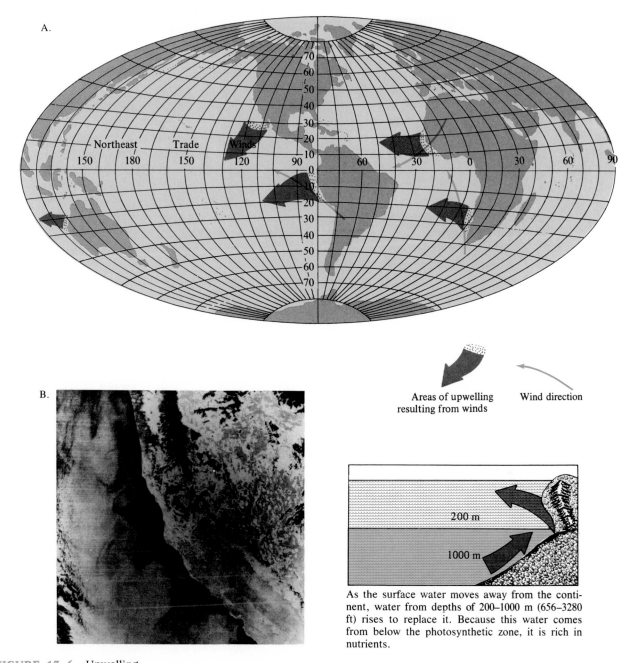

A.

Areas of upwelling resulting from winds   Wind direction

As the surface water moves away from the continent, water from depths of 200–1000 m (656–3280 ft) rises to replace it. Because this water comes from below the photosynthetic zone, it is rich in nutrients.

FIGURE 13–6 Upwelling.
A, the coastal winds usually drive currrents south along the western margins of continents in the Northern Hemisphere, while they are usually driven north in the Southern Hemisphere. The Ekman transport moves water away from the western margins of the continents. Base map courtesy of National Ocean Survey. B, on this infrared satellite image of the California and Oregon coastal water, cold upwelling water appears light while warmer water is darker. Photo courtesy of NOAA.

tinents. How can we explain the fact that marine plants have not followed the path of aggregation (becoming multicellular)?

Assuming the availability of nutrients, the major requirements of marine plants are that they maintain themselves within the uppermost, sunlit, layers of

ocean water to carry on photosynthesis. They must also take in nutrients from the surrounding waters and expel waste materials as efficiently as possible. These requirements lead us to a consideration of size and shape. The ease with which marine plants maintain their position in the upper layers of the ocean is

W hen a wind-driven current moves in a clockwise rotation around a Northern Hemisphere ocean basin, Ekman transport carries the light surface water offshore. The density barrier (pycnocline) between it and the deeper water tilts up toward the shore until it intersects the surface, allowing cold, deeper water to reach the surface near shore—upwelling.

The effect of shoreline features such as capes that jut into the ocean on the behavior of the current was thought to be the production of upwelling on the downstream side of the cape. However, research into this subject has shown that capes have little effect on the dynamics of upwelling. Appearing to be much more significant in their effect on upwelling patterns are bumps on the ocean floor—ridges.

Siavash Narimousa and Tony Maxworthy of the University of Southern California conducted experiments into these effects by use of a cylindrical tank to simulate an ocean basin. The "wind stress" was simulated by a disc that pulled the surface water in a clockwise direction around the tank. As the water begins to rotate, it is blocked by the ridge labeled in figure 13A. The following occurs:

1. The first sign of upwelling occurs as the dense water beneath the pycnocline is forced up over the ridge. A plume of upwelled water migrates away from shore, producing a zone of maximum upwelling (MU) on the downstream edge of the ridge.

2. Directly above the ridge, shoreward of the zone of maximum upwelling, is a trapped cyclone (TC).

FIGURE 13A The Effects of Bumps on Upwelling Patterns. Deep water from beneath the density interface (pycnocline) is dyed so it shows up as white when it reaches the surface as upwelled water. After Siavash Narimousa, University of Southern California.

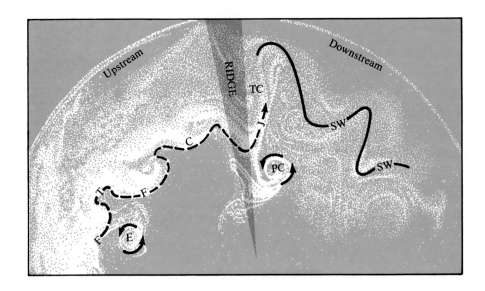

3. Upstream of the ridge, the pycnocline intersects the surface and migrates as a front (F) away from the shore. Depending on the speed of the current, the front between the upwelled water near shore and the surface water stops migrating away from shore and takes on the shape of a wave. There is a clockwise flow in the troughs (T) of the waves near shore and a counterclockwise flow in the crests (C) farther from shore.

4. At the crests of the upstream waves, cyclonic eddies (E) may pinch off and move offshore.

5. When the upstream wave crests reach the ridge, they disappear and their energy is transferred to the jet (J) or pinched-off cyclone (PC) that moves offshore.

6. The jet (J) flows toward shore on the downstream side of the ridge, is deflected, and produces a large standing wave (SW) downstream of the ridge. The upwelling associated with this downstream feature is small compared to that on the ridge and upstream of the ridge.

When these experimental results were applied to events recorded by satellites and field observations, they successfully predicted the relationship of upwelling patterns to bottom topography. This is an indication that the complex phenomena of upwelling are beginning to be understood. Upwelling patterns of the Gulf Stream and California Current where they cross the Charleston Bump and Mendocino Escarpment, respectively, display the responses described here.

Reference: Narimousa, S., and Maxworthy, T. 1985. Two-layer model of shear-driven coastal upwelling in the presence of bottom topography. *Journal of Fluid Mechanics* 159: 503–531.

closely related to the ratio that exists between surface area and body mass. The greater the surface area per unit of body mass, the easier it will be for an organism to maintain its position because there is more surface per unit of body mass in frictional contact with the surrounding water. Greater surface area per unit of body mass amounts to a greater resistance to sinking per unit of body mass. An examination of figure 13–7 will help you understand that this ratio of surface area per unit of mass increases with a decrease in size. It can readily be seen that it would benefit the one-celled plants making up the bulk of marine plant life to be as small as possible. They are, in fact, microscopic.

The efficiency with which the plants take in the nutrients from the surrounding water and expel their waste through their cell membranes is also related to the amount of surface area per unit of mass. The greater the surface area per unit of mass, the easier it is for the organism to carry on these functions. Since both processes are dependent on diffusion, there would be a point at which increased size would reduce the surface area to body mass ratio to the point that the cell could not be serviced and would die. It is a direct result of this condition that we find cells in all plants and animals to be microscopic, regardless of the size of the organism. Each cell must take in nutrients and dispose of wastes by diffusion.

Examining diatoms carefully, we see that it is quite common for them, as well as other members of the microscopic marine community, to have long needle-like extensions from their siliceous *tests,* or protective coverings. These extensions serve to further increase the ratio of surface area to the mass of the organism. The increased ratio of surface area to body mass allows more frictional resistance against the tendency to sink.

Another contrivance used by the small plants and animals to increase their ability to stay in the upper layers of the ocean is to produce an internal drop of oil, which tends to decrease their overall density and increase their buoyancy.

In spite of these adaptations to increase the ability to float, the organisms still have a density in excess of the density of water and therefore tend to sink, if ever so slowly. This is not a serious handicap, since they are so small and are carried quite readily along with the water movements. In these surface layers, there is considerable turbulent action from the mixing that results from wind action on the surface. This turbulence tends to dominate the movement of these small organisms to the point that they are able to spend adequate time bathed in the solar radiation to carry on their role of producing the energy needed by the other members of the marine biological community.

Viscosity

We are familiar with the fact that liquids flow with varying degrees of ease, and we know that syrup flows less readily than water. This condition is indicative of the higher viscosity of syrup as compared to

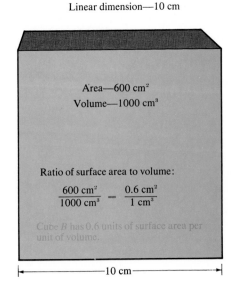

CUBE *A*
Linear dimension—1 cm

←|1 cm|←

Area—6 cm²
Volume—1 cm³

Ratio of surface area to volume:

$$\frac{6 \text{ cm}^2}{1 \text{ cm}^3}$$

Cube *A* has 6 units of surface area per unit of volume.

CUBE *B*
Linear dimension—10 cm

Area—600 cm²
Volume—1000 cm³

Ratio of surface area to volume:

$$\frac{600 \text{ cm}^2}{1000 \text{ cm}^3} = \frac{0.6 \text{ cm}^2}{1 \text{ cm}^3}$$

Cube *B* has 0.6 units of surface area per unit of volume.

←————— 10 cm —————→

Cube *A* has ten times as much surface area per unit of volume as cube *B*.

If both cubes are composed of the same substance, their masses are proportional to their volumes, since mass = volume × density.

If cubes *A* and *B* were plankton, cube *A* would have ten times as much frictional resistance to sinking per unit of mass as cube *B*. It could stay afloat by exerting far less energy than cube *B*.

Cube *A*, were it a planktonic alga, could also take in nutrients and dispose of waste through the cell wall ten times as efficiently as an alga with the dimensions of cube *B*.

FIGURE 13–7   The Importance of Size.

water. *Viscosity,* defined as internal resistance to flow, is a characteristic of all fluids.

The viscosity of ocean water is affected by two variables—temperature and salinity. Viscosity of ocean water increases with an increase in salinity but is affected even more by temperature, a lowering of which increases viscosity. As a result of this temperature relationship, we find that floating plants and animals in colder waters have less need for extensions to aid them in floating than those that occupy warmer waters. It has been observed that members of the same species of floating crustaceans will be very ornate, with featherlike appendages, where they occupy warmer waters; these appendages are missing in the colder, more viscous environments. It appears then that high viscosity would benefit the floating members of the marine biocommunity because it would make it easier for them to maintain their positions in the surface layers.

With increasing size of organisms and a change in the mode of life, viscosity ceases to enhance the ability of an organism to pursue life and becomes an obstacle instead. This is particularly true of the large organisms that swim freely in the open ocean. They must pursue prey and displace water to move ahead. The more rapidly it swims, the greater the stress that is created on the organism. Not only must water be displaced ahead of the animal, but water must move in behind it to occupy the space that it has vacated. The latter is one of the more important considerations in streamlining.

The familiar fusiform shape of the free-swimming fish and the marine mammals manifests the type of adaptations that must be achieved in streamlining an organism to move with a minimum of stress through the water. We see that a common shape achieved to meet this need is a laterally flattened body presenting a small cross section at the anterior end with a gradually tapering posterior. This form, which is characteristic of the bony fishes, reduces stress that results from movement through the water and makes it possible for this movement to be achieved with a minimum of energy (fig. 13–8).

Temperature

Of all the conditions we can measure in the ocean, there are probably no physical characteristics we have discussed that are more important to life in the oceans than temperature. It has been stated that the marine environment is much more stable than that of the land. A comparison of temperature ranges that exist in the sea to those that are found on land will serve to exemplify this condition.

A body with a low degree of streamlining. Its broad structure causes too much resistance area in front and too much wake region behind.

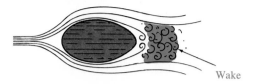

Wake

Teardrop shape (fusiform) front is well rounded and the body tapers gradually to the rear, producing a small wake region.

Wake

FIGURE 13–8 Streamlining.
Due to the viscosity of water, any body moving rapidly through it must be designed properly to produce as little stress as possible as it displaces the water through which it moves. After the water has moved past the body, it must fill in behind the body with as little eddy action as possible.

The minimum temperature observed in the sea is never much less than $-2°C$ ($28.4°F$), and the maximum seldom exceeds $27°C$ ($80.6°F$). This value compares to the continental temperature range from a low of $-88°C$ ($-127°F$) to a high of about $58°C$ ($136°F$). The temperature found on the continent varies considerably on a seasonal and daily basis. Daily variations at the ocean surface rarely exceed more than $0.2°C$ or $0.3°C$ ($0.4°$ or $0.5°F$), although they may be as great as $2°C$ or $3°C$ ($4°$ or $5°F$) in shallower coastal waters. Annual temperature variations are also small. They range from $2°C$ ($4°F$) at the equator to $8°C$ ($14°F$) at $35°$ to $45°$ latitude and decrease again in the higher latitudes. Annual variations of temperature in shallower coastal areas may be as high as $15°C$ ($27°F$).

The temperature variations we are discussing as characteristic of surface waters are reduced in magnitude with increasing depth of observation. In the deep oceans, daily or seasonal variation in temperature becomes a consideration of little or no importance. Throughout the deeper parts of the ocean the temperature remains uniformly low. Where the depth is in excess of 1.5 km (0.9 mi), temperatures will hover at or below $3°C$ ($37°F$), regardless of the latitude.

We previously discussed the fact that a decrease in temperature will increase density, viscosity, and the capacity of water to hold gases in solutions. All of these changes have significant effects on the organisms inhabiting the ocean. A direct consequence of the increased capacity of water to contain dissolved gases is seen in the vast plant communities that develop in the high latitudes during the summer seasons, when solar energy is available to carry on photosynthesis. One of the major factors in this phenomenon is the abundance of dissolved gases—carbon dioxide for the use of plants in carrying on photosynthesis and oxygen needed by the animals that feed on the plants.

It has been observed that floating organisms are larger individually in colder waters than in warmer waters of the tropics, although the tropical populations appear to be characterized by a larger number of species. However, the total biomass of floating organisms in the colder high-latitude planktonic environments greatly exceeds that of the warmer tropics. The fact that organisms in the tropics are smaller than those observed in the higher latitudes could quite possibly be related to the lower viscosity and density found in the lower-latitude waters. Because they are smaller, these tropical species can expose more surface area per unit of body mass, and they are also characterized by ornate plumage to further increase surface area per unit of body mass (fig. 13–9). These plumose adaptations are strikingly absent in the larger cold-water species.

Increases in temperature increase the rate of biological activities, which more than doubles with an increase in the temperature of 10°C (18°F). Tropical organisms apparently grow faster, have a shorter life expectancy, and reproduce earlier and more frequently than their counterparts in the colder waters.

There are some species of animals that can successfully live only in cooler waters and others that can do so only in warmer waters. Many of these can withstand only a very small change in temperature and are called *stenothermal.* Other varieties apparently are little affected by temperature change and can withstand changes over a large range. These are classified as *eurythermal.* Stenothermal organisms are found predominantly in the open ocean and at the greater depths where it is very unlikely that large ranges of temperature will occur. The eurythermal organisms are more characteristic of the shallow coastal waters, where the largest ranges of temperature are found, and the surface waters of the open ocean. See *Gulf Stream Rings and Marine Life* on pages 339–341 for an example of the effect of changing temperature on marine organisms.

DIVISIONS OF THE MARINE ENVIRONMENT

We can readily divide the marine environment into two basic units: the ocean water itself is the *pelagic* environment, and the ocean bottom constitutes the *benthic* environment (fig. 13–10).

Pelagic Environment

The pelagic environment is further divided into two provinces. The *neritic province,* which extends from the shore seaward, includes all water overlying an ocean bottom less than 200 m (656 ft) in depth. Beyond this 200 m depth the *oceanic province* is found.

The oceanic province, which includes water with a very great range in depth from the surface to the bottom of the deepest ocean trenches, is further subdivided on the basis of physical conditions. The oceanic province is subdivided into the *epipelagic* zone from the surface to a depth of 200 m, the *mesopelagic* zone from 200 to 1000 m (3280 ft), the *bathypelagic* zone from 1000 to 4000 m (13,120 ft), and the *abyssopelagic* zone, which includes all of the deepest parts of the ocean below 4000 m depth.

FIGURE 13–9 Water Temperature and Appendages.
A, copepod *(Oithona)* that displays the ornate plumage characteristic of warm-water varieties. *B,* copepod *(Calanus)* that displays the less ornate appendages found on temperate and cold-water forms.

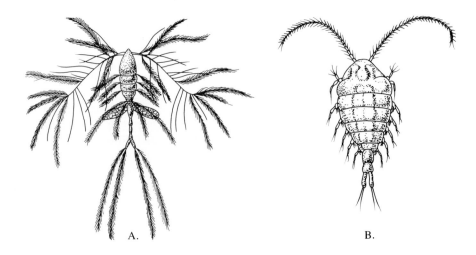

A.   B.

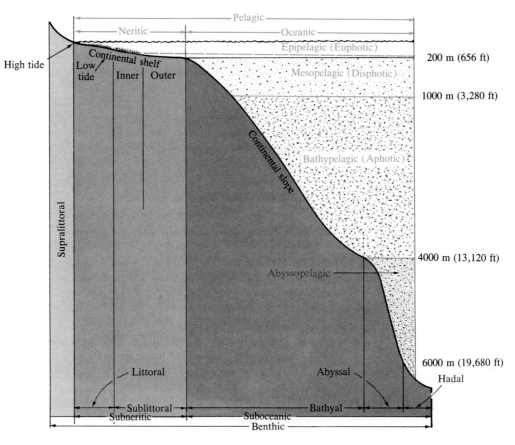

FIGURE 13–10   Biozones.

An important factor in determining the distribution of life in the oceanic province and the boundaries of biozones within it is the availability of light. The *euphotic* region extends from the surface to a depth where there is enough light to support photosynthesis, rarely to more than a depth of 100 m (328 ft). Below this zone, there are small but measurable quantities of light within the *disphotic* zone to a depth of about 1000 m. Below this depth there is no light in the *aphotic* zone (fig. 13–10).

The upper epipelagic zone is the only oceanic biozone in which there is sufficient light to support photosynthesis. The boundary between it and the mesopelagic zone (200 m) is also the approximate depth at which the level of dissolved oxygen begins to decrease significantly. This decrease in oxygen results because no plants are found beneath about 150 m (492 ft) and the dead organic tissue (detritus) descending from the biologically productive upper waters is undergoing decomposition by bacterial oxidation (fig. 13–11). Nutrient content of the water also increases abruptly below 200 m (fig. 13–11). This depth also serves as the approximate bottom of the mixed layer, seasonal thermocline, and surface-water mass.

Within the mesopelagic zone, a dissolved oxygen minimum occurs at a depth of about 700 to 1000 m (2296 to 3280 ft). The intermediate water masses that move horizontally in this depth range often possess the highest levels of plant nutrient content in the ocean. The base of the permanent thermocline and the boundary between the disphotic and aphotic zones characteristically occur at the 1000-m boundary between the mesopelagic and bathypelagic zones.

Within the mesopelagic zone, we see evidence of animals sensing the presence of light in the phenomenon known as the deep scattering layer. During sonar equipment tests by the U.S. Navy early in World War II, a sound-reflecting surface that changed depth on a daily basis was observed. The variation indicated a reflecting mass at a depth of 100 to 200 m (328 to 656 ft) during the night that sank to depths as great as 900 m (2952 ft) during the day. After considerable investigation it was determined that this echo, the deep scattering layer (DSL), was produced by masses of migrating marine life that moved closer to the surface at night and then to a greater depth dur-

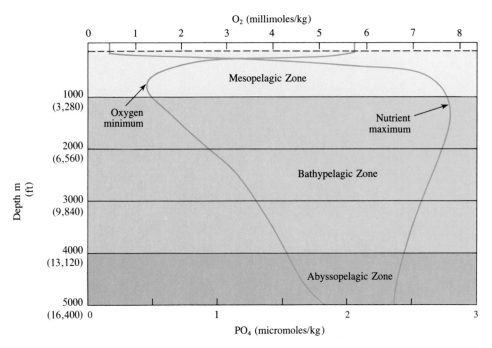

FIGURE 13–11 Distribution of Oxygen (O₂) and Plant Nutrient Phosphate (PO₄) in the Water Column, Mid- to Low Latitudes.

Oxygen is abundant in surface water owing to mixing with the atmosphere and plant photosynthesis. Nutrient content is low in surface water owing to uptake by plants. Oxygen content decreases and nutrient content increases abruptly below the euphotic zone. An oxygen minimum and nutrient maximum are recorded at or near the base of the mesopelagic zone. Nutrient levels remain high to the bottom while oxygen content increases with depth as deep- and bottom-water masses carry oxygen into the deep ocean.

FIGURE 13–12 Deep Scattering Layer.

The deep scattering layer, which scatters and reflects sonar signals well above the bottom, may be caused by euphausids that grow to lengths greater than 2 cm (0.8 in) and lantern fish (myctophid) that reach lengths of 7 cm (2.8 in). They are predators that feed on smaller planktonic organisms migrating vertically in the water column on a daily cycle.

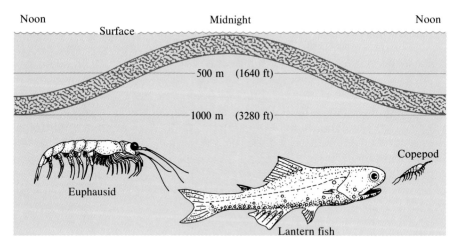

ing the day. DSL appeared to be a response to the changing intensity of solar radiation (fig. 13–12).

The use of plankton nets and submersibles has indicated that the DSL contains layers of siphonophores, copepods, euphausiids, cephalopods, and small fish. Tests that have been done using fish with swim bladders suggest that they may be the primary cause of the reflection since a very small concentration of such fish is sufficient to produce such a phenomenon. Fish, however, are predators and if their presence is responsible for the echoes, the organisms on which they prey are probably responsible for the cyclic nature of their movement. The smaller organisms responding to a changing intensity of light could trigger the vertical movements, while the presence of fish that prey upon them could be primarily responsible for the echoing phenomenon associated with the migration.

GULF STREAM RINGS AND MARINE LIFE

The Gulf Stream separates two distinct water masses. To the north lies the Slope Water with an average temperature of less than 10°C (50°F), and to the south lies the Sargasso Sea water with temperatures ranging from 15°C (59°F) to as high as 25°C (77°F). The average width of the meandering stream is perhaps 100 km (62 mi), and periodically, the meanders cut themselves off and move as individual rings (eddies) on either side of the Gulf Stream (fig. 13B and plates 5 and 6).

When a meander breaks off and forms a ring (fig. 13C) on the north side of the Gulf Stream, it may be up to 200 km (124 mi) in diameter and have a lens of warm Sargasso Sea water trapped inside it. The warm water may go to a depth of 1500 m (4920 ft). Although of great variety, the concentration of life in this "warm ring" is meager and has not been studied in detail. Conversely, the "cold rings" that break off on the south side may be up to 300 km (186 mi) in diameter and contain within them a "plug" of cold Slope Water that extends to the ocean floor. The depth of the ocean in this region may be in excess of 4000 m (13,120 ft). This cold water contains a much smaller number of species

FIGURE 13B Atlantic Ocean Surface Currents.

Path of Gulf Stream separates cold, biologically productive, Slope Water on the north from the warm, less productive Sargasso Sea water to the south. As the Gulf Stream meanders, some meanders pinch off to form rings—warm rings of Sargasso Sea water enter the cold slope water to the north and cold rings of Slope Water enter the Sargasso Sea to the south (see plate 6). Base map courtesy of National Ocean Survey.

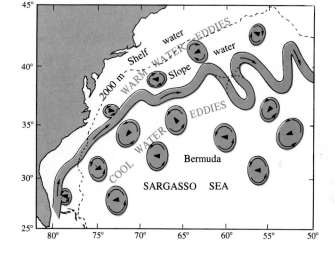

FIGURE 13C Temperature Structure of Warm Ring and Cold Ring.

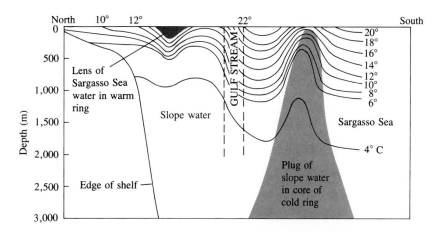

but a much greater biomass than the Sargasso Sea water trapped in the warm rings.

Because of their greater concentration of marine life, the cold rings have been studied extensively by marine biologists. In the following paragraphs, we will describe the life history of the cold rings and the responses of the Slope Water biotic community to physical changes within the ring structure.

Although most cold rings have the energy to remain intact for from 2 to 4 years, the average life of a ring is thought to be about 1½ years. Most cold rings break away from the Gulf Stream between 60° and 75°W longitude and move southwest at an average speed of 4 km/day (2.5 mi/day) through the Sargasso Sea. They rejoin the Gulf Stream off the Florida coast north of the Bahamas.

While the rings are detached, the cold core water is modified. It warms up and becomes more saline, and the depth of the oxygen minimum drops from the 150 m (492 ft) typical of Slope Water to around 800 m (2624 ft), which is characteristic of the Sargasso Sea. A striking feature of this change is the collapse of the dome of cold water extending up into the ring. Figure 13D shows the difference between the temperature structure in a newly formed cold ring and one that has completed its trip across the Sargasso Sea.

As the cold-water environment in the ring decays, the Slope Water life forms in the core become stressed. Most severely affected are the planktonic forms that do not have the ability to swim away to a more favorable environment. The phytoplankton population in the core of newly formed rings is most abundant at a depth of about 30 m (98 ft), as is indicated by maximum chlorophyll concentrations of up to 3.0 micrograms/liter of water. Toward the flanks of the rings, the phytoplankton concentration decreases and is at a maximum at a depth of about 60 m (197 ft). This condition is typical of the Sargasso Sea. Observations in one ring showed a rapid decrease in chlorophyll content. At an age of 7 months, the ring contained only an eighth of the chlorophyll it had contained four months earlier. This rapid reduction in phytoplankton activity results because nutrient supply cannot be replenished in the trapped Slope Water.

The zooplankton population of a young ring core is quite similar in species abundance and depth range to that of Slope Water. It includes some euphausids (small omniverous shrimplike crustaceans) and a herbivorous snail that is found only in the Slope Water. The euphausids (fig. 13E) and other her-

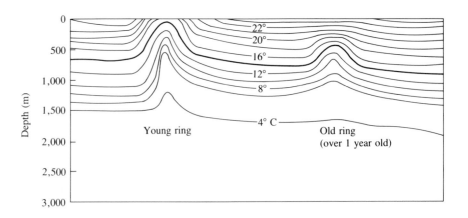

FIGURE 13D Comparison between Temperature Structure in Young and Old Cold Rings.

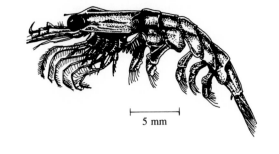

FIGURE 13E Slope-Water Euphausid *Euphausia krohnii.*

Courtesy of Scripps Institution of Oceanography, University of California, San Diego.

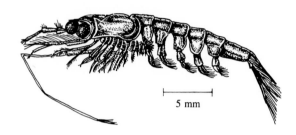

FIGURE 13F Slope-Water Shrimplike Crustacean *Nematoscelis megalops.*

Courtesy of Scripps Institution of Oceanography, University of California, San Diego.

5 mm

5 mm

bivorous zooplankton populations are greatly reduced by the time the rings are six months old because of the rapid decline in phytoplankton.

One of the shrimplike crustaceans, *Nematoscelis megalops,* prefers temperatures ranging between 8° and 12°C (46° and 54°F) (fig. 13F). This population lives almost entirely above a depth of 300 m (984 ft) in the Slope Water when it is initially trapped in the ring. As the core water warms from the top and the isotherms (lines of equal temperature) sink, the depth range of *N. megalops* also descends. As the population sinks deeper into the ring, it becomes separated from the higher concentration of food near the surface. By the time the ring is 6 months old, most of the *N. megalops* population will be found between 300 and 800 m (2624 ft) depth. They begin to starve. Resulting physiological and biochemical changes include a declining rate of respiration, increased water content, and reduced lipid, carbon, and nitrogen content. In a 17-month-old-ring, no *N. megalops* remained. The same fate may await many of the cold-water species trapped in the ring.

Some of the Slope Water life forms may be left behind as the ring moves across the Sargasso Sea. Since the annulus of Gulf Stream water around the cold core extends only to a depth of 1000 to 1500 m (3280 to 4920 ft), deep-living forms may escape beneath this barrier. It is possible that some *N. megalops,* as well as a number of carnivorous species, may be removed from the core water in this way.

As the cold core decays, Sargasso Sea populations move into the ring structure. They gradually replace the Slope Water populations as conditions become favorable to them.

Rings similar to those formed along the Gulf Stream are formed in association with other major ocean current systems. Continued study of these systems will help increase our understanding of the physical processes of the oceans and the relationships of marine organisms to the processes.

References: Weibe, P. 1976. The biology of cold-core rings. *Oceanus* 19:3, 69–76. Weibe, P. 1982. Rings of the Gulf Stream. *Scientific American* 246:3, 60–70.

The mesopelagic zone is also inhabited by species of fish that have unusually large and sensitive eyes, capable of detecting light levels 100 times less than those which the human eye could detect. Another important inhabitant of this zone is the bioluminescent group, especially shrimp, squid, and fish. Approximately 80% of the inhabitants carry light-producing *photophores*. These are glandular cells containing luminous bacteria surrounded by dark pigments. Some contain lenses to amplify the radiation. This cold light is produced by a chemical process involving the compound luciferin. The molecules of luciferin are excited and emit photons of light in the presence of the enzyme luciferase and oxygen. Only a 1% loss of energy is required to produce the illumination. This system is similar to that of the firefly.

The aphotic bathypelagic zone and abyssopelagic zone below the mesopelagic zone represent more than 75% of the living space in the oceanic province. In this region of total darkness many totally blind fish exist. Very bizarre small predaceous species make up the total fish population. Many species of shrimp that normally feed on detritus become predators at these depths, where the food supply has been greatly reduced from what is available in the shallower waters. The animals that live in the aphotic bathypelagic and abyssopelagic zones feed mostly upon one another and have developed impressive warning devices and unusual apparatus to make them more efficient predators. They are characterized by small expandable bodies, extremely large mouths relative to body size, and very efficient sets of teeth (fig. 13–13).

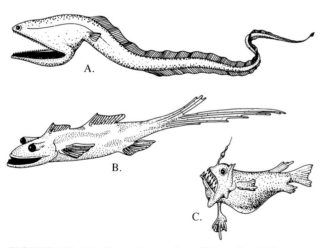

FIGURE 13–13   Some Examples of Deep-Sea Fish.
A, Eurypharynx pelecanoides. Length, 50–60 cm (20–24 in). *B, Gigantura chuni.* Length, to 12 cm (5 in). *C, Linophryne bicornis.* Length, 5–8 cm (2–3 in).

Aside from the depth of light penetration, the primary physical criteria used in defining the boundaries of the bathypelagic zone are the tops and bottoms of deep-water masses. Oxygen content increases with depth, as these water masses carry oxygen from the cold surface waters where they formed to the deep ocean. The abyssopelagic zone is the realm of the bottom-water masses, which commonly are moving in the direction opposite to that of the overlying deep-water masses of the bathypelagic zone.

Benthic Environment

The benthic environment can be subdivided into two larger units, which will then be subdivided on the basis of various criteria (fig. 13–10). These are the *subneritic* province, which extends from the spring high-tide shoreline to a depth of 200 m (656 ft) (approximately the continental shelf), and the *suboceanic* province, which includes all the benthic environment below the 200-m depth.

A transitional region above the high-tide line is the *supralittoral.* The supralittoral zone, commonly called the *spray zone,* is covered with water only during periods of extremely high tides and when tsunami or large storm waves break on the shore.

The intertidal zone is classified as the *littoral* zone, and from low tide to 200 m, the subneritic is called the *sublittoral* or shallow subtidal zone. The *inner sublittoral* includes that part of the sublittoral to a depth of approximately 50 m (164 ft). This seaward limit will vary considerably because it is determined by that depth at which we find no plants growing attached to the ocean bottom. Its extent is determined primarily by the amount of solar radiation that penetrates the surface water. In areas of high turbidity the inner sublittoral will have its seaward limit at shallower depths owing to the decreased penetration of solar radiation, while in areas where the water is unusually clear it may stand well below the 50-m mark. The *outer sublittoral* includes that portion of the sublittoral zone from the seaward limit of the inner sublittoral to a depth of 200 m or the shelf break, the seaward edge of the continental shelf.

With increased depth in the suboceanic system, we find the *bathyal* zone extending from the depth of 200 to 4000 m (656 to 13,120 ft) and corresponding generally to that geomorphic province found beyond the continental shelf—the continental slope. From a depth of 4000 to 6000 m (13,120 to 19,680 ft) stretches the *abyssal* zone, which includes in excess of 80% of the benthic environment. A restricted environment, including all depths below 6000 m, found only in the trenches along the margins of continents, is the *hadal* zone.

Once we have reached the outer sublittoral benthic region, we will see no more attached plants as we go from there into the deep-ocean basin. All of the photosynthesis seaward of the inner sublittoral zone is carried on by the microscopic algae floating in the neritic province above the outer sublittoral zone and in the epipelagic zone of the oceanic province. The *base,* or seaward limit, of the inner sublittoral can be said to coincide with the base of the *euphotic zone,* which is that zone of water near the ocean surface into which enough light penetrates to support photosynthesis. The seaward limit of the outer sublittoral zone approximately coincides with the shelf break. Currents will in general be stronger across the outer sublittoral bottom than along the upper continental slope of the bathyal zone. Sediments will also be coarser on the continental shelf than on the continental slope.

Change in the sediment type of the continental slope to a depth of about 4000 m are subtle. Throughout much of the ocean, the calcium carbonate compensation depth occurs at about this depth, so sediments in the bathyal zone may contain considerable calcium carbonate, while those below will contain little or none. In general, the deposits of neritic sediment found around the continents begin to be replaced by oceanic sediment below 4000 m.

The ocean floor representing the abyssal zone is covered by soft oceanic sediment, primarily abyssal clay. The tracks and burrows of animals that live in this sediment are frequently recorded in bottom photographs. The abyssal zone represents over 60% of the surface area of the benthic environment, or almost 43% of the earth's surface.

The hadal zone below 6000 m is primarily ocean trenches. Isolation in these deep, linear depressions allows the development of faunal assemblages found only in these trenches.

Distribution of Life in the Oceans

It is difficult to describe the degree to which the sea is inhabited because of its immense space and the paucity of our knowledge of it. We do know, however, that some populations fluctuate greatly each season, and this fact increases the difficulty of describing with numbers the extent to which the marine environment is populated. Although it is perhaps meaningless to estimate the number of individual organisms in the ocean, we may derive some means of comparing the marine and terrestrial environments by comparing the number of marine species to land species.

Well over 1.2 million species of animals are known, and only about 17% of these live in the ocean. Many biologists believe there may be from 3 to 10 million unnamed and undescribed animals living on earth. A large number of these may well inhabit the oceans. Yet, the following theory may explain why we can expect fewer species of marine animals to exist than terrestrial animals.

If the ocean is such a prime habitat for life, and if life originated in this environment, why do we now see such a small percentage of the world's animal species living there? This lesser number may well result because the marine environment is more stable than the terrestrial environment. The relatively uniform conditions of the open ocean do not produce pressures for adaptation. Also, once we get below the surface layers of the ocean, the temperatures are not only stable but also relatively low. Chemical reactions are retarded by this lower temperature; this, in turn, reduces the tendency for variation to occur.

If we are to consider briefly the great variety of species of organisms found on the continents, we can assume that this development was the product of an environment more diverse than is found in the ocean, one possessing many opportunities for natural selection to produce new species to inhabit new niches within this terrestrial environment. At least 75% of all land animals are insects that have evolved species capable of inhabiting very restricted environmental niches. If we ignore the insects, the sea does possess in excess of 65% of the remaining animal species living in the marine and terrestrial environments.

In describing the distribution of life within the ocean, we can get a general idea of the relative concentrations in the various marine environments if we consider that of the approximately 200,000 known animal species inhabiting the marine environment, only 4000 (about 2%) live in the pelagic environment. The remainder inhabit the ocean floor (fig. 13–14).

Before proceeding further in our discussion on the organisms of the oceans, we need to define some terms to describe these organisms on the basis of the portion of the ocean they inhabit and the means by which they move.

Plankton

The *plankton* include all those organisms that drift with ocean currents. This does not mean that all are without the ability to move. Many plankters have this capacity but are either able to move only weakly or are so restricted to vertical movement that they can-

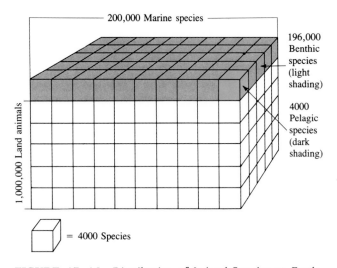

FIGURE 13–14   Distribution of Animal Species on Earth.
The cube represents the 1,200,000 species of animals living on
the earth. The shaded portion indicates the proportion (17%) that
live in the oceans. Of the marine species, 98% live on or in the
ocean floor.

not determine their horizontal position within the
ocean. The plants that follow this life style in the upper layers of the ocean are called *phytoplankton,* and
the animals are *zooplankton* (fig. 13–15). It has recently been discovered that free-living bacteria are
much more abundant in the plankton community
than they were previously thought to be. Having an
average dimension of only 0.5 μm (0.0002 in), the
bacterioplankton were missed in earlier studies because of their small size.

In considering the plankton in more detail, we will
try to develop some basic understanding of this portion of marine life that many of us have never seen
or imagined. The fact is, most of the biomass of the
earth is found adrift in the oceans as plankton. The
volume of the earth's space inhabited by animals that
either drift or swim greatly exceeds that occupied by
all animals that live on land or the ocean's bottom.

The plankton range in size from larger animals and
plants such as jellyfish, salps, and *Sargassum* (macro-
plankton, 2 to 20 cm, or 0.8 to 8 in) to bacteria that
are too small to be filtered from the water by a silk
net and must be removed by other types of microfilters (picoplankton, 0.2 to 2 μm, or 0.00008 to 0.0008
in).

An additional scheme for classifying plankton is
based on the portion of their life cycle spent within
the plankton community. Those organisms such as
planktonic diatoms and copepods, which spend their
entire life as plankton, are *holoplankton.* Many of the
organisms we normally consider as *nekton* or *benthos*
because they spend their adult life in one or the other
of these living modes actually spend a portion of their

life cycle as plankton. Many of the nekton and very
nearly all of the benthos during their larval stages
make the plankton community their home. These organisms, which as adults sink to the bottom to become members of the benthos or begin to swim
freely as members of the nekton, are *meroplankton.*

Nekton

The *nekton* include all those animals capable of moving independently of the ocean currents. Such animals are not only capable of determining their own
positions within relatively small areas of the ocean
but are also capable, in many cases, of long migrations. Included in the nekton are most adult fish and
squid, marine mammals, and marine reptiles (fig.
13–16).

The freely moving nekton, with some exceptions,
are not able to move at will throughout the breadth
of the ocean. By the presence of invisible barriers
they are effectively limited in their lateral range.
Gradual changes in temperature, salinity, viscosity,
and availability of nutrients create impenetrable barriers. The death of large numbers of fish has frequently been caused by temporary lateral shifts of
masses of ocean water. Vertical range is normally determined by pressure for all fish that contain air bladders and mammals that belong to the nekton.

Fish appear to be everywhere but are normally
considered to be more abundant near continents and
islands and in colder waters. Some fish, such as the
salmon, ascend rivers to spawn. Many eels do just the
reverse, growing to maturity in fresh water, then descending the streams to breed in the great depths of
the ocean.

Benthos

The *benthos* live on or in the ocean bottom. There are
those called *infauna* that live buried in the sand,
shells, or mud. Those that live attached to rocks or
move over the surface of the ocean bottom are the
epifauna. Benthos such as some shrimp and demersal
flounder that live in the bottom but move with relative ease through the water above the ocean floor are
the *nektobenthos* (fig. 13–17).

The littoral and inner sublittoral are the only zones
in which we find the macroscopic algae attached to
the bottom because they are the only benthic zones
to which sufficient light can penetrate. There is a
great diversity of physical and nutritive conditions in
these zones. Animal species have developed in great
numbers within this nearshore benthic community as
a result of the variations existing within the habitat.

PHYTOPLANKTON

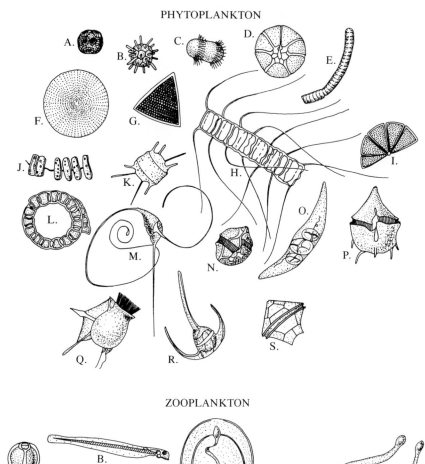

ZOOPLANKTON

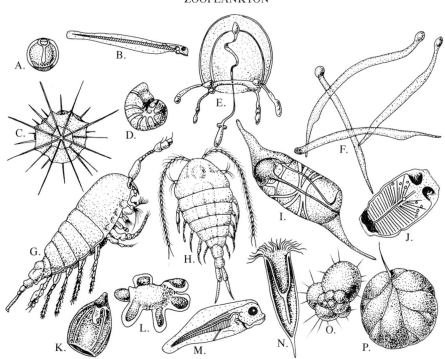

FIGURE 13–15 Phytoplankton and Zooplankton.

Maximum dimensions in parentheses. **Phytoplankton** *A* and *B,* Coccolithophoridae (15 μm, or 0.006 in). *C–L,* diatoms (80 μm, or 0.032 in): *C, Corethron; D, Asteromphalus; E, Rhizosolenia; F, Coscinodiscus; G, Biddulphia favus; H, Chaetoceras; I, Licmophora; J, Thalassiorsira; K, Biddulphia mobiliensis; L, Eucampia.* M–S, dinoflagellates (100 μm, or 0.04 in): *M, Ceratium recticulatum; N, Gonyaulax scrippsae; O, Gymnodinium; P, Gonyaulax triacantha; Q, Dynophysis; R, Ceratium bucephalum; S, Peridinium.* **Zooplankton** *A,* fish egg (1 mm, or 0.04 in); *B,* fish larva (5 cm, or 2 in); *C,* radiolarian (.5 mm, 0.2 in); *D,* foraminifer (1 mm, or 0.04 in); *E,* jellyfish (30 m, or 98 ft); *F,* arrowworms (3 cm, or 1.2 in); *G* and *H,* copepods (5 mm or 0.2 in); *I,* salp (10 cm, or 4 in); *J, Doliolum; K,* jellyfish (30 m, or 98 ft); *L,* worm larva (1 mm, or 0.04 in); *M,* fish larva (5 cm, or 2 in); *N,* tintinnid (50 μm, or 0.02 in); *O,* foraminifer (1 mm, or 0.04 in); *P,* dinoflagellate *(Noctiluca)* (100 μm, or 0.04 in).

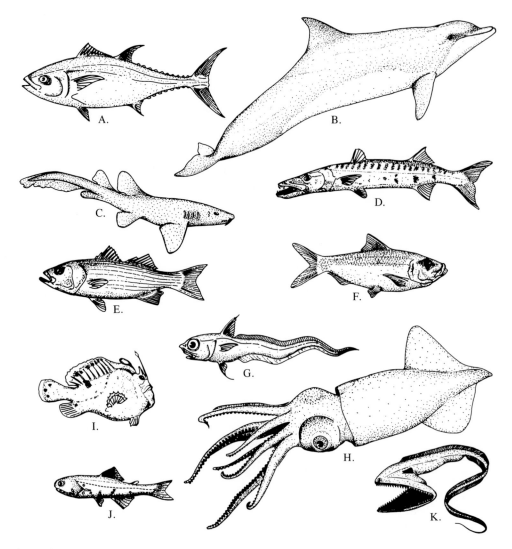

FIGURE 13–16   Nekton.

Not drawn to scale; typical maximum dimension is in parentheses. *A,* bluefin tuna (2 m, or 6.6 ft); *B,* bottlenosed dolphin (4 m, or 13 ft); *C,* nurse shark (3 m, or 10 ft); *D,* barracuda (1 m, or 3.3 ft); *E,* striped bass (0.5 m, or 1.6 ft); *F,* sardine (15 cm, or 6 in); *G,* deep-ocean fish (8 cm, or 3 in); *H,* squid (1 m, or 3.3 ft); *I,* angler fish (5 cm, or 2 in); *J,* lantern fish (8 cm, or 3 in); *K,* gulper (15 cm, or 6 in).

As one moves across the bottom from the littoral into the deeper benthic environments, one commonly observes that an inverse relationship exists between the distance from shore and the number of benthic species that can be found. Some studies, however, indicate a local increase in species diversity at the shelf break.

Throughout most of the benthic environment, animals live in a region of perpetual darkness, where photosynthetic production cannot occur. They must feed on each other or on whatever outside nutrients fall from the productive zone near the surface. The deep-sea bottom is an environment characterized by coldness, stillness, and darkness. Under these condi-tions, it would be expected that life would move at a relatively slow pace. For those animals that move around on the bottom, streamlining, which is very important to the nekton, is of little importance.

The organisms that live in the deep sea normally have quite a wide range of distribution because the physical conditions for life do not vary greatly, even over great distances, on the deep-ocean floor. A few species appear to be extremely tolerant of pressure changes in that members of the same species may be found in the littoral province and at depths of several kilometers.

The 1977 discovery of the first hydrothermal vent biocommunity in the Galapagos Rift has shown us

FIGURE 13–17 Benthos—Some Intertidal and Shallow Subtidal Forms.

Not drawn to scale; typical maximum dimension is in parentheses. *A,* sand dollar (8 cm, or 3 in); *B,* clam (30 cm, or 12 in); *C,* crab (30 cm, or 12 in); *D,* abalone (30 cm, or 12 in); *E,* sea urchins (15 cm, or 6 in); *F,* sea anemones (30 cm, or 12 in); *G,* brittle star (20 cm, or 8 in); *H,* sponge (30 cm, or 12 in); *I,* acorn barnacles (2.5 cm, or 1 in); *J,* snail (2 cm, or 0.8 in); *K,* mussels (25 cm, or 10 in); *L,* gooseneck barnacles (8 cm, or 3 in); *M,* sea star (30 cm, or 12 in); *N,* brain coral (50 cm, or 20 in); *O,* sea cucumber (30 cm, or 12 in); *P,* lamp shell (10 cm, or 4 in); *Q,* sea lily (10 cm, or 4 in); *R,* sea squirt (10 cm, or 4 in).

that high concentrations of deep-ocean benthos are possible. See *Life on the Deep-Ocean Floor,* on pages 432–434, for an interview with Dr. Robert Hessler, one of the foremost authorities on deep-ocean biology. He presents his views on the nature, distribution, and origin of life on the deep-ocean floor.

SUMMARY

The relatively stable marine environment is thought to have given rise to all phyla of organisms. Those organisms that have established themselves in the terrestrial realm have had to develop complex systems for support and for acquiring and retaining water.

Although the relative proportions of the constituents of salinity in the ocean and in the body fluids of the organisms are often very nearly the same, the salinity of one may differ greatly from that of the other. If the body fluids of an organism and ocean water are separated by a membrane that allows water molecules to pass, problems requiring adaptation may develop. Marine invertebrates and sharks are essentially **isotonic,** having body fluids with a salinity similar to that of ocean water. They rarely face adaptation problems. Most marine vertebrates are **hypotonic,** having body fluids with a salinity lower than that of ocean water, and tend to lose water through **osmosis,** the passing of water molecules by **diffusion** from a region in which they are in higher concentration through a semipermeable membrane into a region where they are in lower concentration. Freshwater organisms are essentially all **hypertonic,** having body fluids much higher in salinity than the water in which they live, so they must compensate for a tendency to take water into their cells through osmosis.

For life to flourish in any environment, there must be a sufficient food supply. The basic producers of food are plants, so the requirements of plants must be met if food is to be plentiful. The **availability of nutrients** and **solar radiation** make plant life possible.

Since solar radiation is available only in surface water of the ocean, plant life is restricted to a thin layer of surface water usually no more than 100 m (328 ft) deep. Nutrients derived ultimately from the continents are much more abundant near continental features. Although much is yet to be learned about the distribution of life in the oceans, it appears that the **biomass concentration** of the oceans decreases away from the continents and with increased depth. The color of the oceans ranges from green in highly productive regions to blue in areas of low productivity.

The plants that must stay in surface water to receive sunlight and the small animals that feed on plants do not have effective means of locomotion. They depend, therefore, on their **small size** and other adaptations to give them a **high ratio of surface area per unit of body mass**, which results in a greater frictional resistance to sinking. Large animals that swim freely face an altogether different problem and generally have **streamlined bodies** to reduce frictional resistance to motion.

Compared to life in colder regions, organisms living in warm water tend to be individually smaller, comprise a greater number of species, and constitute a much smaller total biomass. Warm-water organisms also tend to live shorter lives and reproduce earlier and more frequently than their cold-water counterparts.

The marine environment is divided into two basic units—the **pelagic** (water) and the **benthic** (bottom) environments. These regions, which are further divided primarily on the basis of depth, are inhabited by organisms we can classify into three categories on the basis of life style. These categories are the **plankton**, or free-floating forms with little power of locomotion; the **nekton**, or free swimmers; and the **benthos**, or bottom dwellers.

QUESTIONS AND EXERCISES

1. Discuss the major differences between marine plants and land plants, and explain land plants' need for greater complexity.

2. Define the terms euryhaline, stenohaline, eurythermal, and stenothermal. Where in the marine environment will organisms displaying a well-developed degree of each characteristic be found?

3. Describe the relationships among osmotic pressure, salinity, and freezing point of a solution.

4. What is the problem requiring osmotic regulation that is faced by a hypotonic fish in the ocean? How have these animals adapted to meet this problem?

5. An important variable in determining the distribution of life in the oceans is the availability of nutrients. What are the relationships among the continents, nutrients, and the concentration of life in the oceans?

6. Another important determinant of plant productivity is the availability of solar radiation. Why is biological productivity relatively low in the tropical open ocean where the penetration of sunlight is greatest?

7. Discuss the characteristics of the coastal ocean where unusually high concentrations of marine life are found.

8. What factors create the color difference between coastal waters and the less productive open-ocean water?

9. Compare the ability to resist sinking of an organism with an average linear dimension of 1 cm (0.4 in) to that of an organism with an average linear dimension of 5 cm (2 in). Discuss some adaptations other than size used by organisms to increase their resistance to sinking.

10. Changes in water temperature significantly affect the density, viscosity of water, and ability of water to hold gases in solution. Discuss how decreased water temperature changes these variables and may affect marine life.

11. Describe how higher water temperatures in the tropics may account for the greater number of species in these regions compared to low-temperature, high-latitude areas.

12. Construct a table listing the subdivisions of the benthic and pelagic environments and the physical factors used in assigning their boundaries.

13. Describe the vertical distribution of oxygen and nutrients in the oceanic province, and discuss the factors that are responsible for this distribution.

14. Discuss the probable cause and composition of the deep scattering layer.

15. List the relative number of species of animals found in the terrestrial, pelagic, and benthic environments, and discuss the factors that may account for this distribution.

16. Describe the lifestyles of plankton, nekton, and benthos. Why do plankton account for a relatively larger percentage of the biomass of the oceans than the benthos and nekton?

17. List the subdivisions of plankton and benthos and the criteria used for assigning individual species to each.

REFERENCES

Borgese, E. M., and Ginsburg, N. 1980. *Ocean Yearbook 2.* Chicago: The University of Chicago Press.

Coker, R. E. 1962. *This great and wide sea: An introduction to oceanography and marine biology.* New York: Harper and Row.

Darwin, C. 1859. *On the origin of species by means of natural selection, or the preservation of favoured races in the struggle for life.* London: John Murray.

Hedpeth, J., and Hinton, S. 1961. *Common seashore life of southern California.* Healdsburg, Calif.: Naturegraph.

Isaacs, J. D. 1969. The nature of oceanic life. *Scientific American* 221:65–79.

Sieburth, J. M. N. 1979. *Sea microbes.* New York: Oxford University Press.

Sumich, J. L. 1976. *An introduction to the biology of marine life.* Dubuque, Iowa: Wm. C. Brown.

Sverdrup, H.; Johnson, M.; and Fleming, R. 1942. Renewal 1970. *The oceans.* Englewood Cliffs, N.J.: Prentice-Hall.

Thorson, G. 1971. *Life in the sea.* New York: McGraw-Hill.

SUGGESTED READING

Sea Frontiers

Burton, R. 1977. Antarctica: Rich around the edges. 23:5, 287–95.
   The high level of biological productivity around the continent of Antarctica is the topic.

Gruber, M. 1970. Patterns of marine life. 16:4 194–205.
   Many varieties of life in the ocean are discussed in terms of how their form and size fit them for life in a particular environmental niche.

Hammer, R. M. 1974. Pelagic adaptations. 16:1, 2–12.
   A comprehensive discussion of the adaptations of pelagic organisms to reduce the energy required to maintain their position in the open ocean.

Patterson, S. 1975. To be seen or not to be seen. 21:1, 14–20.
   A discussion of the possible role of color in the protection and behavior of tropical fishes.

Schellenger, K. 1974. Marine life of the Galapagos. 20:6, 322–32.
   A discussion of the unique life forms of the Galapagos Islands, 950 km (589 mi) from South America.

Thresher, R. 1975. A place to live. 21:5, 258–67.
   An interesting discussion of how bottom-dwelling animals compete for space on the ocean floor.

Scientific American

Denton, E. 1960. The buoyancy of marine animals. 203:1, 118–29.
    The means by which some marine animals reduce the energy expenditure re-
    quired to live in the ocean water far above the ocean floor are discussed.

Isaacs, J. D., and Schwartzlose, R. A. 1975. Active animals of the deep-sea floor. 233:4,
    84–91.
    A surprisingly large population of large fishes on the deep-sea floor is suggested
    by automatic cameras dropped to the ocean bottom.

Isaacs, J. D. 1969. The nature of oceanic life. 221:3, 146–65.
    A well-developed survey of the conditions for life in the ocean as they relate to the
    variety and distribution of marine life forms.

Palmer, J. D. 1975. Biological clock and the tidal zone. 232:2, 70–79.
    This article investigates the mechanism of biological clocks set to the rhythm of
    the tides which are found in organisms from diatoms to crabs.

Partridge, B. L. 1982. The structure and function of fish schools. 246:6, 114–23.
    Schooling benefits and the means by which fish maintain contact with the school
    are considered.

Vogel, S. 1978. Organisms that capture currents. 239:2, 128–39.
    The manner in which sponges use ocean currents is an important part of this dis-
    cussion.

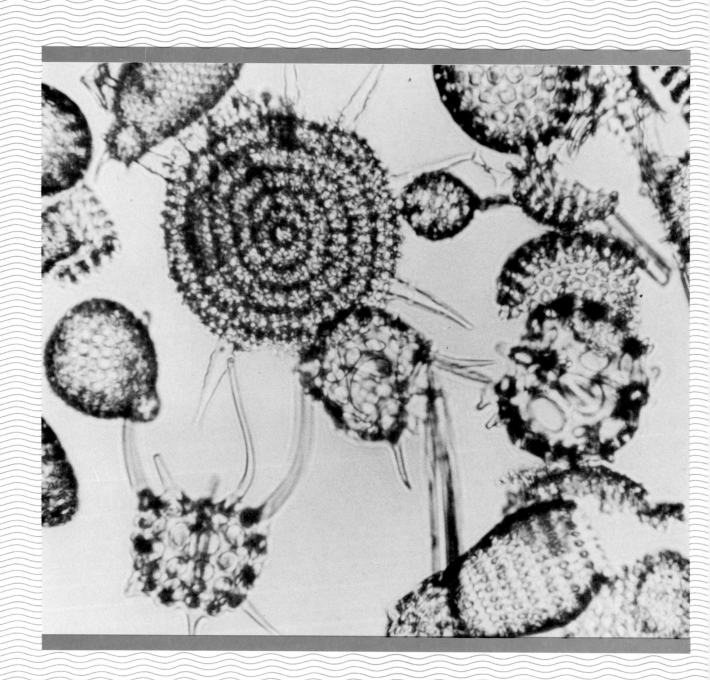

The major primary producers of the oceans, marine algae, capture most of the energy used to support the marine biological community. When we think of marine plants, most of us undoubtedly first consider the large macroscopic plants we see growing near shore in many coastal areas. These large plants, however, play only a minor part in the production of energy for the ocean population as a whole. Instead, marine organisms depend primarily on the small planktonic varieties of marine plant life that inhabit the uppermost sunlit water of the world's oceans. They are not so obvious. All are microscopic, but they are scattered throughout the breadth of the ocean surface layer and represent the largest biomass community in the marine environment—phytoplankton.

In this chapter we will first consider the classification of organisms. We will then discuss the composition of the marine algae, the ecological factors related to plant and bacterial productivity, and the passing of stored chemical energy on to other marine organisms within food webs.

TAXONOMIC CLASSIFICATION

All living things belong to one of the five kingdoms shown in figure 14–1 (*see* appendix V). The simplest of all organisms are classified as *Monera*. These organisms are single-celled and have their nuclear material spread throughout the cell. Included in this kingdom are the blue-green algae and bacteria. The blue-green algae (phylum Cyanophyta, or Cyanobacteria) include the type of algae that, parodoxically, are responsible for the color of the Red Sea, but their overall role in the marine environment is not well known. Bacteria are rapidly being recognized as an important component of the food chain in addition to their well-known role as decomposers of dead organic matter.

Representing a higher stage of evolutionary development is the kingdom *Protista*. It includes those organisms that are single-celled but have their nuclear material contained within a nuclear sheath. Here we find the rest of the algae (simple plants) and the single-celled animals, Protozoa. The Protozoa component of the Protista will be discussed in more detail in the following chapter, but here we will mention members of these

CHAPTER FOURTEEN
BIOLOGICAL PRODUCTIVITY—ENERGY TRANSFER

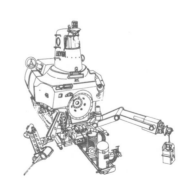

Radiolarians. Photo by Michael DiSpezio.

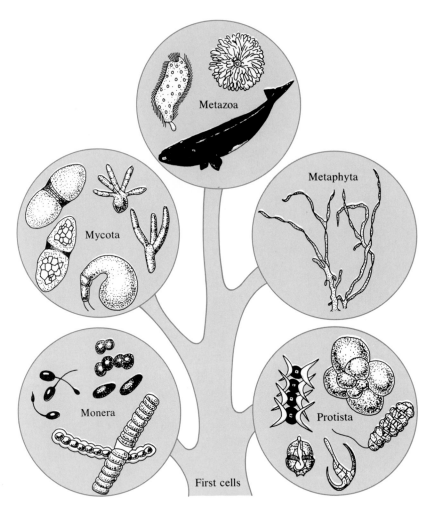

FIGURE 14–1   The Five Kingdoms of Organisms.

Monera—single cells without a nucleus; Protista—single cells with a nucleus; Mycota—the fungi;
Metaphyta—complex, many-celled plants; Metazoa—complex, many-celled animals.

groups as we consider the paths followed in disseminating energy throughout the marine biological community. Most of the marine plant community falls within this kingdom, and our discussions in this chapter will deal primarily with the conditions related to the existence of protistan algae within the marine biological community.

Kingdom *Mycota,* the fungi, appears to be poorly represented in the oceans. Fewer than 1% of the known 50,000 species are found there. Fungi are found throughout the marine environment, but they are much more common in the intertidal zone in a mutualistic relationship with blue-green or green algae. In this relationship, called *lichen,* the fungus provides a protective covering that retains water during periods of exposure, while the algae provides food for the fungus through photosynthesis. Other fungi func-

tion primarily as decomposers in the marine ecosystem: they remineralize organic matter.

Though there is one more kingdom of marine plants—*Metaphyta,* the multicelled plants—its members are mostly restricted to the shallow coastal margins of the ocean and are not an important portion of the marine productive community. They do play a very important role as producers within restricted communities, such as mangrove swamps and salt marshes.

The kingdom *Metazoa* consists of multicelled animals. Metazoans range in complexity from the simple sponges to the vertebrates (animals with backbones).

The kingdoms are divided into increasingly specific groupings *(taxa).* The system of taxonomic classification introduced by Carl von Linné (Linnaeus) in 1758 includes the following major categories:

Kingdom
  Phylum
    Class
      Order
        Family
          Genus
            Species

Every organism's scientific name includes its genus and species. For example, the toothed whale called the common dolphin is the species *Delphinus delphis.*

MACROSCOPIC PLANTS

As we discuss the marine plants, we will first consider those groups of plants with which we are most familiar. These are the attached forms of macroscopic algae and Metaphyta found in shallow waters along the margins of the ocean (fig. 14–2).

Phaeophyta (Brown Algae)

The largest members of the marine plant community that are attached to the bottom wherever suitable substrate is available in the littoral and inner sublittoral zones belong to the Phaeophyta, brown algae. The dominant pigment in these brown algae is *fucoxanthin* and the color may range from a very light brown to black. The brown algae occur primarily in temperate and cold-water areas. Their sizes range from the small black encrusting patch of the *Ralfsia,* found in the upper and middle intertidal zones where it is exposed to an extensive threat of desiccation and may become crisp and dry in the sun without dying, to the *Pelagophycus* (bull kelp), which may grow to the surface from small holdfasts (attachment structures) in water depths in excess of 30 m (98 ft). A genus of brown algae, *Sargassum,* may be attached to the bottom of nearshore waters in a belt between the subtropics or float freely at the ocean's surface. The floating variety found in the Sargasso Sea southeast of Bermuda is buoyed up by small grape-sized air bladders.

Chlorophyta (Green Algae)

The predominant forms of algae found in freshwater environments belong to the Chlorophyta, green algae, and are not well represented in the ocean. Most species are intertidal, or grow in shallow waters of bays. Because of the pigment *chlorophyll* they range in color from yellow green to a very dark green, but most are grass green in color. They grow only to mod-

erate size, seldom exceeding 30 cm (12 in) in the largest dimension. Forms range from finely branched filaments to flat thin sheets. The various species of the genus *Ulva* (sea lettuce), a thin membranous sheet two cell layers thick, may be found widely scattered throughout cold-water areas (fig. 14–2A). The genus *Codium* (sponge weed) is a dichotomously branched form that can be in excess of 6 m (20 ft) in length and is more commonly found in warm waters (fig. 14–2D).

Three genera of green algae, *Acetabularia, Halimeda* and *Penicillus,* secrete calcium carbonate structures and are import-contributors to marine sediment in warm waters.

Rhodophyta (Red Algae)

Red algae belong to the Rhodophyta and are the most abundant and widespread of the marine macroscopic algae. Over 4000 species are found, many of them attached, from the very highest intertidal levels to the outer edge of the inner sublittoral zone. They are very rare in fresh water. The red algae range in size from just visible to the naked eye to lengths of up to 3 m (9 ft). While Rhodophyta are found in both warm- and cold-water areas, the warm-water varieties are relatively small. The characteristic pigment of the red algae is *phycoerythrin.* The color of the red algae will vary considerably depending on their depth in the intertidal or inner sublittoral zones. In the upper well-lighted areas it may be green to black or purplish in color, and it changes through a brown to a pinkish red in the deeper water zones where light concentrations lessen. Although the bulk of marine plant productivity is believed to occur above water depths where the amount of light is reduced to 1% of that available at the surface (approximately 100 m, or 330 ft) a red alga has been observed growing at a depth of 268 m (880 ft) on a seamount near San Salvador, Bahamas. Available light at this sighting was thought to be only 0.05% of the light available at the ocean's surface.

Spermatophyta (Seed-Bearing Plants)

The only Metaphyta observed in the marine environment belong to the highest group of plants, the seed-bearing Spermatophyta. Two seed-bearing plants commonly found in the marine environment are *Zostera* (eelgrass) and *Phyllospadix* (surf grass). *Zostera,* a grasslike plant with true roots, is found primarily in quiet waters of bays and estuaries from the low-tide zone down to a depth of some 6 m (20 ft). *Phyllo-*

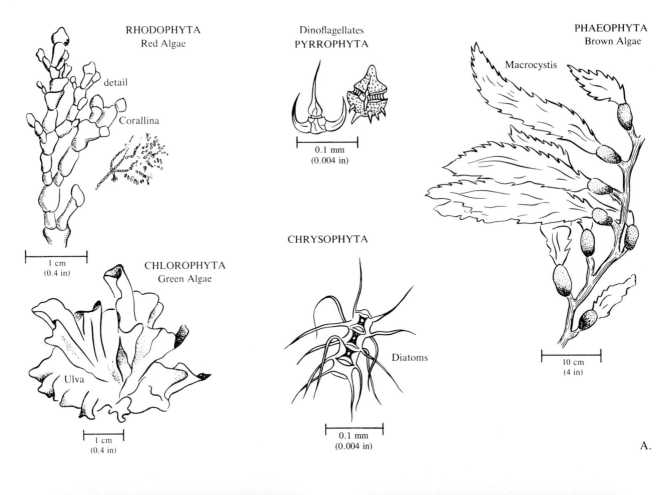

RHODOPHYTA
Red Algae

detail

Corallina

1 cm
(0.4 in)

Dinoflagellates
PYRROPHYTA

0.1 mm
(0.004 in)

PHAEOPHYTA
Brown Algae

Macrocystis

CHLOROPHYTA
Green Algae

Ulva

1 cm
(0.4 in)

CHRYSOPHYTA

Diatoms

0.1 mm
(0.004 in)

10 cm
(4 in)

A.

1 cm.

B.

1 cm.

C.

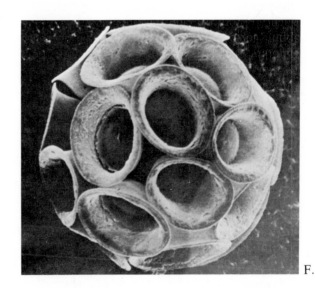

FIGURE 14–2 Algae.

A, detailed drawing of several algae. Line drawing by Phil David Weatherly. *B,* red algae, *Corallina vancouveriensis. C,* brown algae, *Egregia laevigata.* Feather-boa kelp, common in the surf zone. Note the flat axis with featherlike fronds, some of which have air bladders at the base. *D,* green algae, *Codium fragile.* The prominent fingerlike plant is called *sponge weed. B–D,* photos from John D. Roche, Jr. *E,* photomicrograph of *Gonyaulax polyedra* magnified 1100 times. *Gonyaulax* is a large genus of phosphorescent marine dinoflagellates which, in great abundance, cause what is known along the shoreline as red tide. Photo from Scripps Institute of Oceanography, University of California, San Diego. *F,* coccoliths, disc-shaped $CaCO_3$ plates that cover coccolithophore cells. Courtesy of Deep Sea Drilling Project, S.I.O.–U.C.S.D.

FIGURE 14–3 Rocky Coast Exposed During Low Tide.
Phyllospadix (surf grass) is visible in the upper right and left corners of the photo. *Eisenia arborea,* predominant brown algae in the photo, is commonly called *sea palm. Egregia,* feather-boa kelp, is in the center of the lower right quarter of the photo. *Corallina,* a red algae, can be seen as the prominent tuft in the tide pool at the lower left. Photo from John D. Roche, Jr.

spadix prefers the high-energy environment of an exposed rocky coast and can be found from the intertidal zones down to a depth of 15 m (50 ft). Both of these plants are considered to be important sources of the detrital food for the marine animals that inhabit their environment (fig. 14–3). Found in salt marshes are grasses belonging mostly to the genus *Spartina,* while mangrove swamps contain primarily the mangrove genera *Rhizophora* and *Avicennia.*

MICROSCOPIC PLANTS

The following plants include all the members of the

important phytoplankton that produce in excess of 99% of the food supply for marine animals. They are primarily free-drifting forms, although some will live on the bottom in the nearshore environment.

Chrysophyta (Golden Algae)

Containing the yellow pigment *carotene* (or carotin), these microscopic plants store food in the form of *leucosin* (a carbohydrate) and oils.

Diatoms   The diatoms are cells contained in a shell, or *frustule,* composed of opaline silica ($SiO_2 \cdot nH_2O$). These silica housings are important geologically because they accumulate on the ocean bottom and produce a siliceous sediment, *diatomite.* Some deposits of diatomite on land are mined and used primarily in the manufacture of filtering devices.

The frustule of the diatom is similar in structure to a microscopic pill box. The top and the bottom of the box are called *valves;* the larger of the valves is the *epitheca* and the smaller valve is the *hypotheca.* The protoplasm of the plant is contained within this housing and exchanges nutrients and waste products with the surrounding water through slits or pores in the valves.

The reproduction of diatoms is by simple cell division. To achieve this division the valves of the frustule must separate. Each valve then serves as one-half the housing for each newly formed cell. A new valve must form to complete the enclosure of each daughter cell, and each of these newly formed halves forms as a hypotheca.

As can be seen in figure 14–4, this process leads to the formation of smaller and smaller organisms.

FIGURE 14–4   Diatom Reproduction.
Diatom cell division. As the epitheca and hypotheca of a diatom separate, each becomes the epitheca of a new cell. The new frustule is completed by the generation of a new hypotheca by each new diatom. Following the arrow through the formation of three generations of diatoms, it can be seen that some cells will be crowded into ever smaller frustules. When the size of the frustule becomes critically small, an auxospore forms to allow growth of a larger cell.

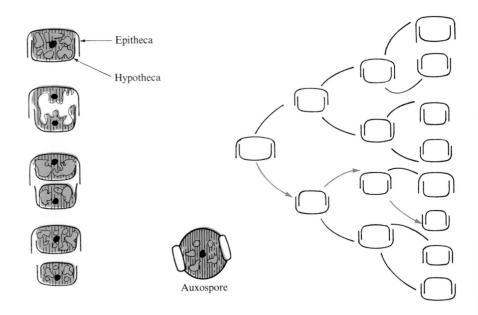

Epitheca

Hypotheca

Auxospore

Eventually, the newly formed daughter cell is going to approach a size so small that if it were decreased further, abnormalities would result. At this point, a change in the process occurs, namely, the formation of an *auxospore*. The auxospore forms between the two separating valves within a membrane that allows the auxospore to grow into a full-sized diatom so that the process of reproduction by the splitting of the frustule can commence again.

Coccolithophores Coccolithophores are for the most part flagellum-bearing organisms (whiplike locomotive structures) covered with small calcareous plates ($CaCO_3$), *coccoliths*. The name of the group means *bearers of coccoliths*. The individual plates are about the size of a bacterium, and the entire organism is too small to be captured in plankton nets. Coccolithophores are included in the nanoplankton, which have dimensions of less than 0.06 mm (0.002 in). The coccolithophores contribute significantly to calcareous deposits in all of the temperate and warmer oceans.

Pyrrophyta (Dinoflagellate Algae)

A second group of phytoplankton important in marine productivity is the dinoflagellates. Possessing *flagella* for locomotion, they have a slight capacity to move into more favorable areas for plant productivity. The dinoflagellates are not important geologically because many have no protective covering. Those that do have coverings made of cellulose, which is easily decomposed by bacterial action when the organism dies. Many of the 1100 species are luminescent.

Red tides result from conditions where up to 2 million dinoflagellates may be found in one liter (1.1 qt) of water. Mainly responsible for this red tide phenomenon are the genera *Gymnodinium* and *Gonyaulax*, which produce water-soluble toxins. *Gonyaulax* toxin is not poisonous to shellfish but concentrates in their tissue and is poisonous to humans who eat the shellfish. *Gymnodinium* toxin kills fish and shellfish. April through September are particularly dangerous months. In most areas there is a quarantine against the taking of shellfish that feed on these microscopic organisms and concentrate the poisons that they secrete to levels that are dangerous to humans. *Gymnodinium brevis* is an important contributor to Gulf of Mexico red tides. *Gonyaulax* is an important participant in the cooler waters of the New England and West Coast areas.

A potentially tragic epidemic of *paralytic shellfish poisoning* (PSP) from a *Gonyaulax tamarensis* red tide occurred in the coastal waters of Massachusetts in the fall of 1972. Fortunately, no deaths occurred, although 30 cases of poisoning were reported. Symptoms of PSP are similar to those of drunkenness, including incoherent speech, uncoordinated movement, dizziness, and nausea. Documented cases throughout the world include 300 deaths and 1750 nonfatal cases. There is no known antidote for the toxin, which attacks the human nervous system, but the critical period usually passes in 24 hours.

PRIMARY PRODUCTIVITY

Primary productivity is defined as the amount of carbon fixed by autotrophic organisms through the synthesis of organic matter from inorganic compounds such as CO_2 and H_2O using energy derived from solar radiation or chemical reactions. The major process through which primary productivity occurs is thought to be photosynthesis. Although new knowledge of the role of chemosynthesis in supporting hydrothermal vent communities on oceanic spreading centers has emerged, chemosynthesis is much less significant in overall marine primary production than is photosynthesis. Because of the far greater importance of photosynthesis in the overall productivity of the oceans and our greater knowledge of the factors that affect marine photosynthesis, the following considerations will relate primarily to this process.

The total amount of organic matter produced by photosynthesis per unit of time represents the *gross primary production* of the oceans. Plants will use some of this organic matter for their own maintenance through respiration. That which remains is the *net primary production,* which is manifested as growth and reproduction products. It is the net primary production that supports the heterotrophic marine populations—animals and bacteria.

In the 1920s, the Gran Method of measuring net primary productivity was developed based on the fact that oxygen is produced by photosynthesis in proportion to the amount of organic carbon synthesized. The method involves having equal quantities of phytoplankton in each of a series of bottles, all of which contain the same amount of dissolved oxygen. The bottles are then arranged in pairs, one being transparent and the other totally opaque. These pairs are suspended on a hydrographic line through the euphotic zone, where they are left for a specific period of time. After the bottles are brought to the surface, the oxygen concentration is determined for each bottle.

Photosynthetic activities, which are confined to the transparent bottles, will add oxygen to the water, whereas respiration will reduce the oxygen content in both the transparent and opaque bottles. Increased oxygen concentration in the transparent bottles is proportional to the amount of photosynthesis that has occurred minus the oxygen consumed by plant respiration and thus represents the net increase in biomass of the plants within the transparent bottle. Decreased oxygen content within the opaque bottles corresponds to the respiration rate. For any depth, an assessment of photosynthesis can be made by adding the oxygen gain in the clear bottles to the oxygen loss in the opaque bottles.

The depth at which the oxygen production and the oxygen consumption are equal is called the *oxygen compensation depth,* which represents that light intensity below which plants do not survive. Since respiration goes on at all times, during the daylight hours plants must produce, through photosynthesis, biomass in excess of that which is consumed by respiration in any 24-h period if the total biomass of the community is to increase (fig. 14–5). An analogy can be made with the common paycheck: Gross primary production (Gross pay earned) = Oxygen change in clear bottle (Take-home pay) + Oxygen loss in dark bottle (Income tax withheld).

During the 1950s, a method involving the use of radioactive carbon (C^{14}) was developed. It has been refined and is currently the most often used method for determining marine primary productivity. The procedure is similar to that described above, because it involves the use of a series of paired clear and opaque bottles. Each bottle contains identical phytoplankton samples, equal amounts of CO_2 containing carbon-14, and equal amounts of CO_2 containing stable carbon.

The phytoplankton sample is filtered from each bottle after the system has been suspended in the ocean for a sufficient period of time. The amount of beta-radiation emitted by each sample is measured by a radiation counting device, and the rate of assimilation of carbon-14 is computed from these measurements. High levels of radioactive emission indicate high levels of productivity.

A third method used in studies of marine primary productivity is the measurement of *chlorophyll a* content of living phytoplankton samples taken from ocean surface waters. Although this method is not as precise as the measurement of net primary productivity by use of carbon-14, there is a direct relationship between the amount of chlorophyll *a* in a phytoplankton sample from a given volume of ocean water and gross primary productivity.

The special feature *Satellites and Plankton,* on pages 378–380, discusses a method being developed to help extrapolate local measurements of chlorophyll content into a global estimate of biological productivity (plates 6 and 21).

Distribution of Productivity

The production of organic compounds such as carbohydrates, proteins, and fats by plants through the process of photosynthesis will vary considerably throughout the areal extent of the ocean as well as in time. Photosynthetic productivity in the oceans depends on (1) the availability of solar radiation and (2) the availability of nutrients. Both variables are related to the seasonal pattern that grows out of the earth's revolution around the sun while rotating on an axis that is tilted relative to the ecliptic.

Primary photosynthetic productivity of the oceans varies from about 0.1 gram of carbon per square meter per day ($gC/m^2/d$) in the open ocean to over $10 gC/m^2/d$ in highly productive coastal areas. This variability is primarily the result of the uneven distribution of nutrients throughout the photosynthetic zone. The causes of this variability are discussed below.

Temperature Stratification and Nutrient Supply

Thermoclines are relatively permanent in low-latitude areas, while a seasonal thermocline may be present during the summer months in the mid-latitudes.

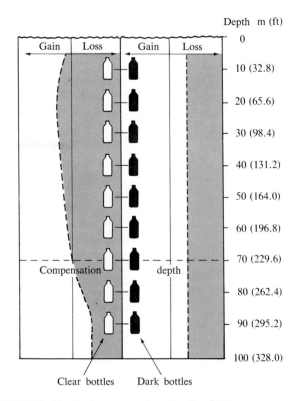

FIGURE 14–5   Compensation Depth of Oxygen.

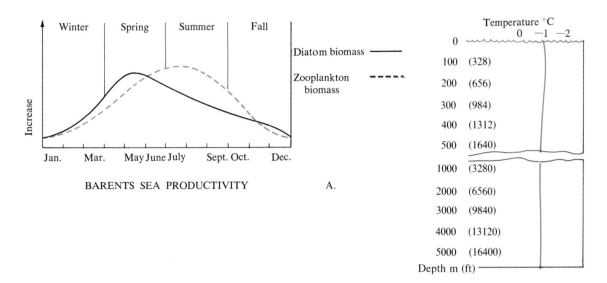

BARENTS SEA PRODUCTIVITY   A.

Polar water shows little temperature stratification.
Nearly isothermal conditions from surface to
bottom are common.   C.

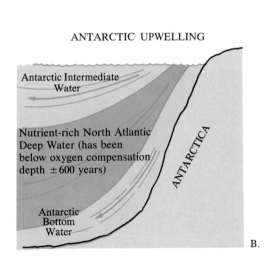

ANTARCTIC UPWELLING

B.

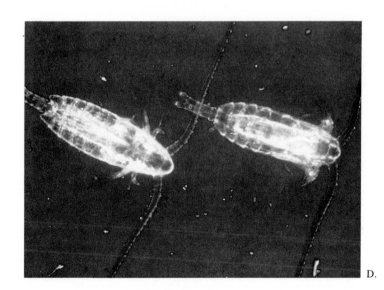

D.

FIGURE 14–6   Productivity in Polar Oceans.
D, photo courtesy of Scripps Institution of Oceanography, University of California, San Diego.

They are generally nonexistent in high-latitude areas. In the following discussion of productivity, we will see that the presence or absence of the thermocline will have an important effect on nutrient availability.

In considering the availability of solar radiation and nutrients, we can rather conveniently divide the oceans into polar, temperate, and tropical regions. We will concern ourselves primarily with the nature of the open ocean in all three regions. The polar re-gion will be discussed first as the productivity considerations are simplest there.

Polar Region   As an example of productivity on a seasonal basis in a polar sea, we will consider the Barents Sea off the northern coast of Europe, where there is peak diatom productivity during the month of May that tapers off through July (fig. 14–6A). This brings about a peak period of zooplankton development, consisting primarily of small crustaceans of the genus *Calanus.* The zooplankton biomass reaches a peak in June and continues at a relatively high level until winter darkness begins in October. In this region above 70°N latitude, there is continuous darkness for about 3 months of winter and continuous illumination for a period of 3 months during summer.

In the Antarctic region, particularly at the southern end of the Atlantic Ocean, productivity patterns are similar to those found in the Barents Sea, except that the seasons are reversed, and productivity is somewhat greater. The most likely explanation for this greater productivity in Antarctic waters is the continual upwelling of water that has sunk in the North Atlantic. Moving southward as a deep-water mass, the North Atlantic Deep Water surfaces hundreds of years later carrying high concentrations of nutrients (fig. 14–6B).

To illustrate the very great productivity that occurs during the short summer season in polar oceans, we can consider the growth rate of baby whales. The largest of all whales, the blue whale, migrates through the temperate and polar oceans at the times of maximum zooplankton productivity. As a result of this excellent timing, they are able to develop and support calves that during a gestation of 11 months reach lengths in excess of 7 m (23 ft) at birth. The mother suckles the calf for 6 months with a teat that actually pumps the youngster full of rich milk. By the time the calf is weaned, it is over 16 m (52 ft) in length and will in a period of 2 yr reach a length approaching 23 m (75 ft). In 3½ yr, a 60-ton blue whale has developed. This phenomenal growth rate gives some indication of the enormous biomass of copepods and krill that these large mammals feed upon.

In polar waters there is little density stratification to prevent the mixing of deeper water with shallow water. Polar waters are relatively isothermal (fig. 14–6C). In most polar areas, the surface waters freely mix with the deeper nutrient-rich water. There can be, however, some density segregation of water masses due to the summer melting of ice, a process that lays down a thin low-salinity layer that does not readily mix with the deeper waters.

There are usually high concentrations of phosphates and nitrates in the surface waters. Thus, plant productivity in the high latitudes is more commonly limited by the availability of solar energy than by the availability of nutrients. The productive season in these waters will be relatively short but will be characterized by an outstandingly high rate of production.

Tropical Region   In direct contrast with high productivity associated with the summer season in the polar seas, low productivity is the rule in the tropical regions of the open ocean. Light penetrates much deeper into the open tropical ocean than into the temperate and polar waters. This produces a very deep compensation depth. In the tropical ocean, however, a permanent thermocline produces a relatively permanent stratification of water masses and prevents mixing between the surface waters and the nutrient-rich deeper waters (fig. 14–7A).

At about 20° latitude we will commonly find the concentrations of phosphate and nitrate to be less than 1/100 of the concentrations of these nutrients in temperate oceans during winter. Nutrient-rich waters within the tropics lie for the most part below 150 m (492 ft), and the highest concentration of nutrients occurs between 500 and 1000 m (1640 and 3280 ft) depth.

There is generally a steady low rate of primary productivity in tropical oceans. Although the rate of tropical productivity is low, when we compare the total annual productivity of tropical oceans with that of the more productive temperate oceans, we find that the tropical productivity is generally at least half of that found in the temperate regions on an annual basis.

Within tropical regions there are three environments where productivity is unusually high—regions of equatorial upwelling, coastal upwelling, and coral reefs (fig. 14–7B). In areas where trade winds drive westerly equatorial currents on either side of the equator, surface water diverges as a result of the Ekman transport. This surface water that moves off toward higher latitudes is replaced by nutrient-rich water that surfaces from depths of up to 100 m (328 ft). This condition of equatorial upwelling is probably best developed in the eastern Pacific Ocean. Where the prevailing winds blow toward the equator and along the western margin of continents, the surface waters are driven away from the coast. They are replaced by nutrient-rich waters from depths of 200–900 m (656–2952 ft). As a result of this upwelling of nutrient-rich water, there is a high rate of primary productivity in these areas, which support large fisheries. Such conditions exist along the southern coast of California and the southwest coast of Peru in the Pacific Ocean. Upwelling also occurs along the northwest coast of Morocco and the southwest coast of Africa in the Atlantic Ocean. The relatively high productivity of coral reef environments is not related to the upwelling process and is discussed in chapter 16.

Temperate Region   We have discussed the general productivity picture in the polar regions, where productivity is limited primarily by the availability of solar radiation, and in the tropical low-latitude areas, where the limiting factor is the availability of nutrients. We will next consider the temperate regions,

TROPICAL REGIONS OF HIGH PRODUCTIVITY
(local areas where nutrients are brought to the surface)

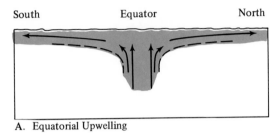

A. Equatorial Upwelling

Nutrient-enriched water from 200 to 300 m
(656 to 984 ft) replaces diverging equatorial
surface water.

B. Coastal Upwelling

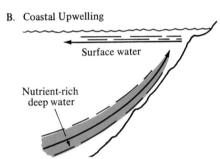

Surface water moved away from the shore
is replaced by deep water that is rich in
nutrients because it comes from below the
euphotic layer.

FIGURE 14–7 Productivity in Tropical Oceans.

NORMAL TROPICAL REGIONS

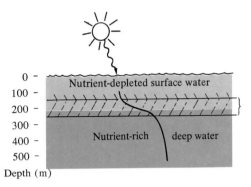

Deep penetration of sunlight produces a deep
compensation depth with a good supply of solar
radiation available for photosynthesis.

Permanent thermocline

Thermocline serves as an effective barrier to mixing of surface and deep water. As plants use
nutrients in surface layer, productivity is retarded because thermocline prevents replenish-
ment from deeper water.

where an alternation of these factors controls productivity in a pattern that is somewhat more complex.

Productivity in temperate oceans is at a very low
level during the winter months, although high concentrations of nutrients are available in the surface
layers. In fact, the nutrient concentration is higher
during the winter season than at any time throughout
the year. The limiting factor on productivity during
the winter season in the temperate ocean is the availability of solar energy. Since the sun is at its lowest
elevation above the horizon during this season, a
higher percentage of solar energy is reflected and a
smaller percentage absorbed into the surface waters.
The compensation depth for basic producers such as
diatoms is so shallow that it does not allow growth of
the diatom population (fig. 14–8).

As the sun rises higher in the sky during the spring
season, the compensation depth deepens as the
amount of solar radiation being absorbed by the surface water increases. Eventually, there is sufficient
water volume included within the water column

above the compensation depth to allow for the expo-
nential growth of the diatom population. This expanding population puts a tremendous demand on
the nutrient supply in the euphotic zone. In most
Northern Hemisphere areas, decreases in the diatom
population will occur by May as a result of insufficient nutrient supply.

As the sun rises higher in the sky during the summer months, the surface waters in the temperate
parts of the ocean are warmed and the water becomes separated from the deeper water masses by a
seasonal thermocline that may develop at depths of
approximately 15 m (49 ft). As a result of this thermocline development, there is little or no exchange
of water across this discontinuity, and the nutrients
that are depleted from surface waters cannot be replaced by those available in the deep waters.
Throughout the summer months the plant population
will remain at a relatively low level but will again increase in some temperate areas during the autumn
months.

FIGURE 14–8 Productivity in Temperate Oceans.

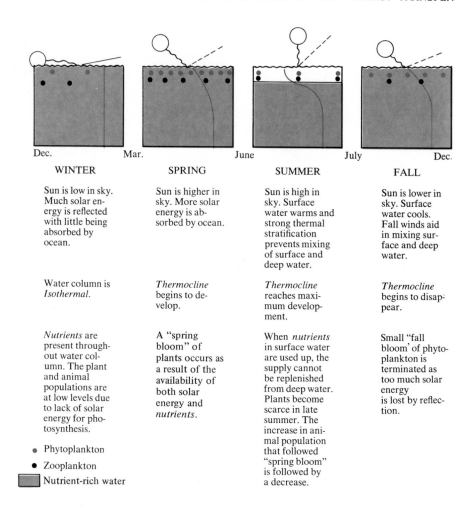

WINTER	SPRING	SUMMER	FALL
Sun is low in sky. Much solar energy is reflected with little being absorbed by ocean.	Sun is higher in sky. More solar energy is absorbed by ocean.	Sun is high in sky. Surface water warms and strong thermal stratification prevents mixing of surface and deep water.	Sun is lower in sky. Surface water cools. Fall winds aid in mixing surface and deep water.
Water column is *Isothermal*.	*Thermocline* begins to develop.	*Thermocline* reaches maximum development.	*Thermocline* begins to disappear.
Nutrients are present throughout water column. The plant and animal populations are at low levels due to lack of solar energy for photosynthesis.	A "spring bloom" of plants occurs as a result of the availability of both solar energy and *nutrients*.	When *nutrients* in surface water are used up, the supply cannot be replenished from deep water. Plants become scarce in late summer. The increase in animal population that followed "spring bloom" is followed by a decrease.	Small "fall bloom" of phytoplankton is terminated as too much solar energy is lost by reflection.

- Phytoplankton
- Zooplankton
- Nutrient-rich water

The autumn increase is much less spectacular than that of the spring because of the decreasing availability of solar radiation resulting from the sun dropping lower in the sky. This causes a decrease in surface temperature and the breakdown of the summer thermocline. A return of nutrients to the surface layer occurs as increased wind strength mixes it with the deeper water mass in which the nutrients have been trapped throughout the summer months. This bloom is very short-lived. The phytoplankton population begins to decrease rapidly. The limiting factor in this case is the opposite of that which reduced the population of the spring phytoplankton. In the case of the spring bloom, solar radiation was readily available, and the decrease in nutrient supply was the limiting factor.

Coastal waters with high nutrient levels are highly productive, or *eutrophic*. Most of the open ocean has a low level of nutrients and productivity. This low productivity is referred to as an *oligotrophic* condition. Figure 14–9 shows the general patterns of ocean productivity based on photosynthetically fixed carbon.

New measurements of photosynthetic productivity of oligotrophic waters in the North Atlantic and North Pacific oceans indicate they may be from 2 to 7 (or even more) times as productive as C^{14} data indicate. Instead of using the small bottles used in the C^{14} measurements and suspending them for a short time in the ocean, some physical oceanographers have used a much larger "bottle." In their studies of the circulation within the subtropical gyres, physical oceanographers use "bottles" that are "capped" by the pycnoclines produced by the thermoclines beneath the warm layers of surface water within the subtropical gyres of the open oceans. These "bottles" contain (depending on the design of the study) from tens to thousands of km^3 of water that record the average results of photosynthetic activity over periods of from a few months to decades.

The new studies analyze the effect of photosynthesis on the oxygen concentrations (1) in the euphotic zone and (2) beneath the euphotic zone. In the North Pacific Ocean, oxygen saturations in *subsurface oxygen maximums* (SOMs) at depths between 50 and 100 m (164 and 328 ft) were from 110% to 120% at

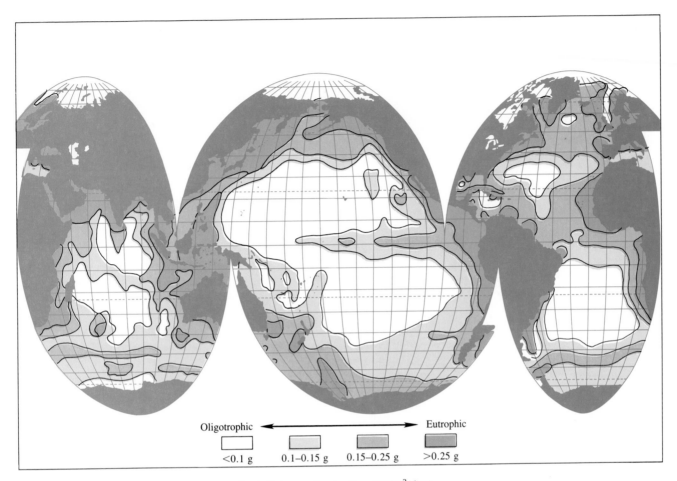

Oligotrophic ⟵——————————————⟶ Eutrophic

<0.1 g	0.1–0.15 g	0.15–0.25 g	>0.25 g

FIGURE 14–9 Distribution of Photosynthetic Primary Production (gC/m²/day).

latitudes of from 30° to 40°N during summer months based on data obtained from 1962 through 1979. The excess oxygen (anything over 100%) may be due to trapped photosynthetically produced oxygen. The North Atlantic study determined the rate at which oxygen is used up beneath the euphotic zone (100 m) by decomposing organic matter falling toward the ocean floor. *Oxygen utilization rates* (OURs) of twice the accepted mean oligotrophic rate indicate there may be much more organic matter produced in the oligotrophic euphotic zone than is indicated by C^{14} data.

The problem with the C^{14} method may be that by using such a small sample and exposing it for such a short time, periods of unusually high productivity may be missed and not averaged into the results. An interesting find may help explain how some of the biological productivity of oligotrophic water may have been missed. Mats of diatoms belonging to the genus *Rhizosolenia* have been found floating in the North Pacific Gyre and the Sargasso Sea. Averaging 7 cm (3 in) in length, these mats possess symbiotic bacteria that fix molecular nitrogen (N_2) into nitrate (NO_3)

useable as a nutrient by the diatoms (fig. 14–10). Such mats are readily broken up by net tows and missed by standard bottle casts. Recent studies in the Pacific and Atlantic oceans indicate that a large percentage (60% in one study) of oligotrophic photosynthesis is achieved by picoplankton (0.2–2.0 μm, or 0.00008–0.0008 in) that pass through many filters used in productivity studies. Continued investigation of this problem may indicate the overall productivity of the oligotrophic ocean waters may have been greatly underestimated.

CHEMOSYNTHETIC PRODUCTIVITY

Another source of potentially significant biological productivity in the oceans is occurring in the rift valleys of the oceanic spreading centers at depths of over 2500 m (8200 ft), where there is no light for photosynthesis. In regions where ocean water seeps down fractures in the ocean crust to depths where it is heated by underlying magma chambers, hydrothermal springs exist. Once the water is heated, it

FIGURE 14–10 Symbiotic Bacterial Cells Fix Nitrogen for Diatoms Living in the Oligotrophic Ocean Waters.

A, typical mat of *Rhizosolenia* about 5 cm (2 in) long. It is composed of intertwining chains of *R. castracanei* (wider cells) and *R. imbricata* (narrower cells). Photo by James M. King. *B,* transmission electron micrograph (TEM) showing cross section of *R. imbricata* containing intracellular bacteria (b) within a vacuole (v). Note the dividing bacterial cells (arrow). Photo by Mary W. Silver.

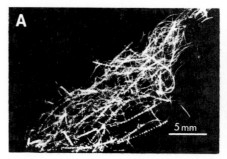
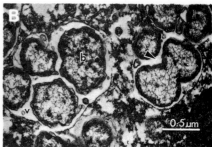

rises to the ocean floor, dissolving minerals from the crustal rocks as it does so. Potentially significant deposits of minerals are associated with biotic communities. A significant component of these communities is the vestimentiferan worm population, a sample of which can be seen in figure 14–11. These 1 m (3.28 ft) tube worms and other chemosynthetic symbionts such as clams and mussels reach unusually large size in association with autotrophic bacteria that derive energy to produce their own food from the hydrogen sulfide (H_2S) gas dissolved in the water of the hydrothermal springs. By oxidizing the gas to form free sulfur (S) and sulfate (SO_4), the bacteria release chemical energy that they use much as plants use solar energy to carry on photosynthesis. Similar communities have been discovered near cold-water seeps at the base of the Florida Escarpment in the Gulf of Mexico. Since the bacterial synthesis of inorganic nutrients into organic molecules depends on the release of chemical energy, it is called *chemosynthesis.* The true significance of bacterial productivity on the deep-ocean floor will not be fully understood until much more research on the phenomenon is conducted. It has the potential, however, of increasing our estimates of the biological productivity of the ocean.

Biochemists have recently discovered that bacteria can obtain the chemical energy needed for the synthesis of organic molecules through the oxidation of a large variety of compounds containing the metals

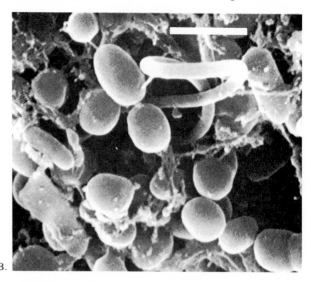

FIGURE 14–11 Chemosynthetic Life of the Galapagos Rift.

A, this photograph taken during the 1979 biological expedition to the hydrothermal biotic community of the Galapagos Rift shows the large chemosynthetic worms found there and at other vents on the East Pacific Rise. These worms possess symbiont sulfur-oxidizing bacteria and chemosynthetically produce their own food by combining inorganic nutrients dissolved in the deep ocean water. *B,* sample of sulfur oxidizing bacteria from the Galapagos Rift with small particles of what is probably free sulfur attached to them. Filter-feeding benthos may feed on these bacteria much as filter feeders in sunlit waters feed on phytoplankton. Magnification 20,000×. Bar is 1 micrometer (0.0004 in). Courtesy of Woods Hole Oceanographic Institution.

iron, manganese, copper, nickel, and cobalt. These microorganisms may well be an important factor in the deposition of ore-quality deposits of the oxides of these metals on the ocean floor in the form of manganese nodules.

ENERGY TRANSFER

We have been discussing the general consideration related to availability of nutrients. Our attention will now be turned to the cycling of specific important classes of nutrients.

Marine Ecosystem

The term *biotic community* refers to the assemblage of organisms that live together within some definable area. An *ecosystem* includes the biotic community and *abiotic* environment with which it interacts in the exchange of energy and chemical substances. Within an ecosystem there are generally three basic categories of organisms—*producers, consumers,* and *decomposers.* Plants and some bacteria are the *autotrophic* producers and have the capacity to nourish themselves through chemosynthetic and photosynthetic processes. The consumers and the decomposers are *heterotrophic* organisms that depend on the organic compounds produced by the autotrophs for their food supply.

Animals may be divided into three categories: *herbivores,* which feed directly on the plants; *carnivores,* which feed only on other animals; and *omnivores,* which feed on both. As the role of bacteria in the marine ecosystem becomes better understood, a fourth category of animals, the *bacteriovores,* which feed on bacteria, may be identified as an important component of the marine ecosystem. The decomposers, such as bacteria, break down the organic compounds of dead plants and animals and animals' excretions while taking some of these decomposition products for their own energy requirements. They characteristically release simple inorganic salts that are used by the plants as nutrients.

Energy Flow

Before considering the biogeochemical cycles that involve the transfer of organic and inorganic matter, we will consider the flow of energy in general. Most energy is put into a biotic community through plants. From the plants, it follows a unidirectional path (although cycles of energy exchange occur at many intermediate biological levels) that leads to a continual degradation of energy culminating in *entropy,* or energy converted to a form where it is no longer available to do work. As can be seen in figure 14–12, which depicts the flow of energy through a plant-supported biotic community, energy enters the system as the high-grade radiant form, solar energy, which is absorbed by the plants. Photosynthesis converts it to a medium-grade chemical energy that is used for plant respiration and passes on to the animals to be used for growth and carrying on their various life-sustaining functions. Energy is expended by the animals as mechanical and heat energy, which are progressively lower forms of energy. Finally, it becomes biologically useless as entropy increases.

Composition of Organic Matter

Having discussed the noncyclic nature of flow of energy through the biotic community and observed that it is a unidirectional flow, let us now consider the biogeochemical cycles involving matter that is not lost to the biotic community but is cycled by being converted from one chemical form to another by the various members of the community.

Biological mass is made up of compounds involving a number of elements that are abundant on earth. All of the elements that are naturally occurring can be assumed to be present in the oceans, at least in extremely small concentrations. However, some of these elements are in concentrations so small that we

FIGURE 14–12 Energy Flow. Energy enters the biological community as solar radiation, a high grade of energy. The energy in the system follows a path of degradation to lower forms of energy until it achieves a high degree of entropy and is no longer available to do work.

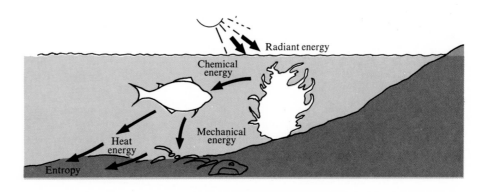

cannot detect their presence. Fewer than 60 have been identified by chemical analysis as being present in the oceans.

In view of the fact that living organisms are composed of carbohydrates, fats, and proteins, let us first concern ourselves with the elements that go into the makeup of these substances. About 20 elements in these basic organic compounds are considered to be essential to their production:

1. The major constituents of organic materials, those that individually make up at least 1% each of organic material on a dry weight basis, are, in the order of their abundance, carbon, oxygen, hydrogen, nitrogen, and phosphorus.
2. Occurring in concentrations from 500 ppm to 10,000 ppm by weight are the elements sulfur, chlorine, potassium, sodium, calcium, magnesium, iron, and copper.
3. A third group occurring in concentrations of less than 500 ppm of dry weight is composed of boron, manganese, zinc, silicon, cobalt, iodine, and fluorine.

Considering the availability of these elements as solutes in ocean water, we find that the second group, which we may refer to as the *secondary constituents* of organic composition, is in sufficient concentration to make it unlikely that members of this group would ever be limiting factors in plant productivity. However, the third group, the *tertiary constituents,* which occur in very low concentrations in organic compounds, is also found to be in very low concentrations in the marine environment. It seems probable that the tertiary constituents might create by their absence conditions that would limit productivity.

Returning to a consideration of the *major constituents* we find that carbon dioxide and water should certainly, on the basis of their concentrations, provide enough carbon, oxygen, and hydrogen to assure that these elements would never limit productivity. Nitrates and phosphates are the nutrients containing nitrogen and phosphorus that we need to consider in more detail since they do, under many conditions, limit marine productivity.

Biogeochemical Cycling

In biogeochemical cycles, elements will follow a pattern where an inorganic form is taken in by an autotrophic organism that synthesizes it into organic molecules. The food is passed through a food web that usually ends in bacterial decomposition of the organ-

ically produced compounds into the inorganic forms that may again be used in plant production.

A New Role for Bacteria Although it has been widely accepted that zooplankton are the primary grazers of phytoplankton, new evidence indicates that free-living bacteria may consume up to 50% of the production of phytoplankton. The bacteria are thought to consume the dissolved organic matter that is lost from the conventional food web by three processes:

1. Phytoplankton exudate. As phytoplankton age they loose some of their cytoplasm directly into the ocean.
2. Phytoplankton "munchates." As phytoplankton are eaten by zooplankton, cytoplasm is "spilled" into the ocean.
3. Zooplankton excretions. The liquid excretions of zooplankton are dissolved into ocean water.

Free-living heterotrophic bacteria absorb this dissolved organic matter and re-enter the conventional food web primarily through the grazing of microscopic flagellates. Other larger zooplankton may be important in this process locally.

The details of the process by which free-living bacteria participate in the marine food web are still under investigation, but microbial ecologists studying the problem are convinced the evidence indicates bacteria have an important role in the transfer of energy through the marine ecosystem. Thus, bacteria may have more varied roles in the biogeochemical cycling of matter in the oceans than had previously been known.

Carbon, Nitrogen, and Phosphorus Although there is no concern that carbon concentrations in the marine environment are a limiting factor in marine productivity, we will examine the carbon cycle since this element is the basic component of organic compounds. To appreciate the excess of carbon dioxide in the world's oceans as it is related to the requirements for plant productivity, we should know that only about 1% of the total carbon present in the oceans is involved in plant productivity.

Comparatively, the soluble nitrogen compounds involved in plant productivity may be 10 times the total nitrogen compound concentration that can be measured as a yearly average. This level implies that the soluble nitrogen compounds must be cycled completely up to 10 times per year. Available phosphates may be turned over up to 4 times per year.

ANTARCTIC SEA STAR LARVAE: UNUSUAL FEEDING BEHAVIOR

Sea star larvae live as plankton in the surface waters and are found throughout the world's oceans. Here they feed on phytoplankton, bacteria, and detritus, as well as absorbing dissolved organic matter before settling to the bottom as adults. In low and mid-latitudes they prefer phytoplankton. Thus their productivity is greatly restricted when phytoplankton are scarce. The feeding habits of high-latitude sea star larvae are only now being studied.

In the Antarctic, sea stars and other echinoderms are a major component of the benthos beneath the permanent ice shelves that prohibit penetration of sufficient sunlight to support phytoplankton. How is this Antarctic benthic community of echinoderms maintained where the availability of phytoplankton is limited to a meager supply carried in occasionally by ocean currents?

In McMurdo Sound, it was determined that the sea star larvae ingest dissolved organic matter and feed only on bacteria. Because bacterial concentrations remain relatively high throughout the year, the sea star larvae appear to have evolved to prosper in this special environment. They totally exclude phytoplankton, which is not a dependable food supply, from their diet. They depend entirely on the available bacteria and dissolved organic matter present in the water beneath the ice shelf. Here is an excellent example of natural selection choosing a few sea star larvae with feeding behavior distinctly different from that of their "cousins" living throughout the world's oceans and allowing them to establish a thriving community in a unique environment.

FIGURE 14A   Sea Star.

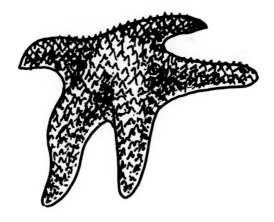

Comparing the ratio of carbon to nitrogen to phosphorus in dry weights of diatoms, we find that proportions are 41:7:1. This ratio is also observed in the zooplankton that feed on the diatoms, as well as in ocean water samples taken from the euphotic zone throughout the world. Thus, phytoplankton take up nutrients in the ratio in which they are available in the ocean water and pass them on to zooplankton in the same ratio. When these planktonic plants and animals die, carbon, nitrogen, and phosphorus are restored to the water in this ratio.

The Carbon Cycle   This cycle involves the uptake of carbon dioxide by plants that use it in the photosynthetic process. Carbon dioxide is returned to the ocean water primarily through respiration of plants, animals, and bacteria and secondarily by autolytic breakdown of dissolved organic materials. *Autolytic decomposition* results from the action of enzymes present in the cells of organic tissue and does not require bacterial action (fig. 14–13).

Carbon dioxide is present in the ocean in the molecular form (CO_2), as carbonic acid (H_2CO_3), as bicarbonate ions (HCO_3^-), and as carbonate ions (CO_3^{--}). In ocean waters of normal pH levels (8.1), bicarbonate ions are about 10 times more abundant than the carbonate ions, which are 10 times more abundant than the carbon dioxide molecules. Carbon dioxide is, in turn, about 100 times more abundant than carbonic acid.

A rapid rate of carbon dioxide consumption resulting from a phytoplankton increase will result in an increase in the pH as bicarbonate ions change to carbonate ions. A high rate of respiration, which produces unusually high concentrations of carbon dioxide, will drive the pH to lower levels, approaching a neutral value of 7. Extreme ranges of pH are rare in the ocean owing to the buffering activity of the car-

bon system that quickly restores the water to values within the 8.1 to 8.3 range.

The Nitrogen Cycle   Nitrogen is primarily important in the production of *amino acids,* the building blocks of proteins that are synthesized by plants. These photosynthetic products are consumed by animals and free-living bacteria. They are then passed on to the saprobic (decomposing) bacteria, along with dead plant tissue, as dead animal tissue and excrement.

The saprobic bacteria gain energy from breaking down these compounds. This breakdown leads to the liberation of inorganic compounds, such as nitrates, that are the basic nutrient salts used by plants. Most nitrogen in the ocean is found as molecular nitrogen (N_2). The most common combined forms are nitrite (NO_2) and nitrate (NO_3). These are highly oxidized states of nitrogen. The most abundant reduced form is ammonia (NH_3).

Different bacteria involved in the nitrogen cycle make it somewhat complicated. Although most of the bacteria play the heterotrophic roles of consuming dissolved organic matter or converting organic compounds into inorganic salts, there are some which are able to fix molecular nitrogen into combined forms. These are *nitrogen-fixing bacteria. Denitrifying bacteria* make up another special group; their metabolism depends upon the breakdown of nitrates and the liberation of molecular nitrogen. The nitrogen cycle is graphically presented in figure 14–14.

Autotrophic organisms of various types can use ammonium nitrogen and nitrites. However, the most important nutrient form of inorganic nitrogen, nitrate, can be utilized more efficiently by plants. Studies of nutrient cycles in various portions of the world's oceans indicate that the availability of nitrogen is clearly a limiting factor in productivity during sum-

FIGURE 14–13   Carbon Cycle.

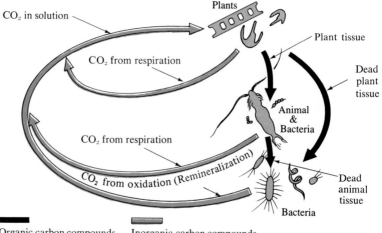

CO₂ in solution

CO₂ from respiration

CO₂ from respiration

CO₂ from oxidation (Remineralization)

Plants

Plant tissue

Dead plant tissue

Animal & Bacteria

Dead animal tissue

Bacteria

Organic carbon compounds   Inorganic carbon compounds

FIGURE 14–14   Nitrogen Cycle.

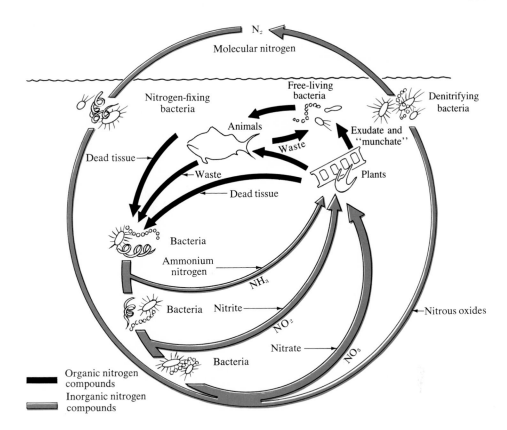

mer months. This condition arises from the fact that the processes involved in converting particulate organic substance to nitrate salts through bacterial action may require up to 3 months. The conversion begins in the lower portions of the photosynthetic zone as the particular matter is sinking toward the ocean bottom.

There are three basic stages in the conversion of particulate organic matter to nitrate salts. The stages involve three distinct bacterial types. The first type oxidizes the particulate organic nitrogen into ammonium nitrogen, which is then acted upon by the second bacterial community, which converts it to nitrite. A third action is required to convert the nitrite to the prime organic salt species, nitrate. By the time this conversion is completed, the nitrogen compounds are usually below the euphotic zone and thus unavailable for photosynthesis. They cannot readily be returned to the euphotic zone during summer because of the strong thermostratification that exists throughout much of the ocean surface. If nitrates are to again become available to plants in these regions, their return must wait until the thermostratification disappears, allowing upwelling and mixing during the winter to effect the movement of the nutrients into the surface waters.

The Phosphorus Cycle   The phosphorus cycle is simpler than the nitrogen cycle primarily because the bacterial action involved in breaking down the organic phosphorus compounds is simpler. This difference can be studied by comparing figures 14–14 and 14–15. The rate at which the organic compounds can be decomposed into inorganic _orthophosphates,_ which are the phosphorus compounds primarily used by plants, is much more rapid than that of nitrogen breakdown. This quicker rate is due to autolytic breakdown of phosphatic organic material by enzymes and the single step required in the bacterial breakdown. As a result of the greater rate of breakdown for organic phosphorus, much of it can be completed above the oxygen compensation depth. It is therefore made available to plants within the photosynthetic zone. Although the concentration of phosphorus in the oceans is only about 1/7 that of the concentration of nitrogen, the fact that the recycling can take place within the photosynthetic zone usually allows adequate quantities of inorganic phosphorus to be available for plant productivity. The lack of phosphorus is rarely a limiting factor of plant productivity.

The Silicon Cycle   As previously discussed, frustules of diatoms are composed of silica (SiO_2). Although the availability of silica can be a limiting factor in the productivity of diatoms, it is rarely a limiting factor of total primary productivity since not

FIGURE 14–15 Phosphorus
Cycle.

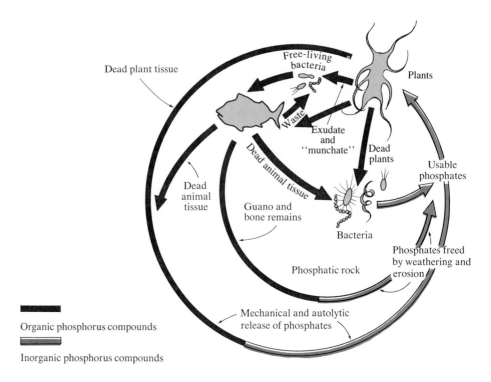

Organic phosphorus compounds

Inorganic phosphorus compounds

all phytoplankton require silica as a protective covering. There is very little likelihood that silicon will ever be a limiting nutrient in total productivity because it is very abundant in the rocks that make up the earth's crust and the small fraction of the earth's crustal silica that will eventually find its way into the sea is very great in relation to the silicon needs of the diatom population. Silicon concentrations range from unmeasurable up to 400 mg/m^3. Fluctuations in concentration roughly coincide with those observed in nitrogen and phosphorus. However, the fluctuation in silica concentrations displays a much greater amplitude than fluctuations for nitrogen and phosphorus. This condition is probably due to the fact that silica does not undergo bacterial decay and is taken directly into solution after undergoing autolytic breakdown.

TROPHIC LEVELS AND BIOMASS PYRAMIDS

Trophic Levels

The transfer of chemical energy stored in the mass of the ocean's plant population to the animal community occurs, in part, through the feeding process. Zooplankton feed as herbivores on the diatoms and other microscopic marine plants, while larger animals feed on the macroscopic algae and "grasses" found growing attached to the ocean bottom near shore. The herbivores will be fed upon by larger animals, carni-

vores, who in turn will be fed upon by another population of larger carnivores, and so on. Each of these feeding levels is a *trophic* level. As was discussed earlier in this chapter and illustrated in figures 14–13, 14–14, and 14–15, an amount of secondary production equal to that of the zooplankton may be achieved by free-living bacteria.

In general, the individual members of a feeding population will be larger but not too much larger than the individual member of the population on which they feed. Although we can consider this to be generally true, there are outstanding exceptions to this condition. The blue whale, which is the largest animal known to have existed on earth, feeds upon the krill that grow to maximum lengths of 6 cm (2.4 in).

We should recall that all consideration of energy transfer must be approached with the understanding that the transfer of energy from one population to another represents a continuous flow of energy. Small-scale recycling and conservation of this energy occurs and slows the process of conversion of potential (chemical) energy to kinetic energy, then to heat energy, and finally to be lost to entropy. However, despite the cycling of energy, all energy that enters the organic community is inevitably lost to entropy in the end.

Transfer Efficiency

In the transfer of energy between feeding, or trophic, levels we are greatly concerned with efficiency. There

FIGURE 14-16 Passage of Energy through a Trophic Level.

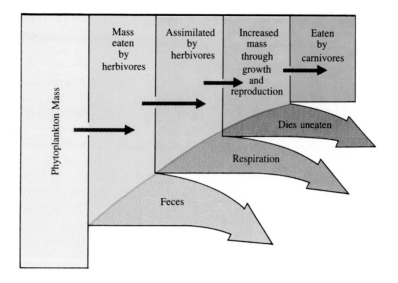

is a relatively high degree of variablility in the efficiency of various plant species under various laboratory conditions. As an average, we consider the percentage of light energy absorbed by plants and ultimately synthesized into the organic substances made available to herbivores to be about 2%.

The *gross efficiency* at any trophic level is the ratio of energy passed on to the next higher trophic level divided by the energy received from the trophic level below. When we examine the transfer of energy from feeding population to feeding population within the ocean, we can readily see that of the biomass representing the food intake of a given population only a portion will be passed on to the next feeding level.

Figure 14-16 shows that some of the energy taken in as food by herbivores is passed by the animal as feces and the rest is assimilated by the animal. Of that portion of the energy that is assimilated, much is quickly converted through respiration to kinetic energy for maintaining life, while what remains is avail-

able for growth and reproduction. Only a portion of this mass is passed on to the next trophic level through feeding. Figure 14-17 represents diagrammatically the passage of energy between trophic levels through an entire ecosystem, from the solar energy assimilated by plants to the mass of the ultimate carnivore.

Many investigations have been conducted into the efficiency of energy transfer between trophic levels. There are many variables, such as the age of the animals involved, to be considered. Young animals display a higher growth efficiency than older animals. The availability of food can also alter efficiency. When food is plentiful, animals are observed to expend more energy in digestion and assimilation than when food is not readily available. Most efficiencies that have been determined range between 6% and 15%. It is well accepted that ecological efficiencies in natural ecosystems average approximately 10%. There is some evidence that in populations important to our

FIGURE 14-17 Ecosystem Energy Flow—Efficiency. Based on probable efficiencies of energy transfer within the ecosystem, one unit of mass equivalent is made available to the fifth trophic level (the tertiary carnivores) for every 50,000 units of radiant energy absorbed by the producers (plants). This value is based on a 2% efficiency of transfer by plants and 10% efficiency at all other levels.

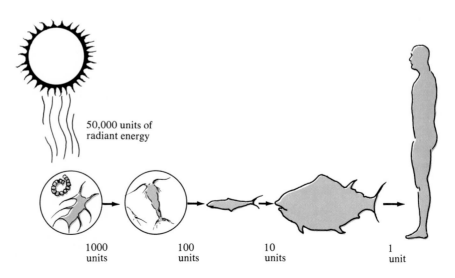

present fisheries this efficiency may run as high as 20%. The true value of this efficiency is of practical importance to us because it determines the fish harvest we can anticipate from the oceans.

Biomass Pyramid

It is obvious from considering the energy losses that occur in each feeding population that there is some limit to the number of feeding populations in a food chain and that each feeding population must necessarily be smaller than the population upon which it feeds. It was previously mentioned that the individual members of a feeding population would be larger than their prey, but it seems they should also be less numerous.

Food Chains   *Food chains* represent a sequence of organisms through which energy is transferred from the primary producers through the herbivore and carnivore feeding populations until one feeding population does not have any predators, which marks the end of the chain. In nature it is rare to find food chains comprising more than five trophic levels. On the basis of our previous considerations of the efficiency of energy transfer between trophic levels, we see that for a population within a food chain it may be beneficial to feed as close to the primary producing population as possible. This arrangement increases the biomass available for food and the number of individuals in the population to be fed upon.

An example of an animal population that is an important fishery and usually represents the third trophic level in a food chain is the herring off the coast of Newfoundland. Although some herring populations are involved in longer food chains, the Newfoundland herring feed primarily on a population of small crustaceans, copepods, that in turn feed upon diatoms (fig. 14–18).

Food Webs   It is, however, uncommon to see simple feeding relationships of the type described above in nature. More commonly the animals representing the last step in a linear food chain feed on a number of animals that have simple or complex feeding relationships, constituting a *food web*. The overall importance of food webs is not well understood, but one consequence for animals that feed through a web rather than a linear chain is their greater likelihood of survival if population extinctions or sharp decreases occurred within the web at or below their feeding level. Those animals involved in food webs, such as the North Sea herring illustrated in figure 14–18, are less likely to suffer from the extinction of one of the populations upon which they depend for food than the Newfoundland herring that feed only on copepods (fig. 14–6D). The extinction of the copepods in the latter case certainly would be expected to have a catastrophic effect on the herring population.

The Newfoundland herring population does, however, have an advantage over its relatives who feed through the broader-based food web. The Newfoundland herring are more likely have a larger biomass to feed on since they are only two steps removed from the producers, while the North Sea herring represent the fourth level in some of the food chains within its web.

The ultimate effect of the transfer of energy between trophic levels can be seen in figure 14–19, which depicts the progressive decrease in numbers of individuals and total biomass at successive trophic levels as a result of decreased amounts of available energy.

THE PERUVIAN ANCHOVY FISHERY

The magnificent anchovy fishery that existed at the north end of the Peru Current can be used as a case

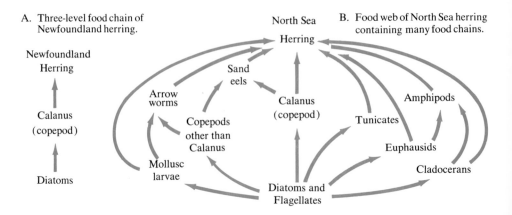

FIGURE 14–18 Food Chains—Food Webs.

A. Three-level food chain of Newfoundland herring.

B. Food web of North Sea herring containing many food chains.

FIGURE 14–19 Biomass Pyramid.

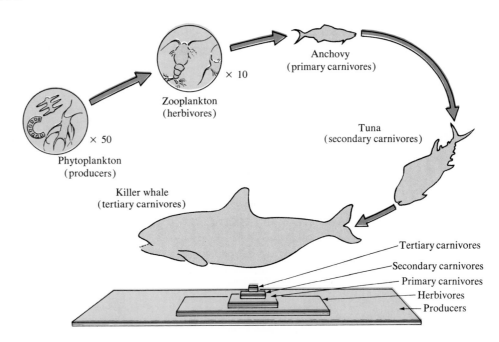

study to show the relationships between physical and biological variables and productivity.

Physical Conditions

Southeasterly winds blow along the coast of Peru and, owing to the earth's rotation, drive the surface waters away from the coast. The surface water is replaced by an upwelling of cold water from depths of 200 to 300 m (656 to 984 ft).

The Peru Current (fig. 14–20) is the northward-flowing eastern boundary of the South Pacific current gyre. Velocity seldom exceeds 0.5 km/h (0.3 mi/h), and the current extends to a depth of 700 m (2296 ft) near its seaward edge, some 200 km (124 mi) from the coast. The Peru Countercurrent knifes its way between the coastal and oceanic regions of the Peru Current near the equator. This southbound surface flow seldom reaches more than 2° or 3° south of the equator, except during the summer, when it reaches its maximum strength. Beneath the surface current is the southward-flowing Peru Undercurrent.

Biological Conditions

Biologically, the significance of the cold water and nutrients provided by upwelling is great. During the years 1964 through 1970, fully 20% of the world fishery was anchovy taken off the coast of Peru.

The food web starts with diatoms and other phytoplankton that support a vast zooplankton population of copepods, arrowworms, fish larvae, and other small animals. Energy is passed on through a series of predators, but most used to stop with the anchovy, which in certain seasons feed directly on large diatoms and once attained a biomass of 20 million met-

FIGURE 14–20 The Peru Current.

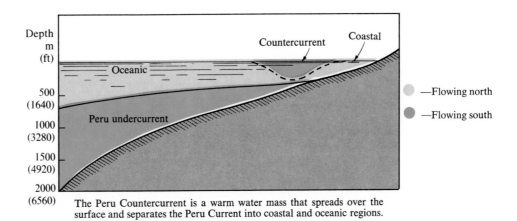

The Peru Countercurrent is a warm water mass that spreads over the surface and separates the Peru Current into coastal and oceanic regions.

ric tn (fig. 14–21). Natural predators that feed on the anchovy are marine birds—boobies, cormorants, and pelicans—and large fishes and squid.

A Natural Hazard

Periodically (every few years) during the summer, the winds become westerly. The northward-flowing current and upwelling slows down, and the Peru Countercurrent may move as far as 1000 km (621 mi) south of the equator, laying down a warm low-salinity layer over the coastal water. Since this event commonly occurs around Christmas, it is called *El Niño*, The Child. Oceanographers and marine biologists have studied this phenomenon extensively in an attempt to learn its cause and the effect of El Niño on marine populations. One of the earlier investigators of this relationship is Warren S. Wooster (fig. 14–22). See *El Niño–Southern Oscillation Events* on pages 312–316 for a discussion of the present knowledge of the physical nature of El Niño.

The increase in water temperature and decrease in nutrients brought about by the slowdown or complete halt of upwelling cause a severe decrease in the anchovy population, the remnants of which seek deeper, cooler water. Marine birds that feed primarily on the anchovy and produce the enormous guano deposits (excrement rich in phosphorus and nitrogen compounds) on the arid islands off the Peruvian coast have their populations cut severely by the decrease in the anchovy.

In 1972, a severe El Niño occurred. The total biomass of anchovy decreased from an estimated 20 million metric tn in 1971 to an estimated 2 million metric tn in 1973. But before blaming this decrease solely on the natural event of El Niño, with which the anchovy have contended for thousands of years, another factor must be evaluated.

A Man-Made Hazard

The Peruvian anchovy fishery began to develop into a major industry in 1957. In 1960, the Instituto del Mar del Peru was established with the aid of the United Nations to study the fishery and recommend a management program to the government. It was hoped that early study would prevent the fishery from going the way of the Japan herring and California sardine fisheries, which had collapsed as a result of overexploitation.

Biologists estimated that 10 million metric tn was the maximum annual take that the anchovy could sustain. Figure 14–23 shows that this limit was exceeded in 1968, 1970, and 1971. The 1970 catch of

A.

B.

FIGURE 14–21 Peruvian Anchovy.

A, these small fish represented one of the world's major fisheries until the collapse of 1972. An active fish meal industry based on anchovy made Peru, for a while, the world's leading fishing nation. Photo courtesy of National Marine Fisheries Service. *B*, a catch of Peruvian anchovy during the early 1960s when the fishery was undergoing rapid development. Photo courtesy of FAO.

12.3 million metric tn was probably 14 million tn if processing losses and spoilage were included.

Dr. Paul Smith, who studies the anchovy fishery problem for the National Marine Fisheries Service in the Southwest Fisheries Center at La Jolla, California, believes, as do other biologists familiar with the problem, that the yearly anchovy catches may have been much larger than the official figures indicate. This belief grows out of the fact that many catches contained large numbers of *pelladilla*, young anchovy low in protein. This underreporting would result from buyers reducing the total tonnage of a catch they are purchasing by a percentage determined by the propor-

FIGURE 14–22 Warren S. Wooster.

In an oceanographic research career that dates back to 1949, Wooster has been concerned primarily with the circulation of the oceans and the application of this knowledge to fishery problems. Wooster conducted early investigations of the effects of El Niño on the region of the Peru Current. In recent years, his interest has been directed toward the problems of conducting oceanographic research in an increasingly complex international political environment. Wooster has recently been Director, Institute for Marine Studies at the University of Washington, and now serves there as Professor of Marine Studies and Fisheries.

tion of the catch accounted for by high-water-content and low-oil-content pelladilla. For example, a 100-tn catch may be recorded as a 70-tn catch because it contained a large number of pelladilla.

Dr. Smith feels the crisis may have resulted from a combination of the effects of El Niño and the industry's lack of understanding of the anchovy fishery. This is reflected in the fact that the January 1972 allotment was set at 1.2 million tn. This figure was reached easily by the industry with about half the expected effort. Because the fishing vessels were able to meet their allotment so easily without going far from shore, the fishery was thought to have developed handsomely, and the February allotment was raised to 1.8 million tn. Much more effort was required to meet this allotment, and in March the fishery collapsed. What had happened?

Apparently, El Niño had come south in January, but it had remained offshore. The anchovy were concentrated in the nearshore portion of the Peru Current that had not been affected (*see* fig. 14–20). In January the nets were filled easily near shore, and it was natural to think the high concentration of anchovy extended over the normal fishery area as well. This one event cannot explain the collapse of the fishery, but it is a symptom common of fisheries, which are poorly understood.

It may be that even with upwelling, ammonia (NH_3) from decomposition of anchovy excrement is required to support enough phytoplankton to allow the anchovy population to expand. Since anchovy populations are now small and scattered, they may swim through great expanses of water without adding much nutrient enrichment. When the population was larger, the whole region received it. It may require a long time, with a gradual increase in the population, until this nutrient supplement is made available to the phytoplankton throughout the entire region.

The anchovy are gradually being replaced by sardines, a more valuable variety of fish. Where the entire anchovy catch was converted to fish meal and ex-

FIGURE 14–23 Annual Anchovy Catch.

After a peak catch of 12.3 million tn in 1970, the fishery declined.

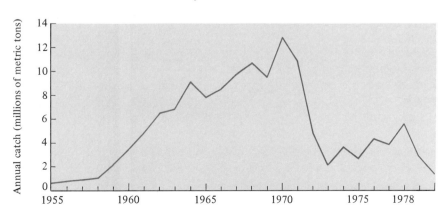

SATELLITES AND PLANKTON

In the past, satellites have been designed specifically to sense components of the earth environment. In 1959, the *TIROS* series began observing the atmosphere to help meteorologists improve weather forecasting. This series was followed by *NIMBUS, SMS, ITOS,* and the *GEOS/NOAA* series. Over 40 satellites have added to our knowledge of the earth's atmosphere since the launching of *TIROS* I. Between 1972 and 1978, three *ERTS* and *LANDSAT* satellites were placed in orbit to help assess terrestrial resources. The only satellite ever launched to specifically sense the ocean surface was *SEASAT.* Launched in 1978, it lasted a little over three months before shorting out. However, during this short time *SEASAT* provided a large amount of data that are only now undergoing a level of analysis that indicates the accuracy of its measurements and the potential of resuming such observations.

Although no existing satellite is designed specifically to sense the ocean surface, most of the meteorological satellites provide some information about the nature of the oceans. The *NIMBUS* 7 carries a Coastal Zone Color Scanner (CZCS) that has significant potential for a marine application: the determination of biological productivity in ocean surface water. Launched on October 7, 1978, the *NIMBUS* 7 CZCS began providing data on ocean color. The Nimbus Experiment Team assigned to the CZCS immediately began to evaluate whether the data were useful. Making them useful required correcting for light scattered by the atmosphere and reflected off the ocean surface.

The idea for using the CZCS for measuring marine biological productivity is based on the fact that the color of ocean water is affected significantly by the light absorption and scattering of suspended particles just below the surface. Near the mouths of rivers such particles are predominantly inorganic sediment, which rules out determining biological productivity by using the CZCS in these waters. Elsewhere, phytoplankton play a major role in determining the ocean's color. The light-scattering and absorption properties of phytoplankton are greatly affected by chlorophyll *a,* the dominant photosynthetic pigment. Chlorophyll *a* and its detrital form, phaeopigment *a,* absorb strongly in the blue and red wavelengths, but they reflect most radiation in the green wavelengths. Thus water with low concentrations of chlorophyll *a* will appear deep blue (the normal color of clear ocean water), while water rich in chlorophyll *a* will appear green.

To correct for atmospheric scattering, about 80% of the radiation detected by the CZCS must be removed by calculations based on atmospheric conditions. The sensor can be tilted up to 20° ahead or behind vertical relative to the earth's surface to avoid the effects of direct reflection off the ocean surface.

The CZCS senses light in four wavelength bands that are used in determining ocean color. They are band 1 (433–53 nm, blue), band 2 (510–30 nm, green-1), band 3 (540–60 nm, green-2), and band 4 (660–80 nm, red). Plate 21A shows a false-color image of the Gulf of Mexico based on the combined sensings of bands 1, 3, and 4 without the removal of atmospheric effects. Plate 21B is derived from the same data, but the effects of atmospheric scattering have been removed. Also, a ratio of band 1 (blue) to band 3 (green) has been used to determine pigment concentration. Because band 1 radiation is essentially zero above concentrations of 1 mg/m^3, all values above that level cannot be calculated by this method and appear reddish-brown. It is clear that coastal concentrations are high and that concentrations throughout much of the open Gulf are below 0.20 mg/m^3.

It appears that the CZCS is capable of observing very subtle variations in water color, but are the pigment concentrations determined using the data and presented in plate 21B correct? One way to answer this question is to determine the concentration of chlorophyll a simultaneously at the same location by CZCS sensing and direct sampling.

Plate 21C shows the distribution of pigment concentrations off the Florida coast. Plate 21C (C_1) has concentrations calculated as they were in plate 21B—band 1/band 3. Plate 21C (C_2) uses the ratio of band 2 to band 3, which provides useable values for concentrations above 1 mg/m^3. The track of R.V. *Athena II* is superimposed on both figures. At the time of the CZCS overpass, the ship was due west of Tampa Bay. Shipboard measurements of chlorophyll a made along the plotted course were compared with plate 21C (C_2) concentrations, as shown in figure 14B.

The CZCS values appear to be consistently lower and more variable than values determined aboard ship. A navigational error may account for some of the difference. If the ship's course was only 10 km (6 mi) east of that indicated, the agreement in the curves of figure 14B would be much better.

The spectacular image shown in plate 6 was produced in a manner similar to that used in plate 21B, so the same color scale can be used to determine the pattern of biological productivity. The brown of the continent and white clouds are not related to the productivity pattern. The low

FIGURE 14B Values of C = Chl a + Phaeo a (in mg/m^3); from Plate 21C Plotted with the Pigment Concentrations Determined Aboard R.V. *Athena II* on November 13 and 14, 1978.

The distances are along the course plot on plate 21C and start at the south end.

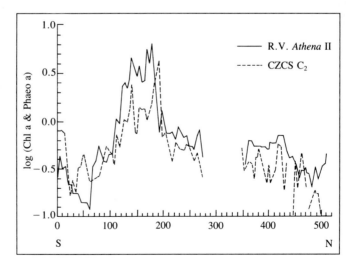

productivity of the Gulf Stream can be identified as the deep blue starting at the lower left corner and meandering across the lower portion of the image. Note the sharp northern boundary of the Gulf Stream indicated by the sharp boundary between the blue of this current and the red and green of more productive waters to the north as it moves out from under the clouds just off Cape Hatteras, N.C. This boundary becomes more diffuse as the current moves east into the North Atlantic. Georges Bank, southeast of Cape Cod, receives a continuous supply of nutrients as currents carry deep nutrient-rich waters up over this shallow feature. High productivity is indicated here by the red color. An interesting feature of this image is the disk of low productivity associated with a ring of warm Sargasso Sea water that is trapped north of the Gulf Stream ring south of Cape Cod. (See *Gulf Stream Rings and Marine Life* on pages 339–341.)

Although precise measurements of the chlorophyll *a* concentrations in ocean water have not yet been achieved using the satellite technique, there is every reason to believe that useful data will become available with further observations by the CZCS. If it does become a dependable means of determining biological productivity in the oceans, it will be a valuable tool in determining the standing crop of phytoplankton, the management of fisheries, and the control of red tide fish kills. It could greatly increase the efficiency with which we exploit the biological resources of the ocean.

References: Gordon, H. R.; Clark, D. K.; Mueller, J. L.; and Hovis, W. A. 1980. Phytoplankton pigments from the *Nimbus-7* Coastal Zone Color Scanner: Comparisons with surface measurements. *Science* 210:63–66. Hovis, W. A.; Clark, D. K.; Anderson, F.; Austin, R. W.; Wilson, W. H.; Baker, E. T.; Ball, D.; Gordon, H. R.; Mueller, J. L.; El-Sayed, S. Z.; Strum, B.; Wrigley, R. C.; and Yentsch, C. S. 1980. *Nimbus-7* Coastal Zone Color Scanner: System description and initial imagery. *Science* 210:60–63.

ported as poultry and hog feed, the sardine fishery will be directed at producing canned and frozen products for direct human consumption.

With anchovy fishing banned since 1980 and an emphasis on species conservation in Peru's future fishing activity, the anchovy fishmeal fishery may be viable in the future. However, it will never reach previous levels. Although it is much smaller in biomass, the sardine fishery may eventually develop into a major source of revenue owing to the higher value of the products that can be produced from it. To improve the present state of affairs, the Ministry of Fisheries will have to conduct a prudent program of fishery assessment and management.

SUMMARY

Organisms are divided into five **kingdoms**: Monera, single-celled organisms without a nucleus; **Protista**, single-celled organisms with a nucleus; **Mycota**, fungi; **Metaphyta**, many-celled plants; and **Metazoa**, many-celled animals. **Taxonomic classification** of organisms involves dividing the kingdoms into the increasingly specific groupings: **phylum**, **class**, **order**, **family**, **genus**, and **species**.

Protistan plants include the macroscopic algae **Phaeophyta** (kelp and *Sargassum*), **Chlorophyta** (green algae), and **Rhodophyta** (red algae). The microscopic algae include **Chrysophyta** (diatoms and coccolithophores) and **Pyrrophyta** (dinoflagellates). The more complex **Spermatophyta** are represented by a few genera of nearshore plants such as *Zostera* (eelgrass), *Phyllospadix* (surf grass), *Spartina* (marsh grass), and mangrove trees.

Plant productivity of the oceans is limited by the availability of solar radiation and of nutrients. The depth to which sufficient light penetrates to allow plants to produce only that amount of oxygen required for their respiration is the **oxygen compensation depth**. Plants cannot live successfully below this depth, which may occur at less than 20 m (65 ft) in turbid coastal waters or at a probable maximum of 150 m (492 ft) in the open ocean. Nutrients are most abundant in coastal areas owing to runoff and upwelling. In **high-latitude areas** thermoclines are generally absent, so upwelling can readily occur, and productivity is commonly limited more by the availability of solar radiation than lack of nutrients. In **low-latitude regions**, where a strong thermocline may exist year-round, productivity is limited except in areas of upwelling. The lack of nutrients is generally the limiting factor. In **temperate regions**, where distinct seasonal patterns are developed, productivity peaks in the spring and fall and is limited by lack of solar radiation in the winter and lack of nutrients in the summer.

In addition to the **primary productivity** through plant photosynthesis, organic biomass is produced through bacterial **chemosynthesis**. Chemosynthesis, recently observed on oceanic spreading centers in association with hydrothermal springs, is based on the release of energy by the oxidation of hydrogen sulfide.

Radiant energy captured by plants is converted to **chemical energy** and passed through the **biotic community**. It is expended as **mechanical** and **heat energy** and ultimately reaches a state of **entropy**, where it is biologically useless. There is, however, no loss of mass. The mass used as nutrients by plants is converted to biomass. Upon the death of organisms, the mass is decomposed to an inorganic form ready again for use as nutrients for plants. Of the nutrients required by plants, compounds of nitrogen are most likely to be depleted and restrict plant productivity. Since the total decomposition of organic nitrogen compounds to inorganic nutrients requires three stages of bacterial decomposition, these compounds may have sunk beneath the photosynthetic zone before decomposition is complete and therefore are unavailable to plants.

As energy is transferred from plants to **herbivores** and the various **carnivore** feeding levels, only about 10% of the mass taken in at one feeding level is passed on to the next. The ultimate effect of this decreased amount of energy that is passed between trophic levels higher in the food chain is a decrease in the number of individuals and total biomass of populations higher in the **food chain**.

An example of the combined effects of natural phenomena and human activities on a marine population is the sharp decline in the population of **Peruvian anchovy** that occurred in 1972. This reduction seems to have been the result of a severe El Niño condition and a poor understanding of the fishery in the coastal waters of Peru.

QUESTIONS AND EXERCISES

1. List the five kingdoms of organisms and the fundamental criteria used in assigning members to them.

2. Compare the macroscopic algae in terms of pigment, maximum depth in which they grow, common species, and size.

3. The Chrysophyta contains two classes of important phytoplankton. Compare their composition and the structure of their hard parts as well as their geological significance.

4. Discuss and compare the contributions of the Pyrrophyta genera *Gymnodinium* and *Gonyaulax* to red tide development.

5. Define oxygen compensation depth, and explain the use of the dark and transparent bottle technique for its determination. Discuss how the quantity of oxygen produced by photosynthesis in each clear bottle is determined.

6. Compare the biological productivity of polar, temperate, and tropical regions of the oceans. Include a discussion of seasonal variables, thermal stratification of the water column, and the availability of nutrients and solar radiation.

7. Discuss chemosynthesis as a method of primary productivity. How does it differ from photosynthesis?

8. Describe the components of the marine ecosystem.

9. Describe the flow of energy through the biotic community, and include the forms to which solar radiation is converted. How does this flow differ from the manner in which mass is moved through the ecosystem?

10. What are the proportions by weight of carbon, nitrogen, and phosphorus in ocean water, phytoplankton, and zooplankton? Suggest how these amounts may support or refute the idea that life originated in the oceans.

11. Explain why nitrogen is much more likely than phosphorus to be a limiting factor in marine productivity.

12. How is the energy taken in by a feeding population lost so that only a small percentage is made available to the next feeding level? What is the average efficiency of energy transfer between trophic levels?

13. If a killer whale is a third-level carnivore, how much phytoplankton mass is required to add each gram of new mass to the whale? Assume 10% efficiency of energy transfer between trophic levels. Include a diagram.

14. Describe the probable advantage to the ultimate carnivore of the food web over a single food chain as a feeding strategy.

15. Discuss the natural and synthetic conditions that may have combined to cause the catastrophic decline of the Peruvian anchovy fishery in 1972.

REFERENCES

Alldredge, A. L., and Cohen, Y. 1987. Can microscale chemical patches persist in the sea? Microelectrode study of marine snow, fecal pellets. *Science* 235:4789, 689–691.

Craig, H., and Hayward, T. 1987. Oxygen supersaturation in the ocean: Biological versus physical contributions. *Science* 235:4785, 199–201.

George, D., and George, J. 1979. *Marine life: An illustrated encyclopedia of intervertebrates in the sea.* New York: Wiley-Interscience.

Grassle, J. F., et al. 1979. Galapagos '79: Initial findings of a deep-sea biological quest. *Oceanus* 22:2, 2–10.

Harvey, H. W. 1966. *Chemistry and fertility of sea waters.* London: Cambridge University Press.

Jackson, D. F., ed. 1963. *Algae and man.* New York: Plenum Press.

Jenkins, W. J. 1982. Oxygen utilization rates in North Atlantic subtropical gyre and primary production in oligotrophic systems. *Nature* 300, 246–48.

Littler, M. M; Littler, D. S.; Blair, S. M.; and Norris, J. N. 1985. Deepest known plant life discovered on an uncharted seamount. *Science* 227:4683, 57–59.

Martinez, L.; Silver, M.; King, J.; and Alldredge, A. 1983. Nitrogen fixation by floating diatom mats: A source of new nitrogen to oligotrophic ocean waters. *Science* 221:4606, 152–54.

Parsons, T. R.; Takahashi, M.; and Hargrave, B. 1984. *Biological oceanographic processes,* 3rd ed. New York: Pergamon Press.

Raymont, J. E. G. 1963. *Plankton and productivity in the oceans.* New York: Pergamon Press.

Rivkin, R.; Bosch, I.; Pearse, J.; and Lessard, E. 1986. Bacteriovory: A novel feeding mode for asteroid larvae. *Science* 233:4770, 1311–13

Russell-Hunter, W. D. 1970. *Aquatic productivity.* New York: Macmillan.

Shulenberger, E., and Reid, J. L. 1981. The Pacific shallow oxygen maximum, deep chlorophyll maximum, and primary productivity reconsidered. *Deep Sea Research* 28A:9, 901–919.

SUGGESTED READING

Sea Frontiers

Arehart, J. L. 1972. Diatoms and silicon. 18:2, 89–94.
A very readable description of the important role of silicon and other elements in the ecology of diatoms, including microphotographs showing the varied forms of diatoms.

Ebert, C. H. V. 1978. El Niño: An unwanted visitor. 24:6, 347–51.
A description of the effects of El Niño and the attempts that are being made to understand its causes.

Idyll, C. P. 1971. The harvest of plankton. 17:5, 258–67.
An interesting discussion of the potential of zooplankton as a major fishery.

Jensen, A. C. 1973. WARNING—Red tide. 19:3, 164–75.
An informative discussion of what is known of the cause, nature, and effect of red tides.

Johnson, S. 1981. Crustacean symbiosis. 27:6, 351–60.
A description of various symbiotic relationships entered into by tropical shrimps and crabs.

Oremland, R. S. 1976. Microorganisms and marine ecology. 22:5, 305–310.
The role of such microorganisms as phytoplankton and bacteria in cycling matter in the oceans is discussed.

Philips, E. 1982. Biological sources of energy from the sea. 28:1, 36–46.
The potential for converting marine biomass to energy sources useful to society is discussed.

Scientific American

Benson, A. A. 1975. Role of wax in oceanic food chains. 232:3, 76–89.
A report on the findings from observations made of the content of wax in the bodies of many marine animals, from copepods to small deep-water fishes, and their implications.

Levine, R. P. 1969. The mechanism of photosynthesis. 221:6, 58–71.
Reveals what is known of the process by which energy is captured by plants and converted to useful forms of chemical energy while freeing oxygen to the atmosphere.

Pettitt, J.; Ducker, S.; and Knox, B. 1981. Submarine pollination. 244:3, 134–144.
Discusses the pollination of sea grasses by wave action.

In the sunlit waters of the pelagic environment live the phytoplankton that provide the basis for nearly all of the marine biomass through their photosynthetic activities. This important marine population was discussed in chapter 14, so our emphasis here will be on the animal populations. The discussion will deal primarily with the adaptations that allow them to live successfully in the pelagic environment.

CHAPTER FIFTEEN
LIFE IN THE OPEN OCEAN

STAYING ABOVE THE OCEAN FLOOR

Since the tissues of the zooplankton and nekton (muscle, cartilage, scales, bone, and shell) are more dense than ocean water, pelagic animals can remain off the bottom only through the application of buoyancy or energy. Chapter 13 discussed how microplankton's decreased size aided in this process by an increased frictional resistance to sinking. However, the larger animals we will discuss in this chapter need additional means to remain off the ocean floor.

Gas Containers

Since air is approximately 0.001 the density of water at sea level, a small amount of it inside an organism makes it easier for that organism to remain in the water. Some cephalopods have rigid gas containers (fig. 15–1). The genus *Nautilus* has an external shell; the cuttlefish *Sepia* and deep-water squid *Spirula* have an internal chambered structure. These animals become neutrally buoyant and can maintain their positions in the water easily. Since the air pressure in their air chambers is always 1 atmosphere, they are limited in the depth to which they may venture. The *Nautilus* must stay above a depth of approximately 500 m (1640 ft), or its chambered shell will collapse as the external pressure approaches 50 atmospheres. The *Nautilus* is rarely observed below a depth of 250 m (820 ft).

Neutral buoyancy is achieved by some bony fish through filling a gas bladder, or *swim bladder,* with gases. The swim bladder is normally not present in very active swimmers, such as the tuna, or in fish

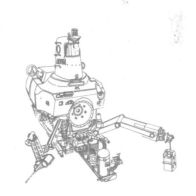

The clown fish lives among the stinging tentacles of the sea anemone. Photo by Christopher Newbert.

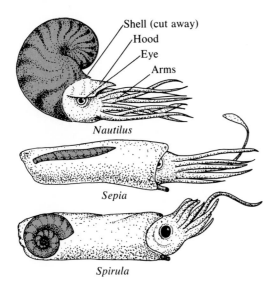

FIGURE 15–1   Gas Containers of the Cephalopods *Nautilus, Sepia,* and *Spirula.*

The *Nautilus* has an external chambered shell, and the *Sepia* and *Spirula* have internal chambered structures that can be filled with gas to provide buoyancy.

The composition of gases in the swim bladders of shallow-water fishes is similar to that of the atmosphere. With increasing depth (fish with swim bladders have been taken from a depth of 7000 m, or 22,960 ft) the concentration of oxygen in the swim bladder gas increases from the 20% common near the surface to more than 90%. At the 700-atmosphere pressure that exists at a depth of 7000 m, gas is compressed to a density of 0.7 gm/cc. This is about the density of fat, so many deep-water fish have air bladders filled with fat instead of compressed gas. This provides almost as much buoyancy as compressed gas and avoids the energy requirements of maintaining a constant volume of gases in the swim bladder.

Floating Forms

Floating at the surface are some relatively large members of the plankton, the evolutionarily primitive siphonophore and scyphozoan *coelenterates,* as well as the pelagic tunicates. Somewhat smaller are the ctenophores and arrowworms. They all have soft gelatinous bodies with little if any hard tissue. A major strategy of this body type is the replacement in body fluids of heavy sulfate ions with chloride ions to maintain osmotic equilibrium.

The siphonophore coelenterates are represented in all oceans by the genera *Physalia* (Portuguese man-of-war) and *Velella* (by-the-wind sailor). Their gas floats, pneumatophores, serve as floats and sails that allow the wind to push these colonial forms across the ocean surface (fig. 15–3). A colony of tiny sea-anemonelike *polyps* and jellyfishlike *medusae* (fig. 15–3B) are suspended beneath the float. The long tentacles of the polyps primarily capture food, and

that live on the bottom. Some fish have a *pneumatic duct* that connects the swim bladder to the esophagus (fig. 15–2). These fish can add or remove air through this duct. In other fish, the gases of the swim bladder must be added or removed more slowly by an interchange with the blood. Since change in depth will cause the gas in the swim bladder to expand or contract, the fish removes or adds gas to the bladder to maintain a constant volume. Those fish without the pneumatic duct are limited in the rate at which they can make these adjustments and, therefore, cannot withstand rapid changes in depth.

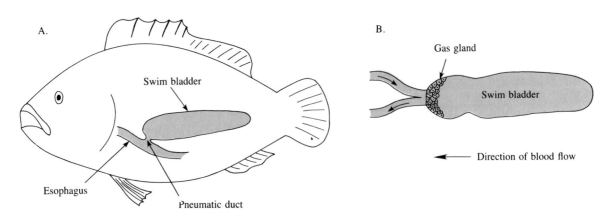

FIGURE 15–2   Swim Bladder of Some Bony Fishes.

A, example of a swim bladder connected to the esophagus by the pneumatic duct, allowing air to be added or removed rapidly. *B,* in fish with no pneumatic duct, all gas must be added or removed through the blood. This exchange requires more time and is achieved by a network of capillaries associated with the gas gland.

FIGURE 15–3 Planktonic Coelenterates.

A, Physalia and Jellyfish. *B, Medusa.* Photo by Larry Ford.

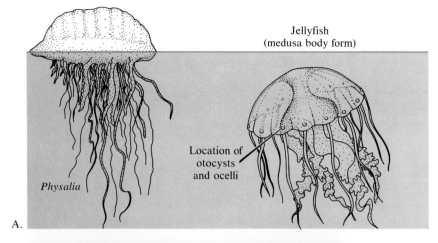

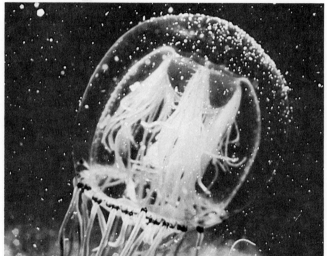

A.

B.

the medusae carry on the sexual reproduction required to create a new colony. The tentacles of the *Physalia* may be many meters long and possess nematocysts long enough to penetrate the skin of humans; they have been known to inflict a painful and occasionally dangerous neurotoxin poisoning. The colonies grow from the initial polyp through asexual budding.

The scyphozoans, or jellyfish, have a medusoid, bell-shaped body with a fringe of tentacles and a mouth at the end of a clapperlike extension hanging down under the bell-shaped float. Ranging in size from near microscopic to 2 m (6.6 ft) in diameter with 60-m (197-ft) tentacles, most jellyfish have bells with a diameter of less than 0.5 m (1.6 ft). Jellyfish are capable of movement through muscular contractions (*see* fig. 15–3B). Water enters the cavity under the bell and is forced out by contractions of muscles that circle the bell, jetting the animal ahead in short spurts. To allow the animal to swim generally in an upward direction, there are

sensory organs spaced around the outer edge of the bell. These may be light-sensitive *ocelli* or gravity-sensitive *otocysts*. This orientation ability is important because the jellyfish feed by swimming to the surface and sinking slowly through the life-rich surface waters.

Pelagic *tunicates* are generally transparent and barrel-shaped, with an anterior incurrent and posterior excurrent opening (fig. 15–4). They move by a feeble form of jet propulsion effected by the contraction of bands of muscles that force water out the excurrent opening. Reproduction of solitary forms is complicated, involving alternate sexual and asexual reproduction. The genus *Salpa* includes solitary forms reaching a length of 20 cm (8 in) and smaller aggregate forms that produce new individuals by budding. Individual members of an aggregate chain may be 7 cm (2.8 in) long, and the chain of newly budded members may reach great lengths (*see* fig. 15–4B). The genus *Pyrosoma* is luminescent and colonial. Individual members of the colony have their

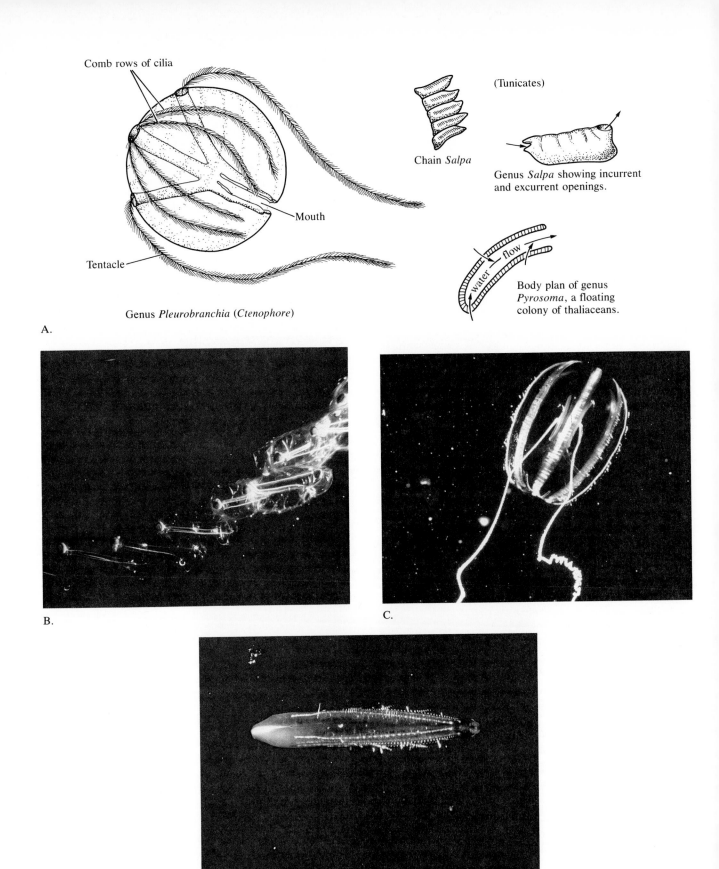

Comb rows of cilia

Mouth

Tentacle

Genus *Pleurobranchia* (*Ctenophore*)

A.

(Tunicates)

Chain *Salpa*

Genus *Salpa* showing incurrent and excurrent openings.

water flow

Body plan of genus *Pyrosoma*, a floating colony of thaliaceans.

B.

C.

D.

FIGURE 15–4 Pelagic Tunicates and Ctenophores.
A, body structure. *B,* chain of salps. Photo by James M. King. *C,* ctenophore, *Pleurobranchia.* Photo by James M. King. *D,* ctenophore, *Beroe.* Photo by James M. King.

incurrent openings facing the outside surface of a tube-shaped colony that may be a few meters long. One end of the tube is closed, and the excurrent openings of the thousands of individuals all empty into the tube. Muscular contraction forces water out the open end to provide a means of propulsion.

Ctenophores are comb-bearing animals closely related to the coelenterates in that their radially symmetrical bodies develop around their mouths. (Some forms, however, also possess an anal opening from the digestive cavity. Often referred to as *comb jellies,* this group of animals is entirely pelagic and confined to the marine environment.) The body form of most members of the phylum Ctenophora is basically spherical, with eight rows of cilia spaced evenly around the sphere. It is from these structures that the name *comb-bearing* is derived; when magnified, they look like miniature combs. If a ctenophore possesses tentacles, there will be two that contain adhesive organs instead of stinging cells to capture prey. The

genera *Pleurobranchia* (sea gooseberries) (fig. 15–4C) and the pink, elongated *Beroe* (fig. 15–4D) range in size from the gooseberry dimension of the former to over 15 cm (6 in) in the latter.

The *arrowworms* (chaetognaths) are transparent and difficult to see, although they may grow to more than 2.5 cm (1 in) in length (fig. 15–5). The name *chaetognath* refers to the hairlike attachments around the mouth, which are used to grasp prey while they devour it. They are voracious feeders, eating primarily the small zooplankton; in turn, they are eaten by fishes and larger planktonic animals such as jellyfish. These exclusively marine, hermaphroditic animals are usually more abundant in the surface waters some distance from shore.

Swimming Forms

This section examines larger pelagic animals, referred to as the *nekton,* with substantial powers of lo-

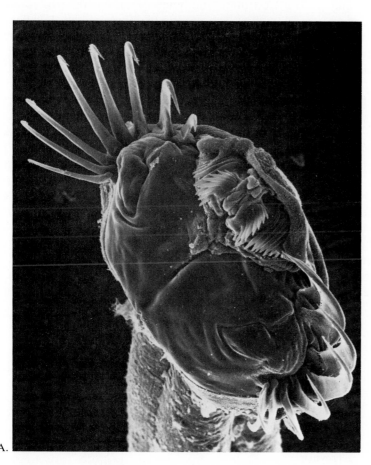

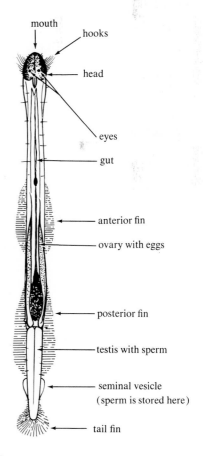

A.

B. General Anatomy of an Arrowworm

FIGURE 15–5 Arrowworm.

A, head of chaetognath *Sagitta tenuis* from Gulf of Mexico, magnified 161 times. Photo by Howard J. Spero, University of South Carolina. *B,* adult chaetognaths reach lengths ranging from 1 to 5 cm (0.4 to 2 in).

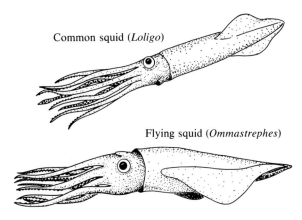

FIGURE 15–6   Squids.

comotion. They include rapid-swimming invertebrate squids, as well as fish and marine mammals.

Swimming squid include the genera *Loligo* (common squid), *Ommastrephes* (flying squid), and the *Architeuthis* (giant squid). Active predators of small fish, the smaller varieties of squid have long slender bodies with paired fins (fig. 15–6). Unlike *Sepia* and *Spirula,* they have no hollow chambers in their bodies and therefore require more energy to remain in the upper water of the oceans. Trapping water in their mantle cavities and forcing it out through a siphon, these invertebrates can swim about as fast as any fish their size. Two long arms with pads containing suction cups at the ends are used to capture prey, and eight shorter arms containing suckers take the prey to the mouth, where it is crushed by a beaklike mouthpiece.

The consideration of locomotion in fish is more complex than that of the squid because of body motion and the role of fins. The basic movement of swimming fish is the passage of a wave of lateral body curvature from the front to the back of the fish. This is achieved by the alternate contraction and relaxation of muscle segments, *myomeres,* along the sides of the body. The backward pressure of the fish's

body and fins produced by the movement of this wave provides the forward thrust (fig. 15–7).

Most active swimming fish have two sets of paired fins, *pelvic* and *pectoral,* used in maneuvers such as turning, braking, or balancing. When they are not in use, they can be folded against the body. Vertically oriented fins, *dorsal* and *anal,* serve primarily as stabilizers. The fin most important in propelling the high-speed fish is the tail, or *caudal,* fin. Caudal fins flare dorsally and ventrally to increase the surface area available to develop thrust. Increased surface area also increases frictional drag. The efficiency of the design of a caudal fin depends on its shape and can be expressed mathematically as an *aspect ratio,* calculated as follows:

$$\frac{(\text{fin height})^2}{\text{fin area}}$$

There are five basic shapes of caudal fins: rounded (as on sculpin), truncate (bass), forked (yellowtail), lunate (tuna), and heterocercal (shark). The rounded fin, with an aspect ratio of approximately 1, is flexible (fig. 15–8) and useful in accelerating and maneuvering at slow speeds. The somewhat flexible truncate (ratio 3) and forked (ratio 5) tails will be found on faster fish and may still be used for maneuvering. The lunate caudal fin, with an aspect ratio of up to 10, is found on the fast-cruising fishes such as tuna, marlin, and swordfish; it is very rigid and useless in maneuverability but very efficient in propelling. The heterocercal caudal fin is asymmetrical, with most of its mass and surface area in the dorsal lobe.

The heterocercal fin produces a significant lift to sharks as it is moved from side to side. This lift is important because sharks have no swim bladder and tend to sink when they stop moving. To aid this lifting, the pectoral fins are large and flat. Located on the anterior of the shark's body like the wings of an airplane, they serve as hydrofoils that lift the front of

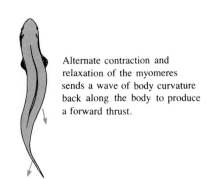

Alternate contraction and relaxation of the myomeres sends a wave of body curvature back along the body to produce a forward thrust.

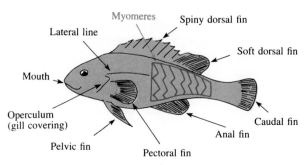

FIGURE 15–7   Swimming Motions and Fins.

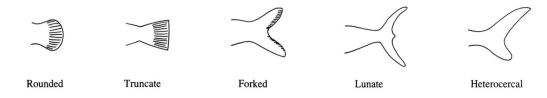

Rounded   Truncate   Forked   Lunate   Heterocercal

Rounded—Aspect ratio 1, typical of bottom-dwelling fish like sculpin and flounder

Truncate—Aspect ratio 3, typical of salmon and bass

Forked—Aspect ratio 5, typical of herring and yellowtail

Lunate—Aspect ratio 7–10, typical of most tuna

Heterocercal—Aspect ratio varies, typical of sharks that need an upward thrust from their swimming motion

FIGURE 15–8 Caudal Fin Shapes and Aspect Ratios.

the shark's body to balance the posterior lift applied by the caudal fin. The shark loses a lot in maneuverability as a result of this modification of the pectoral fins.

Pectoral fins in many fish do more than just help maneuvering (fig. 15–9). Wrass and sculpins use them to "row" themselves through the water in jerky motions while the rest of the body remains motionless. Skates and rays swim by sending undulating motions across the greatly modified pectoral fins, and manta rays flap them like wings. The flying fish, *Exocoetus,* uses greatly enlarged pectoral fins to glide for up to 400 m (1312 ft) across the ocean surface after propelling itself with the enlarged ventral lobe of its caudal fin into the air to escape dolphins and other predators. Other fish have pectoral fins modified into fingerlike structures used to walk on the ocean floor.

Many fish in no great hurry propel themselves by undulation of the dorsal fin. Examples are triggerfish, sea horses, ocean sunfish, and trunkfish. The sunfish also uses its anal fin and stubby caudal fin to aid in the process, still without impressive results.

Modifications in Swimming Behavior

Some fish spend most of their time waiting patiently for prey and exert themselves only in short bursts as they lunge at the prey; others cruise relentlessly through the water seeking out prey. There is a marked difference in the nature of the musculature of fishes with these different styles of seeking out food. A grouper, representative of the *lungers,* has a truncate caudal fin (aspect ratio 3), and almost all its muscle tissue is *white*. Tuna, on the other hand, are *cruisers,* and less than half their muscle tissue is white; it is mostly of a variety referred to as *red* (fig. 15–10).

Red muscle fibers are much smaller in diameter (25–50 μm, or 0.01–0.02 in) than white muscle fibers (135 μm, or 0.05 in) and contain higher concentrations of *myoglobin,* a red pigment with an affinity for oxygen (O_2). These factors allow the red fibers to obtain a much greater oxygen supply than is possible for white fibers and therefore achieve a metabolic rate six times that of white fibers. The red muscle tissue is abundant in cruisers that constantly swim.

Flying fish—
Pectoral fins modified for gliding

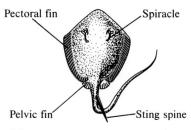

Pectoral fin   Spiracle

Pelvic fin   Sting spine

Stingray—
Pectoral fins modified for swimming

Gurnard—
Pectoral fins modified for walking

FIGURE 15–9 Pectoral Fin Modifications.

FIGURE 15–10 Feeding
Styles, Lunger and Cruiser.
A, lungers like the grouper sit patiently on the bottom and capture
prey with quick, short lunges. Their
muscle tissue is predominantly
white. *B, cruisers* like the tuna swim
constantly in search of prey and
capture it with short periods of
high-speed swimming. Their muscle
tissue is predominantly red. *A,*
photo courtesy of Scripps Institution
of Oceanography, University of California, San Diego. *B,* photo courtesy
of National Marine Fisheries Service.

A.

B.

The lungers can get along quite well with little red tissue because they need not constantly move. White tissue, which fatigues much more rapidly than red tissue, is used by the tuna for short periods of acceleration while on the attack. It is also quite adequate for propelling the grouper and other lungers during their quick passes at prey.

The speed of fish is thought to be closely related to body length. For tuna, well-adapted for sustained cruising and short bursts of high-speed swimming, cruising speed averages about 3 body lengths/second; maximum speed of about 10 body lengths/second can be maintained for only one second. A yellowfin tuna, *Thunnus albacares,* has been clocked

at 74.6 km/hr (46 mi/h) or more than 20 body lengths/second over a period of 0.19 second. Theoretically, a 4-m (13-ft) bluefin tuna, *Thunnus thynnus,* could reach speeds of up to 143.9 km/h (89.2 m/h). Many of the toothed whales are known to be capable of high rates of speed. The porpoise *Stenella* has been clocked at 40 km/h (25 mi/h), and it is believed that the top speed of killer whales may be in excess of 55 km/h (34 mi/h).

As is true with many chemical reactions, metabolic processes can proceed at greater rates at higher temperatures. Although fish are generally considered to be *poikilothermic* (cold-blooded), having body temperatures that conform to the temperature of their environment, there are some fast swimmers with body temperatures above that of the surrounding water. Some are only slightly above the temperature of the water: the mackerel *(Scomber),* yellowtail *(Seriola),* and bonito *(Sarda)* have body temperature elevations of 1.3°, 1.4°, and 1.8°C (2.3°, 2.5°, and 3.2°F), respectively. Other fast swimmers that have much higher elevations are members of the mackerel shark genera, *Lamna* and *Isurus,* as well as the tuna, *Thunnus.* The bluefin tuna has been observed to maintain a body temperature of 30°–32°C (86°–90°F) *(homeothermic)* regardless of the temperature of the water in which it is swimming. Although it is more commonly found in warmer water where the temperature difference is no more than 5°C (9°F), body temperatures of 30°C (86°F) have been measured in bluefin tuna swimming in 7°C (45°F) water.

Why do these fish exert so much energy to maintain their body temperatures at high levels when other fishes do quite well with ambient body temperatures? Their mode of behavior is that of a cruiser, and any adaptation (high temperature and metabolic rate) that can increase the power output of their muscle tissue helps them search out and capture prey.

The mackerel sharks and tuna are aided in conserving the heat energy needed to maintain their high body heats by a modification of the circulatory system (fig. 15–11). While most fish have a dorsal aorta, just beneath the vertebral column, that provides blood to the swimming muscles, the mackerel sharks and tuna have a *cutaneous artery* just beneath the skin on either side of the body. As cool blood flows

into red muscle tissue, its temperature is increased by heat generated by muscle metabolism (muscle contractions). A fine network of tiny blood vessels within the muscle tissue is designed to minimize heat loss. The vessels that return the blood to the *cutaneous vein,* parallel to the cutaneous artery along the side of the fish, are all paired with small vessels carrying blood into the muscle tissue. In this way, the warm blood leaving the tissue helps to heat the cooler blood entering from the cutaneous artery.

MARINE MAMMALS

Ocean mammals include the coast-dwelling herbivores, the sirenans (sea cows), and a variety of carnivorous forms (fig. 15–12). Spending all or most of their lives in coastal waters and coming ashore to breed are the sea otter and pinnipeds (sea lions, seals, and walruses). The truly oceanic mammals are the cetaceans—whales, porpoises, and dolphins (fig. 15–13 and plates 1, 27B, 28, and 29). Since mammals evolved from reptiles on land some 200 million years ago and no marine forms are known earlier than 50 million years ago, it is believed that all marine forms evolved from some ancient land-dwellers.

Cetaceans

Well known as the mammals best adapted to life in the oceans, whales are of two basic types. The toothed whales, the Odontoceti, include 74 species

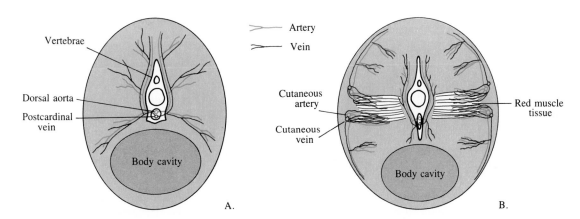

FIGURE 15–11   Circulatory Modifications in Warm-Blooded Fishes.
A, most cold-blooded (poikilothermic) fishes have major blood vessels arranged in the pattern represented on the left. All blood flows to muscle tissue from the dorsal aorta and returns to the postcardinal vein beneath the vertebral column. *B,* warm-blooded (homeothermic) fishes like the tuna have cutaneous arteries and veins that help maintain high blood temperature by using heat energy generated by contracting muscle tissue. Its blood vessel pattern is represented on the right.

A. Sea cow.

B. Elephant seals.

C. Harbor seals.

D. Sea otter.

FIGURE 15–12   Marine Mammals—Sea Cow and Pinnipeds.
A, photo by Marty Snyderman. *B, C,* and *D,* photos from Scripps Institution of Oceanography, University of California, San Diego.

grouped under three common categories: sperm whales, porpoises, and dolphins. The baleen whales, the Mysticeti, probably evolved from the toothed whales some 30 million years ago. In place of teeth, they have baleen, plates of horny material, that hang down from the upper jaw to serve as a sieve. The toothed whales are active predators that feed mostly on smaller fish and squid, although the killer whale is known to feed on a variety of larger animals, including other whales. The baleen whales feed primarily by filtering crustaceans at depths ranging from the surface down to and including the sediment of shallow ocean basins.

The body of a whale is more or less cigar-shaped, nearly hairless, and insulated with a thick layer of blubber. The forelimbs are modified into flippers that

may be moved only at the "shoulder" joint. The hind limbs are vestigial, not attached to the rest of the skeleton and not externally visible. The skull is highly modified, with the nasal opening or openings near the top. Whales propel themselves by vertical movements of a horizontal *fluke* (tail fin).

Modifications to Increase Swimming Speed

Cetaceans' muscles are not vastly more powerful than those of other mammals, so it is believed that their ability to swim at high speed must result from modifications that reduce frictional drag. To illustrate the importance of streamlining in reducing the energy requirements of swimmers, a small dolphin would require muscles five times more powerful than it has

FIGURE 15–13 Baleen and
Toothed Whales.

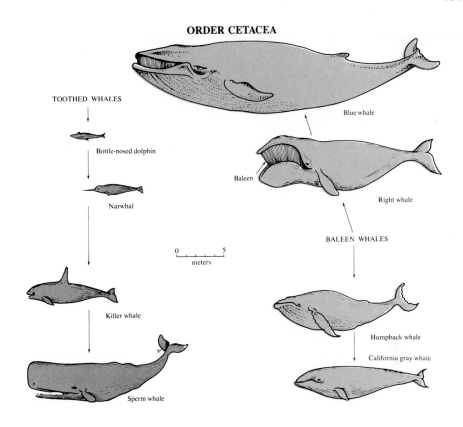

ORDER CETACEA

TOOTHED WHALES

Blue whale

Bottle-nosed dolphin

Baleen

Narwhal

Right whale

BALEEN WHALES

0   5
meters

Killer whale

Humpback whale

California gray whale

Sperm whale

to swim at 40 km/h (25 mi/h) in turbulent flow than in streamlined laminar flow. In addition to having a streamlined body, cetaceans are believed to achieve near-laminar flow of water across their bodies with the aid of a specialized skin structure. The skin is composed of two layers: a soft outer layer that is 80% water and has narrow canals filled with spongy material and a stiffer inner layer composed mostly of tough connective tissue. The soft layer tends to reduce the pressure differences at the skin–water interface by compressing under regions of higher pressure and expanding into regions of low pressure.

Modifications to Allow Deep Diving

Where humans can free-dive to a maximum recorded depth of about 60 m (197 ft) and hold their breath in rare instances for 6 minutes, the sperm whale, *Physeter,* is known to dive deeper than 2200 m (7216 ft), and the North Atlantic bottle-nosed dolphin, *Hyperoodon ampullatus,* can stay submerged for up to two hours.

One reason why whales can make deep dives and stay submerged for long periods is their ability to alternate between periods of *eupnea* (normal breathing) and *apnea* (cessation of breathing). The periods of apnea occur while the animal is submerged. To understand how some cetaceans are able to go for long periods without breathing requires a general knowledge

of the lungs and their associated structures. Air that is inhaled passes through the *trachea,* into the *bronchi,* and through a series of smaller *bronchioles* and *alveolar ducts,* to tiny terminal chambers, the *alveoli* (fig. 15–14). The alveoli are lined by a thin *alveolar membrane* that is in contact with a dense bed of capillaries; the exchange of gases between the inspired air and the blood occurs across the alveolar membrane. Some cetaceans have an exceptionally large concentration of capillaries surrounding the alveoli, which have muscles that move air against the membrane by repeatedly contracting and expanding. Cetaceans take from 1 to 3 breaths per minute while resting, compared to about 15 in humans. Because of the longer time inhaled breath is held, the large capillary mass in contact with the alveolar membrane, and the circulation of the air by muscular action, many cetaceans can extract almost 90% of the O_2 in each breath, compared to the 4% to 20% extracted by terrestrial mammals.

To use this large amount of O_2, which can be taken into the blood efficiently during long periods under water, cetaceans apply two strategies. They (1) store large amounts of O_2 and (2) can reduce oxygen use. The storage of so much O_2 is possible because prolonged divers have a much greater blood volume per unit of body mass than those that dive for only short periods. Compared to terrestrial animals, some ceta-

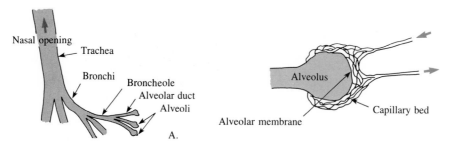

FIGURE 15–14 Cetacean Modifications to Allow Prolonged Submergence.
A, basic lung design. Air enters the lung through the trachea, and oxygen is absorbed into the blood through the walls of the alveoli. *B,* oxygen exchange in the alveolus. A dense mat of capillaries receives oxygen through the alveolar membrane. Because air is normally held in the lungs of cetaceans for up to 1 minute before exhaling, as much as 90% of the oxygen can be extracted. Whales have up to twice as many red blood cells and 9 times as much myoglobin in their muscle tissue as terrestrial mammals in which to store this oxygen.

ceans have twice as many red blood cells per unit of blood volume and up to 9 times as much myoglobin in the muscle tissue. Thus large supplies of O_2 can be chemically stored in the hemoglobin of the red blood cells and the myoglobin.

Additionally, the muscles that start a dive with a significant O_2 supply can continue to function through anaerobic glycolysis when the O_2 is used up. The muscle tissue is relatively insensitive to high levels of carbon dioxide, a product of aerobic respiration, and lactic acid, a product of anaerobic respiration.

Because the swimming muscles can function without oxygen during a dive, they and other organs, such as the digestive tract and kidneys, can be sealed off from the circulatory system by constriction of key arteries. The circulatory system services primarily the heart and brain. Because of the decreased circulatory requirements, the heart rate can be reduced by 20% to 50% of its normal rate. However, recent research has shown that no such reduction in heart rate occurs when the common dolphin *(Delphinus delphis),* white whale *(Delphinapterus leucas),* and the bottle-nosed dolphin *(Tursiops truncatus)* dive.

Another problem with deep and prolonged dives is the absorption of compressed gases into the blood. When humans make dives using compressed air, the prolonged breathing of compressed air, particularly N_2, can result in *nitrogen narcosis,* or *decompression sickness* (the bends). Nitrogen narcosis is an effect similar to drunkenness, and it can occur when a diver either goes too deep or stays too long at depths greater than 30 m (98 ft). If a diver surfaces too rapidly, the lungs cannot remove the excess gases fast enough, and the reduced pressure may cause small bubbles to form in the blood and tissue. The bubbles

interfere with blood circulation, and the resulting decompression sickness can cause excruciating pain, severe physical debilitation, or even death.

Cetaceans and other marine mammals do not suffer from these difficulties. Their main defense against absorbing too much N_2 seems to be a more flexible rib cage. By the time a cetacean has reached a depth of 100 m (328 ft), the rib cage has collapsed under the 11 atmospheres of pressure. The lungs within the rib cage also collapse, removing all air from the alveoli. Since most absorption of gases by the blood occurs across the alveolar membrane, the blood cannot absorb the compressed gases, and the problem of nitrogen narcosis is avoided.

It is, however, possible that the collapsible rib cage is not the main defense against the bends. A recent research effort designed to increase the nitrogen concentration in the lungs of a trained dolphin and give it the bends put enough nitrogen into the tissue of the animal to give a human a severe case of the bends. The dolphin suffered no ill effects. This whale, as well as other species, may have simply evolved an insensitivity to this gas not present in other mammals.

Echolocation Based on present evidence, fully 20% of mammalian species are thought to echolocate. Of the terrestrial forms, bats are the best known, but shrews, flying lemurs, the fat dormouse, and golden hamster are also known to use echolocation. Marine mammals that are known or strongly suspected to use echolocation are all the toothed whales, the Weddell seal, the California sea lion, the walrus, and the gray, blue, and minke whales. These animals can involuntarily calculate the distance of an object by multiplying the velocity with which a sound signal travels to and returns from the object by the time it takes to reach it and dividing this product by 2.

$$D = \frac{V \times T}{2}$$

Although all marine mammals have good vision, conditions within the marine environment often limit its effectiveness. In coastal waters, where suspended sediment and dense plankton blooms make the water turbid, and in the deeper waters, where light is limited or absent, echolocation would surely be beneficial in pursuit of prey or the location of objects in the water.

Cetaceans emit a wide variety of sounds, including what might be called wails, creaks, squeals, moans, clicks, and even songs in the case of the humpback whale, *Megaptera*. Although the other sounds are thought to be of value in communication, only the clicks are thought to be used for echolocation (fig. 15–15).

The common dolphin, *Tursiops truncatus,* has been closely studied by echolocation researchers. For a general scanning of its environment, *Tursiops* emits low-frequency clicks to locate objects at a distance; upon close approach, higher-frequency clicks are produced to determine fine details. Using only their echolocation abilities, the common dolphin has been able to discriminate between such similar objects as fish of similar size and shape, identically shaped plates of different metallic composition, and plates of the same metallic composition with slightly different thickness. The clicks of the small common dolphin are of a frequency range that is partly audible to humans; however, some are of ultrasonic frequencies that can be repeated up to 800 times per second. Large toothed whales emit lower-frequency clicks at lower repetition rates. The killer whale, *Orcinus orca,* produces from 6 to 18 clicks per second, and the sperm whale, *Physeter catodon,* may produce from less than one to more than 40 clicks per second. Each click of the sperm whale contains 9 pulses, or *clickettes.* It has been estimated that the sperm whales can detect their main prey, squid, from a distance of

up to 400 m (1312 ft) by use of their low-frequency scanning clicks.

How the pulses of clicks are produced and how the returning sound is received are not fully understood. However, there are some widely accepted theories that attempt to answer both questions for certain species of toothed whales. Measurements of the intensity of clicks emitted by smaller toothed whales indicate that the sound originates in the forehead region. The source of sound in this area could be the complex of air sacs associated with the nasal passage and functioning in conjunction with the muscular nasal plugs that close off the blowhole (fig. 15–16A). Sounds produced in the blowhole could be reflected off the concave surface at the front of the skull of toothed whales and focused by the lens-shaped *melon.* The melon is a dome-shaped fatty structure located in the center of the forehead of many toothed whales. Studies using hydrophones have indicated that the clicks of dolphins are focused into a forward-directed beam. Although most researchers think it is a less likely source of echolocation clicks, the larynx of toothed whales is well muscled and complex; even without vocal cords, it could be capable of producing sound.

In the sperm whale, the *spermaceti organ,* once prized by whalers for the fine quality of its oil, may play a major role in focusing the clicks (fig. 15–16B). The spermaceti organ is encased in ligaments and rests in the *rostrum,* a forward extension from the base of the frontal skull. Norris and Harvey (1972) suggest that a special complex arrangement, part of the respiratory system, produces the click. The blowhole high on the front of the sperm whale's bulbous head opens into the *distal air sac,* located in front of the spermaceti organ, and the *left nasal passage,* which leads to the trachea. Smaller than the left nasal passage, the *right nasal passage* extends from the *museau du singe,* at the base of the distal air sac, along the base of the spermaceti organ, to the *frontal air sac* at the posterior end of the spermaceti organ, then on to the trachea. The sound-making structure in this system is the museau du singe or *monkey's muzzle.* It is composed of a pair of hard, tightly closed lips.

FIGURE 15–15 Echolocation. Clicking sound signals are generated by whales and bounced off objects in the ocean to determine their size, shape, and distance.

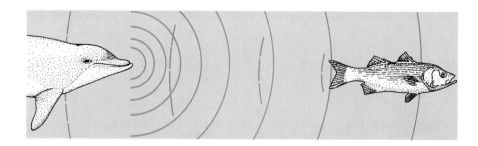

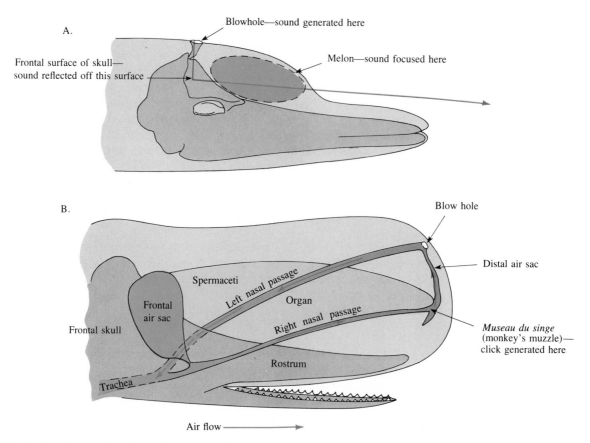

FIGURE 15-16 Generation of Echolocation Clicks in Small-Toothed and Sperm Whales.

A, in small-toothed whales, the clicks may be generated within the blowhole mechanism, reflected off the frontal surface of the skull, and focused by the fatty melon into a forward-directed beam. *B,* structures that may be related to the generation of clicks by sperm whales. Air may pass from the trachea through the right nasal passage across the *Museau du singe,* where clicks are generated. It passes along the distal air sack past the closed blow hole and returns to the lungs along the left nasal passage. The initial click is followed by 8 progressively weaker clicks that may result from sound energy bouncing back and forth between the frontal air sack and the distal air sack.

A means by which this complex system could generate sound involves the forcing of air from the trachea through the right nasal passage. As air is forced past the lips of the monkey's muzzle, the lips separate and snap back together, producing a click. This sound is the first pulse of the 9-pulse click characteristic of sperm whales. The other 8 pulses, of progressively lower intensity, may result from the reverberation of the initial sound back and forth between the distal air sac and the frontal air sac across the spermaceti organ. The spermaceti organ may serve as the sound channel that guides the reflected sound signals. After the air passes through the monkey's muzzle, it is not lost through the blowhole. The blowhole is sealed off, and the air is returned to the trachea through the left nasal passage.

Whereas the human ear is sensitive to frequencies from 16 Hz to 20 kHz, some toothed whales respond to frequencies as high as 150 kHz. In most mammals, the bony housing of the inner ear structure is fused to the skull, so that, when submerged, sounds transmitted through the water are first picked up by the skull and travel to the hearing structure from many directions. This structure makes it impossible for such mammals to locate accurately the source of the sounds. Obviously such an arrangement for the hearing structure would not suit an animal that depended on echolocation to accurately determine the positions of objects in the water. All cetaceans have evolved structural features that insulate the inner ear housing, the *tympanic bulla,* from the rest of the skull. In the toothed whales, the tympanic bullae are separated from the rest of the skull by an appreciable distance and surrounded by an extensive system of air sinuses. The sinuses are filled with an insulating emulsion of oil, mucus, and air and are surrounded by fibrous connective tissue and venous networks.

There is some disagreement on how sound reaches the inner ear to be processed. Some investigators believe it passes through the external auditory canal as it does in other mammals, yet this canal is usually blocked by a plug of ear wax in baleen whales. In toothed whales it may be nearly or completely covered by skin. A thin flaring of the lower jaw, less than 0.1 mm (0.004 in) thick, is directed toward the tympanic bulla on either side of the head and connected with it by a fat- or oil-containing body. It is believed that sound is picked up by the thin, flaring jawbone and passed to the inner ear via the connecting fat body. Tests of acoustical sensitivity have indicated the lower jaw is up to 6 times as sensitive as the external auditory canal.

GROUP BEHAVIOR

Depending on the size, feeding behavior, nature of the reproductive process, and other requirements of maintaining the species, many animals inhabiting the open ocean have developed patterns of group behavior that allow them to most efficiently exploit their environment. Two of these patterns are *schooling* and *migration.*

Schooling

Obtaining food occupies most of the time of many inhabitants of the open ocean. It is achieved by the fast and agile animals through active predation; other animals move more leisurely while filtering small food particles from the water. Examples of predators and filter feeders can be found in populations of pelagic animals from the tiny zooplankton to the massive whales. Patches of zooplankton may occur simply because the nutrients that support the phytoplankton they feed on may be in high concentrations in certain coastal waters. We usually do not refer to these patches as *schools.* The term school is usually reserved for well-defined social organizations of fish, squid, and crustaceans.

The number of individuals in a school can vary from a few larger predaceous fish to hundreds of thousands of small filter feeders. Within the school, individuals of the same size move in the same direction with equal spacing between each. This spacing is probably maintained through visual contact, and in

FIGURE 15–17 Schooling.
School of Pacific sardines, typical of the many small herringlike fishes that form large schools.

the case of fish, by use of the lateral line system that detects vibrations of swimming neighbors. The school can turn abruptly or reverse direction as individuals at the flank or the rear of the school assume leadership positions (fig. 15–17).

The advantage of schooling seems obvious from the reproductive point of view. During spawning, it assures that there will be males to release sperm to fertilize the eggs shed into the water or deposited on the bottom by females. However, most investigators believe the most important function of schooling in small fish is protection from predators. On first consideration, it seems illogical that the formation of schools would serve such a purpose. Aren't the smaller fish making it easier for the predator by forming a large mass? This is a difficult question; in fact, no experimental evidence provides the basis for an unequivocal answer. However, the consensus, based primarily on conjecture, is *no.*

The belief that small fish are safer from predators if they form schools is based in part on the following reasoning: Over four thousand species of fish are known to form schools. This fact alone indicates that this behavior has evolved and is so pervasive today among fish populations because, for fish with no other means of defense, it somehow provides a better chance of survival than swimming alone. How schooling is protective may grow out of the following considerations: (1) If members of a species form schools, they reduce the percentage of the volume of the ocean in which a cruising predator might find one of their kind. (2) Should a predator encounter a large school, it is less likely to consume the entire unit than if it encounters a small school or an individual. (3) The school may appear as a single large and dangerous opponent to the potential predator and prevent

In the past decade, research has disclosed a close relationship between the temperature of incubation of reptile eggs and the sex of the hatchlings. Lizard and alligator eggs will produce females in cool temperatures and males in warm temperatures. Most other reptiles, including marine turtles, produce females at higher incubation temperatures and males at lower temperatures (fig. 15A).

A study done on beaches of South Carolina and Georgia barrier islands showed summer season change in the sex ratios of loggerhead turtle hatchlings from May to August 1982. No females appeared in May, but the hatchlings were 80% female in July. The female percentage dropped to 10% in August. Conservationists are beginning to figure these findings into their management plans for the turtles.

FIGURE 15A Green Turtle *(Chelonia mydas).*
A large marine turtle found in the Atlantic and Pacific oceans. Photo by Alan R. Hargens. Courtesy of Scripps Institution of Oceanography, University of California, San Diego.

some attacks. (4) Predators may find the continually changing position and direction of movement of fish within the school confusing, making attack particularly difficult for predators that can attack only one fish at a time. It is surely possible that there are other more subtle reasons for schooling. Whatever they might be, this gregarious behavior must enhance species survival because it is widely observed among pelagic animals.

Migration

Many oceanic animals undertake migrations of varying magnitudes. This behavior is observed among sea turtles, fish, and mammals. Some animals, for instance the sea turtles, have simply not evolved far enough from their land-based ancestors to carry out reproduction at sea and must return to dry beach deposits to lay their eggs. Other populations, including many of the baleen whales, migrate because the physical environment and nature of available food suitable for adults do not meet the needs of young members of their species.

Migratory routes of commercially important baleen whales have been well known since the mid-nineteenth century. The paths of these and other air-breathing mammals are easy to observe. However, the migratory paths of many less visible fish, even though they may also have high commercial value, have been more difficult to identify. Tagging studies have helped us understand the patterns of movement of these fish. Sampling the distribution patterns of eggs, larvae, young, and adult populations and radio-tracking individuals have also helped to identify migratory routes.

One aspect of migratory behavior that is of interest to investigators is orientation: How do migratory species orient themselves in time and space? To put the problem another way, How do they know where they are in relationship to where they want to go? How do they know when to leave their present location to get where they are going at the proper time?

Although these answers are still being sought, it is believed that all of the migratory species have an innate sense of time referred to as a *biological clock.* External manifestations of the biological clock of a species, called *circadian rhythms,* are physiological variations that occur independent of changes in the external environment. Circadian rhythms include cyclic changes in respiration rate and body temperature. However, some changes in the external environment can alter the circadian rhythms. It is possible that changes in food availability, water temperature, and length of daylight periods may serve to trigger seasonal migrations.

It seems that orientation in space must be a more complex problem than orientation to time. Animals such as mammals and turtles that migrate at the surface could well use their sight as a means of orientation. Gray whales that migrate close to land over a large part of their migratory route could possibly identify landmarks along the shore. When out of sight of land, mammals and turtles could use the relative positions of the sun, moon, and stars to guide them on their way. Fish that migrate beneath the surface are thought to use smell, relative movement within ocean currents, and small induced currents (generated as the ions dissolved in ocean water are carried through the earth's magnetic field by ocean currents) to help them orient themselves in ocean space.

Many of the important food fish of the high northern latitudes deposit pelagic egg masses that are transported by currents. Such fish will actively migrate upcurrent during spring or summer and spawn in a location that will assure the current will carry the eggs back to a nursery ground where the food supply will be appropriate for newly hatched fry. Migrations of this type are usually over distances of a few tens or hundreds of kilometers. This behavior is observed in the population of Atlantic cod, *Gadus morhua,* that spawns along the south shore of Iceland. Adults that occupy feeding grounds along the north and east coasts of Iceland and around the southern tip of Greenland migrate to the spawning grounds in the late winter and early spring. This migration involves swimming against the East Greenland and Irminger currents and includes only mature adults at least 8 years of age. These spawning migrations are undertaken annually until the adults die at an age of 18 to 20 years (fig. 15–18).

The spawning grounds are bathed in the relatively warm water of the Irminger Current, a northward-flowing branch of the Gulf Stream. Each female releases up to 15 million eggs, which float in the surface current. The drifting eggs, carried by the East Greenland and Irminger currents, hatch in about two weeks. The larvae feed at midwater depths while the currents are carrying them to the adult feeding grounds. They willl remain there, undertaking only short onshore-offshore migrations, until they mature and make their first spawning migration.

Longer migrations are undertaken by two Atlantic bluefin tuna populations that spawn in tropical waters near the Azores in the eastern Atlantic and the Bahamas in the western Atlantic. Little is known about where the newly hatched tuna feed, but the adults are known to move north along the seaward

FIGURE 15–18 Migration Routes of the Icelandic Cod, Bluefin Tuna, and Atlantic Eel *(Anguilla)*.

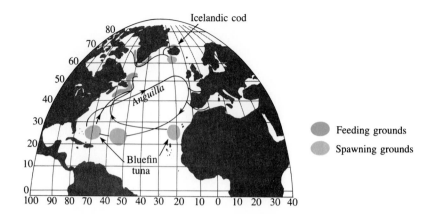

edge of the Gulf Stream in May and June and feed on the herring and mackerel populations off the coasts of Newfoundland and Nova Scotia. They may move even farther north, but the route used by them to return to the spawning grounds is unknown.

Anguilla, the Atlantic eel, undertakes what appears to be a single round trip migration during its lifetime. Its behavior is *catadromous:* after being spawned in the ocean, it enters freshwater streams to spend its adult life and returns near the end of its life to the ocean spawning site.

Spawning is thought to occur in the Sargasso Sea southeast of Bermuda. The spawning has not been observed, and the earliest stage of *Anguilla* that has been observed is the leaf-shaped, transparent *leptocephalus* larva. The 5-mm(0.2-in)-long larvae are carried into the Gulf Stream and north along the North American coast. After one year in the ocean, some of the leptocephali metamorphose into young eels, *elvers,* and move into the streams of North America. Others spend another year migrating to the European coast before metamorphosing and moving into European streams. A third population spends another year moving into the Mediterranean Sea before undergoing metamorphosis. The American and European populations are considered by some to be different species—*A. rostrata* and *A. anguilla,* respectively—based primarily on the fact that the American eel has 8 to 10 fewer vertebrae than the European eel.

After spending up to ten years in the freshwater environment, the mature eels undergo yet another change. They develop a silvery color pattern and enlarged eyes that are typical of fish inhabiting the mesopelagic zone. They swim downriver to the ocean and presumably back to the spawning ground, where they are thought to die.

An opposite behavior pattern is characteristic of a salmon of the north Atlantic and Pacific oceans. This *anadromous* behavior includes spawning in freshwa-

ter streams and spending most of their adult lives in the ocean. Pacific species die after spawning; Atlantic salmon return to the ocean.

Six species of Pacific salmon, *Oncorhynchus,* follow similar paths in their migrations. Spawning occurs during the late summer and early fall. Eggs are deposited and fertilized in gravel beds far upstream. The chinook salmon spawns from as far south as central California to Alaska. The young chinook hatching in the spring may head downstream as a silvery, filter-feeding smolt during its first or second year. By the time it reaches the ocean, the young chinook has become a predator, feeding on smaller fish. After spending about four years in the North Pacific, mature salmon of more than 10 kg (22 lb) in weight and 1 m (3.28 ft) in length return to their home streams to spawn (fig. 15–19). Although it is not known how salmon achieve this homing, most investigators think the most likely sources of information processed by the salmon during their return trip to spawning grounds are odors and currents.

Some of the longest migrations known to occur in the open ocean are the seasonal migrations of baleen whales. The benefit of these migrations to the maintenance of the species seems clear. Essentially all baleen species feed in the colder high-latitude waters and breed and calve in warm tropical waters. Feeding occurs during summer while the long hours of sunlight illuminating the nutrient-rich waters produce a vast population of crustaceans to be fed upon. Only the vast supplies of food available in these waters makes it possible for the whales to maintain their great bulk. The relatively small size and thin layers of blubber characteristic of newborn whales necessitate calving in warmer waters. Although most newborn whales weigh over 2 tn, they are still small enough and their blubber layer thin enough that they would lose body heat at too great a rate in the cold, high-latitude waters. They can survive only if they are born

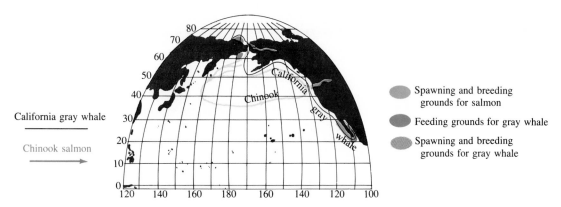

FIGURE 15–19   Migration Routes of the Chinook Salmon and California Gray Whale.

into warm tropical water. Because of the energy demands faced by female baleen whales in producing large offspring (the gestation period is up to 11 months) and providing them with fat-rich milk for several months, it is not uncommon for them to mate only once every two or three years.

The migration route of the *California gray whale* demonstrates how the conditions discussed above are met by whale migration patterns. Gray whales are moderately large whales, reaching lengths of 15 m (49 ft) and weighing over 30 metric tons. Populations of gray whales feed during the summer months in the extreme north Pacific Ocean, Sea of Okhotsk, Bering Sea, and Chukchi Sea. They are unique among baleen whales in that they do not feed by straining pelagic crustaceans and small fish from the water; instead they stir up bottom sediment with their snouts and feed on bottom-dwelling amphipods.

The western populations winter along the coast of Korea, but the population known as the California gray whale migrates from the Chukchi and Bering seas to winter in lagoons of the Pacific coast of Baja California and the Mexican mainland coast near the southern end of the Gulf of California (fig. 15–19). The migration usually begins in September when pack ice begins to form over the continental shelf areas that are their feeding grounds. This migration, which is the longest known migration undertaken by any mammal, may involve a round trip distance of 22,000 km (13,662 mi). First to leave are the pregnant females. They are followed by a procession of mature females that are not pregnant, immature females, mature males, and immature males. After cutting through the Aleutian Islands by way of Unimak Pass, they follow the coast throughout their southern journey. Traveling at an average rate of about 200 km/day (124 mi/day), most of the whales reach the lagoons of Baja California by the end of January.

In these warm-water lagoons, the pregnant females give birth to 2-tn calves. The calves nurse and put on weight quickly during the next two months. While the calves are nursing, the mature males breed with the mature females that did not bear calves. Late in March, the return to the feeding grounds begins, with the procession order reversed. Most of the whales are back in the feeding grounds by the end of June, and they feed on prodigious amounts of amphipods to replenish their depleted store of fat and blubber before the next trip south.

In the case of the gray whales and other mammals that have been studied, the reason for migration is clearly to find an optimum environment for feeding and reproduction. It is probably the same for other migratory species, although it is sometimes more difficult to identify the reason for the choice of environments.

Reproduction

As in the examples just described, reproduction of pelagic animals involves bringing the males and females of the species together for this purpose on some periodic basis. These gatherings usually occur during the warmer spring or summer months, when water temperatures are higher and primary productivity is at its peak.

Most of the invertebrates and fish that inhabit the open ocean are *oviparous:* they lay eggs that hatch in the open water or on the bottom into larval forms that become meroplankters. Animals that reproduce in this way usually produce enormous numbers of eggs—15 million eggs per season in the migrating Icelandic cod—because most will be lost to predators before they hatch or the larvae will be eaten by other members of the plankton community. However,

SALMON RANCHING—A PILOT PROJECT

An interview with William McNeil (fig. 15B), fishery biologist at Oregon State University and one-time director of the Oregon Aqua Foods salmon ranching project.

Q. Dr. McNeil, will you give us an overview of what is being done at the salmon ranching operation in Springfield, Oregon?

A. Oregon Aqua Foods is a subsidiary of The Weyerhaeuser Company, which is a timber products company. Fundamentally, they are trying to take advantage of the heat made available from their paper product plants to increase the rate of growth for the salmon in their project. Here in Oregon, they have built a hatchery adjacent to a paperboard plant. An effort is being made to develop an optimum stock of chum, coho, and chinook (fig. 15C). Oregon, which is the only West Coast state that allows private investment and hatchery operations for profit, has granted Weyerhaeuser permission to release 40 million chum and 20 million each of coho and chinook into the Pacific Ocean each year. The operation opened in 1972 and since 1981 has been able to produce its coho allotment from its own stock. The chum and chinook programs have been greatly reduced and are sustaining themselves at only 1 to 2 million per year.

Q. What are the reasons for the poor development of the chum and chinook populations?

A. Chinook and chum eggs are scarce. Private hatcheries such as this are low on the list of those operations that receive eggs from state hatcheries, which are the primary source of chinook eggs. They had to buy chum eggs from Russia in 1978 and released chum smolts into the ocean in 1979. The return from this release was scheduled for 1983, but El Niño made it impossible for them to recognize their home waters. Private hatcheries had their first chinook return in the fall of 1981 from releases that occurred three to five years before. It will be a few years yet before they are able to meet release allowances for these populations from their own brood stock.

FIGURE 15B Dr. William McNeil.

FIGURE 15C Varieties of Salmon.

Top, chum *(Oncorhynchus keta);* center, coho *(O. kisutch);* bottom, chinook *(O. tschawyscha).* Courtesy U.S. Fish and Wildlife Service.

Q. Could you outline the procedure used in ranching one of the salmon populations?

A. Yes. About 30 million eggs are taken from about 11,000 female coho salmon, fertilized, and incubated in 10°C (50°F) water warmed by the adjacent paperboard plant (fig. 15D). After they hatch, the fry are fed and raised to smolt size in 12°C water. They reach a size in 7 or 8 months that they would require 16 to 18 months to reach in nature. The smolt raised in our warmwater environment double their body weight every 23 days, while it may require 2 or 3 months in the natural streams. At the end of the 7- to 8-month feeding period, the smolt are moved to nearshore ponds, where they learn to recognize the bay water to which it is hoped they will return. They are usually kept in these ponds about 2 weeks. The smolt were initially released directly into the coastal ocean at Newport and Coos Bay. Now they use only the Newport facility and carry the smolt from 5 to 15 miles out to sea in barges to release them.

Q. What do they feed the smolt?

A. They are fed a fish meal made from the scraps of fish-processing plants along the coast. It is fortified with vitamins and formed into pellets by use of a vegetable adhesive that makes up about 10% of the mass of the pellets. Food costs account for about one third of the operating cost.

Q. What percent of the salmon released can be expected to return?

A. Under maximum ocean environment conditions, a 5% to 15% return might be expected. However, the average return to date is about 0.7%. This low figure results from a number of causes. There is of course an ocean fishery that takes 1.5 fish for every one that returns. So the project is contributing to that fishery. The ocean environment has been bad for the last 6 or 7 years. The coastal upwelling has been below normal in intensity, making the nearshore waters warmer. This increases smolt loss to shore birds, especially the murres. The smolt do not like the very cold water brought to the surface by upwelling and will move offshore quickly when upwelling is intense. When the water is not so cold, the smolt will spend too long near shore, where they are preyed upon heavily by the shore birds. There is a population of some 300,000 to 400,000 murres that may eat from 15 to 20 smolt per bird per day. That is the main reason for barging them offshore

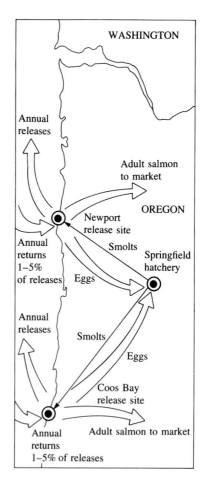

FIGURE 15D A Salmon Ranch Project Design.

for release. Hopefully, the return percentage will increase as the homing acuity of the population is increased by natural selection. The best return was in 1985, when 2.6% of a 1984 release from barges returned to Newport.

Q. What percent of the release needs to return for the operation to be profitable?

A. For the coho, a return of 1.5% to 2% will probably result in profit.

Q. Considering the 14 years of experience that has been gained during the conduct of this project, what do you see in the future of salmon ranching?

A. I am cautiously optimistic. The economic success of ranching is certainly not assured, but I believe when the brood stock reaches the proper level of quality, there is every reason to expect a profit to be made. However, it will require long-term investments at levels that can probably be sustained only by large corporations. Salmon ranching is also an operation that is not free of controversy. There are open-ocean fishery interests and environmental groups that have expressed some legitimate concern about the manipulations of the natural order of things that are required in ranching operations. However, I believe that at some point in the future, mariculture will prove to be superior to open-ocean fisheries as a means of reaping the maximum harvest from the sea. It happened on the land, as agriculture and animal husbandry replaced hunting and gathering as the basis for exploiting the potential yield of the land. It seems inevitable that mariculture will also replace the hunting and gathering technology of our present wild fisheries as the most efficient means of exploiting the potential yield of the oceans.

FIGURE 15–20   Skate Egg Case.
This horny mermaid's purse protects the egg that develops within it until it is ready to hatch. Photo courtesy of James King, Graphic Impressions, Santa Barbara, California.

FIGURE 15–21   Ovoviviparous Behavior in Sea Horses.
The female deposits the eggs in the ventral pouch of the male, where they are protected during incubation. Photo by Larry Ford.

some oviparous fish may produce only one or a few well-protected eggs per season; the cartilaginous sharks, skates, and rays include many such species (fig. 15–20).

The females of other fish species keep their fertilized eggs in their reproductive tract until they hatch. Although the young come into the world live, the process by which the embryos develop is the same as for those that develop from eggs that are laid. This behavior is called *ovoviviparous,* describing a method of reproduction in which the eggs are incubated internally. The sea horse and pipefish exemplify a special twist to internal incubation where the female deposits the eggs in a ventral pouch of the male (fig. 15–21). The male carries the eggs until they hatch, and the young fish enter the ocean directly from this pouch. The additional protection against loss of eggs through pelagic predation results in the reduction of the number of eggs that must be produced; for example, the dogfish shark, *Squalus,* usually produces fewer than a dozen eggs at a time.

Viviparous animals give birth to live young, but the behavior is considered by most to require also that the young be given more than space in the mother's reproductive tract to incubate. Part of the behavior includes providing embryonic nutrient in addition to the yolk of the egg. Some sharks and rays have projections from the wall of the uterus called *villi* that secrete a rich milk. With this additional nutrition, the stringray, *Pteroplatea,* produces young that enter the ocean with a mass up to 50 times that of the mass of

the yolk of the egg from which they came. In the case of the white-tip shark, *Carcharhinus,* this additional nutrition passes to the embryo through a yolk sac attached to the wall of the uterus. Mammals exhibit the highest level of viviparous behavior in that the embryo is encased in the placental sac. The mother's blood flows through the placenta, and the embryo receives nutrition from the mother through an umbilical cord by which it is attached to the placenta; essentially all nutrition is provided by the mother. This behavior provides a high degree of protection to the developing young but puts a high energy demand on the mother. Thus mammalian births typically involve only one or a few young who will remain dependent on their mothers for protection and food for some time after their birth.

SUMMARY

Frictional resistance to sinking—a major factor in helping the tiny plankton stay near the surface—is not a major factor in keeping the larger nekton from sinking; they depend primarily on **buoyancy** or **swimming. Rigid gas containers** found in some cephalopods and **expandable swim bladders** of many bony fishes are adaptations that help increase buoyancy. Many invertebrates such as the jellyfish, tunicates, and arrowworms have soft, **gelatinous bodies** that reduce their density. The Portuguese man-of-war also has a **gas-filled float** that supports this pelagic colony. Many of these invertebrate forms are weak swimmers and depend primarily on buoyancy to maintain their positions near the surface.

Strong **swimmers**, the nekton—squid, fish, and mammals—depend on this expenditure of energy to obtain food in the water column. **Squid** swim by trapping water in their mantle cavities and forcing it out through a **siphon**. Most **fish** swim by creating a **wave of body curvature** that passes from the front to the back of the body and provides a forward thrust. The **caudal fin** is the most important in providing thrust, while the **paired pelvic** and **pectoral fins** are used for **maneuvering**. The vertically oriented **dorsal** and **anal fins** serve primarily as **stabilizers**. The efficiency with which a caudal fin produces thrust is indicated by its **aspect ratio,** equal to the square of the fin height divided by the fin area. The **rounded** caudal fin, found on the sculpin, is flexible, has a low aspect ratio of about 1, and can be used for maneuvering at slow speeds. The **lunate** fin is rigid, has a high aspect ratio of about 10, and is of little use in maneuvering. It is very efficient in producing thrust for fast swimmers such as the tuna.

Fish such as groupers are **lungers** that sit motionless and make short, quick passes at passing prey. They have mostly **white muscle tissue.** Tuna and other **cruisers,** constantly swimming in search of prey, possess mostly **red muscle tissue.** Red fibers have a greater affinity for oxygen and tire less rapidly than white fibers, enabling tuna to maintain a cruising speed of about 3 body lengths/sec. Using their white muscles to help in short periods of rapid swimming, they can reach speeds of up to 10 body lengths/sec. Although most fish are **poikilothermic (cold-blooded),** the tuna, *Thunnus,* is **homeothermic;** that is, it maintains a body temperature well above the temperature of the water.

The best adapted mammals for life in the open ocean are the whales (Cetacea). The **toothed whales** (Odonticeti) and **baleen whales** (Mysticeti) propel their highly streamlined bodies through the water with vertical movement of their **horizontal flukes.**

Whales are able to dive deep and stay submerged for unusually long periods of time because of modifications that allow them to absorb 90% of the O_2 in the air they inhale, store large quantities of oxygen, possibly reduce oxygen use, and collapse their lungs below depths of 100 m (328 ft).

Most whales and many other marine mammals are thought to use **echolocation** in finding their way through the ocean and locating prey. The clicking sounds emitted by whales are bounced off objects, and the animal can determine the size, shape, and distance of the objects by the nature of the returning signals and the time elapsed.

Schooling of active swimmers such as fish, squid, and crustaceans is not fully understood, but it has obvious advantages as a reproductive strategy. Schooling more likely serves a protective function, although the ways in which it may meet this end are speculative.

Migrations are observed among sea turtles, fish, and mammals and are related to the needs of **reproduction** and finding **food.** It is believed that **orientation** is maintained during migrations through use of visible landmarks, smell, and the earth's magnetic field. **Baleen whales** may migrate from their cold-water summer feeding grounds to warm low-latitude lagoons in winter so their young can be born into warm water; they might not be able to survive the cold water of higher latitudes at birth. Fishes such as the Atlantic cod swim upcurrent to deposit their eggs so that when they hatch the cod fry will be in water where suitable food is available.

The North Atlantic eel, *Anguilla,* is **catadromous**: it spawns in the deep waters of the Sargasso Sea and spends its adult life in freshwater streams of North America and Europe. The **Atlantic** and **Pacific salmon** are **anadromous**: they spawn in the freshwater streams of North America, Europe, and Asia and spend their adult life in the open ocean.

Most fish are **oviparous,** depositing their eggs in the ocean. Some sharks and rays maintain their eggs in a body cavity until they hatch; this **ovoviviparous** behavior provides a greater protection of the eggs. Where oviparous fish may produce millions of eggs, ovoviviparous fish may produce fewer than a dozen. The stingray, white-tip shark, and mammals are **viviparous,** not only providing space in their bodies for the eggs to develop but also providing nutrition in addition to the egg yolk. In mammals, the young will still be dependent on their mothers for nutrition for some time after birth.

QUESTIONS AND EXERCISES

1. Discuss how the rigid gas chambers in cephalopods may be more effective in limiting the depth to which they can descend than the flexible swim bladders of bony fish.

2. Describe the body form and lifestyle of the following plankters: *Physalia,* jellyfish, tunicates, ctenophorans, and arrowworms.

3. What are the major structural and physiological differences between the fast-swimming cruisers and the lungers that patiently lay in wait for their prey?

4. List the modifications that are thought to allow some cetaceans to (1) dive to great depths without suffering the bends and (2) stay submerged for long periods of time.

5. Describe the process by which the sperm whale is thought to produce echolocation clicks.

6. Although there is disagreement about how it is achieved, discuss what most investigators believe to be the method by which sound reaches the inner ear of toothed whales.

7. Summarize the reasons some investigators believe schooling increases the safety of fishes from predators.

8. What are the methods believed to be used by migrating animals to maintain their orientation?

9. Why do fishes such as Icelandic cod swim upcurrent to spawn?

10. How are the migrations of the North Atlantic eels and Pacific salmon fundamentally different?

11. Why don't the California gray whales remain in the cold-water feeding grounds during the winter season?

12. Compare the reproductive behavior of oviparous, ovoviviparous, and viviparous animals.

REFERENCES

Carey, F. G. 1973. Fishes with warm bodies. *Scientific American* 228:2, 36–44.

Coker, R. E. 1962. *This great and wide sea: An introduction to oceanography and marine biology.* New York: Harper and Row.

Denton, E. J., and Shaw, T. I. 1962. The buoyancy of gelatinous marine animals. *Journal of Physiology* 161:14P–15P.

George, D., and George, J. 1979. *Marine life: An illustrated encyclopedia of invertebrates in the sea.* New York: Wiley-Interscience.

Herald, E. S. 1961. *Living fishes of the world.* Garden City, N.Y.: Doubleday.

Kanwisher, J. W., and Ridgway, S. H. 1983. The physiological ecology of whales and porpoises. *Scientific American* 248:6, 110–121.

MacGinitie, G. E., and MacGinitie, N. 1968. *Natural history of marine animals.* 2nd ed. New York: McGraw-Hill.

Mrosovsky, N.; Hopkins-Murphy, S. R.; and Richardson, J. I. 1984. Sex ratio of sea turtles: Seasonal changes. *Science* 225:4663, 739–40.

Norris, K. S., and Harvey, G. W. 1972. A theory for the function of the spermaceti organ of the sperm whale (*Physeter catodon* L). Animal orientation and navigation, 397–417. Washington, D.C.: National Aeronautics and Space Administration.

Pike, G. C. 1962. Migration and feeding of the gray whale *(Eschrichtius gibbosus). Journal of the Fisheries Research Board of Canada* 19:815–38.

Royce, W.; Smith, L. S.; and Hartt, A. C. 1968. Models of oceanic migrations of Pacific salmon and comments on guidance mechanisms. *Fishery Bulletin* 66:441–62.

Thorson, G. 1971. *Life in the sea.* New York: McGraw-Hill.

Vaughan, T. A. 1972. *Mammalogy.* Philadelphia: W. B. Saunders.

SUGGESTED READING

Sea Frontiers

Bachand, R. G. 1985. Vision in marine animals. 31:2, 68–74.
An overview of the types of eyes possessed by marine animals.

Bleecker, S. E. 1975. Fishes with electric know-how. 21:3, 142–48.
A survey of fishes that use electrical fields to navigate, to capture prey, and to defend themselves.

Klimley, A. P. 1976. The white shark—a matter of size. 22:1, 2–8.
Describes procedure used by Dr. John E. Randall for determining the size of sharks from the perimeter of the upper jaw and height of teeth.

Lineaweaver, T. H., III. 1971. The hotbloods. 17:2, 66–71.
Discusses the physiology of the bluefin tuna and sharks that have high body temperatures.

Netboy, A. 1976. The mysterious eels. 22:3, 172–82.
An informative discussion describing what is known of the migrations of the catadromous eels and their importance as a fishery.

O'Feldman, R. 1980. The dolphin project. 26:2, 114–18.
A description of the response of Atlantic spotted dolphin to musical sounds.

Reeve, M. 1971. The deadly arrowworm. 17:3, 175–83.
This important group of plankton is described in terms of body structure and life style.

Scientific American

Denton, E. 1960. The buoyancy of marine animals. 303:11, 118–28.
The methods used by various marine animals to use buoyancy to maintain their position in the water column.

Donaldson, L. R. and Joyner, T. 1983. The salmonid fishes as a natural livestock. 249:1, 50–69.
The genetic adaptability of the salmonid fishes may help them adapt to "ranching" operations.

Gosline, J. M., and DeMont, M. E. 1985. Jet propelled swimming in squids. 252:1, 96–103.
By using jet propulsion resulting from expelling water through their siphons, squids can move as fast as the speediest fishes.

Gray, J. 1957. How fishes swim. 197:2, 48–54.
The roles of musculature and fins in the swimming of fishes.

Kooyman, G. L. 1969. The Weddell seal. 221:2, 100–107.
The life style and problems related to this mammal's living in water permanently covered with ice.

Leggett, W. C. 1973. Migration of shad. 228:3, 92–100.
The routes of the anadromous shad in the rivers and Atlantic waters are described, along with factors that may control the migrations.

Rudd, J. T. 1956. The blue whale. 195:6, 46–65.
A description of the ecology of the largest animal that ever lived.

Shaw, E. 1962. The schooling of fishes. 206:6, 128–36.
A consideration of theories seeking to explain the schooling behavior of fishes.

Whitehead, H. 1985. Why whales leap. 252:3, 84–93.
Whales seem to communicate with their spectacular lunges above the oceans surface.

Würsig, B. 1979. Dolphins. 240:3, 136–48.
A summary of the facts concerning dolphin intelligence.

More than 90% of the species of animals found in the ocean live in the two-dimensional world of the ocean floor. Ranging from the rocky, sandy, and muddy environments of the intertidal zone to the muddy deposits of deep-ocean trenches more than 11 km (6.8 mi) in depth, the ocean floor provides a varied environment that is home for a diverse benthic community.

Because it lives at or near the interface of the ocean floor and the ocean water, a benthic organism's success is closely tied to its ability to cope with physical conditions of the water, the ocean floor, and the other members of the biological community. The vast majority of benthic species live on the shallow bottom of the continental shelf, and approximately 400 of the known 157,000 benthic species are found in the hadal zone of the deep-ocean trenches.

One of the most prominent variables affecting species diversification is temperature. We have previously discussed its effect on species diversity on a latitudinal basis; however, even at the same latitude, a significant difference in the number of benthic species is found on opposite sides of an ocean basin because of the effect of ocean currents on the coastal water temperature. Along the European coast where the Gulf Stream warms the water from the northern tip of Norway to the Spanish coast, over three times more species of benthos exist than are found along the similar latitudinal range of the Atlantic coast of North America, where the Labrador Current cools the water as far south as Cape Cod.

ROCKY SHORES

Most of the rocky shore is confined to the supralittoral and littoral zones. The supralittoral corresponds with the backshore above the spring high-tide line and is covered by water only during storms. The littoral zone, also known as the *foreshore* or the *intertidal zone*, lies between the high and low tidal extremes. Along most shores, the intertidal zone can be divided into the high-tide zone, mostly dry—covered by the highest high tide but not the lowest high tides; the middle-tide zone, exposed and covered equally—covered by all high tides and exposed during all low tides; and the low-tide zone, mostly wet—covered

CHAPTER SIXTEEN
LIFE ON THE OCEAN FLOOR

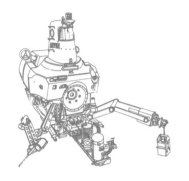

Sea anemones in a tide pool near Laguna Beach, California.

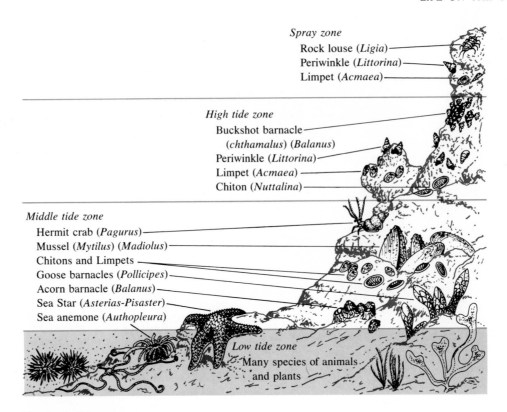

Spray zone
    Rock louse (*Ligia*)
    Periwinkle (*Littorina*)
    Limpet (*Acmaea*)

High tide zone
    Buckshot barnacle
        (*chthamalus*) (*Balanus*)
    Periwinkle (*Littorina*)
    Limpet (*Acmaea*)
    Chiton (*Nuttalina*)

Middle tide zone
    Hermit crab (*Pagurus*)
    Mussel (*Mytilus*) (*Madiolus*)
    Chitons and Limpets
    Goose barnacles (*Pollicipes*)
    Acorn barnacle (*Balanus*)
    Sea Star (*Asterias-Pisaster*)
    Sea anemone (*Authopleura*)

Low tide zone
Many species of animals
and plants

FIGURE 16–1   The Rocky Shore.
A typical rocky intertidal zone and some organisms typically found in its subzones.

during the highest low tides and exposed during the lowest low tides (fig. 16–1).

Along rocky shores these divisions of the intertidal zone are often obvious because of the sharp boundaries between populations of attached organisms. Because each square centimeter of the littoral rocky shore is significantly different from the square centimeter above and below it, evolution has been able to produce organisms with the abilities to withstand very specific degrees of exposure to the atmosphere. This condition results in the most finely defined biozones known in the marine environment.

Supralittoral (Spray) Zone

Throughout the world the most obvious inhabitant of the rocky supralittoral zone is the periwinkle snail. The supralittoral zone can be easily identified if one considers the middle of the periwinkle belt to be the boundary between the littoral and the supralittoral. The genus *Littorina* includes species able to breath air, like land snails. They need the spray of breaking waves as their only tie to the sea since they are viviparous (give birth to live young) rather than depositing their eggs into the water as do more marine periwinkles and other snails (fig. 16–2A).

The periwinkles feed by rasping algae and lichens off the rocks of the spray zone with their filelike *radula*, a calcium carbonate mouthpiece common to many snails. These little snails can be as small as a grain of sand or more than a centimeter long and are a significant factor in the erosion of upper intertidal and spray zone rocks.

Hiding among the cobbles and boulders covering the floors of sea caves well above the high-tide line are isopods belonging to the genus *Ligia*; they are commonly called *rock lice* and *sea roaches* (fig. 16–2B). Neither name is particularly flattering to these little scavengers that reach lengths of 3 cm (1.2 in) and scurry about at night feeding on any type of organic debris.

Another indicator of the spray zone is a distant relative of the periwinkle snails—limpets belonging to the genus *Acmaea* (fig. 16–2C). Having a flattened conical shell, the limpets feed in a manner similar to that of the periwinkles. At low tide, limpets will remain motionless with their shells pulled down tight against the rock on which they make their homes. Some limpets are known to leave their home spots on the rock during high tide and feed for as long as the tide will allow. They will then return to their homes, marked by a depression or discoloration on the rock, to wait out the dry period of low tide.

PLATE 23

A

L ife in the ocean must adapt to a variety of environments ranging from the deep, dark, sediment-filled trenches to the brightly sunlit, wave-battered rocky tidepools of the intertidal zone.

One of the more remarkable ecological units of the marine environment is the coral reef. Above we see the Crown-of-Thorns sea star, which threatened to destroy many coral reefs in the 1960s, being attacked by one of its few predators, the Pacific triton.

Many members of the coral family do not secrete the calcium carbonate of the reef builders. To the right is a view of the feeding polyps extending from the branches of a soft gorgonian coral.

B

PLATE 24

One of the more unusual of the great variety of colorful and strange fishes associated with coral reef and other shallow-water, tropical environments are the puffers. When a potential predator comes too close, they gulp water to fill a ventral stomach pouch, expanding their loose, scaleless skin to produce a spherical shape. Actually, puffers are seldom bothered by predators as they have viscera and skin that often contain deadly neurotoxin. Though they are feeble swimmers, puffers like this swelled-up member of the genus Arothron are virtually free from attack.

Oriental cultures, especially the Japanese, consider fugu, named for the puffer genus Fugu, a delicacy. It is served in public only after being prepared by government-licensed chefs. Those who dine on fugu say the small amount of toxin remaining in the prepared flesh provides a warm, euphoric feeling.

PLATE 25

One of the most colorful members of the benthos is the Cyphoma cowrie and its relatives, most commonly found in coral reef environments, where they browse on a variety of plants and animals such as the sea fan.

This pelagic octopus is a strange and fearsome-looking creature as it jets through the open ocean. Thank goodness it is small enough to fit into the palm of your hand. This specimen was observed off the Kona coast of Hawaii.

A

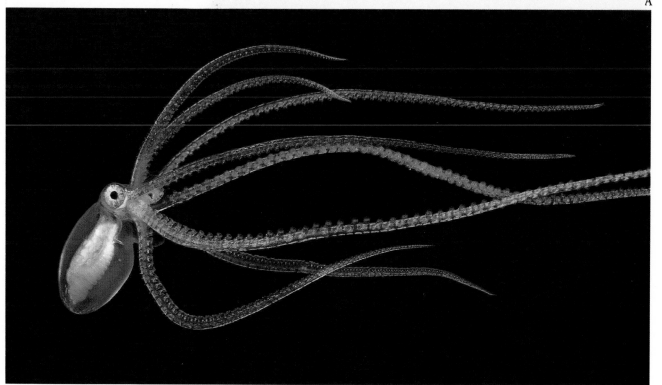

B

PLATE 26

A

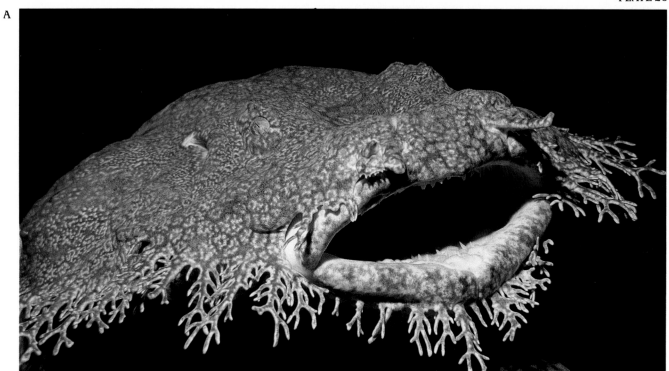

B

*W*hen it comes to ugly, the big-mouthed tasselled wobbegong shark has to be an odds-on favorite to win any competition. In reality, you are very unlikely to recognize one when you see it because it is beautifully camouflaged to lie on the bottom and gulp in unsuspecting prey that swims by. This 5-ft specimen was observed on the Great Barrier Reef.

Feeding more sedately are the tiny and brilliantly colored feather-duster worms that use their tentacled spirals to pluck microscopic plankton and detritus from the gently flowing currents. Like this specimen from the Coral Sea, they live in calcareous tubes tucked safely away in crevices of coral reefs or rocky shores.

PLATE 27

T he identity of this fish, photographed at a depth of 130 ft, remains one of the sea's secrets.

A

N arwhals are unusual toothed whales that may reach 6 m (20 ft) in length. Narwhals possess no dorsal fin and live in the Arctic. They are rarely found south of 65°N latitude. The male has a modified tooth that projects forward from the upper lip as a long tusk with a left-hand spiral. Living in schools of 15 to 20, narwhals feed on cuttlefish, fish, and crustaceans. These narwhals were found in the Beaufort Sea north of Alaska.

B

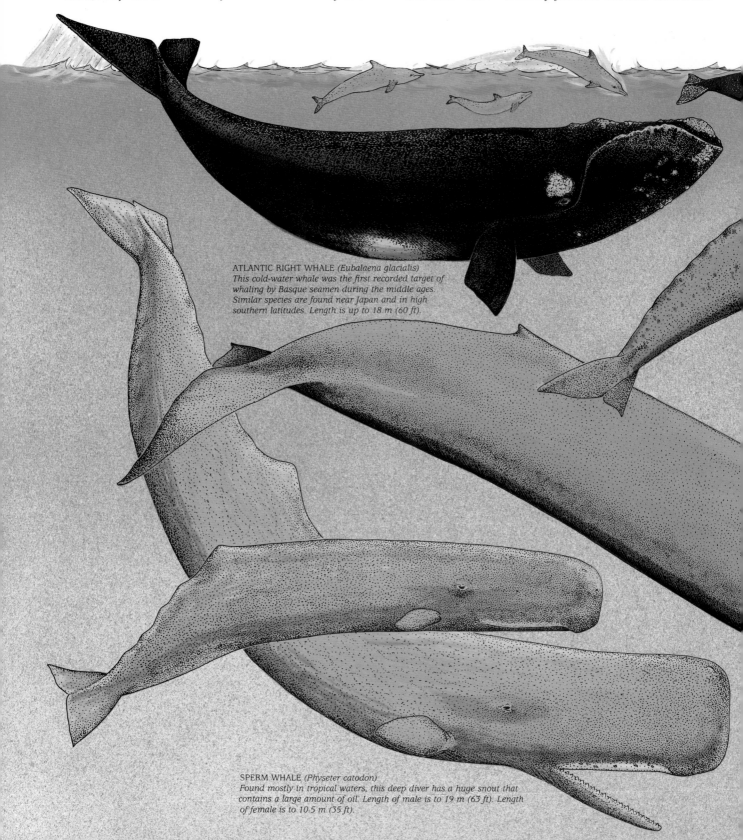

N o creature in the sea captures the human imagination like the whales, huge mammals that returned to the sea about 60 million years ago.

Two suborders are present in today's oceans, the toothed whales (Odontoceti) and baleen whales (Mysticeti). Represented on this plate are members of each

PLATE 28

group drawn to scale.

The toothed whales—bottle-nosed dolphin, narwhal, killer whale, and sperm whale—are active predators and may track down their prey by use of echolocation.

The baleen whales—humpback whale, right whale, and blue whale—leisurely fill their mouths with water

ATLANTIC RIGHT WHALE (Eubalaena glacialis)
This cold-water whale was the first recorded target of whaling by Basque seamen during the middle ages. Similar species are found near Japan and in high southern latitudes. Length is up to 18 m (60 ft).

SPERM WHALE (Physeter catodon)
Found mostly in tropical waters, this deep diver has a huge snout that contains a large amount of oil. Length of male is to 19 m (63 ft). Length of female is to 10.5 m (35 ft).

and force it out between long baleen slats that hang down from the upper jaw. Small fish, krill, and other plankton are trapped behind the baleen. The Pacific gray whale has short baleen slats and feeds by sucking in sediment from the shallow bottom of its North Pacific feeding grounds. Thus it strains out benthic amphipods and other invertebrates.

Although many great whale populations were pushed to near extinction by nineteenth- and twentieth-century whaling efforts, it appears all will survive this threat. The major threat to their survival may now be the disruption of their feeding and breeding grounds.

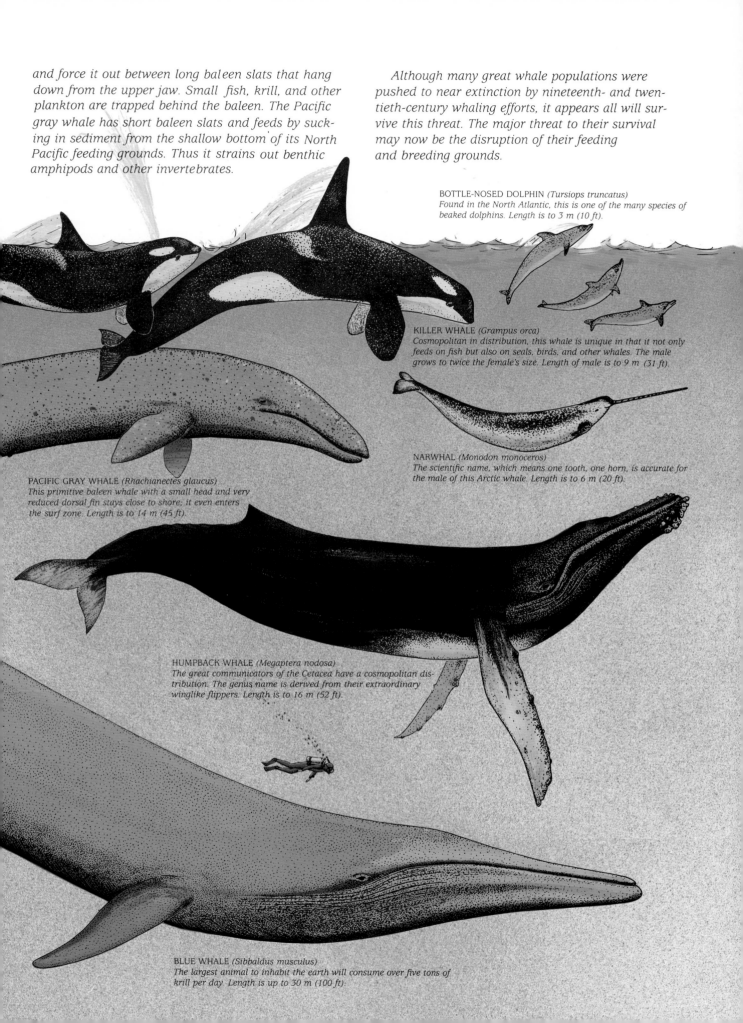

BOTTLE-NOSED DOLPHIN (*Tursiops truncatus*)
Found in the North Atlantic, this is one of the many species of beaked dolphins. Length is to 3 m (10 ft).

KILLER WHALE (*Grampus orca*)
Cosmopolitan in distribution, this whale is unique in that it not only feeds on fish but also on seals, birds, and other whales. The male grows to twice the female's size. Length of male is to 9 m (31 ft).

NARWHAL (*Monodon monoceros*)
The scientific name, which means one tooth, one horn, is accurate for the male of this Arctic whale. Length is to 6 m (20 ft).

PACIFIC GRAY WHALE (*Rhachianectes glaucus*)
This primitive baleen whale with a small head and very reduced dorsal fin stays close to shore; it even enters the surf zone. Length is to 14 m (45 ft).

HUMPBACK WHALE (*Megaptera nodosa*)
The great communicators of the Cetacea have a cosmopolitan distribution. The genus name is derived from their extraordinary winglike flippers. Length is to 16 m (52 ft).

BLUE WHALE (*Sibbaldus musculus*)
The largest animal to inhabit the earth will consume over five tons of krill per day. Length is up to 30 m (100 ft).

A

B

Killer whales frolic in the icy waters of the north Pacific. Leaping whale displays the bold white undermarkings. The erect triangular dorsal fin of a mature male projects from the water. Male dorsal fin may reach a height of more than 1.5 m (5 ft).

The remains of thirteen porpoises and fourteen seals were found in the stomach of one killer whale and give evidence of the animal's voracious appetite.

A view of the underside of the throat of a Bryde's whale reveals the numerous longitudinal grooves that allow the throat to expand to immense proportions when they gulp large volumes of water that contain their primary foods, krill and small fish. The Bryde's whale is a smaller, close relative of the blue whale. This Bryde's whale is "side feeding" on shrimplike krill near the north end of the Sea of Cortez.

PLATE 29

FIGURE 16–2 Animals of the Supralittoral Zone.
A, periwinkles *(Littorina)* nestled in a depression near the upper limit of the high-tide zone. This behavior helps reduce exposure to direct sunlight. *B,* rock lice *(Ligia)* on the roof of a sea cave at Laguna Beach, California. *C,* limpet *(Acmaea). D,* buckshot barnacles *(Chthamalus).* Photos *B* by John D. Roche, Jr., *C* by Greg West.

High-Tide Zone

Unlike the periwinkles, some of whom have given birth to live young, the barnacles cannot abandon the sea for dry land. Their limit is the high-tide shoreline. This *hermaphroditic* animal fertilizes its neighbor's eggs by inserting a penis into a neighboring barnacle. The fertilized eggs hatch within the housing of the barnacle and are released into the water as *nauplii* larvae; these eventually transform into *cypris* larvae

and abandon the planktonic existence to attach to the rocky shore. In addition to their need of water for the planktonic existence of their larval forms, barnacles also are confined to the ocean by their method of feeding.

At the highest intertidal level exists the most prominent inhabitant of this zone, the buckshot barnacle (fig. 16–2D), so called because of its small size, which seldom exceeds 0.5 cm (0.2 in). This unusual arthropod attaches itself to a rock by the back of its neck, secretes a protective calcium carbonate housing that looks like a tiny volcano, and extends its "legs," modified into featherlike cirri, to comb microscopic plankton from the water.

The most conspicuous plants in the high-tide zone are members of the genera *Fucus* (fig. 16–3A) in colder latitudes and *Pelvetia* in warmer latitudes. Both have thick cell walls to reduce water loss during periods of low tide. These plants continue to flourish in the middle-tide zone, where the variety of life forms is much greater than in the upper-tide zone. Not only does the variety increase, but also the total biomass is much greater; there is, thus a greater competition for rock space by *sessile* forms (fig. 16–3).

Middle-Tide Zone

On a clean rocky shore, the *Fucus* or *Pelvetia* (fig. 16–3A) establishes itself before the sessile animal forms; however, it seems doomed once barnacles or mussels

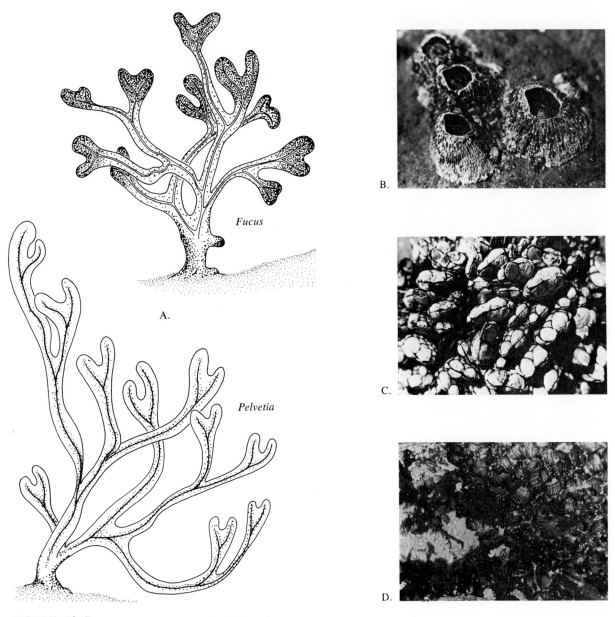

FIGURE 16–3 Life of the High- and Middle-Tide Zones.

A, rock weeds *Pelvetia* and *Fucus.* Sharing the middle-tide zone with these rock weeds are *B*, acorn barnacles *(Balanus)*; *C*, goose barnacles *(Pollicipes)*; and *D*, a mussel bed. Photos *B* and *C* by Greg West.

move in. Although the smaller acorn barnacles usually found in the upper-tide zone can occasionally be found in the upper reaches of the middle-tide zone, larger species of the genus *Balanus* (fig. 16–3B) are the more common acorn barnacles of the middle-tide zone. The barnacle most characteristic of this zone is the goose barnacle, genus *Pollicipes* (fig. 16–3C), with two valves composed of numerous plates that house the animal, which attaches itself to the rock surface by a long muscular neck. The total length of this structure may exceed 10 cm (4 in). Even more successful than barnacles in the competition for space in the middle-tide zone are the various species of mussels (fig. 16–3D) belonging to the genera *Mytilus* and *Modiolus;* they will attach to bare rock, algae, or barnacles. Settling on these surfaces as larval forms, they attach themselves by tough proteinous *byssal threads.* A mussel extends its foot and presses it against the hard surface. It then secretes a fluid that flows down a groove on the surface of the foot to the attachment surface. An attachment thread is soon formed when the ocean water causes the fluid to harden. This process is repeated until many threads hold the mussel firmly in place. Given time, mussels would eventually overgrow all sessile forms.

Life in the intertidal zone is not simple. Mussels are fed upon by predators such as sea stars and carnivorous snails. Two common genera of sea stars are *Pisaster* and *Asterias.*

A sea star feeds by attaching its tube feet to its prey. There are five double rows, which radiate from its mouth at the center of its bottom side. The tube feet are in an oral groove beneath each ray, or arm. To get to the mussel tissue, protected by calcium carbonate bivalve covering, the sea star exerts a continuous pull on the valves by alternating the use of the tube feet so that some are always pulling while others are resting. The mussel is eventually fatigued and can no longer hold the valves closed. The valves open ever so slightly, and the sea star everts its stomach, slips it through the crack, and digests the mussel without having to take it into its mouth (fig. 16–4).

Boring carnivorous snails feed on both barnacles and mussels. Their preferred food, barnacles, can be consumed after the snail envelops it with its fleshy foot and forces the valves open. A narcotic secretion called *purpurin* may also be secreted to make the opening of the valves easier. To feed on the mussels, the snail must use its radula to drill a hole through the shell, through which it can rasp away the flesh. The more widely distributed carnivorous snails are the dogwhelk, *Nucella,* and the similar Pacific coast species *Thais lamellosa.*

FIGURE 16–4   Sea Star Feeding on a Clam.

Even though mussels have survival problems in the intertidal zone, the dominant feature of the middle-tide zone along most rocky coasts is a mussel bed that thickens toward the bottom until it reaches an abrupt bottom limit. This may be so pronounced that it appears as though an invisible horizontal plane has prevented the mussels from growing below this depth. Protruding from the mussel bed will be numerous goose barnacles, and concentrated in the lower levels of the bed will be the sea stars browsing on the mussels. Less conspicuous forms common to the mussel beds are varieties of algae, hydroids, worms, clams, and crustaceans.

Where the rock surface flattens out within the middle-tide zone, tide pools trapping water as the tide ebbs serve as interesting microecosystems containing a wide variety of organisms. The largest member of this community will often be the sedentary relative of the jellyfish, the sea anemone. Shaped like a sack, the anemone has a flat pedal disc that provides a suction attachment to the rock surface. Directed upward, the open end of the sack has the only opening to the gut cavity, the mouth, surrounded by rows of tentacles. The tentacles are covered with *cnidoblast* cells, which contain a stinging threadlike *nematocyst* that automatically is released when any organism brushes against the tentacles. Unlike their pelagic relatives, the tide pool coelenterates do not have nematocysts long enough to penetrate human skin and therefore are not dangerous (fig. 16–5).

Swimming in the tide pools are a variety of small fish. The opaleye reaches a length of 5 cm (2 in) in the pools and grows larger in deeper water offshore.

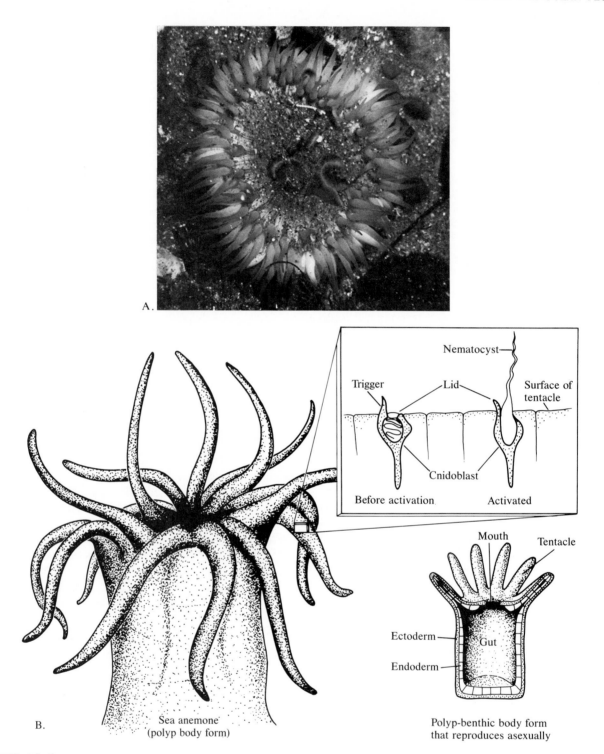

FIGURE 16–5 Sea Anemone.

A, sea anemone. *B,* basic body plan of an anemone and detail of the stinging mechanism.

It is easily identified by the tiny white spots on its back on either side of the dorsal fin. The woolly sculpin is a permanent resident of tide pools and grows longer than 15 cm (6 in). It is covered with hairlike cirri and usually is resting on the bottom or walking on its large pectoral fins. Readily identified by its blunt head and continuous dorsal fin is the 15-cm rockpool blenny (fig. 16–6).

The most interesting inhabitant of the tide pool is the hermit crab, *Pagurus* sp. (fig. 16–7A and B). With a well-armored pair of claws and cephalothorax, hermit crabs have a soft abdomen that is protected by some hard protective container, usually a snail shell. The abdomen has even developed a curl to the right to make it fit properly into snail shells. Once in the snail shell, the opening can be closed off by the crab's large claws. However, one primary enemy is the octopus, which is not bothered by the shell that the hermit crab has adopted. The octopus can cover the crab with the web surrounding its mouth and release a poisonous secretion that paralyzes the crab so that it can be pulled from the shell. Male hermit crabs may be seen fighting over females or carrying them around by grasping the edge of the shell in which the female lives. The male is waiting for the female to molt so he may deposit his sperm on her abdomen. She will later use the sperm to fertilize her eggs.

In tide pools near the lower limit of the middle-tide zone, sea urchins (fig. 16–7C and D), with their five-toothed structure centered on the bottom side of a hard spherical covering that supports many spines, may be found feeding on algae. The hard protective covering is called a *test* and is composed of fused calcium carbonate plates that are perforated to allow tube feet and dermal gills to extend through it.

Low-Tide Zone

Unlike the high- and middle-tide zones, the low-tide zone is dominated by plants rather than animals. A diverse community of animals exists, but its members are less obvious because they are hidden by the great variety of seaweeds and surf grass, *Phylospadix.* The encrusting red alga *Lithothamnion,* which is also seen in middle-zone tide pools, becomes very abundant in the lower-tide pools. In temperate latitudes, moderate-sized red and brown algae provide a drooping canopy beneath which much of the animal life is found (fig. 16–8).

In open tide pools of the low-tide zone, the larger sea anemones will also be found, reaching diameters greater than 25 cm (10 in). *Nudibranchs,* snails with

FIGURE 16–6 Rockpool Blenny.

no shells and exposed gills on their backs, are common in quiet tide pools. They feed on sea anemones, sponges, and hydroids, delicate relatives of the sea anemones. Actually, almost all the phyla of animals are represented in the low-tide zone, where life is less rigorous than in the zones previously discussed. There is no great distinction between the fauna and flora immediately above and below the spring low-tide shoreline.

Scampering from crevice to crevice and in and out of tide pools across the full range of the intertidal zone are various species of shore crabs (fig. 16–9). Scavengers that help keep the shore clean, these creatures can spend long periods of time out of the sea. Shore crabs spend most of the daylight hours hiding in cracks or beneath overhangs. At night they engage in most of their eating activity, shoveling in algae as rapidly as they can tear it from the rock surface with their *chelae,* large front claws. Shore crabs need to return to ocean water only periodically to wet their gills, and females also must deposit the eggs they carry under their abdomen when it is time for them to hatch. A characteristic they share with many crustaceans is that of *autotomy,* the ability to involuntarily shed appendages along a predetermined breaking plane as necessary for survival. It is therefore not unusual to see them with a claw missing or much smaller than its paired counterpart. With each molt the new limb grows until it ultimately becomes full-sized.

A.   B.   C.   D.

FIGURE 16–7   Hermit Crab and Sea Urchin.

A, hermit crab *(Pagurus)* with its protective snail shell home. *B,* hermit crab out of its shell; note the soft curved abdomen. *C,* sea urchins burrowed into the bottom of a lower middle-tide zone tide pool. *D,* the bottom side of this dead sea urchin shows the mouth at the center and the hard calcium carbonate test beneath the spiny exterior.

SEDIMENT-COVERED SHORE

The Sediment

Rocky shores, where the wave energy level is relatively high, are usually dominated by erosion; sediments, if found at all, are composed of large cobbles or boulders. Sediment-covered shore includes *beaches, salt marshes,* or *mud flats* that represent lower-energy environments. As the energy level de-

creases, the particle size and sediment slope decrease, and sediment stability increases; the energy level refers to the strength of the waves and the longshore current. Another characteristic of sediments closely related to particle size is *permeability,* the ease with which fluid can move through the deposit. Because of the strength of cohesive attraction between flat clay-sized particles and the greater angularity of small silt- and fine sand-sized particles compared to the more rounded shape of coarser

FIGURE 16–8 Algal Canopy of the Low-Tide Zone Exposed during an Extremely Low Tide.

sand-sized particles, the permeability of sediments increases with increased particle size. Therefore, breaking waves can more readily percolate down through coarse sands and replace the oxygen used up by animals buried in the sediment. This readily available oxygen also enhances the process of bacterial decomposition of dead tissue.

Because the finer-grained sediment of mud deposits cannot be flushed of organic matter by percolating ocean water, the top 1 cm (0.4 in) of the deposit traps a high concentration of organic matter. Below this surface layer organic matter concentration decreases until, at a depth of about 50 cm (20 in), it becomes negligible in most mud deposits. At the surface, the sediment may be gray or light brown because of the availability of oxygen and the existence of aerobic bacteria that decompose the organic matter. Below the depth of oxygen penetration, anaerobic bacteria and fungi continue the breakdown of organic matter, producing the black color and rotten egg smell of hydrogen sulfide gas. Instead of using oxygen to carry

on their respiration, these anaerobes use sulfate ions (SO_4^{--}), which they reduce to hydrogen sulfide (H_2S) in the process of decomposing the organic matter.

Life in the Sediment

Although sediment-covered shores rarely sustain the relatively large red and brown algae characteristic of the rocky shore, fragments of these plants broken from their rocky substrate by heavy wave action are commonly piled high on beaches during storms. Thus, some of the *suspension feeders* that filter plankton from the clear water of the rocky shore are replaced by *deposit feeders* that eat detritus on many sandy beaches and ingest sediment in mud flats.

Life on and in the sediment requires very different adaptations than on the rocky coast. The sandy beach supports fewer species than the rocky shore, and mud flats fewer still; however, the number of individuals may be as high. In the low-tide zones of some beaches and on mud flats, as many as 5000 to 8000 burrowing clams have been counted in one square meter (10.8 ft^2). Burrowing is the most successful adaptation for life in the sediment-covered shore, so life is less visible. By burrowing a few centimeters beneath the surface, organisms can find a stable environment where they are not bothered by fluctuations of temperature and salinity and the threat of desiccation.

Burrowing in the sediment does not prevent animals from *suspension feeding,* or straining plankton from the clear water above the sediment as do many of the animals of the rocky shore. By the use of various techniques, the water above the sediment surface can be stripped of its plankton content by buried animals. *Deposit feeders* collect debris from the sediment surface and eat their way through the sediment, digesting its organic content and excreting inorganic

A.

B.

C.

FIGURE 16–9 Striped Shore Crab.

A, striped shore crab, *Pachygrapsus,* backed into a rock crevasse. *B,* dorsal view of a preserved *Pachygrapsus. C,* preserved female *Pachygrapsus* showing eggs under the broad abdomen.

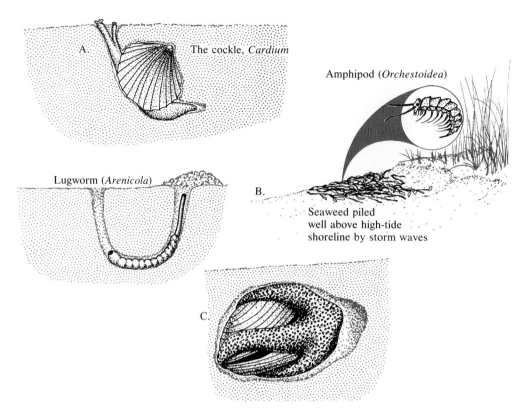

FIGURE 16–10   Modes of Feeding along the Sediment-Covered Shore.

A, suspension feeding. This method is used by clams that bury themselves in sediment and extend siphons through the surface to pump in overlying water. They feed by filtering plankton and other organic matter that is suspended in the water. *B,* deposit feeding. Some deposit feeders such as the segmented worm *Arenicola* feed by ingesting the sediment and extracting organic matter contained within it. Others, like the amphipod *Orchestoidea,* feed on more concentrated deposits of organic mater, detritus, found on the sediment surface. *C,* carnivorous feeding. The sand star, *Astropecten,* cannot climb rocks like its sea star relatives, but it can burrow rapidly into the sand, where it feeds voraciously on crustaceans, mollusks, worms, and other echinoderms.

particles. Burrowing *carnivores* chemically sense the presence of their prey in the darkness of the infaunal habitat (fig. 16–10).

The Sandy Beach

Bivalve Mollusks   Of all the animals that live in the sediment-covered shore, the bivalve mollusks are the best adapted. The greatest variety of clams are found buried beneath the low-tide regions of sandy beaches; their numbers decrease as the sands become muddier. The *pelecypods,* or bivalve mollusks, possess a soft body, a portion of which is called the *mantle;* this part secretes the calcium carbonate lateral valves that hinge together on the dorsal region of the body. The foot extends anteriorly to dig into the sediment and pull the valves and posterior siphons down after it. The tip of the foot is then pumped full of blood to make it swell and anchor the mollusk, while retractor muscles running the length of the foot

contract and pull the animal down. How deep the bivalve can bury itself depends on the length of its siphons; they must reach above the sediment surface to pull in water from which plankton will be filtered. Oxygen is also extracted in the gill chamber before the water is expelled through the excurrent siphon. Indigestible particulate matter will be forced back out the incurrent siphon periodically by quick muscular contractions (fig. 16–11).

Annelid Worms   A variety of *annelids,* or segmented worms, are also well adapted for life in the sediment. Most common of the sand worms is *Arenicola* sp., or lugworm. It lives in a U-shaped burrow, the walls of which are strengthened by mucus. With its finely hooked *parapods* gripping the sides of the tube, it moves forward and backward. The worm moves forward to feed and extends its proboscis up into the head shaft of the burrow to loosen sand with quick pulsing movements. A cone-shaped depression forms at the surface over the head end of the burrow

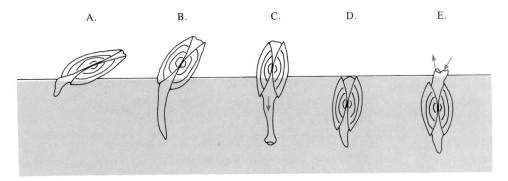

FIGURE 16–11 How a Clam Burrows.
Clams exposed at the sediment surface will quickly burrow into the sediment by *(A)* extending the point-shaped foot into the sediment and *(B)* forcing the foot deeper into the sediment and using this increasing leverage to bring the exposed, shell-clad body toward vertical. When the foot has penetrated deep enough, a bulbous anchor forms at the tip *(C)*, and a quick muscular contraction pulls the entire animal into the sediment *(D)*. The siphons are then pushed up above the sediment to pump in water from which the clam will extract food and oxygen *(E)*.

as sand continually slides into the burrow and is ingested by the worm. As the sand passes through the digestive tract, the organic content is digested, and the processed sand is deposited as castings at the surface surrounding the opening or openings to the tail shaft (fig. 16–10B).

To help prevent asphyxiation resulting from using up the oxygen in the water that fills the tube when the tide is out, lugworms have red, hemoglobin-rich blood that is able to store oxygen. They rarely are forced to leave their tubes since each rising of the tide will result in wave action transporting a new supply of sediment to be ingested. Even reproduction is achieved without leaving the burrow. During October, all the worms release their sex cells into the overlying water, where fertilization occurs, and the larval forms produced settle to the bottom to perpetuate this population of highly specialized sand-dwelling worms.

Also living in fine sand sediment or even silty or muddy sand near the low-tide line are members of the genus *Pectinaria*, species of which are called by the common name of *ice cream cone worm,* derived from the cone-shaped tube composed of very fine sand grains held together by mucus. The body of the worm, no more than 5 to 6 cm (2 to 2.4 in) long, is tapered to fit into the tube and does not attach to the bottom, allowing the worm to burrow. It burrows at an angle into the sediment, leaving the small end of the tube projecting into the overlying water, while the head end creates a small chamber just below the surface. A funnel-shaped tube connects the chamber to the sediment surface. Feeding in a manner similar to that of the lugworm, the ice cream cone worm excretes the sand through an opening at the small end of the tube (fig. 16–12). There is also a variety of detritus-feeding tube worms with permanent tubes

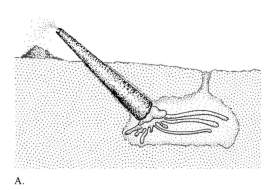

A.

B.

FIGURE 16–12 Ice Cream Cone Worm.
A, Pectinaria, ice cream cone worm, feeding in its chamber. *B, Pectinaria* removed from tube composed of fine sand grains held together by mucus. Photo by Gary Cawthon.

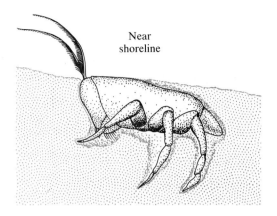

Near
shoreline

FIGURE 16–13   Sand Crab, *Emerita.*

from which they extend tentaclelike *cirri* to search
the sediment surface for food.

Crustaceans   Staying high on the beach and feeding
on kelp cast up by storm or high-tide waves, numerous
amphipod crustaceans called *sand hoppers* are found.
They are known to jump distances more than 2 m (6.6
ft). A common genus is *Orchestoidea,* whose members
usually range from 2 to 3 cm (0.8 to 1.2 in) in length.
Laterally flattened, they usually spend the day buried
in the sand or hidden in the kelp on which they feed.
They become active at night and may form large clouds
above the masses of seaweed on which they are feeding. They are given the name *amphipods* (both legs) because they have one set of appendages specialized for
burrowing and another set that serve well for swimming (fig. 16–10B).

A larger crustacean that may be harder to find on
sandy beaches is a member of the variety of sand
crabs belonging to the genera *Blepharipoda, Emerita,*
and *Lepidopa.* Ranging in length from 2.5 to 8 cm (1
to 3 in), they move up and down the beach near the
shoreline. Burying their bodies into the sand and
leaving their long curved **V**-shaped antennae pointing
up the beach slope, these little crabs filter food particles out of the water (fig. 16–13).

Echinoderms   Echinoderms are represented in the
beach deposits by the sand star, *Astropecten,* and
heart urchins. Members of the genus *Astropecten* are
called *sand stars* because they are well adapted to
prey on invertebrates buried in the low-tide region of
sandy beaches. With five rays and a smooth dorsal
surface, sand stars have sturdy spines along the margins of the rays and tube feet tapered to a point to aid
them in moving through the sediment (fig. 16–10C).

The most widely distributed heart urchins belong
to the genus *Echinocardium.* More flattened and elongated than the sea urchins of the rocky shore, the

heart urchins live buried in the sand near the low-tide
line. Most of the tube feet emerge through a five-pet-
aled flowerlike arrangement of holes on the dorsal
surface of the test. Some tube feet are very long,
forming and maintaining a *respiratory tube* connecting the chamber in which the urchin lives to the
beach surface some 15 cm (6 in) above the animal.
There are a few more tube feet surrounding the ventral mouth; they gather sand grains into the mouth,
where the coating of organic matter is scraped off and
ingested. The posterior anus is surrounded by enlarged spines and long tube feet that create a *sanitary
tube* about 12 cm (4.7 in) long. The feces are passed
into this tube so that the main chamber is not fouled.
The walls of the chamber, respiratory tube, and sanitary tube are supported by mucus applied by the
tube feet (fig. 16–14). The available food at a given
location is usually used up in half an hour, so the
heart urchins are constantly on the move, creating
new chambers and associated features. Each move
involves a distance of about 15 cm (6 in), about twice
the body length, and is accomplished by the short,
fine spines that serve primarily as a means of transporting the heart urchin through the sand.

Meiofauna   Living in the spaces between sediment
particles are tiny organisms ranging in size from 0.1
to 2 mm (0.004 to 0.08 in) in length (fig. 16–15). They
feed primarily on bacteria removed from the surface
of sediment particles. The meiofauna population,
composed primarily of polychaetes, mollusks, arthropods, and nematodes, is found in sediment from the
intertidal zone to the deep-ocean trenches.

Intertidal Zonation   Although it is difficult to see,
there exists a characteristic faunal distribution across
the intertidal range of sediment-covered shores that

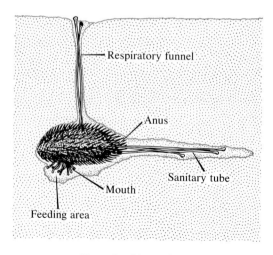

FIGURE 16–14   Heart Urchin, *Echinocardium.*

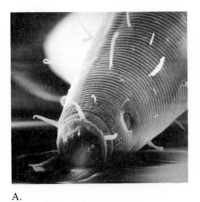

A.

B.

C.

FIGURE 16–15 Meiofauna.
Scanning electron micrographs of *A,* nematode head (×804). The projections and pit on right side are possibly sensory structures. *B,* amphipod (×20). This organism builds a burrow of cemented sand grains. *C,* polychaete (×55) with its proboscis extended. Courtesy of Howard J. Spero, University of South Carolina.

is similar to that observed on rocky shores. Although different species of animals are found in each of the corresponding intertidal zones, the distribution of the life forms across the intertidal rocky and sediment-covered shores is similar in two ways: the maximum number of species and the greatest biomass is found near the low-tide shoreline, and both decrease toward the high-tide shoreline. This zonation, shown in figure 16–16, is best developed on steeply sloping, coarse sand beaches and is less readily identified on the gentler sloping, fine sand beaches. The tiny clay-sized particles of the mud flat produce a deposit with es-

sentially no slope, thus eliminating the possibility of zonation in this protected, low-energy environment.

The Mud Flat

Two widely distributed plants of the mud flat are eelgrass, *Zostera* (fig. 16–17A), and turtle grass, *Thalassia,* which occupy the low-tide zone and adjacent shallow sublittoral regions bordering the flats. Numerous openings at the surface of a mud flat attest to a large population of bivalve mollusks and other invertebrates.

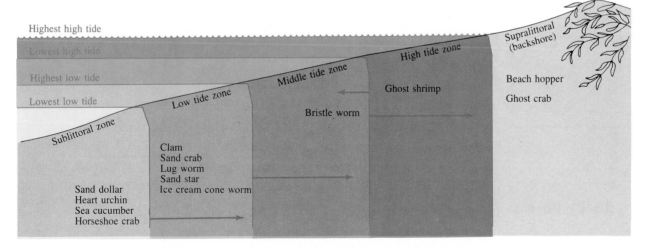

FIGURE 16–16 Intertidal Zonation on the Sediment-Covered Shore.
Zonation is best displayed on coarse sand beaches with steep slopes. As the sediment becomes finer and the beach slope decreases, zonation becomes less distinct. It disappears entirely on flat mud flats.

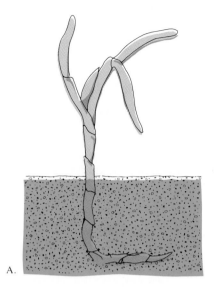

FIGURE 16–17 Life of the Mud Flat.
A, Zostera, eelgrass. *B, Uca,* fiddler crab.

Quite visible and interesting inhabitants of the mud flats are the fiddler crabs, *Uca* (fig. 16–17B), living in burrows that may be more than 1 m (3 ft) deep. Relatives of the shore crabs, they usually measure no more than 2 cm (0.8 in) across the carapace. Fiddler crabs get their name from one out-sized claw, up to 4 cm (1.6 in) long, on the male. This large claw is waved around in such a manner the crab seems to be fiddling. The females have two normal-sized claws. The large claw of the male is used to court females and fight off other competing males. They feed by extracting organic matter from the mud and show a degree of dislike for the sea by jumping into their burrows and closing the mud door when the tide rises to cover the mud flat.

THE SHALLOW OFFSHORE OCEAN FLOOR

Extending from the spring low-tide shoreline to the seaward edge of the continental shelf is an environment that is mainly sediment-covered, although bare rock exposures may be found locally near shore. This region conforms roughly with what has been described as the *sublittoral zone,* the continental shelf.

The Rocky Bottom (Sublittoral)

A rocky bottom within the shallow inner sublittoral region is usually covered with algae. Along the Pacific Coast of North America, the giant bladder kelp, *Macrocystis,* is known to attach to rocks at depths as great as 30 m (98 ft) if the water is clear enough to

allow sunlight to support plant growth. The giant bladder kelp and another fast-growing kelp, *Nereocystis,* often form bands of kelp forest along the Pacific Coast. Smaller tufts of red and brown algae are found on the bottom and living epiphytically (one plant living nonparasitically on another plant) on the kelp fronds. Epifauna commonly found growing along with the algal tufts on the fronds are hydroids and bryozoan colonies. All of these smaller life forms serve as food for many of the animals found living within the kelp forest community. Nudibranchs are important as predators on the hydroids and bryozoan colonies. Surprisingly, very few animals feed directly on the living kelp plant. Among those that do are the large sea hare, *Aplysia,* and sea urchins (fig. 16–18).

Varieties of barnacles, annelid worms, and bivalve mollusks are known to bore into the rocky substrate. Among the bivalves are the date mussels that bear a striking resemblance to date fruit and can be found embedded in soft shale and sandstone. They possess no apparent means of mechanically boring into rock and probably secrete a dissolving chemical. Piddocks belonging to the genus *Pholas* are able to bore into most sedimentary rock after settling onto it in larval form. Developing a filelike rasping surface on the anterior ends of their valves, they mechanically create a small hole that extends deeper into the rock and increases in diameter as the animal grows. Housed inside the rock, the piddock is protected from predators. It feeds by extending siphons into the water and filtering food, much like the clams that bury themselves in soft sediment (fig. 16–19).

Large crustaceans called *lobsters* are common to rocky bottoms. They are a somewhat varied group

B.

A.   C.

FIGURE 16–18   Life of the Sublittoral Zone.
A, Macrocystis and *B,* sea hare *(Aplysia)* in a kelp forest. *C,* sea urchin, *Strongylocentrotus.*

with robust external skeletons. The palinuran, or spiny, lobster is named for its spiny carapace and is characterized by two very large and spiny antennae that have noise-making devices near their bases. The genus *Palinurus* is found at depths greater than 20 m (66 ft) along the coast of Europe, reaches lengths up to 50 cm (20 in), and is considered a delicacy. The species *Panulirus argus,* found in the Caribbean, is sometimes observed to migrate single file for unknown reasons over distances of several kilometers. *Panulirus interruptus* is the spiny lobster of the Amer-

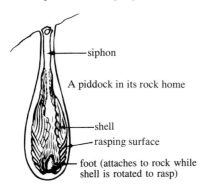

siphon

A piddock in its rock home

shell
rasping surface
foot (attaches to rock while shell is rotated to rasp)

FIGURE 16–19   Rock Borer, *Piddock.*

ican west coast. Other related species are found in the Juan Fernandez Islands and along the coasts of New Zealand, Australia, and South Africa. All of the spiny lobsters are taken for food, but none are as highly regarded as the so-called true lobsters belonging to the genus *Homarus.* Although they are scavengers like their spiny relatives, the true lobsters—which include the American lobster, *Homarus americanus*—also feed on live animals, including mollusks, crustaceans, and members of their own species. Migrating into deep water during winter and returning to near-shore waters in summer, the American lobster is nocturnal. Related species are found along the European coast from Norway to the Mediterranean and the coast of South Africa, and the American lobster is found from Labrador to Cape Hatteras (fig. 16–20).

Oysters are sessile bivalve mollusks found in estuarine environments. They prefer locations where there is a steady flow of clean water to provide plankton and oxygen. Having great commercial importance throughout the world, they have been closely studied. Oyster beds consist of the empty shells of generation after generation cemented to rock bottom or one an-

SPINY LOBSTER (*Panulirus* sp.)

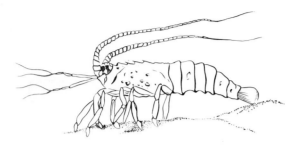

AMERICAN LOBSTER (*Homarus americanus*)

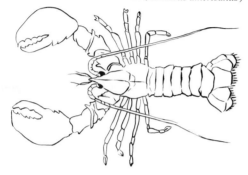

FIGURE 16–20   Spiny and American Lobsters.

other, with the living generation on top. Each female will produce many millions of eggs each year which become planktonic larvae when fertilized. After a few weeks as plankton, the larvae settle to attach themselves to the bottom. The larvae prefer live oyster shells, dead oyster shells, and rock, in that order, as a substrate for attachment. Only a small percentage of the larval forms survive to the attachment stage; after attachment, many oysters perish from predation and competition for food and space before reaching maturity in 1 to 5 years. A great variety of attached forms of sponges, coelenterates, annelids, crustaceans, and tunicates are common to oyster beds; like oysters, most live by filtering plankton from the water. Oysters serve as prey for a variety of sea stars, fishes, crabs, and boring snails that drill through the shell and rasp away the soft tissue of the oyster (fig. 16–21).

Coral Reefs

Although corals are found throughout the ocean, coral accumulations that might be classified as reefs are restricted to the warmer-water regions where the average monthly temperature exceeds 18°C (64°F) throughout the year (fig. 16–22). Such temperature conditions are found primarily between the tropics, although reefs grow to latitudes approaching 35°N and S on the western margins of ocean basins where

warm-water masses move into the high-latitude areas and raise average temperatures.

Not only coral, but algae, mollusks, and foraminifers make important contributions to the reef structure. Reef-building corals are *hermatypic*. They have a symbiotic relationship with the green algae Zooxanthellae that live within the tissue of the coral. Because light is essential for algal photosynthesis, reef-building corals are restricted to shallow bottom waters. The algae could contribute to the calcification capability of corals by extracting carbon dioxide from the animals' body fluids, thus increasing the concentration of the carbonate ion needed for the precipitation of calcium carbonate. This relationship is representative of *mutualism,* in which both parties benefit. The corals contribute to the relationship by providing a supply of nutrients to the Zooxanthellae.

Reef growth also requires that the water have a relatively normal salinity and that it be free from particulate matter. Therefore, we see very little coral reef growth near the mouths of rivers that lower the salinity and carry large quantities of suspended material that would choke the reef colony. Maximum reef growth will occur where a constant flow of water carries food to the waiting tentacles of the polyps, which extrude and feed primarily at night (plate 7). Such conditions are optimum in areas where water circulation through current activity is well developed.

Reefs are widely developed throughout the Pacific and Indian oceans on the flanks of the many volcanic islands rising above the ocean surface from the deep ocean floor. Consisting primarily of the skeletal remains of hermatypic corals and calcareous algae, reefs occur around the margins of all islands and continents where the proper conditions for their existence are found.

Coral reefs actually contain up to three times as much plant as animal biomass. The Zooxanthellae account for less than 5% of the reef's plant mass; most of the rest is filamentous green algae. The concentration of phytoplankton in the waters of coral

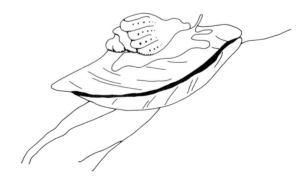

FIGURE 16–21   An Oyster Drill Feeding on an Oyster.

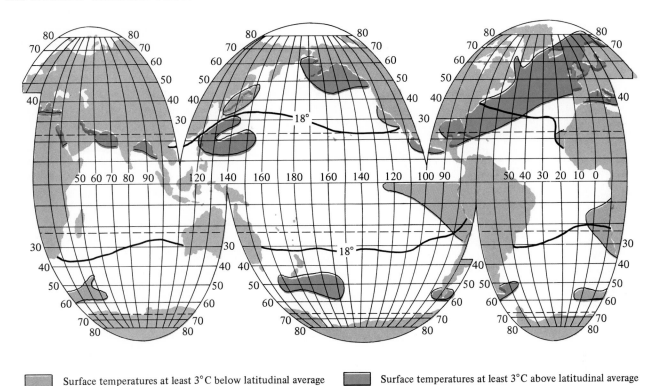

Surface temperatures at least 3°C below latitudinal average   Surface temperatures at least 3°C above latitudinal average

FIGURE 16-22   Coral Reef Distribution.
Minimum water temperatures of 18°C (64°F) in the surface waters of the Northern and Southern hemispheres occur in February and August, respectively. Coral reef development is restricted to the low-latitude area between the two 18°C temperature lines shown on the map. Observing the regions where surface temperatures are 3°C (5.4°F) or more above or below latitudinal averages, the effect of surface currents on coral reef distribution can be seen. Base map courtesy of National Ocean Survey.

reefs is many times greater than in the remarkably nonproductive tropical waters of the open ocean. This is because the reef structure serves to concentrate and hold the nutrients required for plants to prosper.

Because of changes in wave energy, salinity, water depth, temperature, and other less obvious factors, there is a well-developed vertical and horizontal zonation of the reef. These zones are readily identified by the assemblages of plant and animal life found in and near them (fig. 16–23).

The greatest depth to which active coral growth extends is 150 m (492 ft), below which there is not enough sunlight to support hermatypic mutualism. Water motions are not great at these depths, so relatively delicate varieties can live in the outer slope of the reef from 150 m (492 ft) to about 50 m (164 ft). From 50 m to about 20 m (66 ft), the strength of water motion from breaking waves increases on the side of the reef facing into the prevailing current flow. Correspondingly, the mass of coral growth and the strength of the coral structure supporting it increases toward the top of this zone. The genus *Acropora* is representative of the coral found here. The buttress

zone extends from 20 m to the low-tide line; within it more-massive varieties of coral and encrusting algae such as the red alga *Lithothamnion* withstand the crashing waves. The waves cut *surge channels* across the algal ridge, which is inhabited by only a few animals—such as snails, limpets, and the slate pencil urchin, *Heterocentrotus*—that can withstand the constant beating of the surf. The surge channels extend down the reef slope as *debris channels* that carry the products of wave erosion. Small reef fish find protection from larger predators such as sharks, barracuda, and jacks in the debris grooves extending between the buttress ridges. The reef flat extends across the reef beyond the lagoons of atolls and barrier reefs. Here the reef may be under a few centimeters to a few meters of water at low tide. A variety of beautiful reef fish inhabit this shallow water. The sand of reef debris and foraminifer tests fills in the deeper holes and provides a home for sea cucumbers, worms, and a variety of mollusks. In the protected water behind the *Lithothamnion* ridge lagoonal reefs may form, where species of *Porites* and *Acropora* grow into beautiful large colonies (plate 7). Gorgonian coral,

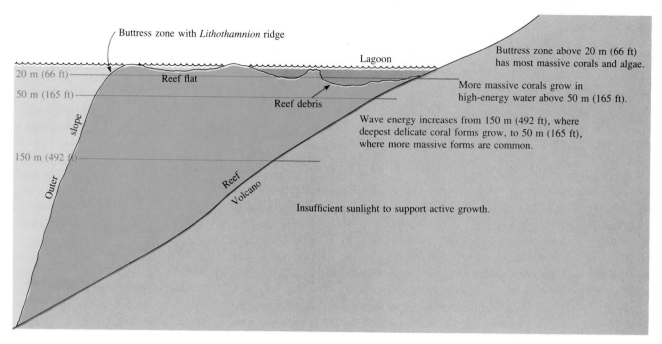

FIGURE 16–23   Coral Reef Zonation.

anemones, crustaceans, mollusks, and echinoderms of great variety also will be found in the lagoon reef (fig. 16–24).

An obvious example of *commensalism* (one organism benefits while the other is unaffected) is the behavior of the shrimpfish that swims head down among the long slender spines of the reef sea urchins. The urchin's spines are a significant deterrent to any predator, and the sea urchin is neither hindered nor aided by the presence of the little fish. The clown fish receives a similar protection by swimming among the tentacles of two species of sea anemones. This relationship is believed to be mutual since the anemones benefit by the clown fish serving as bait to draw other fish within reach of anemone tentacles and actually carrying food to them (fig. 16–25 and plate 8). A variety of cleaner shrimp and fishes that set up cleaning stations on coral reefs are known to be essential to the well-being of many of the larger fishes on the reef by removing parasites and infected tissue.

Sediment-Covered Bottom

Most of the continental shelf is covered with sand or mud deposits. These may be replaced by pebbles,

A.

B.

FIGURE 16–24   Fishes Found in the Coral Reef Lagoon.
A, moray eel. *B,* Moorish idol.

FIGURE 16–25   Clown Fish and Sea Anemone.

FIGURE 16–26   Horseshoe Crab, *Limulus*.
Photo courtesy of James King, Graphic Impressions, Santa Barbara, California.

cobbles, and boulders of glacial origin in the higher latitudes. Generally, similar physical environments may extend over greater distances than in the intertidal region. Although one might expect to find finer sediment near the seaward edge of the shelf than near the shore, this is not always the case. It is not uncommon to find relict beach deposits left behind during periods of glacial advance near the edge of the continental shelf. During these periods, when sea level was 150 m (492 ft) or more lower than today, beaches were created that remain uncovered by later deposits of sediment on the shelf. In general, light availability, temperature, and water movement decrease with increasing depth across the continental shelf.

Below the low-tide line, some of the same species described as burrowing near the low-tide line in the intertidal zone will be found in the sediment. The burrowers usually form similar communities. Although the species may differ throughout the world's temperate and cold-water environments, they are similar. The sandy environments are dominated by bivalve mollusks with a few annelid worms. As the sediment becomes muddier, the bivalves become less varied, and the community contains a greater variety of burrowing annelids. In tropical waters there are many different invertebrate inhabitants of the sediment-covered bottom, and rarely is the community dominated by one or even a few species.

Horseshoe crabs are interesting inhabitants of some shores. These arthropods have not changed substantially from fossilized forms as old as 360 million years. They spend most of their lives in the sandy and muddy offshore waters, where they feed on worms, mollusks, and algae. The common name comes from the horseshoe-shaped carapace, although an equally striking feature is the swordlike tail, which aids in moving through the sediment. Of

the three living species, one is found along the Atlantic Coast of North America and the others inhabit the western Pacific shores. The crabs come ashore to breed. The female deposits the eggs in a shallow nest near the spring high-tide shoreline on sandy beaches, and the male fertilizes them. With the next spring tide two weeks later, the eggs are ready to hatch. The newly hatched crabs quickly enter the ocean, where they undergo a series of molts before achieving adult form (fig. 16–26).

DEEP-OCEAN FLOOR

The Physical Environment

Beyond the edge of the continental shelf, the bathyal zone extends to a depth of 4000 m (13,120 ft) and includes all of the continental slope in most parts of the ocean. The continental rise and deep-ocean floor—between 4000 and 6000 m (19,680 ft) deep, the abyssal zone—comprises over 80% of the benthic environment. The hadal zone of the deep trenches exceeding 6000 m represents the most restricted environment within the benthic habitat. No plant life exists throughout this deep-ocean region. Light is present in only the lowest concentrations above 1000 m (3280 ft) and absent below this depth. Everywhere the temperature is low, rarely exceeding 3°C (37°F) and falling as low as −1.8°C (28.8°F) in the high latitudes. Pressure exceeds 200 atmospheres on the oceanic ridges, ranges between 300 and 500 atmospheres on the deep-ocean abyssal plains, and climbs to over 1000 atmospheres in the deepest trenches. (The pressure is one atmosphere at the ocean surface and increases by one atmosphere for each 10 m, or 33 ft, of depth.) Most of the deep-ocean floor is covered by at least a thin layer of sediment ranging from

LIFE ON THE DEEP-OCEAN FLOOR

FIGURE 16A Dr. Robert R. Hessler.

Interview with Dr. Robert R. Hessler (fig. 16A), professor of oceanography, Scripps Institution of Oceanography, University of California at San Diego.

Q. Dr. Hessler, when does the evidence tell us life first appeared on the deep-ocean floor?

A. There is no absolute evidence for an answer to that question. The conditions for life on the deep-sea floor have probably existed for as long as the deep sea has existed. There has probably always been oxygen and food. The present ocean sediments give us evidence only back to the Cretaceous period (65 to 136 million years ago).

Q. How about the fossil record? Are there any fossils found in rocks that appear to have been deposited in an ancient deep-ocean environment?

A. I have never pursued this much myself. I think there are some. Some of the black shales are thought to have accumulated in deep-basin conditions.

Q. Do you have any ideas about, in an evolutionary sense, where life in the deep ocean came from?

A. My guess would be that ultimately life must have begun in shallow water and penetrated the deep sea from there. One would expect animal life to have evolved closest to the source of primary production, and that can only be in shallow water. Once you get deeper than a few hundred meters there isn't enough light to support primary productivity. Furthermore, considering the variety of physical circumstances in shallow water, the multitude of biogeographic barriers, and so on, I would have expected that to be the area where evolution was the most active. Influx into the deep sea has probably been continuous.

Q. What groups seem to be most effective in achieving this—moving from the shallow water into the deep benthic environment?

A. Some groups have done particularly well. All of the phyla make it into the deep sea. As you go to lower taxonomic levels, the deep-sea fauna is more and more unusual, so that at the level of species, hardly any exist that are common to both the deep sea and shallow water. The percarid crustaceans, macrofauna that brood their young—amphipods, isopods, tanaids, and cumaceans—have been extraordinarily successful in the deep sea. You can get as many as 100 species of isopod in a single sample. Bivalve mollusks and polychaete worms are also very abundant and diverse. In certain environments, other groups tend to be important. Sometimes you find large quantities of sipunculids, sometimes large quantities of pogonophorans. Among the very small animals, another group that is very important is the nematodes. There are incredible quantities of nematodes in the deep sea. The tiny harpacticoid copepods (deposit-feeding copepods less than 2 mm, or 0.08 in, in length) are also very small and abundant. Foraminifers are an interesting group. One doesn't think of the foraminifers as being an important group of benthic organisms, but in the deep sea they are dominant, certainly in terms of biomass and probably also in terms of species diversity.

If you look at the megafauna (the animals that can be photographed), holothurians are fairly important. Glass sponges, tunicates, gorgonians, and certain kinds of decapods are also common. There are a variety of deep-sea families of fishes that are fairly characteristic. The grenadiers are almost always seen.

Q. Which of the physical factors such as pressure, temperature, and the absence of light most limits the vertical migration of life in the deep ocean?

A. Temperature seems to be the most important factor that prevents deep-sea animals from migrating upward. The deep sea is characterized by being isothermal; that is, at any one spot the temperature hardly varies by a measurable amount, and deep-sea animals have become accustomed to this. This means that even a slight change in temperature is likely to have an enormous effect. They can't break through this temperature barrier. However, at north and south high latitudes, you find portions of the deep-sea fauna in shallow water as well. I think the reason is that the temperature barrier isn't there. It's equally cold in the surface waters. With animals penetrating the deep sea from shallow water, the issue is more complex. My guess is that the most important factor in preventing them from penetrating the deep sea is not physical barriers, but food. In order to survive in the deep sea, animals have to be adapted to conditions in which both the rate of supply and the concentration of food are very low. An organism that does not have a life style that can cope with this is not going to make it in the deep sea. Temperature adaptation for a shallow-water animal isn't that difficult. Of the shallow-water species that live through the summer and winter, some are capable of adapting to 20° to 25°C (36° to 45°F) temperature changes on an annual basis. Shallow-water animals could get, or may already be, used to low temperature, but becoming used to low food supply requires major changes in behavior, morphology, and physiology. That's harder to achieve.

Q. The overall consensus is that food is in low supply in general on the deep-ocean floor?

A. That is correct. The reason we think that is so is because one finds so little life there. Any one sample has orders-of-magnitude less life in it than one finds in shallow water. There is only one sensible explanation for that. There's not much food.

Q. We are talking here in terms of biomass?

A. Total biomass—but it is also true in terms of numerical abundance. These two, of course, tend to be correlated.

Q. How about the occasional availability of large supplies of food? Does it have any significant bearing on the distribution of life in the deep ocean?

A. This is called the dead whale hypothesis. The first thing I have to mention is that I don't believe that anyone has written a completely convincing budget for the rate of food supply to the deep sea. So I don't think anyone knows, with any degree of confidence, the rate at which small particles of food are descending to the deep sea—and nobody knows anything at all about the big ones. What we do know about large organic falls is that there seems to be a group of animals that are very well adapted for taking advantage of them. These are the fish and amphipods. When a large organic fall hits the bottom, it isn't long before these animals are there in large quantities consuming it at a fairly rapid rate, and secondarily dispersing it to the rest of the fauna. This dispersal is achieved through their feces, through breaking the food parcel into small particles which are scattered about, and by eating there and dying elsewhere.

Q. What evidence is there that deep-ocean life is moving out of the deep sea and back to the shallow shelf environment from which you think it came?

A. There's not very much. You have to ask yourself, What kind of evidence can I rely on to detect such a phenomenon? It is difficult to come up with really strong criteria, but there is one which isn't bad. It is the absence of eyes. In a group I work on, the isopods, there's a large variety of families that live in the deep sea, but are also found in shallow water at high latitudes. What's interesting about them is that they lack eyes. I won't go into all of the details of the argument, but it looks as though the reason they lack eyes in shallow water is because they originally lost them in the deep sea. When they reinvaded shallow water, they couldn't evolve them again. This is an indication that there are some groups that have come back into shallow water in high latitudes. This apparently happens in other groups as well, but not with every group.

Q. Is there anything else that is of general importance to the nature of life on the deep-ocean floor that you might discuss?

A. Well, there are some very important issues. Let me mention some. First of all, let me emphasize that everything that has been done in the deep sea has been done on soft, mud bottoms. The reason for that is simply that oceanographers have gotten tired of losing their gear while sampling rocky bottoms. But now with the discovery of the deep-sea hydrothermal vents and their wonderful fauna, more and more people are getting interested in studying that hard bottom environment. I think that is going to be the impetus for studying hard bottom environments in general, of which there are an awful lot in the deep sea.

Staying with the hydrothermal vents for a second, I should emphasize that their fascinating fauna is really very different from the normal deep-sea fauna. Not the least of the differences is the fact it is so abundant. This is a nice thing to have found because it seems to be good documentation of the fact that food really is the limiting factor. Here is a source of large quantities of food in the deep sea, and, lo and behold, there is a large amount of biomass taking advantage of it. And these individuals are relatively large, as opposed to most of the animals of the deep sea.

Another point about the deep sea which deserves mention is that almost all deep-sea animals are deposit feeders. Here, I am talking about mud bottom, where suspension feeders and carnivores are rare. This introduces a rather interesting problem. For years now, there has been a debate as to how there can be so many deposit feeders in the same space at the same time. It is a very difficult issue, and probably never will be solved in the deep sea.

Q. With the discovery of large biomass concentrations at the vents, does this give us any hope that further study of the deep-ocean biology will lead to an increase in the food that will be available to us from the ocean?

A. Only to the most limited extent. Except for the vent community, which is inedible because of ambient water chemistry, life is too sparse on the ocean floor to support a fishery. There is a limited fishery for deep, bottom-associated fish, but the turnover rate is so slow in the deep sea that areas could be easily fished out.

the mudlike abyssal clay deposits of the abyssal plains and deep trenches, through the oozes of the oceanic ridges and rises, to some coarse sediment deposited as turbidites on the continental rise. Occasionally, near the crests of the oceanic ridges and rises and down the slopes of seamounts and oceanic islands, basaltic ocean crust is the substrate.

The Deep Fauna

Although the deep-ocean communities have not been extensively sampled, it is becoming apparent that the diversity of deep-ocean benthos is much greater than initially thought. The overriding limiting factor for life on the deep-ocean floor must be the availability of food. Little is known of the nature and availability of food in the deep ocean, but some is apparently cycled through a food chain that exists in the deep environments. Claude ZoBell, a pioneer in the study of marine bacteria, has estimated that 30% to 40% of the organic matter reaching the ocean floor is first used by benthic bacteria at the ocean-sediment interface. The bacteria are then ingested by deposit-feeding infauna. The smaller bacteria-feeding infauna are then preyed upon by larger bottom-dwellers. The carcasses of large fish and mammals may provide food on the bottom, as they would sink rapidly once they die. Although most of the marine phyla may be represented in the deep-ocean fauna, there seems to be a characteristic deep-sea fauna that begins to appear on the lower continental slope. However, they are also frequently found on the continental shelves of high-latitude regions—an indication that temperature is a major factor limiting their distribution. In general, deep-water fauna is differentiated from the shallow-water benthos by a decrease in the variety of sea stars and mollusks and an increase in the diversity of other echinoderms, isopod and amphipod crustaceans, polychaete worms, pycnogonids, and pogonophorans (fig. 16–27).

An exciting discovery of 1977 was the very specialized biological community found near warm-water vents of the Galapagos Rift spreading center. The

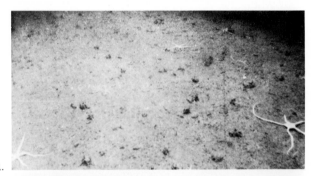

A.

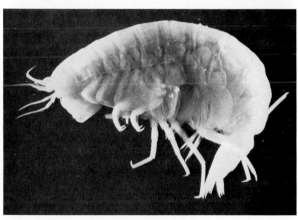

B.

C.

FIGURE 16–27 Deep-Ocean Benthos.
A, although serpent stars such as these are frequently found on the deep-ocean floor, their variety and abundance are considerably less there than in shallow water. *B,* close view of a preserved amphipod recovered in a trap from the floor of the Marianas Trench below 10,400 m (6.46 mi). *C,* amphipods are more abundant on the deep-ocean floor. Here they are attracted to bait in the Philippine Trench at a depth of 9600 m (5.96 mi). Photos from A. A. Yayanos, courtesy Scripps Institution of Oceanography, University of California, San Diego.

first view of the community was observed in pictures taken by the camera sled ANGUS (Acoustically Navigated Underwater Survey System). It showed hundreds of large clams and mussels at depths below 2500 m (8200 ft). The immediate question was, "How do they survive?" Clams and mussels normally feed by filtering phytoplankton out of the water; since there could be no phytoplankton at this depth, an alternative food supply, such as bacteria, must be present. Subsequent investigations found that the organic content of the water, in the form of bacteria, was 500 times greater than normal for this bottom environment and 4 times greater than in productive surface waters. The bacteria are chemosynthetic forms that obtain their energy for the synthesis of organic matter by oxidizing hydrogen sulfide contained in the hydrothermal emissions (fig. 4A, p. 82, and plates 2 and 3).

Continued study of this area and another discovered off the tip of Baja California in 1979 revealed a recognizable zonation of organisms from the 2°C (35.6°F) ambient temperature on the outer edge of the community toward the warmer water near the hydrothermal springs. Spaghettilike acorn worms and benthic siphonophores were found on the outer edge. The siphonophores are relatives of the Portuguese man-of-war and resemble dandelions gone to seed. At intermediate distances, clams, mussels, small anemones, and serpulid worms were found. Thriving in the warm 10° to 15°C (50° to 59°F) water near the vent openings were tube-dwelling pogonophoran worms that are believed to be chemosynthetic in a manner similar to that of the bacteria that support the rest of the colony. The worms have red hemoglobin-filled tentacles that probably have a great capacity for dissolving oxygen extracted from the surrounding water (fig. 16–28A). Living in plasticlike tubes over 1 m (3.3 ft) long, these worms have no mouth or gut. Similar communities have been found in the Gulf of California.

During 1985 and 1986 active hydrothermal vents were found on the Mid-Atlantic Ridge. Associated with two of these vent areas at 23°N and 26°N were biocommunities that differed from those in the Pacific Ocean. The predominant animals at the Atlantic Ocean vents are particulate-feeding shrimp (fig 16–28B).

A relict vent area found on the Galapagos spreading center indicates such communities may have a relatively short life span. This inactive vent was identified by an accumulation of dead clams. It appears that when the vent becomes inactive and the hydrogen sulfide that serves as the source of energy for the community is no longer available, the community dies.

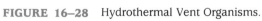

A.

B.

FIGURE 16–28 Hydrothermal Vent Organisms.
A, chemosynthetic tube worms living near the Galapagos Rift hydrothermal vent. *B,* swarm of particulate-feeding shrimp, the predominant animals observed at hydrothermal vents near 23°N and 26°N on the Mid-Atlantic Ridge. This swarm was photographed at 26°N. Courtesy of Peter A. Rona, NOAA.

ADDITIONAL SUBMARINE VENT COMMUNITIES

Three additional submarine vent environments have also been found to have chemosynthetically supported biocommunities. During the summer of 1984, an investigation of an ambient-temperature, hypersaline (46.2‰) seep at the base of the Florida Escarpment in the Gulf of Mexico (fig. 16B) revealed a biocommunity similar in many respects to the hydrothermal vent communities. The seeping water appears to flow from joints at the base of the limestone escarpment and move out across the clay deposits of the abyssal plain at a depth of about 3200 m (10,496 ft).

The hydrogen sulfide-rich waters support a number of white bacterial mats that carry on chemosynthesis in a fashion similar to that of the bacteria of the hydrothermal vents. These and other chemosynthetic bacteria may provide most of the support for a diverse community of animals. The community includes holotheurians, sea stars, shrimp, snails, limpets, brittle stars, anemones, vestimentiferan worms, galatheid crabs, clams, mussels, and zoarcid fish.

Drilling into the limestone rocks of the platform east of the escarpment revealed fluids with temperatures of up to 115°C (239°F) and salinities of 250‰. This dense water apparently seeps deep into the jointed rocks of the platform and is trapped by the impermeable clay deposited on the ocean floor at the base of the escarpment. It then flows out on to the surface clays, mixes with sea water, and produces the spring water with salinity intermediate between that of the platform fluids and ocean water.

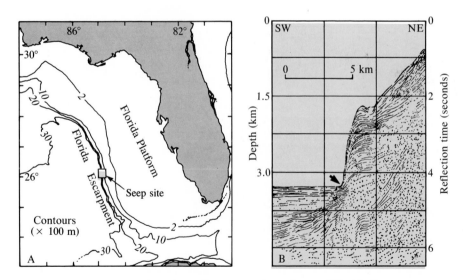

FIGURE 16B Hypersaline Seep Biocommunity at Base of Florida Escarpment. *A,* rectangle is the location of the seep and biocommunity. *B,* seismic reflection profile of limestone Florida Escarpment and abyssal bedded sediments at its base. Arrow marks location of seep. Courtesy of C. K. Paull, Scripps Institution of Oceanography, University of California, San Diego.

Also observed in 1984 were dense biological communities associated with oil and gas seeps on the Gulf of Mexico continental slope (fig. 16C). Trawls at depths of between 600 and 700 m (1968 and 2296 ft) recovered epifauna and infauna similar to those observed at the hydrothermal vents and the hypersaline seep at the base of the Florida Escarpment. Carbon isotope analysis indicates the hydrocarbon-seep fauna is based on a chemosynthetic productivity that derives its energy from hydrogen sulfide and/or hydrocarbons. It is certainly possible that such assemblages will be found throughout the oceans in areas where hydrocarbons are being generated in the underlying sediment.

Finally, a third environment of vent communities was observed from *Alvin* during the summer of 1984. It is located in the subduction zone of the Juan de

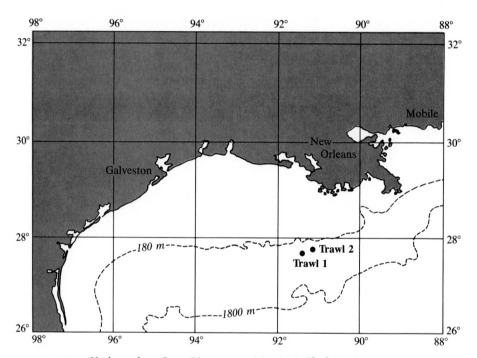

FIGURE 16C Hydrocarbon Seep Biocommunities in Gulf of Mexico.
Locations of trawls that recovered members of biocommunities living in vicinity of hydrocarbon seeps are shown. Courtesy of M. C. Kennicutt II, Texas A&M University.

Fuca Plate at the base of the continental slope off the coast of Oregon (fig. 16D). Here, the trench is filled with sediments. At the seaward edge of the slope, clastic sediments are folded into a ridge. At the crest of the ridge, pore water escapes from the two-million-year-old folded sedimentary rocks into a thin overlying layer of soft sediment. Occurring at a depth of 2036 m (6678 ft) the vents produce water that is only slightly warmer (about 0.3°C, or 0.5°F) than ambient conditions. The vent water contains CH_4 (methane) that is probably produced by decomposition of organic material in the sedimentary rocks. The methane serves as the source of energy for bacteria that oxidize it and chemosynthetically produce food for themselves and the rest of the community, which contains many of the same genera found at other vent sites. During 1985, similar communities were located in subduction zones of the Japan Trench and Peru-Chile Trench. All the vents are located on the landward side of the trenches at depths from 1300 to 5640 m (4264 to 18,499 ft).

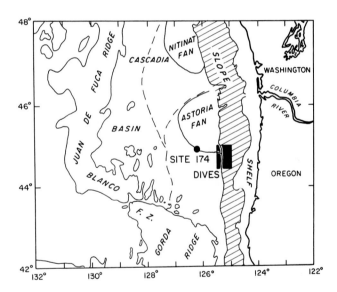

FIGURE 16D Location of Vent Communities off the Coast of Oregon.

The communities are associated with the subduction zone of the Juan de Fuca Plate. Sediment filling the trench is folded into a ridge with vents at its crest.

Based on the observations along spreading centers, volcanic activity (hydrothermal vents are an associated feature) appears to occur in cycles. If this activity is the direct result of the migration of subduction observed in trenches (discussed in chapter 4); the volcanic cycle could be expected to migrate along the median valleys of the spreading centers at rates ranging from a few to 200 km (124 mi) per year.

Subsequent investigations have revealed the presence of other spreading-center hydrothermal activity and associated biocommunities in the Pacific and Atlantic oceans.

SUMMARY

Over 98 percent of the approximately 160,000 species of marine animals live on the ocean floor. Most of these live on the continental shelf, with only 400-odd species found in deep-ocean trenches. The importance of temperature on the development of species diversity is reflected in the fact that there are three times as many species of benthos along the European coast warmed by the Gulf Stream than are along a similar length of North American coast cooled by the Labrador Current.

Because of tidal motions, the littoral (intertidal) zone can be divided into the high-tide zone (mostly dry), middle-tide zone (equally wet and dry), and the low-tide zone (mostly wet). The littoral zone is bounded by the supralittoral zone (covered only by storm waves) and the sublittoral zone, which extends below the low-tide shoreline.

The supralittoral zone along rocky coasts is characterized by the presence of the periwinkle snail, the rock louse, and the limpet. Rock lice stay well above the high-tide line, but snails and limpets may extend down into the high-tide zone.

Within the intertidal zone are found the barnacles, crustaceans that attach themselves to the rocks and comb plankton from the water. The barnacles most characteristic of the high-tide zone are the tiny buckshot barnacles. The most conspicuous plants of the high tide zone are Fucus along colder shores and Pelvetia in warmer latitudes. These plants become more abundant in the middle-tide zone, and the diversity and abundance of the flora and fauna in general increase toward the lower intertidal zone.

Larger species of acorn barnacles are found in the middle-tide zone. A common assemblage of rocky middle-tide zones includes the goose barnacle, the mussels, and sea stars. Joining the sea stars as predators on the barnacles and mussels are carnivorous snails.

Tide pools within the middle-tide zone will commonly house sea anemones, fishes such as the opaleye, woolly sculpin, and rockpool blenny, and hermit crabs. At the lower limit of the middle-tide zone sea urchins become numerous.

The dominance of animals in the upper- and middle-tide zones is ended in the low-tide zone, where plants become dominant. The encrusting Lithothamnion red alga that is present in the middle-tide zone becomes very abundant in tide pools. The low-tide zone in temperate latitudes is characterized by a variety of moderately sized red and brown algae providing a drooping canopy for the animal life.

Scampering across the entire intertidal zone are the scavenging shore crabs.

The only sediments usually found along rocky shores are a few cobbles and boulders. Moving into more protected segments of the shore, lower levels of wave energy allow deposition of sand and mud. Owing to the higher permeability of sand, sand deposits are usually well oxygenated. Mud deposits are anaerobic below a thin surface layer. Hydrogen sulfide (H_2S), a product of anaerobic decomposition of organic matter, gives mud deposits the smell of rotten eggs.

Compared to the rocky shore, the diversity of species is reduced on and in beach deposits and is quite restricted in mud deposits. This does not mean that life is less abundant. Although life is less visible, up to 8000 burrowing clams have been recovered from one square meter (10.8 ft^2) of mud flat. Suspension feeders (filter feeders) characteristic of the rocky shore are still found in the sediment, but there is a great increase in the relative abundance of deposit feeders that ingest sediment and detritus.

Bivalve mollusks are well suited for sediment-covered shores. Clams that burrow into the sediment pump water through siphons and extract food and oxygen. The lugworm and the ice cream cone worm are deposit feeders that ingest sand and extract whatever organic content is available.

Amphipods, called sand hoppers, can be found feeding on the kelp deposited high on the beach by storm waves. Sand crabs filter food from the water with their long, curved antennae while buried in the sand near the shoreline.

Echinoderms are represented in the sandy beach by the sand star, which feeds on buried invertebrates, and the heart urchin, which scrapes organic matter from sand grains.

As is true for the rocky shore, the diversity of species and abundance of life on the sediment-covered shore increases toward the low-tide shoreline. Although it is more difficult to observe on sediment-covered shores, the high-, middle-, and low-tide zonation is also present. It is best developed on steep, coarse sand beaches and is probably missing on mud flats with little or no slope.

Although many burrowing forms inhabit the **mud flats**, the more visible life forms are the **eelgrass, Zostera**, and the **turtle grass, Thalassia**, found at its low-tide margin, and the **fiddler crabs, Uca**, that leave their burrows at low tide.

Attached to the rocky sublittoral bottom just beyond the shoreline is a band of algae including large **kelp** plants. Growing on the large fronds of the kelp are small varieties of **algae, hydroids**, and **bryozoans**. **Nudibranchs** feed on the small hydroids and bryozoans, and the sea hare and sea urchin feed on the kelp and other algae. **Piddocks** and **date mussels** bore into the rocky bottom. **Spiny lobsters** are common to rocky bottoms in the Caribbean and along the United States west coast, and the **American lobster** is found from Labrador to Cape Hatteras.

Oyster beds found in estuarine environments consist of individuals that attach themselves to the bottom or the empty shells of previous generations. Living within these beds is a community consisting of suspension feeders such as **sponges, coelenterates, annelids, crustaceans**, and **tunicates**. **Sea stars, fishes, crabs**, and **boring snails** prey on the oysters.

Above a depth of 150 m (492 ft) in tropical waters, living **coral reef** can be found off the shores of islands and continents. Delicate varieties are found at 150 m, and they become more massive near the surface, where wave energy is higher. The top 20 m (66 ft), the **buttress zone**, is reinforced with calcium carbonate deposited by algae such as *Lithothamnion,* producing an **algal ridge**. Waves cut **surge channels** across the ridge, and **debris channels** extend down the reef front below the surge channels. Many varieties of **commensalism** and **mutualism** are found within the coral reef biological community.

On sediment-covered bottoms, the dominant form of animal life in shallow water is the burrowing **bivalve mollusk**. As sublittoral water depth increases and sediment texture becomes finer, **annelid worms** become dominant. Off the Atlantic Coast of North America and some western Pacific shores, species of **horseshoe crabs** that date back more than 360 million years feed on burrowing mollusks and annelids.

Although little is known of the **deep-ocean benthos**, it is clear that it is much more varied than previously thought. There seems to be a decreased variety of sea stars and mollusks and increased diversity of other echinoderms, isopods, amphipods, polychaetes, pycnogonids, and pogonophorans.

An important discovery of the **hydrothermal vent communities** made in 1977 on the East Pacific Rise has shown that, at least locally, **chemosynthesis** is an important means of primary productivity in the ocean.

QUESTIONS AND EXERCISES

1. Discuss the general distribution of life in the ocean. Include species diversity (variety of types of organisms) between the pelagic and benthic environments and within the benthic environment.

2. Diagram the intertidal zones of the rocky shore and list organisms characteristic of each zone.

3. Describe the dominant feature of the middle-tide zone along rocky coasts, the mussel bed. Include a discussion of other organisms found with the mussels.

4. List and describe crabs found within the rocky shore intertidal zone.

5. Discuss how sediment stability and permeability of sandy and muddy shores differ.

6. Other than predation, discuss the two types of feeding styles that are characteristic of the rocky, sandy, and muddy shores. One of these feeding styles is rather well-represented in all of these environments; name it, and give an example of an organism that uses it in each environment.

7. How does the diversity of species on sediment-covered shores compare with that of the rocky shore? Can you think of any reasons why this should be so? If you can, discuss them.

8. In which intertidal zone of a steeply sloping coarse sand beach would you find the following: clams, sand hoppers, ghost shrimp, sand crabs, and heart urchins?

9. Why do lugworms not need to move frequently, yet heart urchins, which are also deposit feeders, need to move about every half-hour? Consider the specific source of the sediment they ingest.

10. What relationships exist among intertidal zonation, sediment particle size, and beach slope?

11. Discuss the dominant species of kelp, their epifauna, and animals that feed on kelp in the Pacific Coast kelp forest.

12. Where in the world are spiny lobsters and American lobsters (and their related species) found?

13. Discuss the preferred environment, reproduction, and threats to survival of larval and adult forms of oysters.

14. Describe the environment suited to development of coral reefs.

15. What physical factors contribute to vertical zonation of the seaward edge of a coral reef?

16. Why are horseshoe crabs called *living fossils?*

17. As one moves from the shoreline to the deep-ocean floor, what changes in the physical nature of the ocean floor can be expected?

18. Describe the process Claude ZoBell thinks may be important in distributing organic matter throughout the animal populations of the deep-ocean floor.

19. What do you think is the most significant new knowledge gained from the discovery of the hydrothermal vent biological communities of the eastern Pacific Ocean?

REFERENCES

Barnes, R. D. 1968. *Invertebrate zoology.* 2nd ed. Philadelphia: W. B. Saunders.

Childress, J. J.; Fisher, C. R.; Brooks, J. M.; Kennicutt, M. C., II; Bidigare, R.; and Anderson, A. E. 1986. A methanotrophic marine molluscan (Bivalvia, Mytilidae) symbiosis: Mussels fueled by gas. *Science* 233:4770, 1306–8.

Coker, R. E. 1962. *This great and wide sea: An introduction to oceanography and marine biology.* New York: Harper and Row.

George, D., and George, J. 1979. *Marine life: An illustrated encyclopedia of invertebrates in the sea.* New York: Wiley-Interscience.

Hedgpeth, J., and Hinton, S. 1961. *Common seashore life of southern California.* Healdsburg, Calif.: Naturegraph.

Hessler, R. R.; Ingram, C. L.; Yayanos, A. A.; and Burnett, B. R. 1978. Scavenging amphipods from the floor of the Philippine Trench. *Deep-Sea Research* 25:1029–47.

MacGinitie, G. E., and MacGinitie, N. 1968. *Natural history of marine animals.* 2nd ed. New York: McGraw-Hill.

Ricketts, E. F.; Calvin, J.; and Hedgpeth, J. 1968. *Between Pacific tides.* Stanford, Calif.: Stanford University Press.

Rona, P. A.; Klinkhammer, G.; Nelson, T. A.; Trefry, J. H.; and Elderfield, H. 1986. Black smokers, massive sulphides and vent biota at the Mid-Atlantic Ridge. *Nature* 321:6065, 33–37.

Thorson, G. 1971. *Life in the sea.* New York: McGraw-Hill.

Yonge, C. M. 1963. *The sea shore.* New York: Atheneum.

ZoBell, C. E. 1968. Bacterial life in the deep sea. In Proceedings of the U.S.-Japan Seminar on Marine Microbiology, August 1966, Tokyo. *Bulletin Misaki Marine Biology, Kyoto Inst. Univ.* 12:77–96.

SUGGESTED READING

Sea Frontiers

Bellomy, M. D. 1973. Blossoms in the sea. 19:1, 2–13.
An informative discussion of the variety, distribution, and life style of sea anemones.

Coleman, N. 1974. Shell-less molluscs. 20:6, 338–42.
A description of nudibranchs, gastropods without shells.

George, J. D. 1970. The curious bristle-worms. 16:5, 291–300.
The variety of worms belonging to the class Polychaeta of phylum Annelida is described. The discussion includes locomotion, feeding, and reproductive habits of the various members of the class.

Gibson, M. E. 1981. The plight of *Allopora.* 27:4, 211–18.
The reason the author believes the unusual California hydrocoral is headed for the endangered species list is the topic of this article.

Ruggiero, G. 1985. The giant clam: Friend or foe? 31:1, 4–9.
The ecology and behavior of the giant clam *(Tridacna gigas)* is discussed in reference to whether it is a danger to divers.

Shinn, E. A. 1981. Time capsules in the sea. 27:6, 364–74.
The method by which geologists determine past climatic and environmental conditions by studying coral reefs is discussed.

Scientific American

Caldwell, R. L., and Dingle, H. 1976. Stomatopods. 234:1, 80–89.
Presents the ecology of these interesting crustaceans that have appendages specialized for spearing and smashing prey.

Feder, H. A. 1972. Escape responses in marine invertebrates. 227:1, 92–100.
Discusses the surprisingly rapid movements and other responses made by invertebrates to the presence of predators. Some interesting photographs accompany the text, which describes the escape responses of limpets, snails, clams, scallops, sea urchins, and sea anemones.

Wicksten, M. K. 1980. Decorator crabs. 242:2, 146–57.
Describes how species of spider crabs use materials from their environment to camouflage themselves.

Yonge, C. M. 1975. Giant clams. 232:4, 96–105.
The distribution and general ecology of the tridacnid clams, some of which grow to lengths well over 1 m (3.3 ft), are investigated.

CHAPTER SEVENTEEN
HUMANS INTERACT
WITH THE OCEAN

The previous chapters have provided a basis for an appreciation of the marine environment. We can now consider the influence of human activities on its ecology. In the past, extension of technology from the land to the ocean proceeded at a slow rate. However, with the realization that the resources we now take primarily from the land are becoming inadequate to feed our rapidly expanding population and support our industrial growth, the development of technology to exploit the oceans is advancing at an accelerating rate. This development will make it possible in the near future to carry on in the marine environment those activities now restricted to land.

Technology has developed to the point that any fishery without controls could readily be decimated in a short time. The search for petroleum is proceeding at a very rapid rate into the margins of the ocean, while the expansion of drilling to the deep-sea floor only awaits slight improvements in technology. Given the tremendous amount of human and industrial waste that is ultimately disposed of in the ocean, it is time for us to take a close look at the potential consequences of our increasing exploitation of the marine environment.

For thousands of years we have used the ocean as a source of food and minerals, a means of transportation, a recreation area, and a dump for the wastes of society. For a time, the ocean was able to meet all of these demands while suffering very little damage. But eventually the intensity of the assault on the ocean began to show. During the last few decades it has become obvious that the marine environment has been altered measurably, and locally the effects have been such that the lives of marine organisms or of the humans who consume them have been seriously affected or destroyed. See *Subsea Disposal of High-Level Nuclear Waste* on pages 446–449 for a discussion of this disposal option's potential for altering the marine environment.

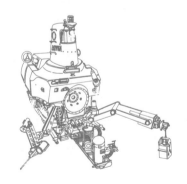

Gill netter fishing off the coast of Nova Scotia. Photo by Michael DiSpezio.

For nearly four decades, high-level nuclear waste has been accumulating from the production of nuclear weapons and commercial power generation. The U.S. alone has more than 75 million gal of waste from weapons production and 12,000 tn of spent reactor fuel that must be safely disposed of until it is no longer dangerous.

This material will be radioactive for more than one million years, but a "safe" disposal system would likely require a shorter time of highly secure confinement. The wastes are composed of over 50 isotopes, each of which has different chemical, half-life, abundance, and radioactive emission characteristics. Considering such factors, investigators believe confinement that allows no more release of radiation into the atmosphere than that of the natural uranium ore from which it was generated may be considered safe and is possible with today's technology (fig. 17A). However, the final determination of the "safe" level of radiation, which is to be made by the Environmental Protection Agency, is yet to be determined.

The major effort is now toward disposal at land sites because the materials remain accessible. However, this accessibility is potentially a major drawback to land disposal. Once the material is safely disposed of, it would seem advisable to have it out of reach of accidental interference and certain acts of intentional human removal. It would also be advisable to put the waste where there is the least likelihood of its being released by natural erosion or tectonic processes. There are few places on the continents that would be as free from all of these threats as parts of the ocean floor.

FIGURE 17A Radiation Doses from Various Sources in Millirems per Year.

The Environmental Protection Agency is responsible for setting the minimum performance requirements (levels of radiation emitted from high-level disposal sites), but has not yet decided on these levels.

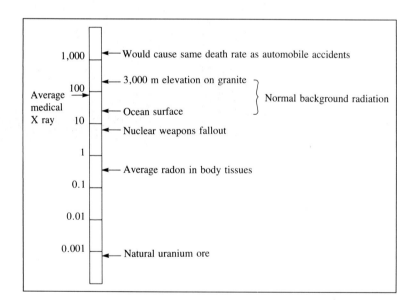

The idea of disposing of nuclear waste in the ocean is not new. However, it is clear the oceans do not have the capacity to dilute the existing supply of high-level waste sufficiently to make it safe.

In 1973, the Sandia National Laboratory in Albuquerque, New Mexico, began consideration of the Subsea Disposal Program. The objective of this investigation is to determine the feasibility of disposing of the high-level nuclear waste in the ocean floor. The form in which the waste will be emplaced is as a glassy mixture that is about 25% radioactive waste. This waste will be housed in a metal container about 3 m (9.8 ft) long and 30 cm (12 in) in diameter and designed to resist corrosion. Although the solid form of the waste will reduce the rate of diffusion of radioactive waste once the container deteriorates, the natural medium in which the waste is buried must be depended upon for the major containment role.

The possibility of drilling into ocean crustal rock is being considered, but any shifting of this rigid material could crush containers and cause premature release of radiation. Initially consideration was given to placing the containers in ocean trenches, but this tectonically active feature of the ocean floor was soon rejected.

In considering the requirements to keep the waste (1) out of the way of human activities, (2) safe from exposure by natural erosion, and (3) away from seismically active regions, it soon became apparent that the centers of oceanic lithospheric plates might be ideal disposal sites. Such regions are beneath at least 5,000 m (16,400 ft) of water, far from the marine activities of humans. Fishing, petroleum production, and mining activities are now conducted primarily on the continental shelves. A potential area of manganese nodule mining between Mexico and the Hawaiian Islands would need to be avoided.

Aside from the fact that such locations meet the requirements listed above, they are covered with up to 1000 m (3280 ft) of fine sediment that has in some regions accumulated uninterrupted for over 100 million years. This pattern of sediment accumulation indicates long periods of stability as the plates move across the earth's surface. It can be expected that this pattern will continue for millions of years. Additionally, the sediment could serve as an ideal medium in which to place the high-level nuclear waste. Many of the radionuclides, after being released from the canisters (they may fail within 1000 yr), would naturally adhere to the clay particles in the sediment. This would slow diffusion of radioactivity away from the burial site.

A number of mid-plate, mid-gyre (MPG) sites have been identified in the North Atlantic and North Pacific oceans (fig. 17B). The initial plan is to place the canisters under 30 to 100 m (98 to 328 ft) of sediment at intervals of at least 100 m. The determination of how to emplace the canisters has not been made, but it may be possible to simply drop them from an appropriate distance above the ocean floor (fig. 17C).

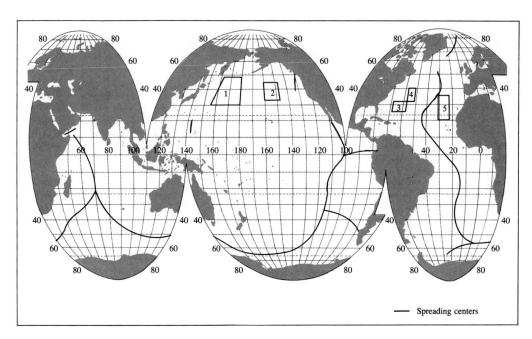

FIGURE 17B MPG Regions.

Mid-plate, mid-gyre (MPG) regions in northern oceans where the environment can be expected to be stable for millions of years.

Before the engineering and social considerations are fully studied, the technical and environment feasibility must be proven. In an effort to investigate this feasibility, computer models have been constructed. To date, those models predict that all short-half-life isotopes and most long-half-life isotopes will be adsorbed to the sediments. These models also indicate that heat release from

FIGURE 17C Canister Placement.

Plan of canister emplacement shows them buried at least 30 m (98 ft) beneath ocean floor spaced 100 m (328 ft) apart. Because of the softness of the sediment, it may be possible to implant the canisters by simply dropping them from an appropriate distance above the sediment surface.

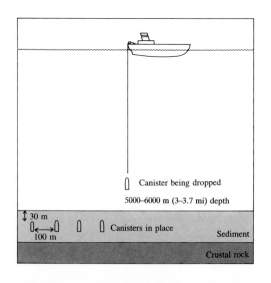

the waste will be transferred almost entirely by conduction. This means that upward convection of radiation carried by sediment pore water will be negligible. This is a crucial consideration. Although these results are promising, models of how well the burial holes will reseal themselves and the means by which radiation will be transported in the water column once it reaches the sediment surface must be completed.

If the models predict the waste can be safely buried in these deep-ocean sediments, they will be verified by actual field tests before emplacement begins. Since the greatest threat of failure will occur during the first 15 years owing to high heat release, real-time tests of the heat effects are feasible. The first such field test will consist of a 400-W heat source buried in MPG 1, located in the North Pacific (fig. 17D). It is expected that technical and environmental feasibility studies will be completed by 1988. If these results are positive, the study of the engineering of transportation and emplacement will be accelerated. Also, the political and legal implications will be clarified, and an environmental impact statement filed.

Should all these phases proceed on schedule and field tests prove the reliability of the disposal of high-level nuclear waste in the seabed, subsea disposal would begin no sooner than the year 2010.

Reference: Hinga, K. R.; Heath, G. R.; Anderson; D. R., and Hollister, C. D. 1982. Disposal of high-level radioactive wastes by burial in the sea floor. *Environmental Science and Technology* 16:1, 28A–37A.

FIGURE 17D Equipment to Test Heated Ocean Sediments.

This equipment for *in situ* verification of physical and chemical properties of heated deep-sea sediments is currently under construction. After depositing the system on the ocean floor at a depth over 5000 m (3 mi), the heat source and sensors will be pushed into the sediment. A ship will periodically visit the site, and data on temperature and other properties will be telemetered to the ship. Cores will be taken in and out of the heated region at the end of the experiment and the entire system will be recovered. Courtesy of EPA.

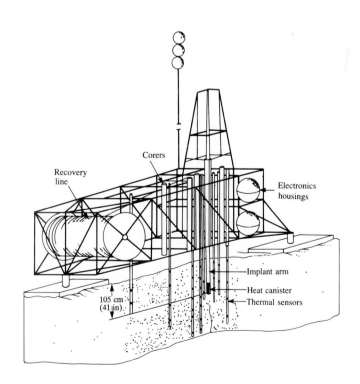

LAW OF THE SEA

Because of the imminence of extensive exploitation of the ocean floor beyond the jurisdiction of the nations that border the seas and the problems developing from international fisheries and pollution, one of the most urgent problems awaiting international agreement has been the establishment of laws by which these activities may be regulated.

Before describing the activities needing regulation, we will discuss some history and some problems facing those who have worked to establish a new law of the sea. In 1609 Hugo Grotius, the Dutch jurist and statesman whose writings helped form international law, established in *Mare Liberum* the doctrine that the ocean is free to all nations. Although controversy continued over whether nations could control portions of the oceans, Cornelius van Bynkershonk produced a satisfactory solution to this problem in *De Dominio Maris,* published in 1702. It provided for national domain over the sea for the distance that could be protected by cannons on the shore. The first official dimension of this zone was established in 1672 when the British determined that they would exercise control over a strip of water one league (3 nautical miles) wide along the shores of the Commonwealth—the *territorial sea.*

Because technology was rapidly developing for drilling beneath the ocean, the first United Nations Conference on the Law of the Sea was held in Geneva in 1958, producing a treaty relating to the continental shelf. It stated that prospecting and mining minerals on the shelf are under the control of the country that owns the nearest land. Unfortunately, the seaward limit of the *continental shelf* was not well defined in the treaty.

A second, less productive United Nations Conference on the Law of the Sea was held in Geneva in 1960; the third had its first meeting in New York in 1973. Subsequent meetings were held at least once per year (fig. 17–1).

The third conference came to a disappointing conclusion in April of 1982, when a large majority—130 to 4, with 17 abstentions—voted in favor of a new Law of the Sea convention. Most of the developing nations who could benefit significantly from the provisions of the treaty voted for its adoption. The opposition was led by the United States, who, along with Turkey, Israel, and Venezuela, voted against it. These countries support private companies planning seabed mining operations and believe certain provisions of the treaty would make it unprofitable. Among the abstainers were the Soviet Union, Britain, Belgium, the

FIGURE 17–1   Law of the Sea Conference.
Third United Nations Conference on the Law of the Sea in Caracas, Venezuela. This gathering was one of a series that has considered the reach of coastal nations' political and economic claims beyond their coastlines, the exploitation of oil and minerals lying beneath the oceans, fishing rights, and curbs on pollution. Carlos Andrés Pérez, former Venezuelan president, is shown giving his speech during the opening session. Photo courtesy of the United Nations.

Netherlands, Italy, and West Germany, all of whom are interested in seabed mining. As of March 12, 1984, 35 nations still had not signed the Treaty, and only 8 nations had ratified it. Because it is unlikely that all dissenting and abstaining nations will ratify the treaty, its ability to facilitate the exploitation of seabed resources will be severely hampered.

The primary features of the treaty are as follows:

Coastal Nations Jurisdiction: A uniform 12-mi (19.3-km) territorial sea and a 200-mi (322-km) Exclusive Economic Zone (EEZ) were established. The coastal nation has jurisdiction over mineral resources, fishing, and pollution within the EEZ and beyond it, up to 350 mi (564 km) from shore, if the continental shelf (defined geologically) extends beyond the EEZ.

Ship Passage: The right of free passage on the high seas is maintained; it is also provided for within territorial seas through straits used for international navigation.

Deep-Ocean Mineral Resources: Private exploitation of seabed resources may proceed under the regulation of the International Seabed Authority (ISA), within which will be a mining company chartered by the United Nations, called *Enterprise.* Mining sites will be operated in pairs: one chosen by the private mining concern, the other by Enterprise, and both approved by ISA. Revenues from the private concern's chosen site will go to that concern; revenues from the other site will be divided among the developing nations.

Arbitration of Disputes: A Law of the Sea tribunal will perform this function.

FISHERIES

Between the end of World War II and 1970, the world catch from marine fisheries steadily increased from around 12 million metric tons (MT) to 50 million MT. Although the capacity of the fishery fleet has increased dramatically since that time, the 1983 catch was only 69.3 million MT.

It has always been understood that fish must be permitted to reproduce if a fishery is to be maintained. However, it has become increasingly difficult to regulate fisheries in a manner that assures sufficient reproduction to maintain the fishery. This has been primarily the result of the great demand that accompanies an increasing human population. Fisheries such as the anchovy, cod, flounder, haddock, herring, and sardine apparently are suffering from overfishing; that is, they are being harvested at a rate above the natural rate of production in such numbers that their population is driven below the size necessary to produce maximum yield. International efforts are now being made to provide worldwide ocean fisheries management. Many regional management authorities have developed over the past few decades.

The establishment of EEZs opens the way to more efficient management of coastal fisheries. It is going to be important for developing nations to establish the capacity to manage and develop their fisheries. Such an ability has the potential of greatly enhancing the economic condition of many nations. It is clear that the exercise of control over a fishery by a single nation will not, of itself, lead to improved fishery management. For example, the Peruvian anchovy fishery was fully under the control of the Peruvian government. Effective management of fisheries depends, of course, on a thorough understanding of the population to be regulated. Obtaining data sufficient for biologists to understand any fishery is extremely difficult. We hope that in the near future more reliable methods for measuring species abundance can be developed.

A method being tested for the assessment of fish stock involves the use of *side-scan sonar*. Figure17–2 shows the plan view and a vertical section of the process through which sonar signals are reflected from fish schools to estimate the dimensions of the schools, making it possible to determine the school's volume. The strength of the reflections is related to the degree of compaction within the school. Possibly, by using this method, the weight of fish schools en-

countered by the sonar signals can be determined. Norwegian investigators have evidence they can determine not only species but also age by sonar.

Figure 17–3 shows a technique that may be adapted to species determination in which resonant reflections at specific frequency ranges are correlated to fishes of a certain size and/or species. The larger jack mackerel with larger air bladders may produce resonance at 800 Hz while 3- and 2-year-old anchovies may be indicated by resonant frequencies at 1300 and 1500 Hz, respectively. A computer-assisted acoustical system developed at the University of Washington has been very successful in actually counting the number of fish within range of the equipment.

Another technique that is valuable in getting a firsthand view of the species of fishes near the deep-ocean floor is the use of *deep-sea free vehicle photography systems*. Figure 17–4 shows the design of such a system, and figure 17–5 shows a clustering of grenadiers and eel pouts gathering around bait cans attached beneath the camera device. Devices of this type have been designed to operate for up to two days completely unattached to surface ships. Still cameras that take periodic exposures or movie cameras that operate at predetermined intervals can be recovered after an automatic timing device releases the camera and float mechanism from the bait.

The effects of inadequate management can be seen in the history of the fisheries in the northwest Atlantic. In this area, regulated by the International Commission for the Northwest Atlantic Fisheries, the fishing capacity of the international fleet increased 500% from 1966 to 1976. The total catch, however, rose by only 15%. This discrepancy indicated a significant decrease in the catch per unit of effort, a good indication that the fishing stocks of the Newfoundland–Grand Banks area are being overexploited. Biologists who had been setting quotas for the major species within this region complained that enforcement was extremely difficult and that the quotas had become international currency used in political bargaining. Politicians were accused of allocating quotas for fish that exceeded the total allowable catch set by the Commission, thus depleting the reserve and preventing stock recovery.

Occurrences like these make it seem that the only answer for regulation with adequate enforcement may be to have fisheries controlled by the nation in whose coastal waters they exist (fig. 17–6). The difficulty of enforcement of regulations by the international commission was largely responsible for the unilateral decision by Canada to extend its right to

FIGURE 17–2 Fish Assessment with Sonar.

A, plan view. *B,* vertical section. Side-scan signal is used to determine area and abundance of fish schools. *C,* a record of a 115-km (71-mi) sonar transect conducted in Southern California Bight from Malibu to Oceanside. Species of fish in schools were not determined, but the major fisheries in this region are northern anchovy and jack mackerel. Courtesy of Paul E. Smith, National Marine Fisheries Service.

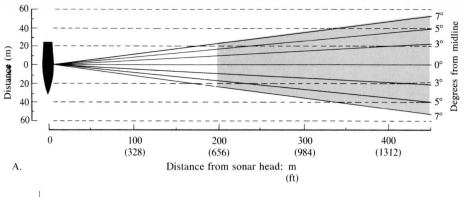

A.

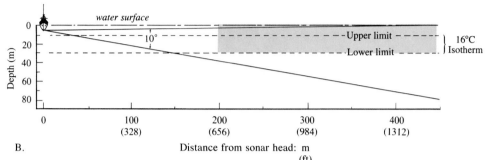

B.

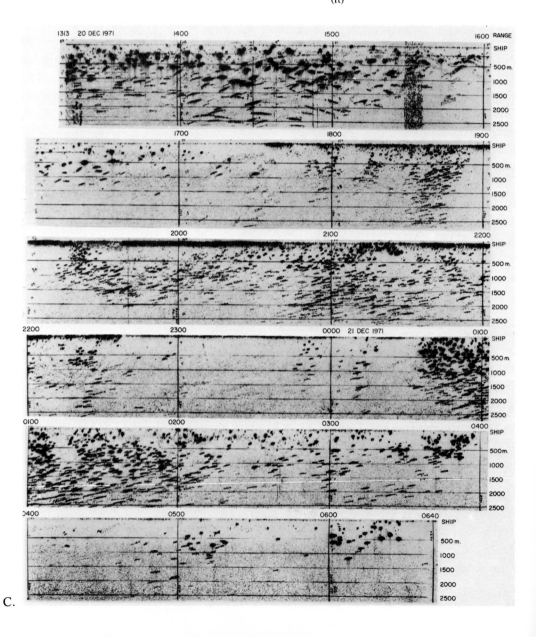

C.

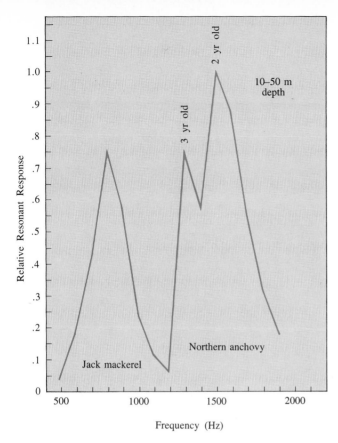

FIGURE 17–3   Swim Bladder Resonance as a Means of Determining Fish Species.

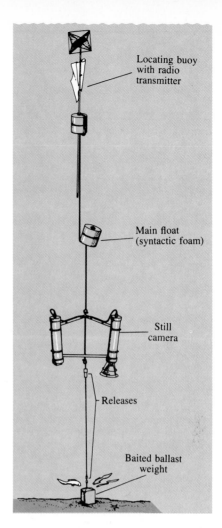

FIGURE 17–4   Deep-Sea Free Vehicle Photography System.

Deep-sea free vehicle cameras are fully automatic, operating for up to two days at a time on the sea floor completely unattached to ship or ocean surface. The cameras were developed by Marine Life Research Group at Scripps Institution of Oceanography. Courtesy of Scripps Institution of Oceanography, University of California, San Diego.

FIGURE 17–5   Fish near Deep-Ocean Floor.

Clustered around a bait can devised to attract ocean-bottom marine life into camera range are grenadiers and eel pouts. Photo from Scripps Institution of Oceanography, University of California, San Diego.

FIGURE 17–6   Soviet Factory Ship.
The development of factory ships serviced by smaller catcher boats such as those seen in this photograph has resulted in a rapid increase in the fishing capacity of the international fleet. Because of the potential catching capacity of such systems, fisheries in which they are involved must be closely monitored and regulated. Photo courtesy of NOAA.

control fish stocks for a distance of 200 mi (322 km) from its shores beginning January 1, 1977. The United States followed with a similar action on March 1, 1977.

Fishing fleets from Russia, Poland, Japan, and Italy fish off the United States coasts, interfering at times with the United States lobster fishery. Occasionally, fishing vessels are boarded, seized, and escorted to port for violation of the fishing regulations established by the United States within their 200-mi limit. Such violations can result in severe fines, imprisonment of the captain, and the confiscation of the ship. U.S. fishermen, particularly from the tuna fleet, occasionally have their vessels seized for fishing within the 200-mi limit of Latin American nations, where they have also been subjected to extremely large fines. Although establishment of the EEZs throughout the world may aid in protecting fisheries, it will certainly create new political and economic problems. U.S. fishermen, happy with the establishment of a 200-mi EEZ by our government, have found they too will be limited by the regulations that go along with its existence. Figure 17–7 shows the area of the ocean covered by the U.S. EEZ. In 1983, 5.9 million MT of fish were taken within the EEZ of the United States; that amount was 8.5% of the world marine fishery for that year.

There is much more to consider in the development of a successful fishery than weight alone. For

instance capelin, anchovy, and menhaden, which are used primarily for the production of fish meal, sold for $60 to $110 per metric ton in 1980. Fisheries such as tuna, flounder, and squid brought $1000 to $2000 per metric ton, and luxury fisheries such as lobster brought up to $8000 per metric ton. Peru will very likely derive more income from its smaller canned sardine fishery than it did from the fish meal anchovy fishery that once made up 20% of the mass of the world fishery.

The direct consumption demand for and value of fish and fish products is increasing rapidly. Over 52 million MT were consumed in 1982, and this demand is projected to reach 93 million MT by the year 2000. The potential for meeting this increased demand, driven by population growth through the expansion of conventional fisheries, is not promising. However, large increases in unconventional fisheries such as krill, mesopelagic fish (lantern fish), and mollusks are thought to have the potential for meeting the increased demand (table 17–1).

MARICULTURE

Marine aquaculture, or mariculture, has been conducted for years throughout the Far East, making major contributions to the available food supply in that part of the world. Most attempts at commercial mariculture in the United States have not been successful because most of the projects were based on intensive production tainted with overambition and impatience.

However, mariculture does have a long history in the United States. In the 1850s, Long Island's natural oyster fishery was depleted and seed oysters were brought in from Chesapeake Bay to be cultivated in the Sound. From this beginning, oyster cultivation spread along the Atlantic Coast into the Gulf of Mexico. Today over 25,000 MT (metric tons), 40% of the U.S. oyster consumption, is produced through mariculture. A significant movement in this direction by the salmon industry is well under way. (See *Salmon Ranching—A Pilot Project* on pages 404–407.) A diversified mariculture industry, along with freshwater aquaculture, is developing in Hawaii. Oyster and shrimp are leading the way in this industry, which is being encouraged by the state government and citizens. The goal is to diversify the state's economy and make it less dependent on tourism.

Organisms chosen for mariculture should be popular marine products that command a high price, are easy and inexpensive to grow, and will reach marketable size within a year or less. Candidates should be

FIGURE 17–7 Exclusive Economic Zone.

Area included in the Exclusive Economic Zone established by the United States. With the establishment of such zones by all nations, 35% of the oceans will be included. Courtesy of Paul Smith, Southwest Fisheries Center, National Marine Fisheries Service, La Jolla, California.

TABLE 17–1 Present Catches and Potential for Increase, Excluding Aquaculture (thousand MT).

Stock	Potential World Catch	World Catch (1966)	World Catch (1977)	Potential Increase By Improved Management	Potential Increase By Increased Effort
Conventional marine resources:					
Salmon	650	453	480	170	. . .
Flounder and cod	6,700	4,692	3,786	2,500	400
Herring and anchovy, etc.	15,600	13,709	1,801	13,800	. . .
Shrimp and lobster	1,670?	830	1,542	25	100
Tuna	2,260	1,031	1,592	100	550
Other demersal	37,100	9,628	17,097	4,000	16,000
Other pelagic	40,200	13,045	28,655	3,000	8,500
Unconventional marine resources:					
Cephalopods	50,000?	833	1,165	. . .	50,000?
Other mollusks	Unknown	2,115	3,011	. . .	Probably large
Krill	50,000?	. . .	123	. . .	50,000?
Mesopelagic fish	50,000?	. . .	. . .	. . .	50,000?

Source: Food and Agricultural Organization of the United Nations

hardy and resistant to disease and parasites. They should also be capable of feeding with a high degree of growth per unit of intake. Severe economic problems can result if the chosen organism is not able to reproduce or cannot be brought to sexual maturity in captivity.

Algae

Algae, in the form of certain seaweeds, are grown and marketed as luxury foods. The life cycle of most of these algae is very complicated, and most of the successful operations are conducted in Japan, where spore-producing plants are cultivated in laboratories. When the plants release their spores into the water, local growers come to the laboratories to dip their nets and ropes, or whatever devices they use for attachment, into the water and allow the spores to attach themselves to these devices. They are next submerged into estuarine waters, where the spores grow into the desired seaweed form using naturally available nutrients.

Bivalves

Cultivation of oysters and mussels is probably the most successful form of mariculture. Commercially, no bivalves have been reared from the larval stage to a marketing size in a controlled environment because of the problems related to producing sufficient phytoplankton to feed them. Usually hatchery-produced juveniles are reared to maturity in natural environments such as bays, estuaries, and other protected coastal waters. They may be set on the bottom or in trays or other suspension devices above the bottom (fig. 17–8). This simple procedure results in annual yields ranging from 10 to over 1000 MT/acre of edible meat per year. There is no attempt at artificial feeding, and the mollusks are fed by the phytoplankton carried to them by the tides and other water movements.

Crustaceans

Crustaceans such as shrimp and lobster are difficult to raise successfully even though the price for such organisms is relatively high (fig. 17–9). The problem with most shrimp operations is that shrimp must be fed extraneously, and so far suitable artificial food has not been developed. American lobsters *(Homarus americanus)* can easily be carried through the complete life cycle in captivity, but they have a cannibalistic nature. To be grown successfully, these lobsters would have to be compartmentalized and maintained

FIGURE 17–8 Bamboo Raft Used in Culturing Oysters. Oysters are suspended from bamboo rafts in nutrient-rich coastal waters off South Korea. This method of culturing off the bottom helps protect oysters from predators such as boring snails and sea stars. Photo from M. Grant Gross.

at an optimum temperature of about 20°C (68°F) for two years. Artificial feeding would also be necessary. Encouraging results have been achieved in rearing American lobster in the warm effluent from an electrical generating plant in Bodega Bay, California. Similar projects will be tried in the coastal waters of Maine. The spiny lobster is not so easily reared through its long complicated larva development, and less effort has been exerted toward developing commercial mariculture with these animals.

Fishes

Mariculture with fishes is also difficult, and only in Japan has a significant marine program been developed. Salmon, yellowtail, tuna, puffers, and a few others are being raised there on an experimental basis. Some estuary fish farming involving mullet and milkfish has been practiced successfully in the Far East for centuries. These fishes are hardy and tolerate salinity ranging from that of fresh water to full sea water. They do not have to be artificially fed. However, since they cannot be spawned and raised to sexual maturity in an artificial setting, the fry must be collected from a natural nursery environment. The costs of operating the farms are relatively low, and yields of about 1 MT/acre are common (fig. 17–10).

What Lies Ahead

Mariculture requiring intensive artificial feeding and high labor costs cannot be expected to help alleviate the world shortage of animal protein. Actually, more

A.

B.

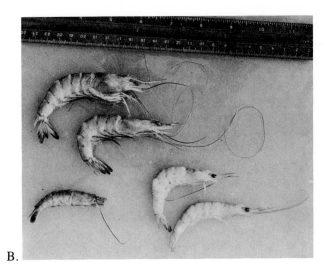

FIGURE 17–9 Shrimp Research.

A, a European biologist and the Philippine students to whom he is teaching research methods examine shrimp kept in concrete ponds to study their molting process. FAO photo by S. Bunnag. *B,* varieties of shrimp, a popular sea food, that are being studied as possible objects of mariculture. FAO photo by Pat Morin.

food is consumed than is produced by such systems, which are typical of the enterprises that have been attempted in the United States. However, those farming enterprises that use natural feeding in estuarine environments have good potential for helping increase the food supply. Most of these efforts require juvenile animals that are obtained from natural nursery areas, which represent the greatest costs to the fish farmer. Potential for increased productivity using the 1 billion acres or so of coastal wetlands in the world could be quite significant. Assuming only 10% of the wetlands would be put into production, it could result in 100 million MT of fish based on the

productivity that has been established for mullet and milkfish (fig. 17–11).

The Food and Agriculture Organization (FAO) of the United Nations has estimated that, based on traditional fisheries, the maximum sustainable yield that might be expected from the oceans is about 120 MT. This estimate may be considered conservative, and could well be exceeded, particularly with the addition of new species to the significant fisheries. A candidate for addition to the list of major fisheries might be the Antarctic krill. Grenadiers and lantern fishes found in deeper waters and the pelagic red crab of the Eastern Pacific and Subantarctic regions

FIGURE 17–10 Milkfish Ponds.

Milkfish are harvested using a gill net. Brackish-water fish-culture techniques applied to ponds such as these in the Philippines are proving successful. FAO photo by P. Boonserm.

FIGURE 17-11 Environment with Potential for Mariculture.

Wetlands such as this salt marsh, if saved from commercial and industrial expansion, may provide the necessary requirements for productive mariculture projects. Photo courtesy of NOAA.

may also be added in the near future. Another significant addition would be cephalopods such as squid, cuttlefish, and octopus, which yield about 80% edible flesh compared to approximately 20% to 50% for the most common representatives of our present fishery.

Surely improved management of existing fisheries, progress in mariculture, and the addition of new spe-

cies to the world fishery will lead to a much needed increase in the supply of marine food.

PETROLEUM

A significant event occurred in the United States during the first week of March 1976: For the first time the United States imported more oil than it produced. Plans were then made to explore and develop petroleum reserves off the Atlantic coast, which had until this time been free of such activities. Test wells were drilled 130 km (81 mi) east of Atlantic City, New Jersey, and 120 km (75 mi) off Cape Cod in the Georges Bank trough. Overcoming the resistance of special interest groups and state governments, the Bureau of Land Management conducted lease sales totalling over $2 billion on Atlantic continental shelf tracts. Testing for petroleum off Baltimore Canyon and on Georges Bank has had discouraging results, and resistance to further exploration continues in the important fishery area of Georges Bank. Similar conflicts have resulted from the federal government's plans to develop petroleum reserves and drill exploratory wells along the California coast. Yet, the urgency to increase domestic petroleum production remains. Figure 17-12 shows the areas of the U.S. continental margin with potential for oil and gas production.

After initial discoveries of North Sea oil and gas in the 1950s, new foreign discoveries continued, and ad-

FIGURE 17-12 Potential Petroleum Production Sites.

Areas of the U.S. continental shelf with the greatest potential for petroleum production. Courtesy of U.S. Department of Energy.

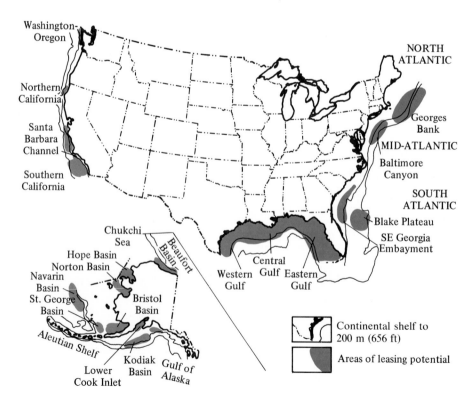

ditional fields were put into production. In the absence of international agreement, one of the more encouraging signs has been the relative ease with which the countries bordering the North Sea, a region of vast reserves, divided the petroleum rights among themselves. In April 1976, the United Kingdom recorded its first exportation of oil, and is now receiving significant income from oil exports as a result of its North Sea reserves. Other offshore discoveries have been made near the Philippines and off the east and west coasts of Spain as well as on the continental shelf off the mouth of the Amazon River. In spite of the occasional glut of oil, the search for petroleum will be the major focus of exploration beneath the oceans in the immediate future.

OTHER MINERALS

The origin of metallic ores that contain rich deposits of copper, lead, zinc, and silver associated with the destructive margins of lithospheric plates is beginning to be understood. Their origin is related to the process of plate destruction (fig. 17–13). Copper sulfide ores associated with ophiolites, sections of oceanic crust that have been pushed up onto continental plates, are part of this process. A good example of this kind of ore is the Troodos Massif along the southern coast of Cyprus, which has been mined since early in the development of the Mediterranean civilizations (see *Ophiolites* on pages 94–97).

The constructive margins of plates associated with oceanic ridges may be the source of this metallic enrichment. The metals found here are predominantly iron and manganese, with significant amounts of copper, nickel, cobalt, zinc, and barium. Analysis of such sediments in a 200-sq-km area of the Atlantis Deep in the Red Sea have shown that they contain more than 3 million tn of zinc, 1 million tn of copper, almost 1 million tn of lead, and 5000 tn of silver. Similar iron- and manganese-rich deposits were recovered by submersibles diving along the Mid-Atlantic Ridge during the French-American Mid-Ocean Undersea Study (FAMOUS) conducted in 1974 (fig. 17–14), as well as the axis of the East Pacific Rise in the late 1970s.

Figure 17–15 shows the general distribution of rich deposits of manganese nodules in the world ocean, and figure 5–14 shows the concentration of manganese, cobalt, nickel, and copper. From the concentrations shown on figure 5–14, it appears that the area of the oceans with the greatest economic potential is in the north equatorial Pacific from Mexico to south of the Hawaiian Islands.

Deposits of phosphorite in nodules or bedded crusts may be found in water close to the edge of the continental shelf (fig. 17–15). These deposits are not likely to be mined in the near future because of the extensive deposits of phosphate on land.

Closer to shore and much easier to mine are placer deposits on the continental shelf. Such deposits are formed through erosion of minerals from the nearby

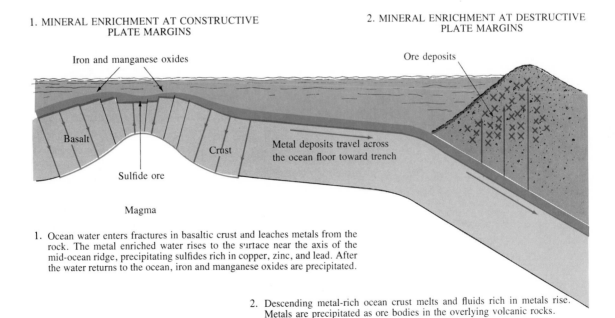

1. MINERAL ENRICHMENT AT CONSTRUCTIVE PLATE MARGINS

2. MINERAL ENRICHMENT AT DESTRUCTIVE PLATE MARGINS

Iron and manganese oxides

Ore deposits

Basalt

Crust

Metal deposits travel across the ocean floor toward trench

Sulfide ore

Magma

1. Ocean water enters fractures in basaltic crust and leaches metals from the rock. The metal enriched water rises to the surface near the axis of the mid-ocean ridge, precipitating sulfides rich in copper, zinc, and lead. After the water returns to the ocean, iron and manganese oxides are precipitated.

2. Descending metal-rich ocean crust melts and fluids rich in metals rise. Metals are precipitated as ore bodies in the overlying volcanic rocks.

FIGURE 17–13 Theory of Metallic Ore Production.

A.

FIGURE 17–14   French-American Mid-Ocean Undersea Study (FAMOUS).

A, the *Alvin,* along with the French submersibles *Archimede* and *Cyana,* descended to depths of 3000 m (9840 ft) during Project FAMOUS to photograph and recover samples from the Mid-Atlantic Ridge rift valley about 320 km (200 mi) southwest of the Azores. *B,* a photo taken by the *Alvin* during Project FAMOUS shows the undulating surface of pahoehoe lava extruded in the rift valley of the Mid-Atlantic Ridge. Photos courtesy of Woods Hole Oceanographic Institution.

B.

landmasses and by concentration in old river channels or along beaches that were subsequently submerged by the rising ocean. These deposits are unconsolidated and can be easily lifted by dredging operations. They include concentrations of diamonds off southwest Africa, gold off Alaska, tin off Indonesia and southwest England, and titanium off Mozambique (fig. 17–15).

With the depletion of sand and gravel deposits on land or their unavailability through urban expansion and other land-use activities, it appears that the continental shelf will have to provide much of these aggregate materials in the near future (fig. 17–16). Economic studies have already shown that marine aggregate is competitive with land-derived sources in many areas. In fact, excluding petroleum, the mining of sand and gravel from the coastal ocean is the largest marine mining operation in existence today. Although environmental problems can result from dredging sand and gravel, they may be less serious than those created by mining land deposits in urban areas.

MARINE POLLUTION

As society engages in many of the activities discussed in the preceding section, it exerts tremendous environmental pressures on various regions of the ocean, particularly the coastal areas.

The ocean is an important recreational resource in the United States as 70% of its population lives within easy access of the coasts of either the oceans or the Great Lakes. Well over 100 million Americans participate in marine recreational activities, spending almost $15 billion annually. In the past it has not been unusual to see local areas closed to recreational activities such as fishing, boating, and swimming as a result of pollution.

We have little difficulty understanding the meaning of marine pollution, which is the introduction into that environment of substances or energy that result in harm to the living resources of the ocean or humans that use these resources. But as far as the marine environment is concerned, there is great difficulty in establishing the degree to which pollution is

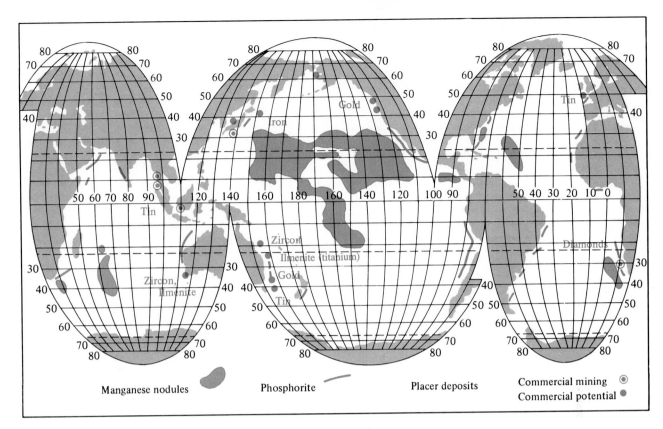

FIGURE 17–15 Manganese Nodule, Phosphorite, and Placer Deposits.
Base map courtesy of National Ocean Survey.

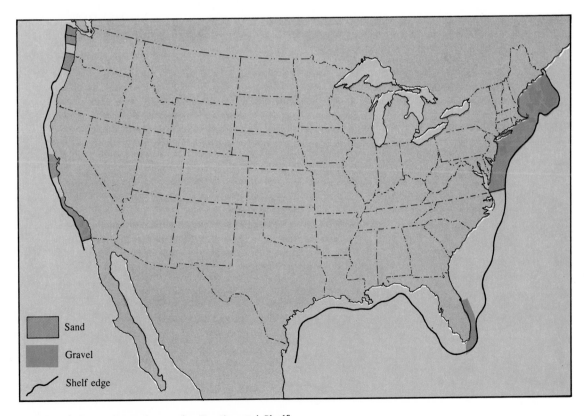

FIGURE 17–16 Sand and Gravel Deposits on the Continental Shelf.
From U.S. Department of Commerce Public Bulletin 188717.

occurring. In most cases, the ocean had not been studied sufficiently prior to society's introduction of pollutants into it. Therefore, we cannot tell how the marine environment has been altered by these activities. In this section we will discuss some examples of marine pollution.

Ocean Dumping in New York Bight

In the past, many substances were routinely dumped in the ocean under the assumption they could be easily absorbed without any damage to the marine environment. Each year in the United States, ocean dumping includes such materials as sewage sludge, industrial waste, explosives, and, particularly, large quantities of dredge spoils. In all, almost 50 million tn of waste materials are dumped into the ocean each year, of which about 80% is composed of dredge spoils (table 17–2). In the United States, the Marine Protection Research and Sanctuaries Act of 1972 provided the first direct legal control over dumping waste in the ocean. About 80% of the dumping that occurs in U.S. waters ends up in the New York Bight off the coasts of New York and New Jersey at six different sites shown in figure 17–17.

Although there are numerous problems related to ocean dumping in such areas, dumping will continue and probably increase until environmentally acceptable alternatives for handling these waste materials can be developed. It must be kept in mind that if it becomes environmentally unacceptable to continue to dump this sludge in the Bight, it must be disposed of elsewhere. Regardless of how it might otherwise be processed, a significant amount of environmental impact will result. The consideration of this problem brings to light a number of the difficult choices society will have to make concerning environmental quality. An acceptable resolution of this problem will require a great amount of cooperation among the public, scientists, and those responsible for administering programs related to public health.

Minamata Disease

Some agricultural and industrial waste products are directly poisonous to marine life and can kill coastal populations directly. Many marine organisms have the ability to concentrate pollutants that are in small concentrations in the ocean waters to very high concentrations in their body fluids and tissue without causing harm to themselves. In some cases these concentrations found in the tissues of marine animals are poisonous to human beings that feed on them.

The most dramatic example of this kind of human poisoning occurred in Minamata Bay, Japan. In 1953 animals were found dying convulsive deaths in streets, and some of the human population of the area also suffered severe symptoms that could not be easily diagnosed. It was finally determined that shellfish in the bay had been filtering mercury compounds from the water and concentrating them in their tissues. The people who had been feeding on shellfish exhibited the symptoms of mercury poisoning. Over 100 people suffered from the disease by 1969 and almost half of them died. Unborn children were later found to have symptoms of the disease. The active poisoning agent in this instance was methyl mercury chloride.

Because of the fish-eating habits of Scandinavians, the Swedish National Institute of Public Health conducted an investigation and found that some persons in Sweden had mercury concentrations in their bodies as high as those in the Japanese who suffered from the Minamata disease. However, the Swedes had no symptoms. This would indicate that populations throughout the world may have varying degrees of sensitivity to the ingestion of foreign substances.

TABLE 17–2   Ocean Dumping in 1968 (U.S.).

Dumped Material (thousands of tons)	Atlantic Ocean	Gulf of Mexico	Pacific Ocean	Total	% of Total
Dredge spoils	15,808	15,300	7,320	38,428	80
Industrial waste	3,013	696	981	4,691	10
Sewage sludge	4,477	0	0	4,477	9
Construction and demolition debris	574	0	0	574	1
Solid waste	0	0	26	26	Trace
Explosives	15	0	0	15	Trace
	23,887	15,996	8,327	48,211	100

Source: U.S. Environmental Protection Agency

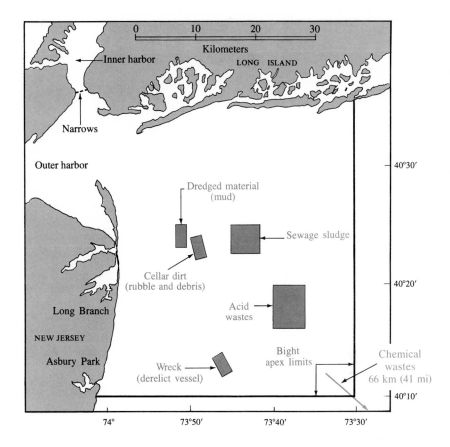

FIGURE 17–17 Dump Sites in New York Bight. Courtesy of Environmental Protection Agency.

An important aspect of the Minamata Bay incident is that the corporation responsible for discharging the mercury into the bay initiated this activity in 1938. Yet the first observance of this disease did not occur for about 15 years. To prevent such occurrences as the Minamata Bay poisoning in the future, the activities in coastal waters must be closely monitored and analyzed for potential polluting effects.

Persistent Compounds

Insecticides such as DDT, Aldrin, and Dieldrin and certain industrial compounds, all of which are valuable commercially because of their great resistance to bacterial breakdown, are serious threats to marine life. Because of their resistance to breakdown, these compounds can become very persistent in the marine environment. These pollutants are commonly carried to the ocean through the atmosphere and are found in higher concentrations in the atmosphere over the coastal ocean than over the open ocean.

However, a study of DDT and PCB (a very stable industrial material) in the North Atlantic showed that their concentrations in coastal waters did not significantly differ from concentrations in the open ocean.

This surprising result has been explained by the fact that the coastal waters contain more particulate matter. Particulate matter has a high capacity to absorb these pollutants from the coastal waters, reducing the amounts found in the water and plankton. Further investigation may find that the high content of particulate matter such as clay and dead plankton may play an important role in reducing the toxic concentrations of many pollutants in the coastal waters, although subsequent settling may introduce them to the sediment.

The use of DDT and PCB has been greatly restricted on a worldwide basis, but the only way to stop pollution by such substances is to stop their use. This would be difficult, however, since the insecticides are necessary in many countries for the control of disease-carrying insects. Any insecticide removed from production would have to be replaced by another to meet these needs.

Oil Pollution

Probably no one can calculate with any great accuracy the amount of petroleum that enters the ocean within a year, but estimates have been made. The

TABLE 17–3 Estimates of Petroleum Annually Entering the Ocean (millions of metric tons per year).

Source	1975 Estimates[1]	1984 Estimates[2]
Natural sources	0.6	0.25 (0.025–2.5)
Production and Distribution	2.413	1.62 (1.05–3.28)
Offshore production and coastal refining	0.28	0.15
Transportation-related operations[3]	1.833	1.05
Accidental spills[4]	0.3	0.42
Use and disposal	2.5	1.08 (0.525–2.52)
Municipal wastes	0.3	0.7
Industrial wastes	0.3	0.2
Urban and river runoff	1.9	0.16
Ocean dumping	NA[5]	0.02
Atmospheric Inputs[6]	0.6	0.3 (0.05–0.5)
Total	6.1	3.3 (1.7–8.8)

[1]NAS, 1975.

[2]NAS, 1985. Ranges are shown in parentheses for some of the best estimates.

[3]Includes tanker operations, drydocking, marine terminals, bilge, and fuel oil bunkering.

[4]Includes transportation and non-transportation accidents.

[5]NA = not available (not estimated).

[6]Atmospheric sources are derived in part from natural and production-related processes, but probably mostly from fuel consumption (combustion).

Source: U.S. Department of Commerce, National Oceanic and Atmospheric Administration. Ocean Assessments Division (Rockville, MD. 1985).

sources of oil pollution most readily documented are those resulting from oil spills at dock facilities and from tanker accidents, offshore drilling rigs, and coastal refineries. Once ships are at sea, it is difficult to determine how much oil enters the ocean from tankers cleaning their tanks and from discharges from other ships. Much oil is known to enter the ocean from natural seeps and, undoubtedly, significant quantities of oil enter the ocean from the atmosphere, but these amounts are open to considerable question (table 17–3). Oil entering the ocean as a result of accidental spills reached a yearly maximum of some 470,000 bbl in 1976. The amount spilled in 1983 was about a third of that, or about 160,000 bbl.

Regardless of the amounts of petroleum that enter the oceans, we need to know what the long-range effects on the marine environment will be. Spills involving refined petroleum products respond much more slowly to nature's cleanup efforts than does natural crude oil. There is much evidence that petroleum spilled in large quantities kills significant numbers of intertidal organisms and birds. It also appears that the ocean environment recovers rather rapidly from most ocean spills, although recovery from severe spills may take years.

Tanker Accidents On March 18, 1967, the first major tanker spill on record occurred off the coast of Cornwall, England, when the *Torrey Canyon* ran aground at Seven Stones Rocks. It spilled 117,000 tn of oil, which washed onto the beaches of Cornwall

and French beaches across the channel. Beginning March 28, the spill was bombed and set afire, and the beaches were cleaned with detergent—a substance probably more toxic to marine life than oil. There is considerable debate about the long-term effects, but a degree of recovery seems to have occurred.

A smaller spill that was well studied occurred when the oil barge *Florida* ran aground in Buzzards Bay, Massachusetts, in 1969 (fig. 17–18). Possibly 700 tn of fuel oil entered the coastal waters and was driven into the productive marshlands by onshore

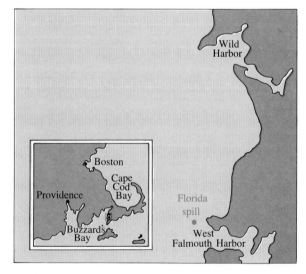

FIGURE 17–18 *Florida* Oil Spill at West Falmouth Harbor.

winds and tides. Immediate study of the affected area in comparison with uncontaminated areas nearby was conducted by scientists from Woods Hole Oceanographic Institution. The ecological damage to local fishing and shell industries has been estimated to have exceeded $1 million, and the oil that penetrated the sediment to a depth greater than 60 cm (24 in) was found to remain toxic for years. Study of this spill points to the fact that ecological damage can be severe. It also shows that the effects can last for years even though the visual appearance of an area after a spill may return to near normal within months.

On December 15, 1976, the tanker *Argo Merchant* ran aground on Fishing Rip, a shoal region 54 km (34 mi) southeast of Nantucket. Heavy weather prevented removing the cargo of 30,000 tn of fuel oil from its tanks (fig. 17–19). On December 21 the *Argo Mer-*

chant began breaking up, spilling virtually all of its oil into the ocean off Massachusetts. Fortunately the oil drifted in a southeasterly direction, doing a minimal amount of environmental damage compared to what could have occurred had it been driven ashore.

Scientific investigation into the effects of the spill were initiated immediately. Because most of it spread out as a slick on the surface, no contamination of more than 250 parts per billion was found in subsurface water samples. Sediment contamination in the immediate area of the grounding was high, and oil concentrations as high as 100 parts per million were found in sediment samples taken 18 km (11 mi) southwest of the grounding point. It is believed this sediment contamination resulted from oil spilling out directly on the bottom at the point of grounding and being carried by subsurface currents. Actually, some

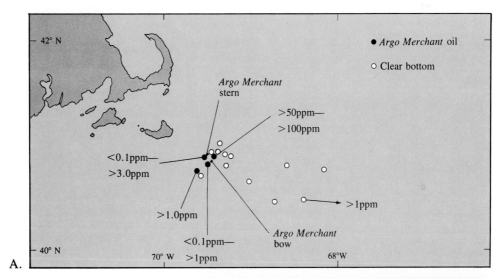

A.

FIGURE 17–19 *Argo Merchant.*

A, location of grounding site and where University of Rhode Island research ship *Endeavor* sampled ocean sediment. Contaminated areas are indicated. From *The* Argo Merchant *oil spill—a preliminary scientific report,* 1977. Courtesy of NOAA. *B,* a Coast Guard helicopter approaches the grounded 196-m (643-ft) vessel to remove crew members prior to its breaking up in heavy seas. Photo courtesy of U.S. Coast Guard.

B.

of the oil contaminating the sediment was not from the *Argo Merchant*. Here again we see the problem of insufficient background data that scientists face when they try to evaluate the damage resulting from an individual event.

Because of the lack of information concerning the biology of the area, it is difficult to assess the biological effect of the *Argo Merchant* spill. Evidence indicates that fuel oil can be expected to do great damage to fish eggs and the zooplankton that provide the basis for fishery populations in the area. Eggs of pollock and cod were found to be significantly affected over the area of the spill. Although the bird popula-

tion in the area is considered small, numerous oiled birds were washed ashore at Nantucket and Martha's Vineyard. Marine mammals apparently suffered little damage as they stayed clear of the oil slick.

On the night of March 16, 1978, the supertanker *Amoco Cadiz* went aground 1.5 km (0.9 mi) off Portsall, France, and the worst tanker spill to date began. During the next weeks, 223,000 tn of crude and bunker oil leaked into the ocean and washed ashore along 320 km (200 mi) of the Brittany coast (fig. 17–20). Spring tides and gale force winds spread the oil, which decimated the intertidal fauna. More than 3200 birds were killed directly and many more died from

FIGURE 17–20 *Amoco Cadiz* Oil Spill.

A, Brittany coast of France showing locations where oil from the *Amoco Cadiz* washed ashore. Courtesy of NOAA. *B, Amoco Cadiz* aground 1.5 km (0.9 mi) north of Portsall, France. 223,000 tn of oil were spilled into the coastal waters of France when it went aground March 16, 1978. Photo courtesy of NOAA.

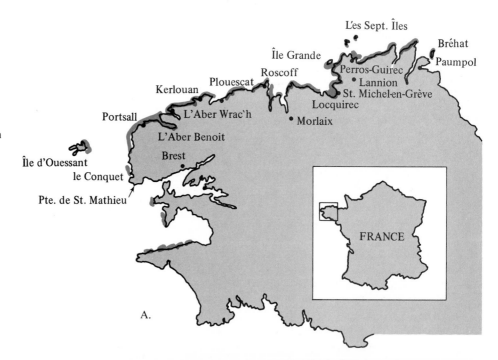

A.

B.

eating contaminated prey. Mariculture operations for oysters and lobsters were severely affected. The total damage will not be known for years. The rocky regions recovered quickly, but the embayments where oil infiltrated the fine sediment may remain contaminated for many years.

In none of these spills, not even in the relatively small spill caused by the *Florida,* was equipment available to control pollution. Oil companies are trying, however, to develop compounds that will jell oil in the tanks of tankers that run aground or are involved in other accidents that might result in spilling their cargo into the ocean.

Offshore Drilling In 1969 a high-pressure well in the Santa Barbara Channel got out of control and about 700,000 gal of oil, representing about 3000 tn, entered the marine environment. There was probably more public reaction to this relatively small spill than to any spill in the history of the United States. Oil leaked into the ocean for more than 300 days. Much of this oil drifted onto the shore, killing intertidal life (fig. 17–21). Many birds were also killed as a result of being coated by the oil. It is difficult to evaluate the environmental effect of this spill because of natural oil seeps in the same area that pour oil into the ocean at a rate greater than the average rate of pollution by the well. Over the duration of the contamination from the leaking well, the average rate of inflow was about 50 barrels per day. Estimates of the inflow from widely scattered natural seeps range between 50 and 75 barrels per day.

A spill 10 times larger than that which occurred in Santa Barbara lasted only 7½ days in the Ekofisk field of the North Sea in April 1977 (fig. 17–21). Even though this blowout was relatively easy to cap, what might the consequences have been had its control required ¹/₁₀ the time needed to stop the Santa Barbara leak, or about 30 days? Flowing at an estimated rate of 4 tn/day, it would have discharged 120,000 tn of oil into the marine environment.

The largest oil spill on record began June 3, 1979, when the Petroleus Mexicanos Ixtoc I blew out and caught fire. Oil began to wash ashore on Padre Island August 17 (fig. 17–22). After a month of cleanup operations along the Texas coast, the offshore current shifted to the south as it usually does during the fall and winter months. The well initially emptied 5000 tn of oil into the Gulf of Mexico per day. The rate of flow was decreased, and the well was finally capped on March 24, 1980, after spewing over 468,000 tn of oil into the Gulf of Mexico. Fortunately, the capping oc-

A.

B.

FIGURE 17–21   Santa Barbara and Ekofisk Oil Spills. *A,* boat sprays detergent into the oil swept onto the California shore after a blowout at an offshore drilling rig in February 1969. Photo courtesy of the U.S. Coast Guard. *B,* fireboat sprays water on 1977 blowout at Ekofisk in the North Sea to lessen fire risk. Photo from Wide World Photos.

curred before the return of northward-flowing currents could carry the oil toward the Texas coast.

Oil is known to be broken down by microorganisms such as bacteria, fungi, and yeast. Certain of these organisms are effective only in breaking down a particular variety of hydrocarbon; none are effective against all forms. Dr. Ananda M. Chakrabarty, a microbiologist at the General Electric Research and Development Center in Schenectady, New York, has produced a microorganism capable of breaking down about two-thirds of the hydrocarbons in most crude oil spills. By manipulating special rings of DNA called *plasmids,* he has been able to combine the consumptive characteristics of numerous natural strains of bacteria into what might be considered a "superbug" that could be very effective in helping clear up oil spills. However, before this superstrain can be used

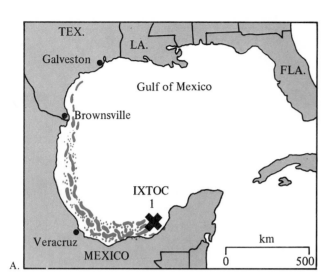

FIGURE 17–22 Ixtoc I Blowout.
Location of Gulf of Campeche blowout and oil slick that threatened Texas coast during the summer of 1979.

against actual oil spills, much testing must be done to ensure that it does not have adverse effects on the environment.

One of the most pervasive agruments concerning the supply of petroleum for our national needs revolves around whether we should develop the potential reserves on the U.S. continental shelf or continue to transport oil from foreign sources, using the large tankers that have been developed. Arguments against the development of the continental shelf reserves generally refer to the damage that might be done by the potential increase in the number of accidents similar to that which occurred at Santa Barbara. However, data collected on oil spills from continental shelf petroleum production from 1964 to 1971 show that there were 16 major spills from over 10,000 producing wells, which released 46,000 tn of oil into the ocean. Even with the blowout of the Ixtoc I, offshore drilling appears to be less of a pollution threat than tanker transportation.

Although new techniques for decreasing the quantities of oil spilled and for cleaning the oil once it is spilled indicate that we may be able to reduce the amount of damage from spills, the ideal is to prevent such disasters in the first place. Of course, this cannot be totally achieved. But with much greater attention than is presently given to ensuring the seaworthiness and safety of operation of tankers, the number of such accidents can be reduced significantly. Increased attention to improving offshore drilling and production techniques should result in significantly less spilling from these operations as well.

As research on alternative sources of energy continues, we are assured that at some time in the future we will have to replace petroleum as our major energy source. However, until this comes about, it seems that the exploitation of our continental shelf reserves is an environmentally safer direction to follow than that of transporting foreign oil. There are, of course, other economic and political benefits that would result from following such a plan.

OCEAN RESEARCH

Freedom for Research

If international regulation of the ocean is to be developed, it can be effective only in proportion to the knowledge regulators have of the marine environment. This will, of course, require widespread and intensive scientific research by scientific institutions of many nations. In the past there has been little constraint to handicap the conduct of research throughout the ocean environment. About the only restriction to investigating the oceans was the 3-mi (4.8-km) territorial sea, but permission was generally granted for work even within this zone. It now seems likely that the coastal state will control research in the EEZ, which includes all of the continental margin.

Only a few of the wealthier nations have the ability to do extensive ocean research. Some of the developing coastal nations are very suspicious of the activities that might be undertaken in their coastal waters and have a fear of being exploited by these larger nations. Other nations are concerned about their national security, since they believe ocean research might be related to espionage.

At present there is considerable uncertainty about the degree of freedom for scientific research that may ultimately exist in the open ocean as well as in coastal waters. Whatever the outcome, scientific research must remain a viable activity. Even though it may require greater cooperation among nations than was necessary in the past, it surely would not seem prudent to increase international regulation of ocean resources without making a very strong effort also to increase our understanding of the resources under regulation.

Remote Sensing

A possible source of information that would not require vessels entering territorial waters of other nations is *remote sensing*. Satellites fitted with sensors such as cameras; infrared, visible, and microwave scanners; laser reflectors; altimeters; and radar have proven valuable in providing information on surface and near-surface ocean phenomena. Satellites that have already proven their value for the study of oceans are the *TIROS N; NIMBUS G* and *NIMBUS 7; NOAA 2, 3, 4,* and *6; ERTS 1; LANDSAT 2* and *3; SMS/GOES 1–5;* and *SEASAT.*

Although data obtained through remote sensing may never be as precise as desired by scientists, it should be of great help in increasing understanding of the distribution of temperature with an absolute accuracy of 2°C (3.6°F) and relative accuracy of ±0.5°C (0.9°F) by sensing visible and infrared wavelengths. Surface winds at an elevation of 20 m (65.6 ft) can be measured at speeds up to 125 km/h (78 mi/h) with accuracies up to ±7 km/h (4.3 mi/h) or 25% of the actual value, whichever is larger. The wind's direction can be found to ±20°. Radiation absorbed and reflected by the ocean can be determined to within ±5 ly/day.

It is very desirable to determine wave heights so forecasts of wave conditions can be made beyond the areas where large waves are being produced by high winds. Significant wave heights, the highest one-third of the waves in a sea area, can be determined to within ±0.5 m (1.6 ft) of 10% of the actual height for waves up to 20 m (66 ft) high using a short-pulse radar altimeter. (See *SEASAT Measures the Ocean Surface* on pages 180–182.) Gravitational anomalies in the height of the ocean surface can even be used to map the surface of the ocean floor. (See *SEASAT Maps the Ocean Floor* on pages 52–54.) Visual images showing wave diffraction, reflection, and refraction patterns will aid in understanding how waves move sediment on shoals and in channels and how they damage ocean structures and will be of importance in planning and designing shoreline protection structures. It appears that such data can be obtained for waves with wavelengths greater than 50 m (164 ft).

Speed of surface currents determined from measurements of ocean surface slope may be possible to within ±0.7 km/h (0.4 mi/h). For lower-speed currents, buoys may be used as tracers. Buoys equipped to collect data are valuable adjuncts to the satellite sensors. Thermal infrared images can be used to supplement the slope data obtained by altimeters (fig. 17–23). Areas of upwelling (fig. 13–6B) can be iden-

FIGURE 17–23 Visible Light and Infrared Images of Eastern United States and Gulf Stream.

From an altitude of 1457 km (905 mi), the *NOAA 2* satellite took these two images of the eastern United States. A Very High Resolution Radiometer recorded the image at left from reflected visible light and the image at right from infrared radiation, which indicates differences in the temperature of the earth, oceans, and clouds. In both images, the Gulf Stream is distinctly darker than the colder near-shore water. Photo courtesy of NOAA.

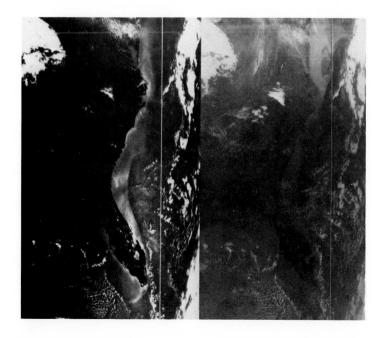

tified on the basis of temperature and, using a combination of temperature and color sensing, the chlorophyll concentration could be measured and used to quantify biological productivity. Open-ocean tides that seldom exceed 1 m (3.28 ft) and are difficult to measure may be measured to within ±25 cm (10 in) relative to mean sea level and to within ±20° of the alignment of the rotating crest and trough with a year of data collecting by altimeters.

Storm surges with amplitudes approaching 10 m (32.8 ft) are responsible for much of the property damage and loss of life from hurricanes. Should altimeter-bearing satellites record a storm surge, it could be measured to within ±1 m (3.28 ft). There is, however, a low probability this would occur. Also, it is unlikely a tsunami would be observed by a satellite. However, should one be passed over, its height could be determined to within ±25 cm (10 in) and wavelength to within 20% of its actual value.

Nearshore wave and current action can produce significant changes in shallow-water features and

shorelines. Using multispectral optical imagers, it is possible to image shoals to a depth of 30 m (98 ft) with vertical resolution of up to 2 m (6.56 ft) and horizontal resolution of 70 m (230 ft). Images obtained for identifying shoal areas are shown in figure 17–24. Sediment transport by currents can be measured using several wavelengths with a resolution of 800 m (2624 ft) over swaths as wide as 700 km (435 mi). Particulate concentrations from 0.2 to 100 mg/m^3 may be determined.

Using a sensor called synthetic aperture radar (SAR), sea ice and very large icebergs can be detected with a resolution of 25 m (82 ft) and swath width of up to 100 km (62 mi). With a scanning multichannel microwave radiometer (SMMR), ice cover over the entire polar caps can be imaged with a low resolution of 20 km (12 mi) along a swath 1000 km (621 mi) wide (*see* plate 4).

Changes in physical or chemical composition of ocean water lead to variations in its color and reflectivity. Such changes, which are more pronounced

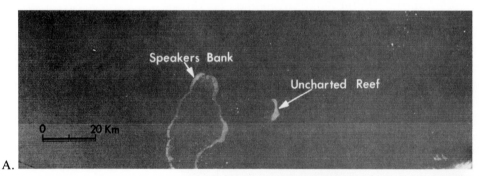

FIGURE 17–24 Shoal Areas Observed by Satellite.
A, a high gain setting on the multispectral scanner was used to increase the water-penetrating capability of Band 4 (0.5–0.6 μm) radiation. Revealed in this *Landsat 2* image is the existence of an uncharted reef in the Chagos Archipelago of the Indian Ocean. *B,* this image was also obtained using Band 4 and has been processed to enhance underwater detail. It shows the bottom to a depth of 30 m (98 ft). The east end of Grand Bahama Island is in the upper left, Great Abaco Island in the upper right, and Berry Islands at lower center. *A,* photo courtesy of A.N. Kover, U.S. Geological Survey. *B,* photo courtesy of A. P. Colvocoresses, U.S. Geological Survey.

near continents, make it possible to identify water masses. A coastal zone color scanner (CZCS) on *NIMBUS 7* can image ocean surface and near surface in visible light and thermal infrared wavelengths with 0.8-km (0.5-mi) resolution over a 700-km (435-mi) swath. It is designed to make quantitative measurements of chlorophyll and sediment concentrations (plates 6 and 21). (See *Satellites and Plankton* on pages 378–380.)

Other potential uses include fish stock location and pollutant distribution and concentration. By viewing reflected sunlight, variations in surface roughness caused by oil spills and internal waves can also be detected. For these and the other limited aspects of ocean study discussed in this chapter, satellites will surely provide the first measuring stations to supply oceanographers with an opportunity to study vast areas of ocean surface as a single unit.

A major disappointment in the development of remote sensing capability to study the oceans was the failure of the *SEASAT* satellite. Launched June 28, 1978, the satellite provided a great amount of data, but electrical problems ended the flow of data on October 10, 1978.

Ocean-sensing satellite launchings that will greatly increase our ability to observe ocean processes worldwide are scheduled for the early 1990s. The Navy Remote Sensing System (NROSS) will measure winds at the ocean surface. The Ocean Topography Experiment (TOPEX) will measure the ocean topography with a precision of 2 cm (0.8 in). The Ocean Color Imager (OCI) will measure the concentration of phytoplankton pigments with greater precision than the CZCS, and the Geopotential Research Mission (GRM) will measure the earth's gravity field and magnetic field.

This rapidly increasing capability to observe and measure the oceans can help our generation make wise decisions that will protect and conserve the ocean resources for generations who will follow us (fig. 17–25).

SUMMARY

As a result of a rapidly expanding human population and a growing need for food and minerals, knowledge and technology have developed to the point where extensive exploitation of the ocean's resources is near at hand. Already the ocean has been significantly affected by the search for food and minerals. Also, use of the ocean as a means of transportation and a dump for the wastes of society has contributed to the deterioration of the marine environment.

Because of the imminence of extensive exploitation of the ocean, the United Nations has conducted conferences on the **Law of the Sea** to develop international agreement on how the resources of the ocean can be shared and managed.

Increased effort in some of the **world's fisheries** has not significantly increased catches. In some cases, the result has been a dramatic decrease in takes, indicating **overexploitation** of these fisheries. An increased emphasis on understanding of fisheries has resulted in development of **sonar, photography,** and other techniques of **fishery assessment.** There has been an agreement to extend coastal nations' control of the ocean to 200 mi (320 km) beyond their coasts to protect coastal fisheries.

Mariculture projects around the world have, in some cases, indicated potential for increasing the food taken from the ocean. The most successful mariculture to date is directed at **oysters, mussels, mullet,** and **milkfish.** These activities, combined with adding new species such as krill, grenadiers, lantern fish, red crabs, and cephalopods to our open-ocean fisheries, may help increase the amount of food we take from the ocean.

Inorganic resources such as petroleum, metal ores, phosphorite, and placer deposits of diamonds, gold, and other minerals and aggregates are targets of exploration and exploitation. **Petroleum** is already being extensively exploited, and plans are well developed for the mining of **manganese nodules.**

FIGURE 17–25 They Will Need to Use the Ocean Resources Too.

A representative sample of the future generations we must consider in making decisions about the use of ocean resources are these young oceanographers busy exploring the marine world at the Aquarium-Museum of Scripps Institution of Oceanography. Photo by Pat Kampman, courtesy of Scripps Institution of Oceanography, University of California, San Diego.

All of these activities exert pressure on the marine environment. Combined with the dumping of society's wastes, they have created the problem of **marine pollution**. Because of our lack of knowledge of the original condition of the ocean, it is difficult to determine the degree to which marine pollution has occurred. Some of the major sources of pollution are (1) **ocean dumping** of sewage sludge, industrial wastes, explosives, and dredge spoils, (2) the entry of **insecticides** from runoff and from the atmosphere, and (3) the **spilling of oil** as a result of tanker and offshore drilling accidents.

For better regulation of marine exploitation activities, research to increase knowledge of the oceans must continue. There is some concern that the **freedom to do research** will decrease as coastal nations increase their control over coastal waters to 200 mi beyond their coastline.

The use of **remote sensing** from orbiting satellites promises to make it possible to gather information from near the ocean surface without entering the territorial waters of coastal nations to take direct measurements. This approach has great potential for measuring temperature, surface winds, wave height, wave length and patterns, current speed, storm surge, tsunami, shoal water, sea ice, icebergs, physical and chemical changes in surface water, primary productivity, fish stocks, and pollution, as well as for mapping the surface of the deep ocean.

QUESTIONS AND EXERCISES

1. Summarize the significant events in the development of international sea law.

2. Describe techniques being developed to aid in the assessment of fish stocks so that fisheries can be managed more knowledgeably.

3. Why did the Marine Mammal Protection Act of 1972 create a problem for the U.S. eastern Pacific tuna fleet? What has the industry done to help solve the problem?

4. List the desirable characteristics of organisms chosen for mariculture projects.

5. What do the procedures for cultivating algae and oysters have in common?

6. What is the major problem in mariculture of crustaceans that makes economic success difficult?

7. How is the world food supply affected by mariculture operations that require supplemental feeding? Name some animals that have potential as new major fisheries.

8. Make a table listing the regions of the (1) Atlantic Coast, (2) Gulf Coast, and (3) Pacific Coast of the conterminous United States and (4) Alaska, where potential for petroleum production from the continental shelf is greatest.

9. Describe the relationship of metallic ore deposits in the lithosphere to the plate tectonics process.

10. Discuss the positive and negative aspects of gem and metal deposits and sand and gravel deposits on the continental shelves.

11. Study the relative quantities of waste materials dumped in the Atlantic Ocean, Gulf of Mexico, and Pacific Ocean, shown in table 17–2. Discuss factors that may explain the range of quantities dumped in each region.

12. Why does the coastal ocean not have higher concentrations of DDT and PCB than the open ocean?

13. What is the apparent relationship between sediment particle size and the time the sediment remains toxic following an oil spill?

14. Compare the threat of oil pollution in the marine environment from oil transportation to the dangers posed by offshore production.

15. Discuss possible reasons the less-developed nations might prefer the concept of the open ocean being the common heritage of all nations, while the more-developed nations might prefer the concept of the open ocean's resources belonging to those who recover them.

16. Given recent developments, do you think marine research will continue to be a viable activity?

REFERENCES

Borgese, E. M., and Ginsburg, N. (editors). *1985 Ocean Yearbook,* Vol 5. Chicago: University of Chicago Press.

Fisheries of the United States, 1974. 1975. Current Fishery Statistics No. 6700. Washington, D.C.: National Oceanic and Atmospheric Administration, U.S. Department of Commerce.

Inter-university program of research on ferromanganese deposits of the ocean floor. 1973. Washington, D.C.: National Science Foundation.

Knauss, J. A. 1974. Marine science and the 1974 Law of the Sea Conference—science faces a difficult future in changing Law of the Sea. *Science* 184:1335–41.

Oil spills and spills of hazardous substances. 1974. Washington, D.C.: U.S. Environmental Protection Agency.

Petroleum in the marine environment. 1975. Washington, D.C.: National Academy of Sciences.

Ross, W. M. 1973. *Oil pollution as an international problem.* Seattle: University of Washington Press.

Smith, P. E. 1969. *The horizontal dimensions and abundance of fish schools in the upper mixed layer as measured by sonar.* La Jolla, Calif.: Bureau of Commercial Fisheries.

SUGGESTED READING

Sea Frontiers

Ehmann, J. 1983. Man-made island—solution to economic and environmental woes? 29:3, 170–79.
Describes a structure proposed for construction off the east coast as a solution to the problems of a deep-water port, pollution, and industrial development.

Engle, M. 1986. Oceanography's eye in the sky. 32:1, 37–43.
The shuttle is used to photograph oceanographic phenomena.

Goldstein, R. J. 1983. Dredge spoil: Not always a waste. 29:4, 241–47.
Some of the beneficial uses of dredge spoils at Cape Hatteras and Pamlico Sound, North Carolina, are discussed.

Holloway, T. 1977. Progress of marine fish farming in Britain. 23:1, 48–54.
An evaluation of the potential for mariculture in the coastal areas of the British Isles is presented.

Hull, E. W. S. 1978. Oil spills: The causes and the cures. 24:6, 360–69.
A general discussion of oil spills, their frequency, and methods to clean them up.

Iverson, E., and Jory, D. 1986. Shrimp culture in Ecuador: Farmers without seed. 32:6, 442–453.
Ecuador strives to maintain its position as the world's number one producer of farmed shrimp.

Land, T. 1982. "Fingerprinting" offending tankers. 28:2, 102–104.
A discussion of techniques for marking oil cargoes so the source of oil spills can be determined.

Nowadnick, J. 1977. There is but one ocean. 23:3, 130–40.
Problems related to exploitation of the ocean in terms of environmental impact and international politics are discussed.

Parry, J. 1983. Nations unite to fight pollution. 29:3, 143–50.
A discussion of the United Nations Environment Program.

Schlais, J. 1982. Deepwater shrimps—a new Pacific fishery. 28:6, 364–69.
Development of a new fishery is chronicled.

Scientific American

Bascom, W. 1974. Disposal of waste in the ocean. 231:2, 16–25.

An informative overview of people's misconceptions concerning the disposal of waste in the oceans and the methods that may be used to safely use the oceans for such a purpose.

Beddington, J. R., and May, R. M. 1982. The harvesting of interacting species in a natural ecosystem. 247:5, 70–127.
The effect of changing populations of krill-feeding baleen whales on Antarctic food chains is discussed.

Bonatti, E. 1978. The origin of metal deposits in the oceanic lithosphere. 238:2, 54–61.
A description of the geochemical processes occurring at spreading centers of oceanic ridges and rises, which are the prime source of metal-rich minerals found throughout the lithosphere.

Donaldson, L. R., and Joyner, T. 1983. The salmonid fishes as a natural livestock. 249:1, 51–58.
A discussion of how the genetic adaptability of this family of fishes is making it possible to replenish depleted stocks, establish new stocks, and adapt them for ranching operations.

Pinchot, G. B. 1970. Marine farming. 223:6, 14–21.
Presents a brief history and description of efforts and potential for future marine farming, with an emphasis on the means by which nutrients may be economically introduced into low-nutrient waters.

Rona, P. A. 1973. Plate tectonics and mineral resources. 229:1, 86–95.
The economically important concentrations of minerals are controlled by plate tectonics.

APPENDIX I
SCIENTIFIC NOTATION

To simplify writing very large and very small numbers, scientists indicate the number of zeros in them by scientific notation. One integer is placed to the left of the decimal, and a multiplication times a power of ten tells which direction and how far the decimal is moved to write the number out in its long form. For example:

$$2.13 \times 10^5 = 213,000$$
$$\text{or}$$
$$2.13 \times 10^{-5} = 0.0000213$$

Further examples showing numbers that are powers of ten are:

$$1,000,000,000 = 1.0 \times 10^9, \quad \text{or } 10^9$$
$$1,000,000 = 1.0 \times 10^6, \quad \text{or } 10^6$$
$$1,000 = 1.0 \times 10^3, \quad \text{or } 10^3$$
$$100 = 1.0 \times 10^2, \quad \text{or } 10^2$$
$$10 = 1.0 \times 10^1, \quad \text{or } 10^1$$
$$1 = 1.0 \times 10^0, \quad \text{or } 10^0$$
$$0.1 = 1.0 \times 10^{-1}, \text{or } 10^{-1}$$
$$0.01 = 1.0 \times 10^{-2}, \text{or } 10^{-2}$$
$$0.001 = 1.0 \times 10^{-3}, \text{or } 10^{-3}$$
$$0.000001 = 1.0 \times 10^{-6}, \text{or } 10^{-6}$$
$$0.000000001 = 1.0 \times 10^{-9}, \text{or } 10^{-9}$$

To add or subtract numbers written as powers of ten, they must be converted to the same power:

Addition		Subtraction	
2.1×10^3	0.021×10^5	3.4×10^4	3.4×10^4
$+1.0 \times 10^5 =$	$+1.000 \times 10^5$	$-2.0 \times 10^3 =$	-0.2×10^4
	1.021×10^5		3.2×10^4

To multiply or divide numbers written as powers of ten, the exponents are added or subtracted:

Multiplication	Division
6.04×10^2	3.0×10^3
$\times\ 2.1\ \times 10^4$	$\div 1.5 \times 10^2$
12.684×10^6	2.0×10^1

APPENDIX II
THE METRIC SYSTEM AND
CONVERSION FACTORS

Units of Length
1 micrometer (μm) = 10^{-6} m = 0.000394 in
1 millimeter (mm) = 10^{-3} m = 0.0394 in = 10^3 μm
1 centimeter (cm) = 10^{-2} m = 0.394 in = 10^4 μm
1 meter (m) = 10^2 cm = 39.4 in = 3.28 ft = 1.09 yd = 0.547 fath
1 kilometer (km) = 10^3 m = 0.621 statute mi = 0.540 nautical mi

Units of Volume
1 liter = 10^3 cm^3 = 1.0567 liquid qt = 0.264 U.S. gal
1 cubic meter (m^3) = 10^6 cm^3 = 10^3 l = 35.3 ft^3 = 264 U.S. gal
1 cubic kilometer (km^3) = 10^9 m^3 = 10^{15} cm^3 = 0.24 statute mi^3

Units of Area
1 square centimeter (cm^2) = 0.155 in^2
1 square meter (m^2) = 10.7 ft^2
1 square kilometer (km^2) = 0.292 nautical mi^2 = 0.386 statute mi^2

Units of Time
1 day = 8.64 × 10^4 s (mean solar day)
1 year = 8765.8 h = 3.156 × 10^7 s

Units of Mass
1 gram (g) = 0.035 oz
1 kilogram (kg) = 10^3 g = 2.205 lb
1 metric ton = 10^6 g = 2205 lb

Units of Speed
1 centimeter per second (cm/s) = 0.0328 ft/s
1 meter per second (m/s) = 2.24 statute mi/h = 1.94 kn
1 kilometer per hour (km/h) = 27.8 cm/s = 0.55 kn
1 knot (1 nautical mile per hour, kn) = 1.15 statute mi/h = 0.51 m/s

Units of Temperature

Celsius (°C)	Fahrenheit (°F)	Kelvin (K)	
−237.2	−459.7	0	Absolute zero (lowest possible temperature)
0	32	273.2	Freezing point of water
100	212	373.2	Boiling point of water

Conversions: °F = (1.8 × °C) + 32
°C = (°F − 32)/1.8

APPENDIX III
PERIODIC TABLE OF THE ELEMENTS

```
 1  ◄——— Atomic number
 H  ◄——— Symbol of element
1.0080 ◄——— Atomic weight
```

☀ Radioactive state at 1 atmos. and 20°C (68°F)

Inert gas ▢

Gas ▨

Liquid ◯

Solid—all others

Light Metals

Nonmetals

Transitional Elements

Heavy Metals

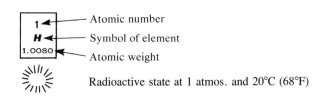

Period	I A	II A	III B	IV B	V B	VI B	VII B		VIII B		I B	II B	III A	IV A	V A	VI A	VII A	VIII A
1	1 H 1.0080																	2 He 4.003
2	3 Li 6.939	4 Be 9.012											5 B 10.81	6 C 12.011	7 N 14.007	8 O 15.9994	9 F 18.998	10 Ne 20.183
3	11 Na 22.990	12 Mg 24.31											13 Al 26.98	14 Si 28.09	15 P 30.974	16 S 32.064	17 Cl 35.453	18 Ar 39.948
4	19 K 39.102	20 Ca 40.08	21 Sc 44.96	22 Ti 47.90	23 V 50.94	24 Cr 52.00	25 Mn 54.94	26 Fe 55.85	27 Co 58.93	28 Ni 58.71	29 Cu 63.54	30 Zn 65.37	31 Ga 69.72	32 Ge 72.59	33 As 74.92	34 Se 78.96	35 Br 79.909	36 Kr 83.80
5	37 Rb 85.47	38 Sr 87.62	39 Y 88.91	40 Zr 91.22	41 Nb 92.91	42 Mo 95.94	43 Tc (99)	44 Ru 101.1	45 Rh 102.90	46 Pd 106.4	47 Ag 107.870	48 Cd 112.40	49 In 114.82	50 Sn 118.69	51 Sb 121.75	52 Te 127.60	53 I 126.90	54 Xe 131.30
6	55 Cs 132.91	56 Ba 137.34	57 TO 71	72 Hf 178.49	73 Ta 180.95	74 W 183.85	75 Re 186.2	76 Os 190.2	77 Ir 192.2	78 Pt 195.09	79 Au 197.0	80 Hg 200.59	81 Tl 204.37	82 Pb 207.19	83 Bi 208.98	84 Po (210)	85 At (210)	86 Rn (222)
7	87 Fr (223)	88 Ra 226.05	89 TO 103															

Rare Earth Elements

Lanthanide series

57 La 138.91	58 Ce 140.12	59 Pr 140.91	60 Nd 144.24	61 Pm (147)	62 Sm 150.35	63 Eu 151.96	64 Gd 157.25	65 Tb 158.92	66 Dy 162.50	67 Ho 164.93	68 Er 167.26	69 Tm 168.93	70 Yb 173.04	71 Lu 174.97

Actinide series

89 Ac (227)	90 Th 232.04	91 Pa (231)	92 U 238.03	93 Np (237)	94 Pu (242)	95 Am (243)	96 Cm (247)	97 Bk (249)	98 Cf (251)	99 Es (254)	100 Fm (253)	101 Md (256)	102 No (254)	103 Lw (257)

APPENDIX IV
THE GEOLOGIC TIME TABLE

Era	Period	Epoch & Age 10^6 Yr		Development of Life and Orogenic Events	
Cenozoic	Quaternary	Recent			
		Pleistocene	0.01 —	Pacific Coast	
		Pliocene	1.6 —	Ranges form	
	Tertiary	Miocene	5 —		
		Oligocene	24 —		
		Eocene	37 —		
		Paleocene	58 —		
Mesozoic	Cretaceous		66 —	First primates	Rocky Mts. form
	Jurassic		144 —	First birds	Sierra Nevada
	Triassic		208 —	First mammals	Ranges form
Paleozoic	Permian		245 —	First flowering plants	
	Pennsylvanian		286 —	First reptiles	
	Mississippian		320 —	First insects	Appalachian Mts. form
	Devonian		360 —	First amphibians	
	Silurian		408 —	First land plants	
	Ordovician		438 —	First vertebrates (fish)	
	Cambrian		505 —	First abundant animal fossils	
	Eocambrian		570 —		
Precambrian			640 —	First algae, bacteria	
			4500 —		

Dates based on Geological Society of America Geological Time Scale, 1983.

APPENDIX V
TAXONOMIC CLASSIFICATION OF COMMON MARINE ORGANISMS (with appropriate number of known species)

Kingdom Monera

Organisms without nuclear membranes; nuclear material is spread throughout cell; predominantly unicellular.

Phylum Schizophyta. Smallest known cells; bacteria (1500 species).

Phylum Cyanophyta. Blue-green algae; chlorophyll *a,* carotene and phycobilin pigments (200 species).

Kingdom Protista

Organisms with nuclear material confined to nucleus by a membrane.

Phylum Chrysophyta. Golden-brown algae; includes diatoms, coccolithophores and silicoflagellates; chlorophyll *a* and *c,* xanthophyll, and carotene pigments (6000 + species).

Phylum Pyrrophyta. Dinoflagellate algae; chlorophyll *a* and *c,* xanthophyll, and carotene pigments (1100 species).

Phylum Chlorophyta. Green algae, chlorophyll *a* and *b,* and carotene pigments (7000 species).

Phylum Phaeophyta. Brown algae; chlorophyll *a* and *c,* xanthophyll, and carotene pigments (1500 species).

Phylum Rhodophyta. Red algae; chlorophyll *a,* carotene, and phycobilin pigments (4000 species).

Phylum Protozoa. Nonphotosynthetic, heterotrophic protists (27,400 species).
Class Mastigophora. Flagellated; dinoflagellates (5200 species).
Class Sarcodina. Amoeboid; foraminifers and radiolarians (11,500 species).
Class Ciliophora. Ciliated (6000 species).

Kingdom Fungi

Phylum Mycophyta. Fungi, lichens; most fungi are decomposers found on the ocean floor, whereas lichens inhabit the upper intertidal zones (3000 species of fungi, 160 species of lichen).

Kingdom Metaphyta

Multicellular, complex plants.

Phylum Tracheophyta. Vascular plants with roots, stems, and leaves that are serviced by special cells that carry food and fluids (287,200 species).
Class Angiospermae. Flowering plants with seeds contained in a closed vessel (275,000 species).

Kingdom Metazoa

Multicellular animals.

Phylum Porifera. Sponges; spicules are the only hard parts in these sessile animals that do not possess tissue (10,000 species).
Class Calcarea. Calcium carbonate spicules (50 species).
Class Desmospongiae. Skeleton may be composed of siliceous spicules or spongin fibers or be nonexistent (9500 species).

Class Sclerospongiae. Coralline sponges; massive skeleton composed of calcium carbonate, siliceous spicules, and organic fibers (7 species).
Class Hexactinellida. Glass sponges; six-rayed with siliceous spicules (450 species).

Phylum Cnidaria. Radially symmetrical, two-cell, layered body wall with one opening to gut cavity; polyp (asexual, sexual, benthic) and medusa (sexual, pelagic) body forms (10,000 species).
Class Hydrozoa. Polypoid colonies such as pelagic Portuguese man-of-war and benthic *Obelia* common; medusa present in reproductive cycle but reduced in size (3000 species).
Class Scyphozoa. Jellyfish; medusa up to 1 m (3.3 ft) in diameter is dominant form; polyp is small if present (250 species).
Class Anthozoa. Corals and anemones possessing only polypoid body form and reproducing asexually and sexually (6500 species).

Phylum Ctenophora. Predominantly planktonic comb jellies; basic eight-sided radial symmetry modified by secondary bilateral symmetry (80 species).

Phylum Platyhelminthes. Flatworms; bilateral symmetry; hermaphroditic (25,000 species).

Phylum Nemertea. Ribbon worms; as long as 30 m (98 ft); benthic and pelagic (800 species).

Phylum Nematoda. Roundworms; marine forms are primarily free-living and benthic; most 1 to 3 mm (0.04 to 0.12 in) in length (5000 marine species).

Phylum Rotifera. Ciliated, unsegmented forms less than 2 mm (0.08 in) in length (1500 species, only a few marine).

Phylum Bryozoa. Moss animals; benthic, branching or encrusting colonies; lophophore feeding structure (4500 species).

Phylum Branchiopoda. Lamp shells; lophophorate benthic bivalves (300 species).

Phylum Phoronida. Horseshoe worms; 24-cm(9-in)-long lophorate tubeworms that live in sediment of shallow and temperate shallow waters (15 species).

Phylum Sipuncula. Peanut worms, benthic (325 species).

Phylum Echiura. Spoon worms; sausage-shaped with spoon-shaped proboscis; burrow in sediment or live under rocks (130 species).

Phylum Pogonophora. Tube-dwelling, gutless worms 5 to 80 cm (2 to 32 in) long; absorb organic matter through body wall (100 species).

Phylum Tardigrada. Marine meiofauna that have the ability to survive long periods in a cryptobiotic state (diversity is poorly known).

Phylum Mollusca. Soft bodies possessing a muscular foot and mantle that usually secretes calcium carbonate shell (75,000 species).
Class Monoplacophora. Rare trench-dwelling forms with segmented bodies and limpetlike shells (10 species).
Class Polyplacophora. Chitons; oval, flattened body covered by eight overlapping plates (600 species).
Class Gastropoda. Large, diverse group of snails and their relatives; shell spiral if present (64,500 species).
Class Bivalvia. Bivalves; includes mostly filter-feeding clams, mussels, oysters, and scallops (7500 species).
Class Aplacophora. Tusk shells; sand-burrowing organisms that feed on small animals living in sand deposits (350 species).
Class Cephalopoda. Octopus, squid, and cuttlefish that possess no external shell except in the genus *Nautilus* (600 species).

Phylum Annelida. Segmented worms in which musculature, circulatory, nervous, excretory, and reproductive systems may be repeated in many segments; mostly benthic (10,000 marine species).

Phylum Arthropoda. Jointed-legged animals with segmented body covered by an exoskeleton (30,000 marine species).

Subphylum Crustacea. Calcareous exoskeleton, two pairs of antennae; cephalon, thorax, and abdomen body parts; includes copepods, ostracods, barnacles, shrimp, lobsters, and crabs (26,000 species).

Subphylum Chelicerata. Horseshoe crabs (4 species).
Class Merostomata.
Class Pycnogonida. Sea spiders.

Subphylum Uniramia. Insects; genus *Halobites* is the only truly marine insect.

Phylum Chaetognatha. Arrowworms; mostly planktonic, transparent and slender; up to 10 cm (4 in) long (50 species).

Phylum Hemichordata. Acorn worms and pterobranchs; primitive nerve cord; gill slits; benthic (90 species).

Phylum Echinodermata. Spiny-skinned animals; benthic animals with secondary radial symmetry and water vascular system (6000 species).
Class Stelleroidea. Star-shaped animals.
Subclass Asteroidea. Starfishes; free-living, flattened body with five or more rays with tube feet used for locomotion; mouth down (1600 species).
Subclass Ophiuroidea. Brittle stars and basket stars; prominent central disc with slender rays; tube feet used for feeding; mouth down (200 species).
Class Concentricycloidea. Sea daisies; tiny disc-shaped form found in a sample taken from crevice in driftwood at a depth of 1000 m (3280 ft) near New Zealand (1 species).
Class Echinoidea. Sea urchins, sand dollars, heart urchins; free-living forms without rays; calcium carbonate test; mouth down or forward (860 species).
Class Holothuroidea. Sea cucumbers; soft bodies with radial symmetry obscured; mouth forward (900 species).
Class Crinoidea. Sea lilies, feather stars; cup-shaped body attached to bottom by a jointed stalk or appendages; mouth up (630 species).

Phylum Chordata. Notochord; dorsal nerve cord and gills or gill slits (55,000 species).

Subphylum Urochordata. Tunicates; chordate characteristics in larval stage only; benthic sea squirts and planktonic thaliaceans and larvaceans (1375 species).

Subphylum Cephalochordata. *Amphioxus,* or lancelets; live in coarse temperate and tropical sediment (25 species).

Subphylum Vertebrata. Internal skeleton; spinal column of vertebrae; brain (52,000 species).
Class Agnatha. Lampreys and hagfishes; most primitive vertebrates with cartilaginous skeleton, no jaws and no scales (50 species).
Class Chondrichthyes. Sharks, skates, and rays; cartilaginous skeleton; 5 to 7 gill openings; placoid scales (625 species).
Class Osteichthyes. Bony fishes; cycloid scales; covered gill opening; swim bladder common (30,000 species).
Class Amphibia. Frogs, toads, and salamanders; Asian mud flat frogs are the only amphibians that tolerate marine water (2600 species).
Class Reptilia. Snakes, turtles, lizards, and alligators; orders Squamata (snakes) and Chelonia (turtles) are major marine groups (6500 species).
Class Aves. Birds; many live on and in the ocean but all must return to land to breed (8600 species).
Class Mammalia. Warm-blooded; hair; mammary glands; bear live young; marine representatives found in the orders Sirenia (sea cows, dugong), Cetacea (whales) and Carnivora (sea otter, pinnipeds) (4100 species).

APPENDIX VI
CALCULATIONS RELATIVE TO SOURCE OF OCEAN WATER

The Mantle Mass of the mantle = density × volume = $(4.5 \text{ g/cm}^3) \times (1.0 \times 10^{27} \text{ cm}^3)$
$= 4.5 \times 10^{27} \text{ g}$

Mass of the oceans = $(1.025 \text{ g/cm}^3) \times (1.4 \times 10^{24} \text{cm}^3) = 1.4 \times 10^{24} \text{ g}$
Percent of mass of mantle represented by mass of oceans:
$$\frac{\text{Mass of oceans}}{\text{Mass of mantle}} = \frac{1.4 \times 10^{24} \text{ g}}{4.5 \times 10^{27} \text{ g}} = 0.00031 = 0.031\%$$

Percent of mass of stony meteorites made up of water = 0.5%

Ratio of percent of meteorite mass composed of water to minimum percent of mantle mass that must be composed of water to allow mantle to be considered source of oceans:
$$\frac{\text{Meteorite}}{\text{Mantle}} = \frac{0.5\%}{0.031\%} = 16.1$$

If the mantle has the same water content as the meteorites, it contains 16 times the mass of water needed to account for the earth's oceans.

APPENDIX VII
ROOTS, PREFIXES, AND SUFFIXES*

a- not, without
ab-, abs- off, away, from
abysso- deep
acanth- spine
acro- the top
aero- air, atmosphere
aigial- beach, shore
albi- white
alga- seaweed
alti- high, tall
alve- cavity, pit
amoebe- change
annel- ring
annu- year
anomal- irregular, uneven
antho- flower
apex- tip
aplysio- sponge, filthiness
aqua- water
arachno- spider
arena- sand
arthor- joint
asthen- weak, feeble
astro- star
auto- self
avi- bird
bacterio- bacteria
balaeno- whale
balano- acorn
barnaco- goose
batho- deep
bentho- the deep sea
bio- living
botryo- bunch of grapes
brachio- arm

branchio- gill
broncho- windpipe
bryo- moss
bysso- a fine thread
calci- limestone ($CaCO_3$)
calori- heat
capill- hair
cara- head
carno- flesh
cartilagi- gristle
caryo- nuclus
cen-, ceno- recent
cephalo- head
chaeto- bristle
chiton- tunic
chlor- green
choano- funnel, collar
chondri- cartilage
cilio- small hair
cirri- hair
clino- slope
cnido- nettle
cocco- berry
coelo- hollow
cope- oar
crusta- rind
cteno- comb
cyano- dark blue
cypri- Venus, lovely
-cyst bladder, bag
-cyte cell
deca- ten
delphi- dolphin
di- two, double
dino- whirling

*Source: D. J. Borror, 1960. *Dictionary of Word Roots and Combining Forms.* Palo Alto, Calif.: National Press Books.

diplo-   double, two
dolio-   barrel
dors-   back
echino-   spiny
eco-   house, abode
ecto-   outside, outer
edrio-   a seat
en-   in, into
endo-   inner, within
entero-   gut
epi-   upon, above
estuar-   the sea
eu-   good, well
exo-   out, withou
fec-   dregs
fecund-   fruitful
flacci-   flake
flagell-   whip
fluvi-   river
fossili-   dug up
fuco-   red
geno-   birth, race
geo-   the earth
giga-   very large
globo-   ball, globe
gnatho-   jaw
guano-   dung
gymno-   naked, bare
halo-   salt
haplo-   single
helio-   the sun
helminth-   worm
hemi-   half
herbi-   plant
herpeto-   creeping
hetero-   different
hexa-   six
holo-   whole
homo-   alike
hydro-   water
hygro-   wet
hyper-   over, above, excess
hypo-   under, beneath
ichthyo-   fish
-idae   members of the animal
family of
infra-   below, beneath
insecti-   cut into
insula-   an island
inter-   between
involute-   intricate
iso-   equal
-ite   rock
juven-   young

juxta-   near to
kera-   horn
kilo-   one thousand
lacto-   milk
lamino-   layer
larvi-   ghost
latent   hidden
latero-   side
lati-   broad, wide
lemni-   water plant
limno-   marshy lake
lipo-   fat
litho-   stone
litorial   near the seashore
lopho-   tuft
lorica-   armor
luci-   light
luna   moon
lux   light
macro-   large
mala-   jaw, cheek
mamilla-   teat
mandibulo-   jaw
mantle-   cloak
mari-   the sea
masti-   chewing
mastigo-   whip
masto-   breast, nipple
maxillo-   jaw
madi-   middle
mega-   great, large
meio-   less
meridio-   noon
meso-   middle
meta-   after
meteor-   in the sky
-meter   measure
-metry   science of measuring
mid-   middle
milli-   thousandth
mio-   less
moll-   soft
mono-   one, single
-morph   form
nano-   dwarf
necto-   swimming
nemato-   thread
neo-   new
nerito-   sea nymph
noct-   night
-nomy   the science of
nucleo-   nucleus
nutri-   nourishing
o-, oo-   egg

ob- reversed
octa- eight
oculo- eye
odonto- teeth
oiko- house, dwelling
oligo- few, scant
-ology science of
omni- all
ophi- a serpent
opto- the eye, vision
orni- a bird
-osis condition
ovo- egg
pan- all
para- beside, near
pari- equal
pecti- comb
pedi- foot
penta- five
peri- all around
phaeo- dusky
pholado- lurking in a hole
-phore carrier
photo- light
phyto- plant
pinnati- feather
pisci- fish
plani- flat, level
plankto- wandering
pleisto- most
pleuro- side
plio- more
pluri- several
pneuma- air, breath
pod- foot
poikilo- variegated
poly- many
poro- channel
post- behind, after
-pous foot

pre- before
pro- before, forward
procto- anus
proto- first
pseudo- false
ptero- wing
pulmo- lung
pycno- dense
quadra- four
quasi- almost
radi- radial
rhizo- root
rhodo- rose-colored
sali- salt
schizo- split, division
scyphi- cup
semi- half
septi- partition
siphono- tube
-sis process
spiro- spiral, coil
stoma- mouth
strati- layer
sub- below
supra- above
symbio- living together
taxo- arrangement
tecto- covering
terra- earth
terti- third
thalasso- the sea
trocho- wheel
tropho- nourishment
turbi- disturbed
un- not
vel- veil
ventro- underside
xantho- yellow
xipho- sword
zoo- animal

GLOSSARY

Abiotic Without life.

Abyssal clay Deep-ocean (oceanic) deposits containing less than 30% biogenous sediment.

Abyssal hill A volcanic extension above the ocean floor less than 1 km (0.62 mi).

Abyssal zone Benthic environment from 4000 to 6000 m (13,120 to 19,680 ft).

Abyssopelagic zone Open-ocean (oceanic) environment below 4000 m (13,120 ft) depth.

Acid A solution in which the H^+ concentration is greater than the OH^- concentration. In a water solution, the hydrogen ion concentration is greater than one part per 10^7.

Acoelomate Without a secondary body cavity (coelom).

Adiabatic Pertaining to a change in the temperature of a mass resulting from compression or expansion. Requires no addition or loss of heat from the mass.

Algae One-celled or many-celled plants that have no root, stem, or leaf systems; simple plants.

Alkaline Pertaining to a solution in which the OH^- concentration is greater than the H^+ concentration. In a water solution, the hydrogen ion concentration is less than one part per 10^7.

Amphidromic point A nodal or no-tide point in the ocean or sea around which the crest of the tide wave rotates during one tidal period.

Amphineura Class of mollusks with eight dorsal calcareous plates; includes chitons.

Amphipoda Crustacean order containing laterally compressed members such as the "sand hoppers."

Amplitude Height of wave crest above or trough below still water.

Anadromous Pertaining to a species of fish that spawns in fresh water, then migrates into the ocean to grow to maturity.

Anaerobic respiration Respiration carried on in the absence of free oxygen (O_2). Some bacteria and protozoans carry on respiration this way.

Anion A negatively charged ion.

Annelida Phylum of elongated, segmented worms.

Antinode Zone of maximum vertical particle movement in standing waves where crest and trough formation alternate.

Aphelion The point farthest from the sun in the orbit of a planet or comet.

Aphotic Without light.

Apogee The point farthest from the earth in the orbit of the moon or a human-made satellite.

Aschelminthes Phylum of wormlike pseudocoelomates.

Aspect ratio The index of propulsive efficiency obtained by dividing the square of fin height by the area of the fin.

Asthenosphere A plastic layer in the upper mantle from depths as shallow as 10 km (6.2 mi) to as deep as 800 km (497 mi) which may allow lateral movement of lithospheric plates and isostatic adjustments.

Atlantic-type margin The passive trailing edge of a continent that is subsiding because of lithospheric cooling and increasing sediment load.

Atom The smallest particle of an element that can combine with similar particles of other elements to produce compounds.

Atomic number A number representing the relative position of an element in the periodic table of elements. It is equal to the number of positive charges in the atom's nucleus.

Augite A dark mineral, usually black, rich in iron and magnesium. An important constituent of basalt, the rock that is characteristic of the ocean crust.

Authigenic minerals Minerals that form in place, such as phosphorite and those found in manganese nodules.

Autolytic decomposition The breakdown of organic matter by enzymes activated when an organism dies.

Autotroph Plant or bacterium that can synthesize organic compounds from inorganic nutrients.

Auxospore A diatom cell that has shed its frustule to allow growth.

Backshore The inner portion of the shore, lying landward of the mean spring-tide high-water line. Acted upon by the ocean only during exceptionally high tides and storms.

Bacterioplankton Bacteria that live as plankton.

Bacteriovore An organism that feeds primarily on bacteria.

Barnacle *See* Cirripedia.

Barrier island A long, narrow, wave-built island separated from the mainland by a lagoon.

Basalt A dark-colored volcanic rock characteristic of the ocean crust. Contains minerals with relatively high iron and magnesium content.

Base A compound that releases hydroxide ions in aqueous solution.

Bathyal Benthic environment from 200 to 4000 m (656 to 13,120 ft).

Bathymetry The study of ocean depth.

Bathypelagic Open-ocean environment of approximately 1000 to 4000 m (3280 to 13,120 ft).

Bay barrier A marine deposit attached to the mainland at both ends and extending entirely across the mouth of a bay, separating the bay from the open water.

Beach Sediment seaward of the coastline through the surf zone that is in transport along the shore and within the surf zone.

Benthic Pertaining to the ocean bottom.

Benthic nepheloid layer A layer of turbid water adjacent to the ocean floor; created by bottom currents.

Benthos The forms of marine life that live on the ocean bottom.

Biogenous sediment Sediment containing material produced by plants or animals; e.g., coral reefs, shell fragments, and housings of diatoms, radiolarians, foraminifers, and coccolithophores.

Biomass Total weight of the organisms in a particular habitat, species, or group of species.

Biotic community All the organisms that live within some definable area.

Bore A steep-fronted tide crest that moves up a river in association with high tide.

Breaker zone Region where waves break at the seaward margin of the surf zone.

Bryozoa Phylum of colonial animals that often share one coelomic cavity. Encrusting and branching forms secrete a protective housing (zooecium) of calcium carbonate or chitinous material. Possess lophophore feeding structure.

Buffering Of chemical or other processes, any process that reduces impact.

Calcareous Containing calcium carbonate.

Calcium carbonate compensation depth The depth at which ocean water dissolves calcium carbonate as rapidly as it is deposited from above. There is no calcium carbonate in sediment deposited beneath this depth (usually between 4000 and 5000 m, or 13,120 and 16,400 ft).

Calorie Unit of energy defined as the amount of heat required to raise the temperature of 1 g of water 1°C (1.8°F).

Capillary wave Ocean wave whose wavelength is less than 1.74 cm (0.7 in). The dominant restoring force for such waves is surface tension.

Carapace Chitinous or calcareous shield that covers the cephalothorax of some crustaceans. Dorsal portion of a turtle shell.

Carbohydrate An organic compound consisting of carbon, hydrogen, and oxygen. Sugars and starches are examples.

Carbon cycle A process occurring in stars that releases energy. It involves the conversion of hydrogen to helium with carbon acting as a catalyst. During the process, a C^{12} atom grows to become a N^{15} atom. N^{15} combines with H^1 (proton) to produce $C^{12} + He^4 + \gamma$.

Carnivore An animal that depends solely or chiefly on other animals for its food supply.

Carotene (carotin) A red to yellow pigment found in plants.

Catadromous Pertaining to a species of fish which spawns at sea, then migrates into a freshwater stream or lake to grow to maturity.

Cation A positively charged ion.

Celsius temperature scale Scale in which 0°C = 273.16°K; 0°C = freezing point of water; 100°C = boiling point of water.

Center of gravity The point where the entire mass of a body may be considered to be concentrated.

Centrifugal force   A force that seems to make an object move away from the center of a curved path it is following. It results from the application of a centripetal force acting against the inertia of the object.

Centripetal force   A center-seeking force that tends to make rotating bodies move toward the center of rotation.

Cephalopoda   A class of the phylum Mollusca with a well-developed pair of eyes and a ring of tentacles surrounding the mouth. The shell is absent or internal on most members. The class includes the squid, octopus, and *Nautilus*.

Cetacea   An order of marine mammals that includes the whales.

Chaetognatha   A phylum of elongate, transparent, worm-like pelagic animals commonly called *arrowworms*.

Chela   Arthropod appendage modified to form a pincer.

Chemosynthesis   The formation of organic compounds from inorganic substances using energy derived from oxidation.

Chiton   Common name for any member of the Amphineura class of mollusks with eight dorsal plates.

Chlorinity   The chloride content of seawater expressed in grams per kilogram (g/kg) or parts per thousand (‰) by weight. It includes all the halide ions (F^+, Cl^+, Br^+, I^+, and At^+).

Chlorophyll   A group of green pigments that make it possible for plants to carry on photosynthesis.

Chlorophyta   Green algae; characterized by the presence of chlorophyll and other pigments.

Chrysophyta   An important phylum of planktonic algae, including the diatoms. The presence of chlorophyll is masked by the pigment carotin, which gives the plants a golden color.

Cilia   Short hairlike structures common on lower animals. Beating in unison, they may be used for locomotion or to create water currents that carry food toward the mouth of the animal.

Cirripedia   An order of crustaceans with up to six pairs of thoracic appendages that strain food from the water. They are barnacles that attach themselves to a substrate and secrete an external calcareous housing.

Clastic   Pertaining to a rock or sediment composed of broken fragments of pre-existing rocks. Two common examples are beach deposits and sandstone.

Clay   A term relating to particle size between silt and colloid. Clay minerals are hydrous aluminum silicates with plastic, expansive, and cation exchange properties.

Cnidoblast   Stinging cell of the phylum Coelenterata; contains a stinging mechanism (nematocyst) used in defense and capturing prey.

Coast   A strip of land that extends inland from the coastline as far as marine influence is evidenced in the landforms.

Coastline   Landward limit of the highest storm waves' effect on the shore.

Coccolithophore   A microscopic planktonic form of algae, encased by a covering composed of calcareous discs (coccoliths).

Coelenterata   Phylum of radially symmetrical animals that includes two basic body forms, the medusa and the polyp. Includes jellyfish (medusoid) and sea anemones (polypoid).

Coelom   Secondary body cavity (gut cavity is primary cavity). Forms within the mesoderm in higher animals, is lined with peritoneum, and contains vital organs.

Cold current   A current carrying cold water into areas of warmer water. It generally travels toward the equator.

Colloid   Substance having particles of a size smaller than clay.

Colonial animals   Animals that live in groups of attached or separate individuals. Groups of individuals may serve special functions.

Comb jelly   Common name for members of the phylum Ctenophora. (*See* Ctenophora.)

Commensalism   A symbiotic relationship in which one organism benefits at no expense to its host.

Compensation depth, $CaCO_3$ The depth at which the amount of calcium carbonate ($CaCO_3$) produced by the organisms in the overlying water column is equal to the amount of $CaCO_3$ the water column can dissolve. There will be no $CaCO_3$ deposition below this depth.

Compensation depth, O_2   The depth where the oxygen produced by photosynthesis is equal to the oxygen requirements of plant respiration. A plant population cannot be sustained below this depth, which will be greater in the open ocean than near the shore owing to the relatively deeper light penetration in the open ocean.

Compound   A substance containing two or more elements combined in fixed proportions.

Condensation   The conversion of water from the vapor to the liquid state. When it occurs, the energy required to vaporize the water is released. This is about 585 cal/g of water at 20°C (68°F).

Conduction   The transmission of heat by the passage of energy from particle to particle.

Conjunction   An apparent closeness of two or more heavenly bodies. During the new moon, the sun and moon are in conjunction on the same side of the earth.

Conservative property   A property of surface ocean water that is changed only by mixing and diffusion after the water sinks below the surface.

Consumer   A heterotrophic organism that consumes an external supply of organic matter.

Continent   About one-third of the earth's surface that rises above the deep-ocean floor to be exposed above sea level. Continents are composed primarily of granite, an igneous rock of lower density than the basaltic oceanic crust.

Continental borderland   A highly irregular portion of the continental margin that is submerged beneath the ocean and is characterized by depths greater than those characteristic of the continental shelf.

Continental drift   A term applied to early theories supporting the possibility that the continents are in motion over the earth's surface.

Continental rise   A gently sloping depositional surface at the base of the continental slope.

Continental shelf   A gently sloping depositional surface extending from the low-water line to a marked increase in slope around the margin of a continent.

Continental slope   A relatively steeply sloping surface lying seaward of the continental shelf.

Convection   Process by which, in a fluid being heated, the warmer part of the mass will rise and the cooler portions will sink. If the heat source is stationary, cells may develop as the rising warm water cools and sinks in regions on either side of the axis of rising.

Convergence   Coming together of water masses in polar, tropical, and subtropical regions of the ocean. Along these lines of convergence, the denser mass will sink beneath the others.

Copepoda   An order of microscopic to nearly microscopic crustaceans that are important members of the zooplankton in temperate and subpolar waters.

Coral   A group of benthic coelenterates that exist as individuals or in colonies and may secrete external skeletons of calcium carbonate. Under the proper conditions, corals may produce reefs composed of their external skeletons and the $CaCO_3$ material secreted by varieties of algae associated with the reefs.

Corange lines   Lines on a map joining points of equal tide range.

Coriolis effect   An effect resulting from the earth's rotation that causes particles in motion to be deflected to the right in the Northern Hemisphere and to the left in the Southern Hemisphere.

Cosmogenous sediment   All sediment derived from outer space.

Cotidal lines   Lines connecting points where high tides occur simultaneously.

Crust   Unit of the earth's structure that is composed of basaltic ocean crust and granitic continental crust. The total thickness of the crustal units may range from 5 km (3.1 mi) beneath the ocean to 50 km (31 mi) beneath the continents.

Crustacea   A class of the phylum Arthropoda that includes barnacles, copepods, lobsters, crabs, and shrimp.

Crystalline rock   Igneous or metamorphic rocks. These rocks are made up of crystalline particles with orderly molecular structures.

Ctenophora   A phylum of gelatinous organisms that are more or less spheroidal with biradial symmetry. These exclusively marine animals have eight rows of ciliated combs for locomotion, and most have two tentacles for capturing prey.

Current   A horizontal movement of water.

Cypris   The advanced, free-swimming larval stage of barnacles. After attaching to a substrate, it metamorphoses into the adult.

Decapoda   1. An order of crustaceans with five pairs of thoracic "walking legs," including crabs, shrimp, and lobsters. 2. Suborder of cephalopod mollusks with ten arms that includes squids and cuttlefish.

Decomposers   Primarily bacteria that break down nonliving organic material, extract some of the products of decomposition for their own needs, and make available the compounds needed for plant production.

Deep boundary current   Relatively strong deep current flowing across the continental rise along the western margin of an ocean basin.

Deep scattering layer   A layer of marine organisms in the open ocean that scatters signals from an echo sounder. The organisms migrate daily from near the surface at night to more than 800 m (2624 ft) during the day.

Deep sea system   System that includes all benthic environments beneath the littoral (sublittoral, bathyal, abyssal, and hadal).

Deep water   That portion of the water column from the base of the permanent thermocline or pycnocline to the ocean floor. Water temperatures are relatively uniform throughout.

Deep-water wave   Ocean wave traveling in water that has a depth greater than one-half the average wavelength. Its velocity is independent of water depth.

Delta   A low-lying deposit at the mouth of a river.

Density   Mass per unit volume of a substance. Usually expressed as grams per cubic centimeter. For ocean water with a salinity of 35‰ at 0°C, the density is 1.028 g/cm^3. (*See* sigma-tee.)

Desiccation   Process of drying out.

Detritus   Any loose material produced directly from rock disintegration. (Organic: material resulting from the disintegration of dead organic remains.)

Diatom   Member of the class Bacillariophyceae of algae; possesses a wall of overlapping silica valves.

Diffraction   Any bending of a wave around an obstacle that cannot be interpreted as refraction or reflection.

Diffusion   The transfer of material or a property by random molecular movement. The movement is from a region in which the material or the property is high in concentration to regions of low concentration.

Dinoflagellates   Single-celled microscopic organisms that may possess chlorophyll and belong to the plant phylum Pyrrophyta (autotrophic) or may ingest food and belong to the class Mastigophora of the animal phylum Protozoa (heterotrophic).

Diploblastic   Pertaining to body structure composed of two cell layers: ectoderm and endoderm. Diploblastic phyla are Porifera, Coelenterata, and Ctenophora.

Discoaster   A tiny star-shaped plate that may have formed on a coccolithophorelike algal cell which died out at the start of the Pleistocene epoch.

Discontinuity   An abrupt change in a property, such as temperature or salinity, at a line or surface.

Disphotic zone   The dimly lit zone corresponding approximately with the mesopelagic in which there is not enough light to carry on photosynthesis. Sometimes called the *twilight zone*.

Distributary   A small stream flowing away from a main stream. Such streams are characteristic of deltas.

Diurnal inequality   The difference in the heights of two successive high or low waters during a lunar (tidal) day.

Diurnal tide A tide with one high water and one low water during a tidal day. Tidal period is 24 hrs 50 min.

Divergence A horizontal flow of fluid from a central region, as occurs in upwelling.

Doldrums A belt of light, variable winds 10° to 15° north and south of the equator, resulting from the vertical flow of low-density air within this equatorial belt. Doldrums is the common name for the Intertropical Convergence Zone.

Dolomite A common rock-forming mineral; a calcium-magnesium carbonate $[CaMg(CO_3)_2]$.

Dolphin 1. A brilliantly colored fish of the genus *Coryphaena*. 2. The name applied to the small, beaked members of the cetacean family Delphinidae.

Dorsal Pertaining to the back or upper surface of most animals.

Drifts Thick sediment deposits on the continental rise produced where the western boundary undercurrent slows and loses sediment as it changes direction to follow the base of the continental slope.

Dynamic topography A surface configuration resulting from the geopotential difference between a given surface and a reference surface of no motion. A contour map of this surface is useful in estimating the nature of geostrophic currents.

Earthquake A sudden motion or trembling in the earth caused by the sudden release of slowly accumulated strain by faulting (movement along a fracture in the earth's crust) or volcanic activity.

Ebb current Seaward-flowing current during a decrease in the height of the tide.

Echinodermata Phylum of animals that have bilateral symmetry in larval forms and usually a five-sided radial symmetry as adults. Benthic and possessing rigid or articulating exoskeletons of calcium carbonate with spines, this phylum includes sea stars, brittle stars, sea urchins, sand dollars, sea cucumbers, and sea lilies.

Echo sounding Determining the depth of water by measuring the time required for a sonic or ultrasonic signal to travel to the bottom and back to the ship that emitted the signal.

Ecological efficiency Efficiency with which energy is transferred from one trophic level to the next, or the ratio of the amount of protoplasm added to a trophic level to the amount of food required to produce it.

Ecosystem All the organisms in a biotic community and the abiotic environmental factors with which they interact.

Ectoderm Outermost layer of cells in an animal embryo. In vertebrates it gives rise to the skin, nervous system, sense organs, etc.

Eddy A current of any fluid forming on the side of or within a main current. It usually moves in a circular path and develops where currents encounter obstacles or flow past one another.

Edreobenthos Benthos that are attached to the bottom of the ocean.

Ekman spiral A theoretical consideration of the effect of a steady wind blowing over an ocean of unlimited depth and breadth and of uniform viscosity. The result is a surface flow at 45° to the right of the wind in the Northern Hemisphere. Water at increasing depth will drift in directions increasingly to the right until at about 100 m (328 ft) depth it is moving in a direction opposite to that of the wind. The net water transport is 90° to the wind, and velocity decreases with depth.

Ekman transport The net transport of surface water set in motion by wind. Owing to the Ekman spiral phenomenon, it is theoretically in a direction 90° to the right and 90° to the left of the wind direction in the Northern Hemisphere and Southern Hemisphere, respectively.

Electromagnetic energy Energy that travels as waves or particles with the speed of light. Different kinds possess different properties based on wavelength. The longest wavelengths (up to 10 km, or 6.21 mi, in length) belong to radio waves. At the other end of the spectrum are cosmic rays with great penetrating power and wavelengths of less than .000001 μm.

Electron A negatively charged particle in orbit around the nucleus of an atom.

Electron cloud The organized assemblage of electrons surrounding the nucleus of an atom.

Element One of a number of substances, each of which is composed entirely of like atoms.

El Niño A southerly-flowing warm current that generally develops off the coast of Ecuador shortly after Christmas. Occasionally it will move farther south into Peruvian coastal waters and cause the widespread death of plankton and fish.

Emergent shoreline A shoreline resulting from the emergence of the ocean floor relative to the ocean surface. It is usually rather straight and characterized by marine features usually found at a greater depth.

Endoderm Innermost cell layer of an embryo. Develops into the digestive and excretory systems and forms the lining for the respiratory system, etc., in vertebrates.

Endothermic reaction A chemical reaction that absorbs energy. For example, energy is stored in the organic products of the chemical reaction photosynthesis.

Entropy A quantity reflecting the degree of uniform distribution of heat energy in a system. It increases with time and represents a state in which energy is unrecoverable for work.

Environment The sum of all physical, chemical, and biological factors to which an organism or community is subjected.

Epicenter The point on the earth's surface that is directly above the focus of an earthquake.

Epifauna Animals that live on the ocean bottom, either attached or moving freely over it.

Epipelagic zone The upper region of the oceanic province, extending to a depth of 200 m (656 ft).

Equatorial tide A semimonthly tide occurring when the moon is over the equator. It displays a minimal diurnal inequality.

Equilibrium tide theory A tidal hypothesis that considers the ocean to be of uniform nature and depth throughout the earth's surface. It is further assumed that this ocean will respond instantly to the gravitational forces of the sun and moon.

Equinox The times when the sun is over the equator, making day and night of equal length throughout the earth. *Vernal equinox* occurs about March 21 as the sun is moving into the Northern Hemisphere. *Autumnal equinox* occurs about September 21 as the sun is moving into the Southern Hemisphere.

Estuarine circulation Circulation characteristic of an estuary and other bodies of water having restricted circulation with the ocean; this results from an excess of runoff and precipitation as compared to evaporation. Surface flow is toward the ocean with a subsurface counter flow.

Estuary The mouth of a river valley, or a bay or lagoon receiving fresh water, where marine influence is manifested as tidal effects and increased salinity of the fresh water.

Euphausiacea An order of planktonic crustaceans ranging in length from 5 to 30 cm (1.97 to 11.82 in). Most possess luminous organs, and some are the principal food for baleen whales.

Euphotic zone The surface layer of the ocean that receives enough light to support photosynthesis. The bottom of this zone, which is marked by the oxygen compensation depth, varies and reaches a maximum value of around 150 m (492 ft) in the very clearest open ocean water.

Euryhaline Pertaining to the ability of a marine organism to tolerate a wide range of salinity.

Eurythermal Pertaining to the ability of a marine organism to tolerate a wide range of temperature.

Eutrophic Characterized by an abundance of nutrients.

Evaporation The physical process of converting a liquid to a gas. Commonly considered to occur at a temperature below the boiling point of the liquid.

Excess volatiles Volatile compounds found in the oceans, sediments, and atmosphere in quantities greater than the chemical weathering of crystalline rock could produce. They are considered to have been produced by volcanic action.

Exothermic reaction A chemical reaction that liberates energy. For example, the energy stored in the products of photosynthesis is released by the chemical reaction respiration.

Fahrenheit temperature scale (°F) Scale in which the freezing point of water is 32°; boiling point of water is 212°.

Fan A gently sloping, fan-shaped feature normally located at the lower end of a canyon.

Fast ice Sea ice that is attached to the shore and therefore remains stationary

Fat A colorless, odorless organic compound consisting of carbon, hydrogen, and oxygen; insoluble in water.

Fathom A unit of ocean depth commonly used in countries using the English system of units. It is equal to 1.83 m, or 6 ft.

Fault A fracture or fracture zone in the earth's crust along which displacement has occurred.

Fault block A crustal block bounded on at least two sides by faults. Usually elongate; if it is down-dropped it produces a graben; if uplifted, it is a horst.

Fauna The animal life of any particular area or of any particular time.

Fecal pellet Organic excrement found in marine sediment and produced primarily by invertebrates. Usually of an ovoid form that is less than 1 mm (0.04 in) in length.

Fetch 1. Area of the open ocean over which the wind blows with constant speed and direction, thereby creating a wave system. 2. The distance across the fetch (wave-generating area) measured in a direction parallel to the direction of the wind.

Fjord A long, narrow, deep, U-shaped inlet that usually represents the seaward end of a glacial valley that has become partially submerged after the melting of the glacier.

Flagellum A whiplike living process used by some cells for locomotion.

Floe A piece of floating ice other than fast ice or icebergs. May range in dimension from about 20 cm (7.9 in) across to more than a kilometer.

Flood current A tidal current associated with increasing height of the tide, generally moving toward the shore.

Flora The plant life of any particular area or of any particular time.

Food chain The passage of energy materials from producers through a sequence of herbivores and carnivores.

Food web A group of interrelated food chains.

Foraminifera An order of planktonic and benthic protozoans that possess protective coverings usually composed of calcium carbonate.

Forced standing wave A wave that is generated and maintained by a periodic force, such as the gravitational attraction of the moon.

Foreshore The portion of the shore lying between the normal high- and low-water marks—the intertidal zone.

Fortnight Half a synodic month (29.5 days), or about 14.75 days. Normally used in reference to a period of time equal to two weeks. The time that elapses between the new moon and full moon.

Fossil Any remains, trace, or imprint of an organism that has been preserved in rocks.

Fracture zone An extensive linear zone of unusually irregular ocean floor topography, characterized by large seamounts, steep-sided or asymmetrical ridges, troughs, or long, steep slopes. Usually represents ancient, inactive transform fault zones.

Free standing wave A wave, created by a sudden rather than continuous impulse, that continues to exist after the generating force is gone.

Freezing point The temperature at which a liquid becomes a solid under any given set of conditions. The freezing point of water is 0°C (32°F) under atmospheric pressure.

Fringing reef A reef that is directly attached to the shore of an island or continent. It may extend more than 1 km (0.62 mi) from shore. The outer margin is submerged and often consists of algal limestone, coral rock, and living coral.

Frustule The siliceous covering of a diatom, consisting of two halves (epitheca and hypotheca).

Fucoxanthin The reddish-brown pigment that gives brown algae its characteristic color.

Furrows Parallel, troughlike structures cut into mud waves by bottom currents. They are aligned in the direction of current flow.

Gastropoda A class of mollusks, most of which possess an asymmetrical spiral one-piece shell and a well-developed flattened foot. A well-developed head will usually have two eyes and one or two pairs of tentacles. Includes snails, limpets, abalone, cowries, sea hares, and sea slugs.

Geostrophic current A current that develops out of the earth's rotation and is the result of a near balance between gravitational force and the Coriolis effect.

Gill A thin-walled projection from some part of the external body or the digestive tract; used for respiration in a water environment.

Glacial Epoch The Pleistocene Epoch, the earlier of two divisions of the Quaterary Period of geologic time. During this time, high-latitude continental areas now free of ice were covered by continental glaciers.

Glacier A large mass of ice formed on land by the recrystallization of old compacted snow. It flows from an area of accumulation to an area of wasting, where ice is removed from the glacier by melting or calving.

Gondwanaland A hypothetical protocontinent of the Southern Hemisphere named for the Gondwana region of India. It included the present continental masses of Africa, Antarctica, Australia, India, and South America.

Graded bedding Stratification in which each layer displays a decrease in grain size from bottom to top.

Gradient The rate of increase or decrease of one quantity or characteristic relative to a unit change in another. For example, the slope of the ocean floor is a change in elevation (a vertical linear measurement) per unit of horizontal distance covered. Commonly measured in m/km.

Granite A light-colored igneous rock characteristic of the continental crust. Rich in nonferromagnesian minerals such as feldspar and quartz.

Gravitational stability The degree to which segments of the water column tend to remain stationary. The portion of the water column with the greatest degree of gravitational stability is called the pycnocline, where water at the top is much less dense than water at the bottom. This zone of density stratification normally coincides with the thermocline.

Gravity wave A wave for which the dominant restoring force is gravity. Such waves have a wavelength of more than 1.74 cm (0.7 in), and their velocity of propagation is controlled mainly by gravity.

Groin A low, artificial structure projecting into the ocean from the shore to interfere with longshore transportation of sediment. It usually has the purpose of trapping sand to cause the buildup of a beach.

Gross primary production The total amount of organic material produced by autotrophs.

Guyot A tablemount; a conical volcanic feature on the ocean floor that has had the top truncated to a relatively flat surface.

Gyre A circular motion. Used mainly in reference to the circular motion of water in each of the major ocean basins centered in subtropical high-pressure regions.

Habitat A place where a particular plant or animal lives. Generally refers to a smaller area than environment.

Hadal Pertaining to the deepest ocean environment, specifically that of ocean trenches deeper than 6 km (3.7 mi).

Half-life The time required for half the atoms of a radioactive isotope sample to decay to atoms of another element.

Halocline A layer of water which has a high rate of salinity change in the vertical dimension.

Headland A steep-faced irregularity of the coast that extends out into the ocean.

Heat Energy moving from a high-temperature system to a lower-temperature system. The heat gained by the one system may be used to raise its temperature or to do work.

Heat budget (global) The equilibrium that exists on the average between the amount of heat absorbed by the earth and its atmosphere in one year, and the amount of heat radiated back into space in one year.

Heat capacity Usually defined as the amount of heat required to raise the temperature of 1 g of a substance 1°C.

Herbivore An animal that relies chiefly or solely on plants for its food.

Hermaphroditic Pertaining to the possession of both functional male and functional female reproductive organs by an animal. It is rare for both systems to function at the same time.

Hermatypic coral Reef-building coral that has symbiotic algae in its ectodermal tissue. Cannot produce a reef structure below the euphotic zone.

Herpetobenthos Benthos that walk or crawl across the ocean floor.

Heterotroph Animals and bacteria that depend on the organic compounds produced by other animals and plants as food. Organisms not capable of producing their own food by photosynthesis or chemosynthesis.

Higher high water (HHW) The higher of two high waters occurring during a tidal day where tides are mixed.

Higher low water (HLW) The higher of two low waters occurring during a tidal day where tides are mixed.

High water (HW) The highest level reached by the rising tide before it begins to recede.

Holoplankton Organisms that spend their entire life as members of the plankton.

Homeotherm An animal which maintains a precisely controlled internal body temperature using its own heating and cooling mechanisms.

Homologous Pertaining to a basic similarity of structures in different organisms resulting from a similar embryonic origin and development, e.g., the foreflippers of a seal and the arms and hands of a man are homologous.

Hook A spit or narrow cape of sand or gravel with an end that bends landward to form a "hook."

Horse latitudes The latitude belts between 30° and 35° north and south where winds are light and variable since the principal movement of air masses at these latitudes is one of vertical descent.

Hurricane A tropical cyclone in which winds reach velocities above 120 km/h (73mi/h). Generally applied to such storms in the North Atlantic Ocean, eastern North Pacific Ocean, Caribbean Sea, and Gulf of Mexico. Such storms in the western Pacific Ocean are called *typhoons.*

Hydrocarbon An organic compound consisting solely of hydrogen and carbon. Petroleum is a mixture of many hydrocarbon compounds.

Hydrogenous sediment Sediment that forms from ocean water precipitation or ion exchange between existing sediment and ocean water. Examples are manganese nodules, phosphorite, glauconite, phillipsite, and montmorillonite.

Hydrologic cycle The cycle of water exchange among the atmosphere, land, and ocean through the processes of evaporation, precipitation, runoff, and subsurface percolation.

Hydrophilic Pertaining to the property of attracting water.

Hydrophobic Pertaining to the property of being impossible or difficult to wet with water.

Hydrozoa A class of coelenterates that characteristically exhibit alternation of generations, with a sessile polypoid colony giving rise to a pelagic medusoid form by asexual budding.

Hypertonic Pertaining to the property of an aqueous solution having a higher osmotic pressure (salinity) than another aqueous solution separated by a semipermeable membrane allowing osmosis. The hypertonic fluid will gain water molecules through the membrane from the other fluid.

Hypotonic Pertaining to the property of an aqueous solution having a lower osmotic pressure (salinity) than another aqueous solution separated by a semipermeable membrane allowing osmosis. The hypotonic fluid will lose water molecules through the membrane to the other fluid.

Iceberg A massive piece of glacier ice that has broken from the front of the glacier (calved) into a body of water. It floats with its tip at least 5 m (16.4 ft) above the water's surface and with at least 4/5 of its mass submerged.

Ice floe *See* floe.

Ice shelf A thick layer of ice with a relatively flat surface which is attached to and nourished by a continental glacier from one side. The shelf, which is for the most part afloat, may extend above water level by more than 50 m (164 ft) along its seaward cliff (formed by the break-off of large tabular chunks of ice that become icebergs).

Igneous rock One of the three main classes into which all rocks are divided, i.e., igneous, metamorphic, and sedimentary. Rock that forms from the solidification of molten or partly molten material (magma).

Inertia Tendency that a body at rest will stay at rest and a body in motion will remain in a uniform motion in a straight line unless acted on by some external force. Stated in Newton's first law of motion.

Infauna Animals that live buried in the soft substrate (sand or mud).

Infrared radiation Electromagnetic radiation between the wavelengths of 0.8 μm (0.00032 in) and about 1000 μm (0.394 in). It is bounded on the shorter-wavelength side by the visible spectrum and on the long side by microwave radiation.

Inner sublittoral zone The section of ocean floor from the low-tide shoreline to the point where attached plants stop growing.

In situ In place, e.g., *in situ* density of a sample of water is its density at its original depth.

Insolation The rate at which solar radiation is received per unit of surface area at any point at or above the earth's surface.

Interface A surface separating two substances of different properties, i.e., density, salinity, or temperature. In oceanography, it usually refers to a separation of two layers of water with different densities caused by significant differences in temperature and/or salinity.

Internal wave A wave that develops below the surface of a fluid, the density of which changes with increased depth. This change may be gradual or occur abruptly at an interface.

Intertidal zone Littoral zone; the foreshore. The ocean floor covered by the highest normal tides and exposed by the lowest normal tides, and the water environment of the tide pools within this region.

Intertropical Convergence Zone Zone where northeast trade winds and southeast trade winds converge. Averages about 5°N in the Pacific and Atlantic oceans and 7°S in the Indian Ocean.

Ion An atom that becomes electrically charged by gaining or losing one or more electrons. The loss of electrons produces a positively charged cation, and the gain of electrons produces a negatively charged anion.

Ionic bond A chemical bond resulting from the electrical attraction that exists between cations and anions.

Island arc system A linear arrangement of islands, many of which are volcanic, usually curved so the concave side faces a sea separating the islands from a continent. The convex side faces the open ocean and is bounded by a deep-ocean trench.

Isobar Lines connecting values of equal pressure on a map or graph.

Isohaline Of the same salinity.

Isopoda An order of dorsoventrally flattened crustaceans that are mostly scavengers or parasites on other crustaceans or fish.

Isostasy A condition of equilibrium, comparable to buoyancy, in which the rigid crustal units float on the underlying mantle.

Isotherm A line connecting points of equal temperature.

Isothermal Of the same temperature.

Isotonic Pertaining to the property of having equal osmotic pressure. If two such fluids were separated by a semipermeable membrane that allows osmosis to occur, there would be no net transfer of water molecules across the membrane.

Isotope One of several atoms of an element that has a different number of neutrons, and therefore a different atomic mass, than the other atoms, or isotopes, of the element.

Jellyfish 1. A free-swimming, umbrella-shaped medusoid member of the coelenterate class, Scyphozoa. 2. Also frequently applied to the medusoid forms of other coelenterates.

Jet stream An easterly moving air mass at an elevation of about 10 km (6.2 mi). Moving at speeds that can exceed 300 km/h, the (186 mi/h) jet stream follows a wavy path in the midlatitudes and influences how far polar air masses may extend into the lower latitudes.

Jetty A structure built from the shore into a body of water to protect a harbor or a navigable passage from being shoaled by deposition of longshore (littoral) drift material.

Juvenile water Water that is derived directly from magma, being released for the first time at the earth's surface as the magma crystallizes to igneous rock.

Kelp Large varieties of Phaeophyta (brown algae).

Kelvin temperature scale (°K) Scale in which 0°K = −273.16°C. One degree on the Kelvin scale equals the same temperature range as one degree on the Celsius scale. 0°K is the lowest temperature possible.

Kelvin wave Wave that results when a progressive tide wave moves from the open ocean into and out of a relatively narrow body of water during a tidal cycle; the tidal range will be greater on the right side of the narrow body of water during flood tide. This results from the fact that the channel rotates in a counterclockwise direction as the earth rotates, while the wave tends to move in a straight line.

Key A low, flat island composed of sand or coral debris that accumulates on a reef flat.

Kinetic energy Energy of motion. It increases as the mass or velocity of the object in motion increases.

Knot (kt) Unit of speed equal to 1 nautical mile per hour (approximately 51 centimeters per second).

Krill A common name frequently applied to members of the crustacean order Euphausiacea (euphausids).

Lagoon A shallow stretch of seawater partly or completely separated from the open ocean by an elongate narrow strip of land, such as a reef or barrier island.

Lamina A layer.

Laminar flow Flow in which a fluid flows in parallel layers or sheets. The direction of flow at any point does not change with time; nonturbulent flow.

Langmuir circulation A cellular circulation set up by winds that blow consistently in one direction with velocities above 12 km/h (7.5 mi/h). Helical spirals running parallel to the wind direction are alternately clockwise and counterclockwise.

Larva An embryo that is on its own before it assumes the characteristics of the adult of the species.

Latent heat The quantity of heat gained or lost per unit of mass as a substance undergoes a change of state (liquid to solid, etc.) at a given temperature and pressure.

Lateritic soil A red subsoil rich in secondary iron and aluminum oxides. It is characteristic of intertropical soils, may contain much quartz and kaolinite, and can harden to a bricklike substance.

Latitude Location on the earth's surface based on angular distance north or south of the equator. Equator, 0°; North Pole, 90° N; South Pole, 90° S.

Laurasia A hypothetical protocontinent of the Northern Hemisphere. The name is derived from Laurentia, pertaining to the Canadian Shield of North America, and Eurasia, of which it was composed.

Lava Fluid magma coming from an opening in the earth's surface, or the same material after it solidifies.

Leeward Direction toward which the wind is blowing or waves are moving.

Levee Natural (resulting from deposition during flooding) or man-made low ridges on either side of a river channel.

Lichen Organism involving a photosynthetic, mutualistic relationship between an alga and a fungus. The alga is protected by the fungus, which is dependent on the alga for photosynthetically produced food.

Limestone A class of sedimentary rock composed of at least 50% calcium or magnesium carbonate. Limestone may be either biogenous or hydrogenous.

Limpet A mollusk of the class Gastropoda that possesses a low conical shell exhibiting no spiraling in the adult form.

Lithogenous sediment Sediment composed of mineral grains derived from the rock of continents and islands and transported to the ocean by wind or running water.

Lithosphere The outer layer of the earth's structure, including the crust and the upper mantle to a depth of about 200 km (124 mi). It is this layer that breaks into the plates that are the major elements of global plate tectonics.

***Lithothamnion* ridge** A feature common to the seaward edge of a reef structure, characterized by the presence of the red alga, *Lithothamnion*.

Littoral zone The benthic zone between the highest seaward and lowest normal water marks; the intertidal zone.

Lobster Large marine crustacean used as food. *Homarus americanus* (American lobster) possesses two large chelae (pincers) and is found off the New England coast. *Panulirus* sp. (spiny lobsters or rock lobsters) have no chelae but possess long spiny antennae effective in warding off predators. *P. argus* is found off the coast of Florida and in the West Indies, while *P. interruptus* is common along the coast of southern California.

Longitude Location on the earth's surface based on angular distance east or west of the Greenwich Meridian (0° longitude). 180° longitude is the International Date Line.

Longitudinal wave A wave in which particle vibration is parallel to the direction of energy propagation.

Longshore current A current located in the surf zone and running parallel to the shore as a result of breaking waves.

Longshore drift The load of sediment transported along the beach from the breaker zone to the top of the swash line in association with the longshore current.

Lophophore Horseshoe-shaped feeding structure bearing ciliated tentacles characteristic of the phyla Bryozoa, Brachiopoda, and Phoronidea.

Lower high water (LHW) The lower of two high waters occurring during a tidal day where tides are mixed.

Lower low water (LLW) The lower of two low waters occurring during a tidal day where tides are mixed.

Low water (LW) The lowest level reached by the water surface at low tide before the rise toward high tide begins.

Lunar day The time interval between two successive transits of the moon over a meridian (approximately 24 hours and 50 minutes of solar time).

Lunar hour One twenty-fourth of a lunar day (about 62.1 minutes).

Lunar tide The part of the tide caused solely by the tide-producing force of the moon.

Magma Fluid rock material which solidifies to form igneous rock.

Magnetic anomaly Distortion of the regular pattern of the earth's magnetic field, resulting from the various magnetic properties of local concentrations of ferromagnetic minerals in the earth's crust.

Manganese nodules Concretionary lumps containing oxides of iron, manganese, copper, and nickel found scattered over the ocean floor.

Mangrove swamp A marshlike environment that is dominated by mangrove trees. They are restricted to latitudes below 30°.

Mantle The zone between the core and crust of the earth; rich in ferromagnesium minerals. In pelecypods, the portion of the body that secretes shell material.

Marginal sea A semienclosed body of water adjacent to a continent.

Marsh An area of soft, wet land. Flat land periodically flooded by saltwater; common in portions of lagoons.

Meander A sinuous curve, bend, or turn in the course of a current.

Mean high water (MHW) The average height of all the high waters occurring over a 19-year period.

Mean low water (MLW) The average height of the low waters occurring over a 19-year period.

Mean sea level (MSL) The mean surface water level determined by averaging all stages of the tide over a 19-year period, usually determined from hourly height observations along an open coast.

Mean tidal range The difference between mean high water and mean low water.

Mediterranean circulation Circulation characteristic of bodies of water with restricted ocean circulation that results from an excess of evaporation as compared to precipitation and runoff. Surface flow is into the restricted body of water with a subsurface outflow, as exists between the Mediterranean Sea and the Atlantic Ocean.

Medusa A free-swimming, bell-shaped coelenterate body form with a mouth at the end of a central projection and tentacles around the periphery. Reproduces sexually.

Meridian of longitude Half a great circle terminating at the North and South poles.

Meroplankton Planktonic larval forms of organisms that are members of the benthos or nekton as adults.

Mesoderm A primitive cell layer of the embryo that develops between the endoderm and ectoderm. In vertebrates, it gives rise to the skeleton, muscles, circulatory and excretory systems, and most of the reproductive system.

Mesoglea Jellylike substance found between the endoderm and ectoderm in coelenterates and ctenophores. Ranges from noncellular and nonliving to a highly cellular nature, which leads some to believe it represents a stage in the development of the mesoderm.

Mesopelagic zone That portion of the oceanic province from about 200 to 1000 m (656 to 3280 ft) depth. Corresponds approximately with the disphotic (twilight) zone.

Metamorphic rock Rock that has undergone recrystallization while in the solid state in response to changes of temperature, pressure, and chemical environment.

Metaphyta Kingdom of many-celled plants.

Metazoa Kingdom of many-celled animals.

Microcontinent A submarine plateau that is an isolated fragment of the continental crust. It usually has linear features; found primarily in the Indian Ocean.

Microplankton Net plankton. Plankton not easily seen by the unaided eye, but easily recovered from the ocean with the aid of a fine-mesh plankton net.

Mineral An inorganic substance occurring naturally in the earth and having distinctive physical properties and a chemical composition that can be expressed by a chemical formula.

Mixed tide A tide having two high and two low waters per tidal day with a marked diurnal inequality. Such a tide may also show alternating periods of diurnal and semidiurnal components.

Mohorovicic discontinuity A sharp seismic discontinuity between the crust and mantle of the earth. It may be as shallow as 5 km (3.1 mi) below the ocean floor or as deep as 60 km (37 mi) beneath some continental mountain ranges. Also known as the Moho.

Mole The weight of a substance in grams numerically equal to its molecular weight (gram molecule). 1 mole of water (H_2O) is 18 g.

Molecule The smallest particle of an element or compound that, in the free state, retains the characteristics of the substance.

Mollusca Phylum of soft, unsegmented animals usually protected by a calcareous shell and having a muscular foot for locomotion; includes snails, clams, chitons, and octopuses.

Molt The periodic shedding of the exoskeleton by arthropods to permit growth.

Monera The kingdom of organisms that do not have nuclear material confined within a sheath but spread throughout the cell, such as bacteria and blue-green algae.

Mononodal Pertaining to a standing wave with only one nodal point or nodal line.

Monsoons A name for seasonal winds derived from the Arabic word for season, *mausim*. The term was originally applied to winds over the Arabian Sea that blow from the southwest during summer and the northeast during winter.

Moraine Unsorted material deposited at the margins of glaciers. Many such deposits have become economically important as fishing banks after being submerged by the rising level of the ocean.

Mud Sediment consisting of silt- and clay-sized particles smaller than 0.06 mm (0.002 in). Actually, small amounts of larger particles will also be present.

Mud waves Wave features with lengths of 2 to 3 km (1.2 to 1.9 mi) that bottom currents produce on the surfaces of drifts or ridges.

Mutualism A symbiotic relationship in which both partners benefit.

Mycota Kingdom of fungi. Marine fungi obtain nutrition primarily as decomposers, parasites, or in mutualistic relationships with algae.

Nadir The point on the celestial sphere directly opposite the zenith and directly beneath the observer.

Nanoplankton Plankton less than 50 μm (0.02 in) in length that cannot be captured in a plankton net and must be removed from the water by centrifuge or special microfilters.

Nansen bottle A device used by oceanographers to obtain samples of ocean water from beneath the surface.

Nauplius A microscopic, free-swimming larval stage of crustaceans such as copepods, ostracods, and decapods. Typically has three pairs of appendages.

Neap tide Tide of minimal range occurring when the moon is in quadrature.

Nearshore zone The seaward zone from the shoreline to the line of breakers.

Nektobenthos Those members of the benthos that actively swim and spend much time off the bottom.

Nekton Pelagic animals such as adult squids, fish, and mammals that can determine their position in the ocean by swimming.

Nepheloid layer Well-mixed, turbid layer of water at the base of the oceanic water column. It is particularly well developed in the western boundary undercurrent.

Neritic province That portion of the pelagic environment from the shoreline to a depth of 200 m (656 ft).

Neritic sediment That sediment composed primarily of lithogenous particles and deposited relatively rapidly on the continental shelf, continental slope, and continental rise.

Net primary production The remaining amount of organic material produced by autotrophs after they have met their respiration needs.

Neutron An electrically neutral particle found in the nucleus of most atoms. It has a mass approximately equal to that of a proton.

Niche The ecological role of an organism and its position in the ecosystem.

Nitrogen fixation Conversion by bacteria of atmospheric nitrogen (N_2) to oxides of nitrogen (NO_2, NO_3) usable by plants in primary production.

Node The point on a standing wave where vertical motion is lacking or minimal. If this condition extends across the surface of an oscillating body of water, the line of no vertical motion is a nodal line.

Nonconservative property A property of ocean water attained at the surface and changed after the water sinks below the surface by processes other than mixing and diffusion. For example, dissolved oxygen content will be altered by biological activity.

Nucleus 1. A central, membrane-bound mass in eukaryotic cells; containing chromosomes. 2. The central, positively charged part of an atom; containing protons and neutrons.

Nudibranch Sea slug. A member of the mollusk class Gastropoda that has no protective covering as an adult. Respiration is carried on by gills or other projections on the dorsal surface.

Nutrient Any organic or inorganic compound used by plants in primary production. Nitrogen and phosphorus compounds are important examples.

Obduction The reverse of subduction. In the case of ophiolites, the rock is pushed up on the continent instead of subducting beneath it.

Oceanic crust A mass of rock with basaltic composition that is about 5 km (3.1 mi) thick and that forms the earth's crust beneath ocean basins.

Oceanic province That division of the pelagic environment where the water depth is greater than 200 m (656 ft).

Oceanic ridge A linear, seismic mountain range that extends through all the major oceans, rising from 1 to 3 km (0.62 to 1.9 mi) above the deep-ocean basins. Averaging 1500 km (932 mi) in width, rift valleys are common along the central axis. A source of new oceanic crustal material.

Oceanic sediment The inorganic abyssal clays and the organic oozes that accumulate particle by particle on the deep-ocean floor.

Offshore The comparatively flat, submerged zone of variable width extending from the breaker line to the edge of the continental shelf.

Oligotrophic Characterized by a low level of nutrients.

Omnivore An animal that feeds on both plants and animals.

Ooze A pelagic sediment containing at least 30% skeletal remains of pelagic organisms, the balance being clay minerals. Oozes are further defined by the chemical composition of the organic remains (siliceous or calcareous) and by their characteristic organisms (diatom ooze, foraminiferan ooze, radiolarian ooze, pteropod ooze).

Ophiolite A suite of rocks representing what is believed to be a piece of oceanic lithosphere embedded in a continent.

Opposition The separation of two heavenly bodies by 180° relative to the earth. The sun and moon are in opposition during the full moon phase.

Organic chemistry The branch of chemistry dealing with carbon compounds.

Orthogonal lines Lines drawn perpendicularly to wave fronts and spaced uniformly so that equal amounts of energy are contained by the segments of the wave front lying between any two orthogonal lines in a series. The areas where energy is concentrated as the waves break on the shore can be identified by the convergence of the orthogonal lines.

Osmosis Passage of water molecules through a semipermeable membrane separating two aqueous solutions of different solute concentration. The water molecules pass from the solution of lower solute concentration to the higher.

Osmotic pressure A measure of the tendency for osmosis to occur. It is the pressure that must be applied to the more-concentrated solution to prevent the passage of water molecules from the less-concentrated solution.

Ostracoda An order of crustaceans that are minute and compressed within a bivalve shell.

Outer sublittoral zone The section of ocean floor from the seaward edge of the inner sublittoral zone to a depth of 200 m (656 ft). No attached plants grow here.

Oviparous Pertaining to an animal that releases eggs which develop and hatch outside its body.

Ovoviviparous Pertaining to an animal that incubates eggs inside the mother until they hatch.

Oxygen compensation depth The depth at which marine plants photosynthesize at a rate which exactly meets their respiration needs (the base of the euphotic zone).

Oxygen utilization rate (OUR) The rate of use of dissolved oxygen caused by the respiration of animals and bacterial decomposition of dead organic matter descending to the bottom. The OUR is highest just beneath the euphotic zone (from about 100 to 1000 m, or 328 to 2280 ft) and gradually decreases with depth as the mass of dead organic matter and living animals decreases. Beneath 1000 m, oxygen values may increase owing to the inflow of oxygen carried by deep- and bottom-water masses that descend from the high-latitude ocean surface.

Pacific-type margin Leading edge of a continent that undergoes tectonic uplift as a result of lithospheric plate convergence.

Pack ice Any area of sea ice other than fast ice. Less than 3 m (9.8 ft) thick, it covers the ocean sufficiently to make navigation possible only by icebreakers.

Pancake ice Circular pieces of newly formed sea ice from 30 cm to 3 m (0.98 to 9.8 ft) in diameter that form in early fall in polar regions.

Pangaea A hypothetical supercontinent of the geologic past that contained all the continental crust of the earth.

Panthalassa A hypothetical proto-ocean surrounding Pangaea.

Parapodia Flat protuberances on each side of most segments of polychaete worms. Most possess cirri and setae (bristlelike projections); may be modified for special functions such as feeding, locomotion, and respiration.

Parasite An organism that takes its nutrients from the tissues of another organism and benefits at the host's expense.

Parasitism A symbiotic relationship in which the parasite harms the host from which it takes its nutrition.

Partial tide One of the harmonic components comprising the tide at any location. The periods of the partial tides are derived from the various combinations of the angular velocities of the earth, sun, and moon relative to one another.

Pedicellariae Minute stalked or unstalked pincerlike structures around the base of spines and dermal branchiae in certain echinoderms, especially Asteroidea, Echinoidea, and Ophiuroidea. They snap shut on debris and small organisms to keep the surface of the echinoderm clean.

Pelagic environment The open-ocean environment which is divided into the neritic province (water depth 0 to 200 m, or 0 to 656 ft) and the oceanic province (water depth greater than 200 m, or 656 ft).

Pelecypoda A class of mollusks characterized by two fairly symmetrical lateral valves with a dorsal hinge. These filter feeders pump water through the filter system and over gills through posterior siphons. Many possess a hatchet-shaped foot used for locomotion and burrowing. Includes clams, oysters, mussels, and scallops.

Perigee The point on the orbit of an earth satellite (moon) that is nearest the earth.

Perihelion That point on the orbit of a planet or comet around the sun that is closest to the sun.

Peritoneum Thin membrane lining the coelom and covering all organs in the coelom.

Permeability A condition which allows the passage of liquids through a substance.

Phaeophyta Brown algae characterized by the carotinoid pigment fucoxanthin. Contains the largest members of the marine plant community.

Photosynthesis The process by which plants produce carbohydrate from carbon dioxide and water in the presence of chlorophyll, using light energy and releasing oxygen.

Phycoerythrin A red pigment characteristic of the Rhodophyta (red algae).

Phytoplankton Plant plankton. The most important community of primary producers in the ocean.

Plankton Passively drifting or weakly swimming organisms that are dependent on currents. Includes mostly microscopic algae, protozoans, and larval forms of higher animals.

Plankton bloom A very high concentration of phytoplankton, resulting from a rapid rate of reproduction as conditions become optimum during the spring in high-latitude areas. Less obvious causes produce blooms that may be destructive in other areas.

Plankton net Plankton-extracting device that is cone-shaped and typically of synthetic material. It is towed through the water or lifted vertically to extract plankton down to a size of 50 μm (0.02 in).

Pogonophora A phylum of entirely marine tube worms that have no gut and are found only in water deeper than 20 m (65.6 ft).

Poikilotherm An organism whose body temperature varies with and is largely controlled by the temperature of its environment.

Polychaeta Class of annelid worms that includes most of the marine segmented worms.

Polynya A nonlinear opening in sea ice.

Polyp A single individual of a colony or a solitary attached coelenterate.

Population A group of individuals of one species living in an area.

Porifera Phylum of sponges. Supporting structure composed of $CaCO_3$ or SiO_2 spicules or fibrous spongin. Water currents created by flagella-waving choanocytes enter tiny pores, pass through canals, and exit through a larger osculum.

Primary productivity The amount of organic matter organisms synthesize from inorganic substances within a given volume of water or habitat in a unit of time.

Prime meridian The meridian of longitude 0° used as a reference for measuring longitude; the Greenwich Meridian.

Progressive wave A wave in which the waveform progressively moves.

Propagation The transmission of energy through a medium.

Protein A complex organic compound that contains nitrogen. It is composed principally of amino acids, and is an essential part of an organism's food requirement.

Protista A kingdom of organisms that includes all one-celled forms with nuclear material confined to a nuclear sheath. Includes the animal phylum Protozoa and the phyla of algal plants.

Proton A nuclear particle of all atoms, containing one unit of electrical charge.

Protoplasm The complicated self-perpetuating living material making up all organisms. The elements carbon, hydrogen, and oxygen constitute more than 95%; water and dissolved salts make up from 50% to 97% of most plants and animals, with carbohydrates, lipids (fats), and proteins constituting the remainder.

Protozoa Phylum of one-celled animals with nuclear material confined within a nuclear sheath.

Pseudocoelomate Phyla of animals with a coelom that forms by the separation of the endoderm and ectoderm. It is not lined with peritoneum.

Pseudopodia Extensions of protoplasm in broad, flat, or long needlelike projections used for locomotion or feeding. Typical of amoeboid forms such as foraminifers and radiolarians.

Pteropoda An order of pelagic gastropods in which the foot is modified for swimming and the shell may be present or absent.

Pycnocline A layer of water in which a high rate of change in density in the vertical dimension is present.

Pycnogonid A spiderlike arthropod found on the ocean bottom at all depths. The more commonly observed nearshore varieties are usually less than 1 cm (0.4 in) across, while deeper water varieties may reach spreads of over 1 m (3.28 ft).

Pyrrophyta A phylum of microscopic algae that possesses flagella for locomotion—the dinoflagellates.

Quadrature Configuration in which two heavenly bodies are 90° apart when viewed from the center of a third body. During the 1st- and 3rd-quarter phases of the moon, the moon is in quadrature relative to the sun when viewed from the earth.

Radiata A grouping of phyla with primary radial symmetry—phyla Coelenterata and Ctenophora.

Radioactivity The spontaneous breakdown of the nucleus of an atom resulting in the emission of radiant energy in the form of particles or waves.

Radiolaria An order of planktonic and benthic protozoans that possess protective coverings usually composed of silica.

Ray A cartilaginous fish in which the body is dorso-ventrally flattened, eyes and spiracles are on the upper surface, and gill slits are on the bottom. The tail is reduced to a whiplike appendage. Includes electric rays, manta rays, and stingrays.

Red tide   A reddish-brown discoloration of surface water, usually in coastal areas, caused by high concentrations of microscopic organisms, usually dinoflagellates. It probably results from increased availability of certain nutrients for various reasons. Toxins produced by the dinoflagellates may kill fish directly, or large populations of animal forms that spring up to feed on the plants, along with decaying plant and animal remains, may use up the oxygen in the surface water to cause asphyxiation of many animals.

Reef   A consolidated rock (a hazard to navigation) with a depth of 20 m (65.6 ft) or less.

Reef flat   A platform of coral fragments and sand that is relatively exposed at low tide.

Reef front   The upper seaward face of a reef from the reef edge (seaward margin of reef flat) to the depth at which living coral and coralline algae become rare (16 to 30 m, or 52 to 98 ft).

Reflection   The process in which a wave has part of its energy returned seaward by a reflecting surface.

Refraction   The process by which the part of a wave in shallow water is slowed down to cause the wave to bend and tend to align itself with the underwater contours.

Relict beach   A beach deposit laid down and submerged by a rise in sea level. It is still identifiable on the continental shelf, indicating no deposition is presently taking place at that location on the shelf.

Residence time   The average length of time a particle of any substance spends in the ocean. It is calculated by dividing the total amount of the substance in the ocean by the rate of its introduction into the ocean or the rate at which it leaves the ocean.

Respiration   The process by which organisms utilize organic materials (food) as a source of energy. As the energy is released, oxygen is used and carbon dioxide and water are produced.

Reversing current   The tide current as it occurs at the margins of landmasses. The water flows in and out for approximately equal periods of time, separated by slack water when the water is still at high and low tidal extremes.

Rhodophyta   Phylum of algae composed primarily of small encrusting, branching, or filamentous plants that receive their characteristic red color from the presence of the pigment phycoerythrin. With a worldwide distribution, they are found at greater depths than other algae.

Ridge   *See* drift; oceanic ridge.

Rip current   A strong narrow surface or near-surface current of high velocity (up to 4 km/h, or 2.5 mi/h) flowing seaward through the breaker zone at nearly right angles to the shore. It represents the return to the ocean of water that has been piled up on the shore by incoming waves.

Ripples   Capillary waves. Ten-to 15 cm (4 to 6 in)-long waves. They are found on the sides of furrows cut into mud waves by current action.

Rise   A long, broad elevation that rises gently and rather smoothly from the deep-ocean floor.

Rotary current   Tidal current as observed in the open ocean. The tidal crest makes one complete rotation during a tidal period.

Sabellid   A member of the annelid family Sabellidae that lives in a tube composed of shell fragments, sand, and agglutinous material. Featherlike gills and feeding structures filter food from the water above the tube opening.

Salinity   A measure of the quantity of dissolved solids in ocean water. Formally, it is the total amount of dissolved solids in ocean water in parts per thousand by weight after all carbonate has been converted to oxide, the bromide and iodide to chloride, and all the organic matter oxidized. It is normally computed from conductivity, refractive index, or chlorinity.

Salinometer   A conductance-measuring device that measures the salinity of ocean water to a precision of 0.003 ‰.

Salpa   Genus of pelagic tunicates that are cylindrical, transparent, and found in all oceans.

Salt   Any substance that yields ions other than hydrogen or hydroxyl. Salts are produced from acids by replacing the hydrogen with a metal.

Salt marsh   A relatively flat area of the shore where fine sediment is deposited and salt-tolerant grasses grow. One of the most biologically productive regions on the earth's surface.

Sand   Particle size ranging from 1/16 to 2 mm (0.0025 to 0.08 in). It pertains to particles that lie between silt and granules on the Wentworth scale of grain size.

Sargasso Sea   A region of convergence in the North Atlantic lying south and east of Bermuda where the water is a very clear deep blue in color and contains large quantities of floating *Sargassum.*

Sargassum   A brown alga characterized by a bushy form, substantial holdfast when attached, and a yellow-brown, green-yellow, or orange color. Two species, *S. fluitans* and *S. natans,* make up most of the macroscopic vegetation in the Sargasso Sea.

Scaphopoda   A class of mollusk commonly called *tusk shells.* The shell is an elongate cone open at both ends. The conical foot surrounded by threadlike tentacles extends from the larger end to aid the animal in burrowing.

Scarp   A linear, steep slope on the ocean floor separating gently sloping or flat surfaces.

Scavenger   An animal that feeds on dead organisms.

Schizocoelomate   Animal that has a coelomic cavity called a *schizocoelom* that develops from a split in the mesoderm and is lined with peritoneum. It is present in members of the phyla Annelida, Arthropoda, Bryozoa, Mollusca, and Phoronida.

Scyphozoa   A class of coelenterates that includes the true jellyfish, in which the medusoid body form predominates and the polyp is reduced or absent.

Sea 1. A subdivision of an ocean. Two types of seas are identifiable and defined. They are the *mediterranean seas,* where a number of seas are grouped together collectively as one sea, and *marginal seas* that are connected individually to the ocean. 2. A portion of the ocean where waves are being generated by wind.

Sea anemone A member of the class Anthozoa whose bright color, tentacles, and general appearance resemble a flower.

Sea arch An opening through a headland caused by wave erosion. Usually develops as sea caves are extended from one or both sides of the headland.

Sea cave A cavity at the base of a sea cliff; formed by wave erosion.

Sea cow An aquatic, herbivorous mammal of the order Sirenia that includes the dugong and manatee.

Sea cucumber A common name given to members of the echinoderm class Holotheuroidea.

Sea-floor spreading A process producing the lithosphere when convective upwelling of magma along the oceanic ridges moves away at rates of from 1 to 10 cm (0.4 to 4 in) per year.

Sea ice Any form of ice originating from the freezing of ocean water.

Seamount An individual peak extending over 1000 m (3280 ft) above the ocean floor.

Sea snake A reptile belonging to the family Hydrophiidae with venom similar to that of cobras. Found primarily in the coastal waters of the Indian Ocean and the western Pacific Ocean.

Sea state A description of the ocean surface that includes the average height of the highest one-third of the waves observed in a wave train; referred to a numerical code.

Sea turtle Any turtle of the reptilian order Testudinata; widely found in warm water.

Sea urchin An echinoderm belonging to the class Echinoidea; possessing a fused test (external covering) and well-developed spines.

Sediment Particles of organic or inorganic origin that accumulate in loose form.

Sedimentary rock A rock resulting from the consolidation of loose sediment, or a rock resulting from chemical precipitation, e.g., sandstone and limestone.

Seiche A standing wave of an enclosed or semienclosed body of water that may have a period ranging from a few minutes to a few hours, depending on the dimensions of the basin. The wave motion continues after the initiating force has ceased.

Seismic Pertaining to an earthquake or earth vibration, including those that are artificially induced.

Seismic sea wave *See* tsunami.

Semidiurnal tide Tide having two high and low waters per tidal day, with small inequalities between successive highs and lows. Tidal period is about 12 hours and 25 minutes solar time; semidaily tide.

Serpulid A polychaete worm belonging to the family Serpulidae that builds a calcareous or leathery tube on a submerged surface.

Sessile Permanently attached to the substrate and not free to move about.

Seta Hairlike or needlelike projections on the exoskeletons of arthropods. Similar structures are found on some annelids.

Shallow-water wave A wave on the surface of the water whose wavelength is at least 20 times water depth. The bottom affects the orbit of water particles, and velocity is determined by water depth.
$$V(m/s) = 3.1\sqrt{\text{water depth (m)}}.$$

Shoal A shallow.

Shore The section of land seaward of the coast; extends from highest level of wave action during storms to the low-water line.

Shoreline The line marking the intersection of the water surface with the shore. Migrates up and down as the tide rises and falls.

Sialic Pertaining to the composition of the granitic continental crust, which is rich in silica and aluminum.

Sigma-tee (σ_t) A term used in place of density, but derived from the density of a water sample after the pressure has been reduced to one atmosphere without temperature change. It is computed as $\sigma_t =$ (specific gravity $- 1$) 1000.

Silica Silicon dioxide (SiO_2).

Sill A submarine ridge partially separating bodies of water such as fjords and seas from one another or from the open ocean.

Silt A particle size ranging from 1/128 to 1/16 mm (0.003 to 0.0025 in). It is intermediate between sand and clay.

Simatic Pertaining to the portion of the earth's crust underlying the oceans and the sialic continents; rich in magnesium and iron and therefore more dense than the sialic continental crust.

Siphonophora An order of hydrozoan coelenterates that forms pelagic colonies containing both polyps and medusae. Examples are *Physalia* and *Velella*.

Slack water Condition existing when a reversing tidal current changes direction at high or low water. Current velocity is zero.

Slick A smooth patch on an otherwise rippled surface caused by a monomolecular film of organic material that reduces surface tension.

Slope current Another name for the western boundary undercurrent; derived from the fact that these currents flow along the base of the continental slope.

Solar tide The partial tide caused by the tide-producing forces of the sun.

Solstice The time when the sun is directly over one of the tropics. In the Northern Hemisphere, the summer solstice occurs on June 21 or 22, when the sun is over the Tropic of Cancer, and the winter solstice occurs on December 21 or 22, when the sun is over the Tropic of Capricorn.

Solute A substance dissolved in a solution. Salts are the solute in salt water.

Solution   A state in which a solute is homogeneously mixed with a liquid solvent. Water is the solvent in ocean water.

Sonar   An acronym for *sound navigation and ranging.* A method by which objects may be located in the ocean.

Sounding   Measuring the depth of water beneath a ship.

Specific gravity   The density ratio of a given substance to pure water at 4°C (39.2°F) and at atmospheric pressure.

Specific heat   The quantity of heat required to raise the temperature of 1 g of a given substance 1°C (1.8°F). For water it is 1 cal.

Spermatophyta   Seed-bearing plants.

Spicule   A minute needlelike calcareous or siliceous form found in sponges, radiolarians, chitons, and echinoderms that acts to support the tissue or provide a protective covering.

Spit   A small point, low tongue, or narrow embankment of land having one end attached to the mainland and the other terminating in open water. It commonly consists of sand deposited by longshore currents.

Sponge   *See* Porifera.

Spray zone   A zone extending above the highest high-tide shoreline to the coastline. Water covers it only during storms.

Spring tide   Tide of maximum range occurring every fortnight close to the times when the moon is new and full.

Stack   An isolated mass of rock that has been detached by wave erosion from the headland and projects from the ocean.

Standing crop   The biomass of a population at any given time.

Standing wave   A wave whose form oscillates vertically without progressive movement. The region of maximum vertical motion is an *antinode.* On either side are *nodes* where there is no vertical motion but maximum horizontal motion.

Stenohaline   Pertaining to organisms that can withstand only a small range of salinity change.

Stenothermal   Pertaining to organisms that can withstand only a small range of temperature change.

Storm surge   A rise above normal water level resulting from wind stress and reduced atmospheric pressure during storms. Consequences can be more severe if it occurs in association with high tide.

Stromatolite   A calcium carbonate sedimentary structure in which algal assemblages trap sediment and bind it into forms that are often dome-shaped. They are known to form only in shallow-water environments.

Subduction   A process by which one lithospheric plate descends beneath another. The surface expression of such a process may be an island arc-trench system or a folded mountain range.

Sublimation   The transformation of the solid state of a substance to a vapor without going through the liquid phase, and vice versa.

Sublittoral zone   That portion of the benthic environment extending from low tide to a depth of 200 m (656 ft). Some consider it to be the surface of the continental shelf.

Submarine canyon   A steep V-shaped canyon cut into the continental shelf or slope.

Submergent shoreline   Shoreline on landforms developed under subaerial processes; formed by the relative submergence of a landmass. It is characterized by bays and promontories and is more irregular than a shoreline of emergence.

Subneritic province   The ocean floor underlying the neritic water column.

Suboceanic province   The ocean floor underlying the oceanic water column.

Substrate   The base on which an organism lives and grows.

Subsurface current   A current usually flowing below the pycnocline, generally at a slower speed and in a different direction than the surface current.

Subsurface oxygen maximum (SOM)   Due to the fact the oxygen produced by photosynthesis is trapped by thermal stratification, dissolved oxygen maximums occur in the low-latitude and temperate oceans. SOM dissolved-oxygen values at depths of 50 to 100 m (164 to 328 ft) are often 20% above saturation.

Supralittoral zone   The spray zone above the high-tide mark.

Surf zone   The region between the shoreline and the line of breakers, where most wave energy is released.

Sverdrup (sv)   A unit of volume transport equal to 1 million m^3/s.

Swash   The rush of water up on the beach following the breaking of a wave.

Swell   One of a series of regular long-period waves that travels out of a wave-generating area.

Swim bladder   An elongate, gas-filled sack dorsal to the digestive tract of most bony fishes. It serves chiefly as a hydrostatic organ to help achieve neutral buoyancy.

Symbiosis   A relationship between two species in which one or both benefit and neither or one is harmed. Examples are commensalism, mutualism, and parasitism.

Tablemount   A flat-topped seamount; a guyot.

Tectonics   The study of the origin and history of structural deformation of the earth's crust.

Temperature gradient   The rate of temperature change within the water column. It is steep within the thermocline, where water temperature changes rapidly with changing depth, but less steep in the deep water, where distribution patterns are more uniform.

Territorial sea   A zone extending seaward from the shore or internal waters of a nation for a distance of 12 mi (19.3 km) as defined by the UNCLOS. The coastal state has full authority over this zone but must allow rights of innocent passage.

Tethys   An oceanic body of water that separated the protocontinents Gondwanaland and Laurasia along the alignment of the present Alpine-Himalayan mountain belt.

Thermocline   A layer of water in which a rapid change in temperature can be measured in the vertical dimension.

Thermohaline circulation The vertical movement of ocean water driven by density differences resulting from the combined effects of variations in temperature and salinity.

Tidal bore *See* bore.

Tidal day *See* lunar day.

Tidal period Elapsed time between successive high or low waters.

Tidal range The difference in height between consecutive high and low waters. The time frame of comparison may also be a day, month, or year.

Tide Periodic rise and fall of the ocean surface and connected bodies of water resulting from the unequal gravitational attraction of the moon and sun on different parts of the earth.

Tide wave The long-period gravity wave generated by the tide-generating forces described above and manifested in the rise and fall of the tide.

Tintinnid A ciliate protozoan of the family Tintinnidae with a tubular to vase-shaped outer shell.

Tissue An aggregate of cells and their products developed by organisms for the performance of a particular function.

Tombolo A sand or gravel bar that connects an island with another island or the mainland.

Topography The configuration of a surface. In oceanography, it refers to the ocean bottom or the surface of a mass of water.

Trade winds The air masses moving from subtropical high-pressure belts toward the equator. They are northeasterly in the Northern Hemisphere and southeasterly in the Southern Hemisphere.

Transform fault A fault characteristic of oceanic ridges along which the ridges are horizontally offset.

Transitional crust Thinned section of continental crust at the trailing edge of a continent; created by the breaking apart of an ancient continent over a newly formed spreading center.

Transitional wave A wave moving from deep water to shallow water that has a wavelength more than 2 but less than 20 times the water depth. Particle orbits are beginning to be influenced by the bottom.

Transverse wave A wave in which particle motion is at right angles to energy propagation.

Trawl A sturdy bag or net that can be dragged along the ocean bottom or at various depths above the bottom to catch fish.

Trench A long, narrow, and deep depression on the ocean floor with relatively steep sides.

Triploblastic Pertaining to the embryonic structure characterized by three layers of cell tissue—endoderm, mesoderm, and ectoderm.

Trophic level A nourishment level in a food chain. Plant producers constitute the lowest level, followed by herbivores and carnivores at the higher levels.

Tropical tide A tide occurring bimonthly when the moon is at its maximum declination north and south of the equator. It is during tropical tides that the greatest diurnal inequalities occur.

Tsunami Seismic sea wave. A long-period gravity wave generated by a submarine earthquake or volcanic event. Not noticeable on the open ocean but builds up to great heights in shallow water.

Tube worms. *See* sabellid and serpulid.

Tunicates Members of the chordate subphylum Urochordata, which includes sacklike animals. Some are sessile (sea squirts) while others are pelagic (salps).

Turbidite A sediment or rock formed from sediment deposited by turbidity currents; characterized by both horizontally and vertically graded bedding.

Turbidity A state of reduced clarity in a fluid; caused by the presence of suspended matter.

Turbidity current A gravity current resulting from a density increase brought about by increased water turbidity. Possibly initiated by some sudden force such as an earthquake, the turbid mass continues under the force of gravity down a submarine slope.

Turbulence A disorderly flow of ocean water.

Turbulent flow Flow in which the flow lines are confused owing to random velocity fluctuations.

Typhoon A severe tropical storm in the western Pacific.

Ultraplankton Plankton smaller than 5 μm (0.002 in). Very difficult to separate from the water.

Ultrasonic Pertaining to sound frequencies above human range (above 20,000 cycles per second).

Ultraviolet radiation Electromagnetic radiation shorter than visible radiation and longer than X-rays. The approximate range is from 1 to 400 nanometers (nm).

Upper water That portion of the water column from the ocean surface to the base of the permanent thermocline or pycnocline. Surface temperatures are relatively high and uniform to the base of the mixed layer (100 to 200 m, or 328 to 656 ft) and decrease rapidly to the base of the thermocline (about 1000 m, or 3280 ft).

Upwelling The process by which deep, cold, nutrient-laden water is brought to the surface, usually by the wind divergence of equatorial currents or coastal winds push water away from the coast.

Valence The combining capacity of an element measured by the number of hydrogen atoms with which it will combine.

Van der Waals force Weak attractive force between molecules; a result of the interaction between the nuclear particles of one molecule and the electrons of another.

Vector A physical quantity that has magnitude and direction. Examples are force, acceleration, and velocity.

Veliger A planktonic larval stage of many gastropods with two ciliated lobes.

Ventral Pertaining to the lower or under surface.

Vertebrata Subphylum of chordates that includes those animals with a well-developed brain and a skeleton of bone or cartilage; includes fish, amphibians, reptiles, birds, and mammals.

Viscosity The property of a substance to offer resistance to flow; internal friction.

Viviparous   Pertaining to an animal that gives birth to living young.

Warm current   A current carrying warm water into areas of colder water. It generally moves poleward.

Water mass   A body of water identifiable by its temperature, salinity, or chemical content.

Wave   A disturbance that moves over or through a medium with a speed determined by the properties of the medium.

Wave-cut bench   A gently sloping surface produced by wave erosion and extending from the base of the wave-cut cliff out under the offshore region.

Wave-cut cliff   A cliff produced by wave erosion cutting landward.

Wave height   Vertical distance between a crest and the preceding trough.

Wavelength   Horizontal distance between two corresponding points on successive waves, such as from crest to crest.

Wave period   The elapsed time between the passage of two successive wave crests past a fixed point.

Wave steepness   Ratio of wave height to wavelength.

Wave train   A series of waves from the same direction.

Weathering   A process by which rocks are broken down by chemical and mechanical means.

Westerly winds   The air masses moving away from the subtropical high-pressure belts toward higher latitudes. They are southwesterly in the Northern Hemisphere and northwesterly in the Southern Hemisphere.

Western boundary undercurrent   Relatively strong bottom current flowing across the continental rise along the western boundary of ocean basins.

Windrows   Rows of floating debris aligned parallel to the direction of the wind; resulting from Langmuir circulation.

Windward   Pertaining to the direction from which the wind is blowing.

Zenith   That point on the celestial sphere directly over the observer.

Zooplankton   Animal plankton.

WE VALUE YOUR OPINION—PLEASE SHARE IT WITH US

Merrill Publishing and our authors are most interested in your reactions to this textbook. Did it serve you well in the course? If it did, what aspects of the text were most helpful? If not, what didn't you like about it? Your comments will help us to write and develop better textbooks. We value your opinions and thank you for your help.

Text Title _____ Edition _____

Author(s) _____

Your Name (optional) _____

Address _____

City _____ State _____ Zip _____

School _____

Course Title _____

Instructor's Name _____

Your Major _____

Your Class Rank _____ Freshman _____ Sophomore _____ Junior _____ Senior

_____ Graduate Student

Were you required to take this course? _____ Required _____ Elective

Length of Course? _____ Quarter _____ Semester

1. Overall, how does this text compare to other texts you've used?

_____ Superior _____ Better Than Most _____ Average _____ Poor

2. Please rate the text in the following areas:

	Superior	Better Than Most	Average	Poor
Author's Writing Style	_____	_____	_____	_____
Readability	_____	_____	_____	_____
Organization	_____	_____	_____	_____
Accuracy	_____	_____	_____	_____
Layout and Design	_____	_____	_____	_____
Illustrations/Photos/Tables	_____	_____	_____	_____
Examples	_____	_____	_____	_____
Problems/Exercises	_____	_____	_____	_____
Topic Selection	_____	_____	_____	_____
Currentness of Coverage	_____	_____	_____	_____
Explanation of Difficult Concepts	_____	_____	_____	_____
Match-up with Course Coverage	_____	_____	_____	_____
Applications to Real Life	_____	_____	_____	_____

3. Circle those chapters you especially liked:
1   2   3   4   5   6   7   8   9   10   11   12   13   14   15   16   17   18   19   20
What was your favorite chapter? _____
Comments:

4. Circle those chapters you liked least:
1   2   3   4   5   6   7   8   9   10   11   12   13   14   15   16   17   18   19   20
What was your least favorite chapter? _____
Comments:

5. List any chapters your instructor did not assign. _____

6. What topics did your instructor discuss that were not covered in the text?_____

_____

7. Were you required to buy this book? _____ Yes   _____ No

Did you buy this book new or used? _____ New   _____ Used

If used, how much did you pay? _____

Do you plan to keep or sell this book? _____ Keep   _____ Sell

If you plan to sell the book, how much do you expect to receive? _____

Should the instructor continue to assign this book? _____ Yes   _____ No

8. Please list any other learning materials you purchased to help you in this course (e.g., study guide, lab manual).

_____

9. What did you like most about this text? _____

_____

10. What did you like least about this text? _____

_____

11. General comments:

May we quote you in our advertising? _____ Yes   _____ No

Please mail to: Boyd Lane
                  College Division, Research Department
                  Box 508
                  1300 Alum Creek Drive
                  Columbus, Ohio 43216

Thank you!